# Design of Polymeric Hydrogels Biomaterials

# Design of Polymeric Hydrogels Biomaterials

Diana C. Silva
Ana Paula Serro
Maria Vivero-Lopez

Basel • Beijing • Wuhan • Barcelona • Belgrade • Novi Sad • Cluj • Manchester

*Editors*

Diana C. Silva
Chemical Engineering
Department
Instituto Superior Técnico-
University of Lisbon
Lisbon
Portugal

Ana Paula Serro
Chemical Engineering
Department
Instituto Superior Técnico-
University of Lisbon
Lisbon
Portugal

Maria Vivero-Lopez
Advanced Materials and
Healthcare Technologies Division
School of Pharmacy
University of Nottingham
Nottingham
United Kingdom

*Editorial Office*
MDPI AG
Grosspeteranlage 5
4052 Basel, Switzerland

This is a reprint of articles from the Special Issue published online in the open access journal *Gels* (ISSN 2310-2861) (available at: www.mdpi.com/journal/gels/special_issues/TJUV9EZ395).

For citation purposes, cite each article independently as indicated on the article page online and as indicated below:

Lastname, A.A.; Lastname, B.B. Article Title. *Journal Name* **Year**, *Volume Number*, Page Range.

**ISBN 978-3-7258-2234-8 (Hbk)**
**ISBN 978-3-7258-2233-1 (PDF)**
**doi.org/10.3390/books978-3-7258-2233-1**

Cover image courtesy of Ana Paula Serro, Diana C. Silva and Maria Vivero-Lopez.
Hydrogel suffering slow biodegradation and releasing bioactive molecules.

© 2024 by the authors. Articles in this book are Open Access and distributed under the Creative Commons Attribution (CC BY) license. The book as a whole is distributed by MDPI under the terms and conditions of the Creative Commons Attribution-NonCommercial-NoDerivs (CC BY-NC-ND) license.

# Contents

**About the Editors** . . . . . . . . . . . . . . . . . . . . . . . . . . . . . . . . . . . . . . . . . . . . . . . . vii

**Ana Paula Serro, Maria Vivero-Lopez and Diana C. Silva**
Editorial for the Special Issue Titled "Design of Polymeric Hydrogels Biomaterials"
Reprinted from: *Gels* 2024, *10*, 344, doi:10.3390/gels10050344 . . . . . . . . . . . . . . . . . . . . . . 1

**Francesca Della Sala, Mario di Gennaro, Pooyan Makvandi and Assunta Borzacchiello**
A Covalently Cross-Linked Hyaluronic Acid/Carboxymethyl Cellulose Composite Hydrogel as a Potential Filler for Soft Tissue Augmentation
Reprinted from: *Gels* 2024, *10*, 67, doi:10.3390/gels10010067 . . . . . . . . . . . . . . . . . . . . . . 5

**Jun Woo Lim, Sang Jin Kim, Jimin Jeong, Sung Gyu Shin, Chaewon Woo and Woonggyu Jung et al.**
Regulated Self-Folding in Multi-Layered Hydrogels Considered with an Interfacial Layer
Reprinted from: *Gels* 2024, *10*, 48, doi:10.3390/gels10010048 . . . . . . . . . . . . . . . . . . . . . . 20

**Aikaterini Gialouri, Sofia Falia Saravanou, Konstantinos Loukelis, Maria Chatzinikolaidou, George Pasparakis and Nikolaos Bouropoulos**
Thermoresponsive Alginate-Graft-pNIPAM/Methyl Cellulose 3D-Printed Scaffolds Promote Osteogenesis In Vitro
Reprinted from: *Gels* 2023, *9*, 984, doi:10.3390/gels9120984 . . . . . . . . . . . . . . . . . . . . . . 30

**Mehtap Sahiner, Aynur S. Yilmaz, Ramesh S. Ayyala and Nurettin Sahiner**
Carboxymethyl Chitosan Microgels for Sustained Delivery of Vancomycin and Long-Lasting Antibacterial Effects
Reprinted from: *Gels* 2023, *9*, 708, doi:10.3390/gels9090708 . . . . . . . . . . . . . . . . . . . . . . 48

**Pouya Dehghani, Aliakbar Akbari, Milad Saadatkish, Jaleh Varshosaz, Monireh Kouhi and Mahdi Bodaghi**
Acceleration of Wound Healing in Rats by Modified Lignocellulose Based Sponge Containing Pentoxifylline Loaded Lecithin/Chitosan Nanoparticles
Reprinted from: *Gels* 2022, *8*, 658, doi:10.3390/gels8100658 . . . . . . . . . . . . . . . . . . . . . . 65

**Tooru Ooya and Jaehwi Lee**
Hydrotropic Hydrogels Prepared from Polyglycerol Dendrimers: Enhanced Solubilization and Release of Paclitaxel
Reprinted from: *Gels* 2022, *8*, 614, doi:10.3390/gels8100614 . . . . . . . . . . . . . . . . . . . . . . 79

**Kamil Sghier, Maja Mur, Francisco Veiga, Ana Cláudia Paiva-Santos and Patrícia C. Pires**
Novel Therapeutic Hybrid Systems Using Hydrogels and Nanotechnology: A Focus on Nanoemulgels for the Treatment of Skin Diseases
Reprinted from: *Gels* 2024, *10*, 45, doi:10.3390/gels10010045 . . . . . . . . . . . . . . . . . . . . . . 89

**Irina Negut and Bogdan Bita**
Exploring the Potential of Artificial Intelligence for Hydrogel Development—A Short Review
Reprinted from: *Gels* 2023, *9*, 845, doi:10.3390/gels9110845 . . . . . . . . . . . . . . . . . . . . . . 143

**Bana Shriky, Maksims Babenko and Ben R. Whiteside**
Dissolving and Swelling Hydrogel-Based Microneedles: An Overview of Their Materials, Fabrication, Characterization Methods, and Challenges
Reprinted from: *Gels* 2023, *9*, 806, doi:10.3390/gels9100806 . . . . . . . . . . . . . . . . . . . . . . 165

**Hossein Omidian and Sumana Dey Chowdhury**
Advancements and Applications of Injectable Hydrogel Composites in Biomedical Research and Therapy
Reprinted from: *Gels* **2023**, *9*, 533, doi:10.3390/gels9070533 . . . . . . . . . . . . . . . . . . . . . **200**

**Caroline S. A. de Lima, Justine P. R. O. Varca, Victória M. Alves, Kamila M. Nogueira, Cassia P. C. Cruz and M. Isabel Rial-Hermida et al.**
Mucoadhesive Polymers and Their Applications in Drug Delivery Systems for the Treatment of Bladder Cancer
Reprinted from: *Gels* **2022**, *8*, 587, doi:10.3390/gels8090587 . . . . . . . . . . . . . . . . . . . . . **237**

# About the Editors

**Diana C. Silva**

Diana Cristina Silva is a Junior Researcher and Invited Assistant Professor at Instituto Superior Técnico and Atlântica, Instituto Universitário, Portugal. She holds a degree in Bioengineering (Biomedical) (Instituto Politécnico de Portalegre, Portugal), as well as a Master's in Bioengineering and Nanosystems and a Ph.D. in Advanced Materials and Processing from Instituto Superior Técnico, Portugal. During her Ph.D., she developed new drug delivery systems based on ophthalmic lenses, collaborating with academic partners (University of Coimbra, Portugal, and University of Santiago de Compostela USC, Spain), industry (Bioceramed), and clinicians (Hospital das Forças Armadas). She spent 3 months at USC to develop molecularly imprinted hydrogels for ocular delivery. Diana's main research work is focused on biomaterials, such as in drug release, material characterization, surface modification, and interaction with biomolecules. She is also involved in the study of sterilization procedures and related biological analysis (e.g., microbiological and sterility tests, cell viability, etc.). Diana has published several peer-reviewed articles, is a member of the Controlled Release Society, and is currently guest editing two Special Issues for *Gels*: (i) "Design of Polymeric Hydrogel Biomaterials" and (ii) "Hydrogel-Based Novel Biomaterials: Achievements and Prospects". She has served as a reviewer for four different journals and supervised several Master and Bachelor students. She started a position as Junior Researcher in 2023 after she placed first in the Individual Call to Scientific Employment Stimulus (CEEC) 5th Edition with the project titled "Multi-functional natural-based wound dressings with healing properties for symptom management of malignant fungating wounds". In 2024, she was awarded a laboratory mobility action for the University of Strasbourg (France) under the guidance of Professor Eric Pollet.

**Ana Paula Serro**

Ana Paula Serro holds a degree in Chemical Engineering and a Ph.D. in Biomaterials, both from Instituto Superior Técnico (IST), University of Lisbon. Currently, she is an Associate Professor at the Chemical Engineering Department of IST and researcher at Centro de Química Estrutural (CQE). She is the Coordinator of the Advanced Materials Group of CQE and Head of the Biomaterials Research Group at IST. Her main research interests fall on the development and characterization of biomaterials, mainly for controlled drug release. She has extensive work in hydrogels for therapeutic applications, including drug-loaded ophthalmic lenses (contact lenses and intraocular lenses), wound dressings, and cartilage substitute materials, addressing topics related to sterilization, biotribology, and biomolecule adsorption, among others. She was principal researcher/institutional leader of several national and international projects and has a vast experience in graduated students' supervision (12 PhD and 45 MSc). She is the author of more than 125 publications in indexed journals and of some book chapters and was the guest editor of several special issues in recognized scientific journals. The excellence of her research work was recognized with several prizes, including a distinction by the University of Lisbon in 2023 and the nomination of one of her projects as a "success story" of the MERA-NET international program.

**Maria Vivero-Lopez**

Maria Vivero-Lopez is a Ramón Areces Postdoctoral Fellow working on the rational design of medicated contact lenses (CLs) to combat biofilm formation at the University of Nottingham (UK) under the supervision of Prof. Morgan Alexander. She obtained her Ph.D. in Pharmacy with Honors from the University of Santiago de Compostela in 2022, supervised by Prof. Carmen Alvarez-Lorenzo and Prof. Angel Concheiro. Her Ph.D. Thesis focused on the development of new techniques to endow hearing aids and CLs with the ability to host and sustainably release substances capable of preventing biofilm formation or addressing already established infections under a Xunta de Galicia predoctoral fellowship. During her Ph.D., she carried out a 3-month predoctoral secondment at Instituto Superior Técnico (Lisbon, Portugal) to evaluate in vitro the release profiles of different active compounds from medicated CLs using microfluidics systems. Additionally, she conducted various secondments at Ocupharm Company (Madrid, Spain) for the in vivo assessment of different ocular drug delivery systems. She is an active member of the Controlled Release Society, where she serves on the Executive Team of the Young Scientist Committee (YSC), the Ocular Delivery Focus Group, and the YSC of the Spanish–Portuguese Local Chapter. She is also a member of the European Society for Biomaterials (ESB) and the Spanish Society for Industrial and Galenical Pharmacy (SEFIG). Her research has garnered recognition, including the VIII Julián Francisco Suárez Freire Award from the Royal Academy of Pharmacy of Galicia and the 2023 Julia Polak European Doctoral Award from the European Society for Biomaterials.

*Editorial*

# Editorial for the Special Issue Titled "Design of Polymeric Hydrogels Biomaterials"

**Ana Paula Serro [1], Maria Vivero-Lopez [2,*] and Diana C. Silva [1,*]**

1. Centro de Química Estrutural, Institute of Molecular Sciences, Instituto Superior Técnico, Universidade de Lisboa, Av. Rovisco Pais, 1049-001 Lisboa, Portugal; anapaula.serro@tecnico.ulisboa.pt
2. School of Pharmacy, University of Nottingham, University Park, Nottingham NG7 2RD, UK
* Correspondence: maria.vivero@nottingham.ac.uk (M.V.-L.); dianacristinasilva@tecnico.ulisboa.pt (D.C.S.)

**Citation:** Serro, A.P.; Vivero-Lopez, M.; Silva, D.C. Editorial for the Special Issue Titled "Design of Polymeric Hydrogels Biomaterials". *Gels* **2024**, *10*, 344. https://doi.org/10.3390/gels10050344

Received: 9 May 2024
Accepted: 13 May 2024
Published: 18 May 2024

**Copyright:** © 2024 by the authors. Licensee MDPI, Basel, Switzerland. This article is an open access article distributed under the terms and conditions of the Creative Commons Attribution (CC BY) license (https:// creativecommons.org/licenses/by/ 4.0/).

Hydrogels have attracted great interest in the biomedical applications field in recent years. Their biocompatibility, similarity with biological tissues, and ability to be tailored with specific properties make them one of the most promising groups of biomaterials [1]. They can be manipulated to obtain controlled structures and dynamic functionalities or mimic the biological complexity of live tissues. They can also be produced as films, 3D printable structures, fibers, and nanoparticles, making them quite versatile. Hydrogels can also load different compounds within their polymeric network to be used as delivery platforms capable of providing sustained drug release [2]. Some hydrogels behave as smart stimuli-responsive materials since their network can suffer modifications in response to external triggers (e.g., pH, temperature, electrical and magnetic fields, light, or the presence of different biomolecules) and change their hydrophilicity, swelling capacity, physical properties, or molecule permeability [3]. They also present self-healing or shape memory properties. Hydrogels currently have a large number of applications that range from the ophthalmic area (contact lenses, intraocular lenses, and ocular implants) to the cardiovascular area (catheter coatings and valves) or skin healing/substitution area (suture threads, wound dressings, and skin grafts). Their role has also become increasingly important in areas such as tissue engineering and regenerative medicine (where they can be used as cell scaffolds) and biosensing [4].

Although we have made several advances in the design of hydrogels for biomedical applications in recent years, many challenges remain in obtaining new materials that, in addition to being safe, ensure an efficient performance. This Special Issue brings together six original research works and five reviews focusing on the most diverse topics related to this theme. Examples of new materials intended for soft tissue augmentation, scaffolds, or drug delivery can be found. Reviews addressing specific hydrogel applications, such as the treatment of bladder cancer or the production of microneedles, and current issues of great interest, like the use of artificial intelligence (AI) for hydrogel development, are presented. Some of the most recent achievements in the area illustrating different aspects of the synthesis, characterization, and application of this type of materials that will certainly continue at the forefront of biomaterial applications are also shown.

Della Salla et al. developed a composite hydrogel based on hyaluronic acid (HA)/ carboxymethyl cellulose (CMC), crosslinked with 1,4-butanediol diglycidyl ether (BDDE), intending it to be used in soft tissue augmentation. To obtain a good-performing HA-based hydrogel filler, they prepared materials with different HA/CMC ratios and reaction conditions (different polymerization temperatures during different times) and evaluated their viscoelastic properties, thermal stability upon sterilization in an autoclave, and swelling capacity. The hydrogel containing HA/CMC at a ratio of 1/1, which was prepared at room temperature for 24 h, presented the highest viscoelastic moduli before and after thermal treatment. It showed a dense crosslinking network that explained its rheological properties and thermal resistance. In tests carried out with fibroblasts, the hydrogels led to a cell viability of 90%, and there were no significant changes in cell morphology.

In another study, Lim et al. developed a new strategy to design multi-layered hydrogels for soft hydrogel actuators. They studied the effect of using diffusion to produce an interfacial layer between each layered hydrogel on the enhancement of the design and fabrication precision. The presence of this interfacial layer reduced the degree of mismatch in the self-folding process. The results show a direct relation between the interfacial layer's thickness and its curvature radius during the self-folding process of the multi-layered hydrogel. Such a layer ensures the integrity of the system in operation as it prevents the separation of layers in the multi-layered hydrogel during actuation.

Gialouri et al. grafted a sodium alginate-based copolymer using thermoresponsive poly(N-isopropylacrylamide) (PNIPAM) chains and combined it with methylcellulose (MC) to be used in scaffolds. The material was achieved via a dual crosslinking mechanism including ionic interactions among $Ca^{2+}$ and carboxyl groups and secondary hydrophobic associations of PNIPAM. The results demonstrate that MC significantly enhanced the mechanical properties. The dynamic moduli of the resulting gels make them suitable for the 3D printing of scaffolds. Adhered pre-osteoblastic cells showed a high viability promoting osteogenic potential, as evidenced by the increased alkaline phosphatase activity, calcium, and collagen production.

Carboxymethyl chitosan (CMCh) microgels were synthesized by Sahiner and coworkers with a tailored size and zeta potential for drug delivery purposes by using a microemulsion environment and divinyl sulfone (DVS) as a crosslinker. The microgels presented a spherical structure and a size in the range of 1–10 μm. The materials showed high biocompatibility in cytotoxicity tests with L929 fibroblasts. The antibiotic drug Vancomycin (Van) was used as the model drug to verify its drug-carrying abilities. The MIC values of the drug released from the Van@CMCh microgels after 24 h were 68.6 and 7.95 μg/mL for *E. coli* and *S. aureus*, respectively. The results demonstrate that Van@CMCh microgels have an effective antibiotic effect against *S. aureus* up to 72 h.

In turn, a lignocellulose sponge containing pentoxifylline (PTX)-loaded lecithin/chitosan nanoparticles (LCNs) was developed by Dehghani et al. to be used in wound dressings. They functionalized lignocellulose hydrogels by oxidation/amination, freeze-dried them, and loaded them with nanoparticles. Drug release assays showed that PTX was released in a sustained way. In vivo wound healing studies were performed in rats to which full-thickness excisional wound models were induced. Histological examination confirmed that the PTX-loaded hydrogels performed better and were more suitable for treating chronic wounds compared to the unloaded hydrogels and those that underwent normal treatment with saline solution.

Polyglycerol dendrimers (PGDs) have demonstrated remarkable properties for drug delivery and solubilization, bioimaging, and diagnostics. Ooya and Lee produced PGD hydrogels crosslinked with ethylene glycol diglycidylether (EGDGE) at various concentrations and evaluated their potential for controlling the release of poorly soluble drugs. Paclitaxel (PTX), an anti-cancer drug, was loaded by soaking hydrogels in the drug solution. The increase in the swelling capacity enhanced PTX loading. No evidence of PTX crystallization was observed as the hydrogels remained transparent, and an FTIR analysis revealed a good dispersion of the drug. About 60% of the loaded PTX was released in sink conditions within 90 min. The results show the potential of these hydrogels for the fast release of hydrophobic drugs, e.g., for oral administration.

In their review, Sghier et al. gathered information on the latest advances in the development of nanoemulgels for wound healing, skin appendage infections, inflammatory skin diseases, skin cancer, neuropathy, or anti-ageing purposes, encapsulating a wide range of molecules, including commercial drugs, repurposed drugs, and other natural and synthetic compounds. All developed formulations showed more advantageous characteristics than those that are currently marketed, with adequate droplet size, PDI, pH, stability, viscosity, spreadability, drug release, and drug permeation and/or retention capacity. Their safety and efficacy were confirmed in vitro and/or in vivo, demonstrating their potential to be used as platforms to replace current therapies or as potential adjuvant treatments, which can one day effectively reach the market to help combat high-incidence skin or systemic diseases and conditions.

Cutting-edge topics, including artificial intelligence (AI) and machine learning (ML), were also addressed. Negut et al. explored their integration in hydrogel development, highlighting their importance in improving the design, characterization, and optimization of hydrogels for various applications. The concept of using AI for train hydrogel design is introduced, underlining its potential to decipher intricate relationships between hydrogel compositions, structures, and properties from complex datasets. Classic physical and chemical techniques in hydrogel design are described to lay the foundation for advances in AI/ML along with AI/ML-enhanced numerical and analytical methods. ML techniques, such as neural networks and support vector machines, which accelerate pattern recognition and predictive modeling using large datasets and advance the discovery of hydrogel formulations, are also presented. In sum, this review shows how AI and ML have transformed hydrogel design by accelerating material discovery, optimizing properties, reducing costs, enabling precise customization, and offering innovative solutions for drug delivery, tissue engineering, wound healing, and more.

Hydrogels have also gained attention in the field of transdermal microneedles thanks to their tunable properties, which allow them to be exploited as delivery systems and extraction tools. However, since hydrogel microneedles are a new emerging technology, their manufacture faces several challenges that need to be addressed for them to be redeemed as a viable pharmaceutical option. Shriky et al. reviewed hydrogel microneedles from a material standpoint, independent of their mechanism of action, citing advances in their formulation, presenting relevant manufacturing and characterization methods, and discussing the regulatory and manufacturing challenges faced by these emerging technologies before their approval.

In their review, Omidian et al. emphasize the adaptability and promise of injectable hydrogel nanocomposites in biomedical research. Injectable hydrogels have become popular due to their ability for controlled release, targeted delivery, and improved mechanical properties. These materials exhibit potential in many areas, including joint ailments, cardiac regeneration, eye disease treatment, and post-operative analgesia. They are also useful in tissue regeneration, cardiovascular issues, ischemic brain injury, and personalized cancer immunotherapy. Moreover, nano-hydroxyapatite-enriched hydrogels offer promise in bone regeneration, tackling bone defects, osteoporosis, and tumor-associated recovery challenges. In wound care and cancer treatment, they facilitate controlled release, expedite wound healing, and target drug release. Their review also includes a perspective section that delves into future possibilities, underscores interdisciplinary collaboration, and emphasizes the bright prospects of injectable hydrogel nanocomposites in biomedical research and applications.

The last review, prepared by Lima et al., focuses on describing the current situation of bladder cancer, the tenth most common type of cancer worldwide. After describing the disease and available treatments, they present a report on the main mucoadhesive polymer-based drug delivery systems (DDSs) that were developed in recent years. These DDSs have an increased ability to improve the drug residence time, permeation capacity, and target release, which may prevent the need for frequent catheter insertions with reduced intervals between doses that are followed by current intravesical therapies and which are highly demotivating for patients. A brief review of the methods used for assessing mucoadhesion properties is also shown, along with a discussion of the different polymers suitable for this application.

**Acknowledgments:** Ana Paula Serro and Diana C. Silva acknowledge Fundação para a Ciência e Tecnologia (FCT) for providing funding through Milk4WoundCare (https://doi.org/10.54499/2022.03408.PTDC), SOL (https://doi.org/10.54499/PTDC/CTM-CTM/2353/2021), and through Centro de Química Estrutural Research Unit projects (https://doi.org/10.54499/UIDB/00100/2020 and the Institute of Molecular Sciences project (https://doi.org/10.54499/UIDP/00100/2020)). Diana C. Silva acknowledges FCT for her Junior Research contract (https://doi.org/10.54499/2022.08560.CEECIND/CP1713/CT0016). Maria Vivero-Lopez acknowledges the Ramón Areces Foundation for a postdoctoral research fellowship [BEVP35A7118].

**Conflicts of Interest:** The authors declare no conflict of interest.

**List of Contributions**

1. Della Sala, F.; di Gennaro, M.; Makvandi, P.; Borzacchiello, A.A. Covalently Cross-Linked Hyaluronic Acid/Carboxymethyl Cellulose Composite Hydrogel as a Potential Filler for Soft Tissue Augmentation. *Gels* **2024**, *10*, 67. https://doi.org/10.3390/gels10010067.
2. Lim, J.W.; Kim, S.J.; Jeong, J.; Shin, S.G.; Woo, C.; Jung, W.; Jeong, J.H. Regulated Self-Folding in Multi-Layered Hydrogels Considered with an Interfacial Layer. *Gels* **2024**, *10*, 48. https://doi.org/10.3390/gels10010048.
3. Gialouri, A.; Saravanou, S.F.; Loukelis, K.; Chatzinikolaidou, M.; Pasparakis, G.; Bouropoulos, N. Thermoresponsive Alginate-GraftpNIPAM/Methyl Cellulose 3D-Printed Scaffolds Promote Osteogenesis In Vitro. *Gels* **2023**, *9*, 984. https://doi.org/10.3390/gels9120984.
4. Sahiner, M.; Yilmaz, A.S.; Ayyala, R.S.; Sahiner, N. Carboxymethyl Chitosan Microgels for Sustained Delivery of Vancomycin and Long-Lasting Antibacterial Effects. *Gels* **2023**, *9*, 708. https://doi.org/10.3390/gels9090708.
5. Dehghani, P.; Akbari, A.; Saadatkish, M.; Varshosaz, J.; Kouhi, M.; Bodaghi, M. Acceleration of Wound Healing in Rats by Modified Lignocellulose Based Sponge Containing Pentoxifylline Loaded Lecithin/Chitosan Nanoparticles. *Gels* **2022**, *8*, 658. https://doi.org/10.3390/gels8100658.
6. Ooya, T.; Lee, J. Hydrotropic Hydrogels Prepared from Polyglycerol Dendrimers: Enhanced Solubilization and Release of Paclitaxel. *Gels* **2022**, *8*, 614. https://doi.org/10.3390/gels8100614.
7. Sghier, K.; Mur, M.; Veiga, F.; Paiva-Santos, A.C.; Pires, P.C. Novel Therapeutic Hybrid Systems Using Hydrogels and Nanotechnology: A Focus on Nanoemulgels for the Treatment of Skin Diseases. *Gels* **2024**, *10*, 45. https://doi.org/10.3390/gels10010045.
8. Negut, I.; Bita, B. Exploring the Potential of Artificial Intelligence for Hydrogel Development—A Short Review. *Gels* **2023**, *9*, 845. https://doi.org/10.3390/gels9110845.
9. Shriky, B.; Babenko, M.; Whiteside, B.R. Dissolving and Swelling Hydrogel-Based Microneedles: An Overview of Their Materials, Fabrication, Characterization Methods, and Challenges. *Gels* **2023**, *9*, 806. https://doi.org/10.3390/gels9100806.
10. Omidian, H.; Chowdhury, S.D. Advancements and Applications of Injectable Hydrogel Composites in Biomedical Research and Therapy. *Gels* **2023**, *9*, 533. https://doi.org/10.3390/gels9070533.
11. de Lima, C.S.A.; Varca, J.P.R.O.; Alves, V.M.; Nogueira, K.M.; Cruz, C.P.C.; Rial-Hermida, M.I.; Kadłubowski, S.S.; Varca, G.H.C.; Lugão, A.B. Mucoadhesive Polymers and Their Applications in Drug Delivery Systems for the Treatment of Bladder Cancer. *Gels* **2022**, *8*, 587. https://doi.org/10.3390/gels8090587.

## References

1. Cao, H.; Duan, L.; Zhang, Y.; Cao, J.; Zhang, K. Current hydrogel advances in physicochemical and biological response-driven biomedical application diversity. *Signal Transduct. Target. Ther.* **2021**, *6*, 426. [CrossRef] [PubMed]
2. Kesharwani, P.; Bisht, A.; Alexander, A.; Dave, V.; Sharma, S. Biomedical applications of hydrogels in drug delivery system: An update. *J. Drug Deliv. Sci. Technol.* **2021**, *66*, 102914. [CrossRef]
3. Chakrapani, G.; Zare, M.; Ramakrishna, S. Intelligent hydrogels and their biomedical applications. *Mater. Adv.* **2022**, *3*, 7757–7772. [CrossRef]
4. Correa, S.; Grosskopf, A.K.; Lopez Hernandez, H.; Chan, D.; Yu, A.C.; Stapleton, L.M.; Appel, E.A. Translational applications of hydrogels. *Chem. Rev.* **2021**, *121*, 11385–11457. [CrossRef] [PubMed]

**Disclaimer/Publisher's Note:** The statements, opinions and data contained in all publications are solely those of the individual author(s) and contributor(s) and not of MDPI and/or the editor(s). MDPI and/or the editor(s) disclaim responsibility for any injury to people or property resulting from any ideas, methods, instructions or products referred to in the content.

Article

# A Covalently Cross-Linked Hyaluronic Acid/Carboxymethyl Cellulose Composite Hydrogel as a Potential Filler for Soft Tissue Augmentation

Francesca Della Sala [1,†], Mario di Gennaro [1,†], Pooyan Makvandi [2] and Assunta Borzacchiello [1,*]

1. Institute of Polymers, Composites and Biomaterials, National Research Council (IPCB-CNR), Viale J.F. Kennedy 54, 80125 Naples, Italy; francesca.dellasala@cnr.it (F.D.S.); mariodigennaro5@gmail.com (M.d.G.)
2. Centre of Research Impact and Outcome, Chitkara University Institute of Engineering and Technology, Chitkara University, Rajpura 140401, Punjab, India; pooyanmakvandi@gmail.com
* Correspondence: bassunta@unina.it or assunta.borzacchiello@cnr.it
† These authors contributed equally to this work.

**Abstract:** The use of fillers for soft tissue augmentation is an approach to restore the structure in surgically or traumatically created tissue voids. Hyaluronic acid (HA), is one of the main components of the extracellular matrix, and it is widely employed in the design of materials with features similar to human tissues. HA-based fillers already find extensive use in soft tissue applications, but are burdened with inherent drawbacks, such as poor thermal stability. A well-known strategy to improve the HA properties is to reticulate it with 1,4-Butanediol diglycidyl ether (BDDE). The aim of this work was to improve the design of HA hydrogels as fillers, by developing a crosslinking HA method with carboxymethyl cellulose (CMC) by means of BDDE. CMC is a water soluble cellulose ether, whose insertion into the hydrogel can lead to increased thermal stability. HA/CMC hydrogels at different ratios were prepared, and their rheological properties and thermal stability were investigated. The hydrogel with an HA/CMC ratio of 1/1 resulted in the highest values of viscoelastic moduli before and after thermal treatment. The morphology of the hydrogel was examined via SEM. Biocompatibility response, performed with the Alamar blue assay on fibroblast cells, showed a safety percentage of around 90% until 72 h.

**Keywords:** hyaluronic acid; carboxymethyl cellulose; BDDE; hydrogels; fillers

**Citation:** Della Sala, F.; di Gennaro, M.; Makvandi, P.; Borzacchiello, A. A Covalently Cross-Linked Hyaluronic Acid/Carboxymethyl Cellulose Composite Hydrogel as a Potential Filler for Soft Tissue Augmentation. *Gels* **2024**, *10*, 67. https://doi.org/10.3390/gels10010067

Academic Editor: Wei Ji

Received: 7 December 2023
Revised: 12 January 2024
Accepted: 13 January 2024
Published: 16 January 2024

**Copyright:** © 2024 by the authors. Licensee MDPI, Basel, Switzerland. This article is an open access article distributed under the terms and conditions of the Creative Commons Attribution (CC BY) license (https:// creativecommons.org/licenses/by/ 4.0/).

## 1. Introduction

Disease, trauma, and aging result in the loss of dermal collagen and fat, leading to deficits in soft tissue [1]. As a consequence, there is a need to develop materials that safely and effectively restore areas of deficiency. Soft tissue fillers have been used for decades for reconstructive and aesthetic procedures [2]. HA is a naturally occurring biopolymer, mainly concentrated in soft connective tissue extracellular matrix (ECM), dermis, vitreous body of the eye, hyaline cartilage, synovial joint fluid, intervertebral disc nucleus pulpous, and umbilical cord. HA consists of repeating disaccharide units composed of N-acetyl-D-glucosamine and D-glucuronic acid linked by a β-1,4 glycosidic, bond, whereas the disaccharides are linked by β-1,3 glycosidic bonds [3]. Its unique characteristics, such as biocompatibility, biodegradability, and mucoadhesiveness, as well as its viscoelastic properties, have led it to be used in a versatile manner in various biomedical applications [4]. HA-based hydrogels hold great promise for soft tissue engineering to replace damaged or lost tissues, since these biomaterials provide an environment close to the native ECM. It has been found that HA-based filler materials are useful in corrective and surgical fields, for example in aesthetic applications (such as facial contouring and in products for soft tissue augmentation), surgery, such as in sutures, drug administration, and moist wound dressing [5–7]. Moreover, HA-based hydrogels can be considered a potential implantable biomaterial for soft tissue augmentation or replacement [8]. Concentrated HA aqueous solutions are characterized by the presence of a self-aggregate polymer network in which intramolecular and intermolecular interactions are present due to the establishment of hydrogen bonds and hydrophobic interactions. [9]. When the material undergoes a shear stress, the so-formed physical network confers to these solutions the properties of a viscoelastic solid, if the strain or the time of application are small enough. Nevertheless, the weak intra- and interchain interactions are not able to maintain the structure upon prolonged stress, which causes the disentanglement of the network and the flow of the solution, which behaves as a viscoelastic fluid. In this frame, to improve the performance of the material as a filler, the chemical crosslinking of HA is necessary in order to increase the rigidity of the polymer network, extend its permanence in the site of application, and reduce its susceptibility to enzymatic degradation [10]. Several crosslinkers have been used to reticulate HA, such as 1,4-Butanediol diglycidyl ether (BDDE), 1,2,7,8-diepoxyoctane (DEO), divinyl sulfone (DVS), hexamethylenediamine (HMDA), and polyethylene glycol diglycidyl ether (PEGDE) [11]. A frequently used method today for crosslinking HA is the reaction with BDDE under alkaline conditions to yield a stable covalent ether linkage between HA and the cross-linker [12,13]. Further studies have evaluated the opportunity to prepare composite hydrogels by crosslinking HA with other polymers in order to obtain hydrogels with enhanced performances. For example, HA crosslinked with lactate-modified chitosan was observed to increase the elasticity of the material, because of the electrostatic interaction between the two polymers [14]. In another work, BDDE was used to crosslink HA and bacterial cellulose to obtain a wound dress with improved surface properties and mechanical and thermal resistance [15]. Indeed, the use of cellulose-based materials is also advantageous in the preparation of biomedical devices with a structural function, such as scaffolds and fillers. Cellulose is a linear polysaccharide of glucose, linked by β-(1,4)-glycosidic bonds, that has a structural task in plants. Cellulose is biocompatible and widely abundant in nature, and is stable in a physiological environment because amylases present in animals are not able to hydrolyze β-glycosidic bonds. At the same time, despite cellulose not being intrinsically hydrophobic, it is almost insoluble in water owing to the high degree of crystallinity [16]. This hindrance can be overcome by means of the chemical modification of the hydroxyl groups of cellulose, such as etherification or esterification, in order to make the polymer soluble in water and enlarge its field of application. Many water-soluble derivatives of cellulose are currently on the market, among which the most widespread is carboxymethyl cellulose (CMC). CMC's cellulose-derived polysaccharide is available in high-purity forms and has found, since it is FDA-approved, several biomedical applications due to its biocompatibility and low cost. CMC has a plant origin, which represents a key advantage over other natural fillers of animal origin, such as collagen, since it is less likely to elicit an immune response. Furthermore, the absence of the cellulase enzyme in humans, which digests cellulose, allows for adequate mechanical stability of CMC in vivo, compared to other natural biomaterial fillers that are susceptible to enzymatic activity. [17]. Several commercially available fillers such as Laresse, Radiesse, and Sculptra incorporate CMC with other materials, such as poly(ethylene oxide), hydroxyapatite, and poly(lactic acid). Most of these fillers use non-cross-linked formulations of CMC, which can potentially reduce their mechanical stability and in vivo retention time. Furthermore, the presence of synthetic components in these fillers can make selective removal following adverse reactions, migration, or placement very challenging, requiring invasive surgical procedures [18]. However, the right compromise between viscoelastic properties, the persistence of fillers, and biocompatibility must be achieved. The combination of HA and CMC is expected to improve CMC as a biomaterial while increasing HA's mechanical properties and their thermal stability. In this frame, a composite hydrogel based of HA/CMC crosslinked with BDDE has been developed for soft tissue augmentation for the first time with the main aim to obtain a better-performing HA-based hydrogel filler. HA/CMC composite hydrogels were developed at different ratios and their viscoelastic properties, thermal stability, and swelling ratio were investigated in order to optimize composition and reaction conditions. The morphology and the biocompatibility response of the hydrogel that possessed the best performance were then investigated.

## 2. Results and Discussion

### 2.1. The Design of the Protocol

In this work a protocol was developed for the preparation of a HA/CMC composite hydrogel, crosslinked with BDDE, for application in soft tissue augmentation. To achieve this goal, two protocols for crosslinking HA with the same concentration of HA and BDDE were compared, investigating the effect of time and temperature (Table 1) [19]. Briefly, HA was dissolved in a NaOH 1% $w/w$ solution, then BDDE was added and the mixture was left to react at 25 °C for 24 h and at 50 °C for 2 h, respectively. The materials prepared with the two protocols were labeled HA1 and HA2. In alkaline solution, the crosslinking of HA occurs by means of the nucleophilic addition of the primary hydroxyl groups present at the C-6 position on the N-acetyl-D-glucosamine to the epoxydic groups of BDDE, forming an ether bond (Figure 1) [20].

**Table 1.** Experimental conditions for the preparation of HA hydrogels.

| Sample | C HA (mg/mL) | C BDDE (μL/mL) | NaOH% | T, °C | Time, h | SR (w/w) |
|---|---|---|---|---|---|---|
| HA1 | 133.3 | 8.33 | 1 | 25 | 24 | 80 |
| HA2 | 133.3 | 8.33 | 1 | 50 | 2 | 54 |

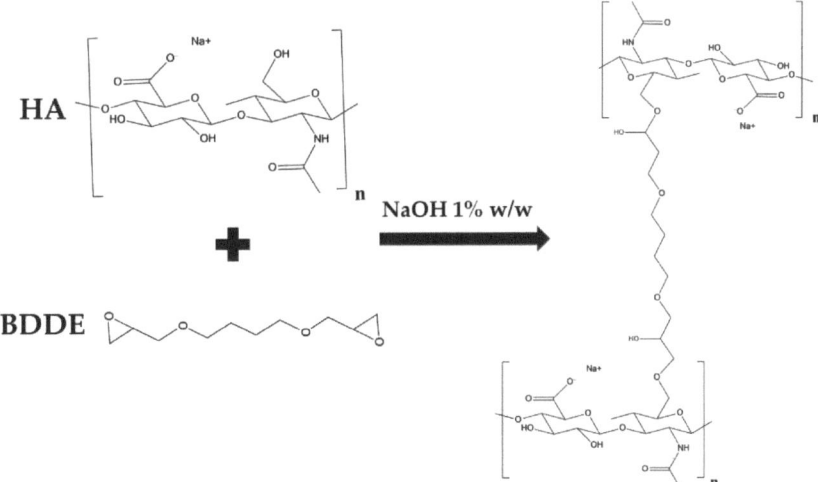

**Figure 1.** Scheme of the crosslinking reaction of HA by means of BDDE.

After, for each protocol, the material obtained was put in 200 mL of bi-distilled water for three days, in order to remove the unreacted BDDE, which is known for being toxic [21]. During this purification process, each hydrogel was periodically removed, dried from excess water and weighed, in order to calculate the swelling ratio (SR) (Table 1). For all of the hydrogels examined in this work, the equilibrium was reached after 24 h. After the purification was completed, the viscoelastic moduli of the two materials were measured. The storage modulus $G'$, the loss modulus $G''$ and the loss factor tanδ as a function of the frequency were reported in Figure 2. $G'$ and $G''$ represent the elastic and viscous response of the material, respectively, and their ratio tanδ expresses the behavior of the viscoelastic materials.

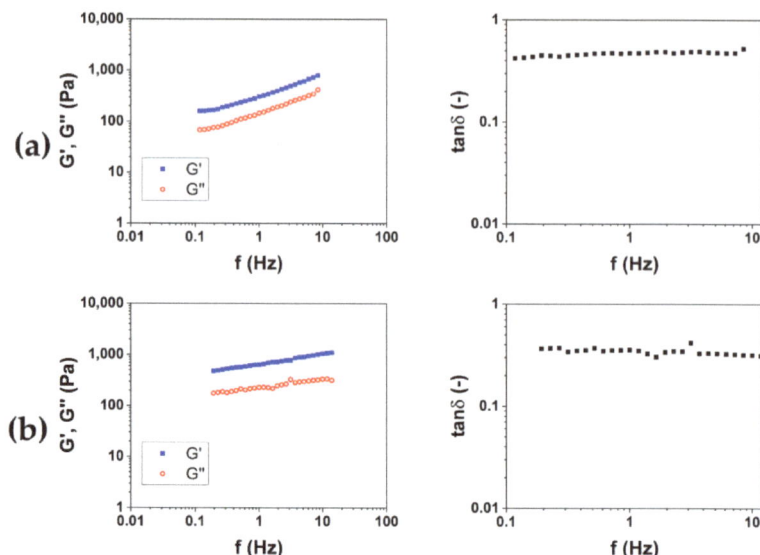

**Figure 2.** Representative images of mechanical spectra and tanδ of HA-based hydrogels crosslinked with BDDE (HA1) at 25 °C for 24 h (**a**) and (HA2) at 50 °C for 2 h (**b**).

The results of rheological analysis are expressed in terms of the value of the storage modulus G' and tanδ at 1 and 10 Hz as reported in Table 2 Both the materials exhibit a gel-like behavior, for which G' > G'' and tanδ < 1 in all of the frequency ranges investigated. The quantitative analysis shows how the HA2 hydrogel, prepared at 50 °C for 2 h, exhibits higher values of the viscoelastic modulus G' (640 ± 40 at 1 Hz and 1050 ± 40 at 10 Hz), lower values of tanδ (0.4 at 1 Hz and 0.32 at 10 Hz) and lower value of SR (54 $w/w$), compared to HA1 hydrogels. In order to evaluate the best protocol to prepare HA/CMC composite hydrogels, CMC-based hydrogels, crosslinked with BDDE, were prepared with the two methods reported in Table 3. CMC is able to form hydrogels with BDDE (Figure 3) by reacting with its carboxylate groups, which act as nucleophiles [22]. Representative mechanical spectra of the two hydrogels, labeled CMC1 and CMC2, respectively, are reported in Figure 4.

**Table 2.** Storage modulus G' and loss factor of the HA-based and CMC-based hydrogels.

| Sample | 1 Hz | | 10 Hz | |
|---|---|---|---|---|
| | G' (Pa) | tanδ (-) | G' (Pa) | tanδ (-) |
| HA1 | 310 ± 30 | 0.5 ± 0.02 | 420 ± 30 | 0.52 ± 0.02 |
| HA2 | 640 ± 40 | 0.4 ± 0.02 | 1050 ± 40 | 0.32 ± 0.02 |
| CMC1 | 130 ± 30 | 0.4 ± 0.02 | 220 ± 30 | 0.35 ± 0.02 |
| CMC2 | 55 ± 30 | 0.7 ± 0.02 | 120 ± 30 | 0.52 ± 0.02 |

**Table 3.** Experimental conditions for the preparation of CMC hydrogels.

| Sample | C CMC (mg/mL) | C BDDE (μL/mL) | NaOH% | T, °C | Time, h | SR ($w/w$) |
|---|---|---|---|---|---|---|
| CMC1 | 133.3 | 8.33 | 1 | 25 | 24 | 90 |
| CMC2 | 133.3 | 8.33 | 1 | 50 | 2 | 18 |

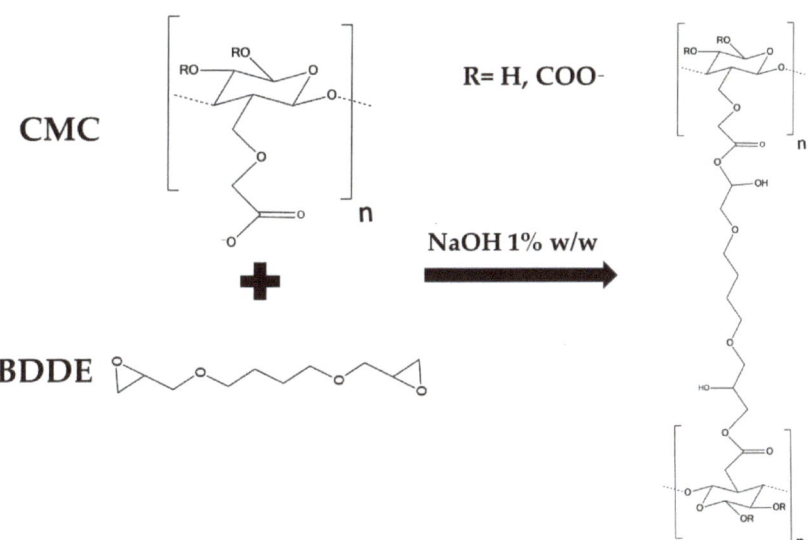

**Figure 3.** Scheme of the crosslinking reaction of CMC by means of BDDE.

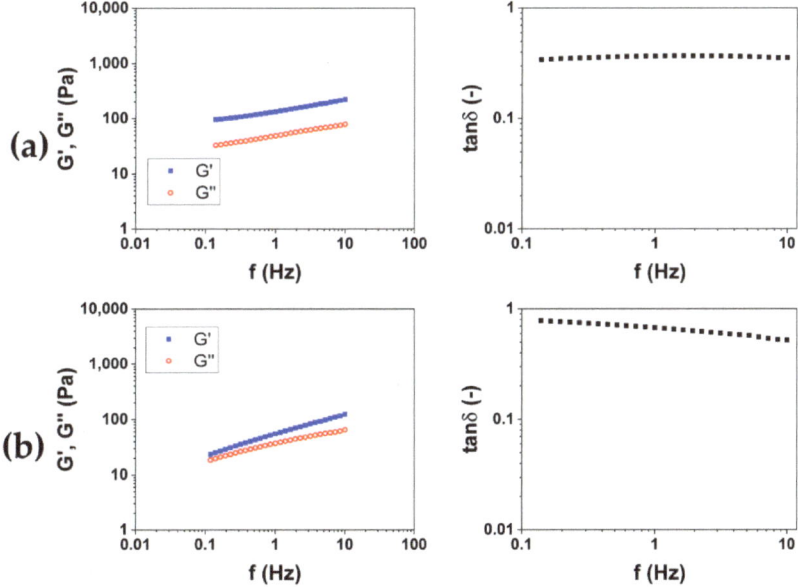

**Figure 4.** Representative images of mechanical spectra and tanδ of CMC-based hydrogels crosslinked with BDDE (CMC1) at 25 °C for 24 h (a) and (CMC2) at 50 °C for 2 h (b).

Differently from HA-based hydrogels, a higher stiffness is observed for the hydrogel CMC1, prepared at 25 °C for 24 h. Furthermore, the mechanical spectrum of the hydrogel CMC2, obtained at 50 °C for 2 h, almost does not present a trend of the moduli typical of a hydrogel (Table 2). Contrariwise, a rising trend of the viscoelastic moduli is observed (Figure 4b), and it appears that an intersection of the curves could be present for lower values of frequency. This mechanical behavior is more assimilable to entanglement polymer solutions than a hydrogel [23]. Also, the SR calculated for CMC2 was lower (18 $w/w$)

compared to CMC1 (90 $w/w$), despite it being a stiffer hydrogel. It follows that, for CMC, reacting for short times at a higher temperature could not lead to a crosslinking of the material to form a hydrogel. For this reason, the protocol at 25 °C for 24 h was chosen for the preparation of the HA/CMC composite hydrogels.

### 2.2. The Optimization of the Composition

The HA/CMC composite hydrogels were prepared following the same procedure used for HA-based and CMC-based hydrogels. As reported in Table 4, hydrogels prepared with the HA/CMC weight ratios of 1:3, 1:1, and 3:1, named HCM1, HCM2, and HCM3, respectively, were examined.

**Table 4.** Composition, protocol and swelling ratio (SR) of the HA/CMC composite hydrogels prepared.

| Sample | HA/CMC Weight Ratio | C HA (mg/mL) | C CMC (mg/mL) | C BDDE (µL/mL) | T (°C) | Time (h) | SR ($w/w$) |
|---|---|---|---|---|---|---|---|
| HCM1 | 3/1 | 100 | 33.3 | 8.33 | 25 | 24 | 58 |
| HCM2 | 1/1 | 66.6 | 66.6 | 8.33 | 25 | 24 | 76 |
| HCM3 | 1/3 | 33.3 | 100 | 8.33 | 25 | 24 | 50 |

After the purification was completed, and SR was assessed, in order to investigate the viscoelastic properties of the hydrogels, the Frequency Sweep (FS) tests were performed at 20 °C and 37 °C on the three samples. The results of the frequency sweep tests of the HCM1, HCM2, and HCM3 were reported in Figure 5a, b, and c, respectively. The mechanical spectra were analyzed in terms of the dependence of the storage modulus G′ and of the loss factor tanδ as a function of the frequency for the two temperatures (Table 5). All of the HA/CMC-based materials exhibit a trend of the viscoelastic moduli and tanδ proper of the hydrogels. The data collected at 20 °C and 37 °C was nearly overlapping, and the irrelevant effects of temperature were observed. The composite hydrogels HCM1 and HCM2 exhibited higher viscoelastic moduli and lower loss factor, compared to the HA1 hydrogel, at the same overall polymer concentration. HA and CMC appear, therefore, to form a synergical network with an increase of rheological properties after the crosslinking. In particular, the hydrogel HCM2, prepared with an HA/CMC weight ratio of 1/1, is the hydrogel with the higher values of the moduli and simultaneously higher SR. Finally, according to the data collected for the CMC-based hydrogels, the sample HCM3, with the highest content of CMC, was the hydrogel with the lowest mechanical properties among the three prepared.

**Table 5.** Viscoelastic properties of the HA/CMC composite hydrogels measured at 20 °C and 37 °C.

| Sample | 20 °C, 1 Hz | | 20 °C, 10 Hz | | 37 °C, 1 Hz | | 37 °C, 10 Hz |
|---|---|---|---|---|---|---|---|
| | G′ (Pa) | tanδ (-) | G′ (Pa) | tanδ (-) | G′ (Pa) | tanδ (-) | G′ (Pa) |
| HCM1 | 1700 ± 130 | 0.13 ± 0.03 | 2000 ± 120 | 0.13 ± 0.02 | 1100 ± 200 | 0.19 | 1400 ± 200 |
| HCM2 | 1900 ± 100 | 0.15 ± 0.05 | 2300 ± 100 | 0.2 ± 0.04 | 1800.0 ± 200 | 0.14 | 2200 ± 100 |
| HCM3 | 380 ± 90 | 0.25 ± 0.05 | 510 ± 150 | 0.25 ± 0.03 | 360 ± 120 | 0.2 | 500 ± 100 |

In order to obtain a product suitable for withstanding high-impact treatments like, for example, thermal sterilization, the rheological characterization of the hydrogels has been completed by measuring the viscoelastic moduli as a function of the frequency, at 20 °C and 37 °C, after sterilization in an autoclave (AC) at 121 °C for 20 min. The representative mechanical spectra of the three autoclaved hydrogels were reported in Figure 6a–c, and the rheological properties were reported in Table 6.

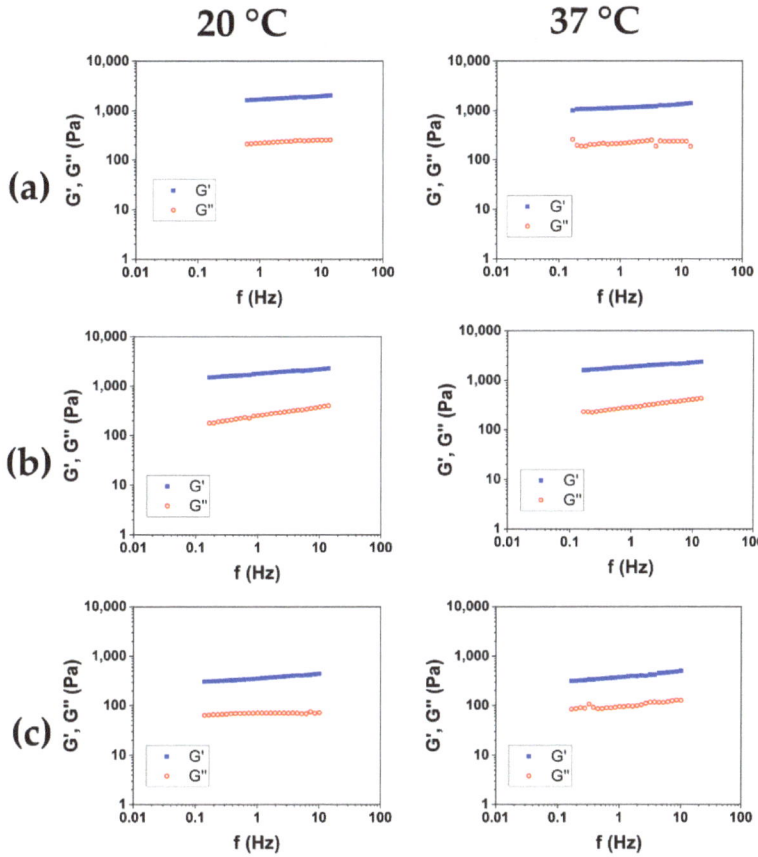

**Figure 5.** Representative mechanical spectra of HCM1 (**a**), HCM2 (**b**), and HCM3 (**c**) at 20 °C and 37 °C.

**Table 6.** Viscoelastic properties of the HA/CMC hydrogels after sterilization in an autoclave, measured at 20 °C and 37 °C.

| Sample | 20 °C, 1 Hz | | | 37 °C, 1 Hz | | |
|---|---|---|---|---|---|---|
| | G' (Pa) | tanδ (-) | G'/G'$_{AC}$ (-) | G' (Pa) | tanδ (-) | G'/G'$_{AC}$ (-) |
| HCM1 | 550 ± 90 | 0.15 ± 0.03 | 3.1 | 550 ± 110 | 0.13 ± 0.04 | 2.1 |
| HCM2 | 870 ± 120 | 0.13 ± 0.05 | 2.2 | 800 ± 150 | 0.09 ± 0.02 | 2.3 |
| HCM3 | 140 ± 90 | 0.30 ± 0.02 | 2.7 | 190 ± 90 | 0.30 ± 0.03 | 1.9 |

Among the three hydrogels, after the autoclave sterilization, HCM2 has the highest storage modulus, and the lower value of tanδ. Furthermore, compared to the material that was not autoclaved, expressed as G'/G'$_{AC}$, HCM2 has the lower loss of mechanical properties at 20 °C, while at 37 °C the loss of mechanical properties was comparable, probably because of the thermal degradation, which caused a reduction in the structuration of the materials.

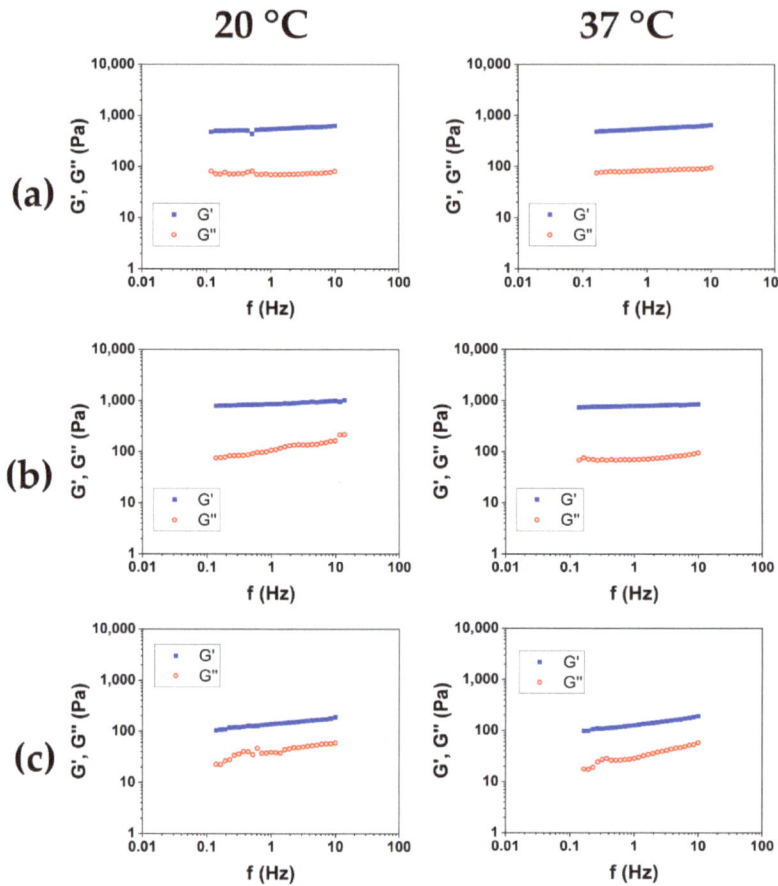

**Figure 6.** Mechanical spectra of HCM1 (**a**), HCM2 (**b**), and HCM3 (**c**) after sterilization in an autoclave at 20 °C and 37 °C.

### 2.3. Fourier-Transformed Infrared (FT-IR) Analysis

The chemical modification of HA/CMC composite hydrogels has been confirmed by FTIR. The ATR-FTIR spectra acquired from the various samples HCM1, HCM2, and HCM3 and the single components of the native HA and CMC are shown in Figure 7a, along with the comparison between HCM1 and HA/CMC 3/1 (b), HCM2 and HA/CMC 1/1 (c), and HCM3 and HA/CMC 1/3 (d) before and after the addition of BDDE. It can be possible to identify the typical polysaccharide –OH signals at the region between the 3000 cm$^{-1}$–3700 cm$^{-1}$ [24,25]. The peaks in the region between the 3000 cm$^{-1}$ and 2700 cm$^{-1}$ are associated with the stretching of –CH$_2$ and –CH$_3$, while the peaks at 1600 cm$^{-1}$ and 1412 cm$^{-1}$ are associated with the symmetric and asymmetric stretching of –COO groups. Finally, at 1030 cm$^{-1}$ the C-O-C symmetric stretching ether bands are observed. The comparison between the spectra of the polymers mixtures after and before the addition of the crosslinking agent BDDE provided information about the chemical modification occurring in the composite hydrogels. Indeed, the peak associated with the stretching of –CH$_2$ and –CH$_3$ changes in shape and intensity after the crosslinking with BDDE occurs. Moreover, the presence of the BDDE covalently cross-linked to polysaccharides is suggested by the absence of the peaks 1256 cm$^{-1}$ and 908 cm$^{-1}$ that, according to the literature, belong to the asymmetric and symmetric stretching vibrations of the epoxy groups of BDDE [26].

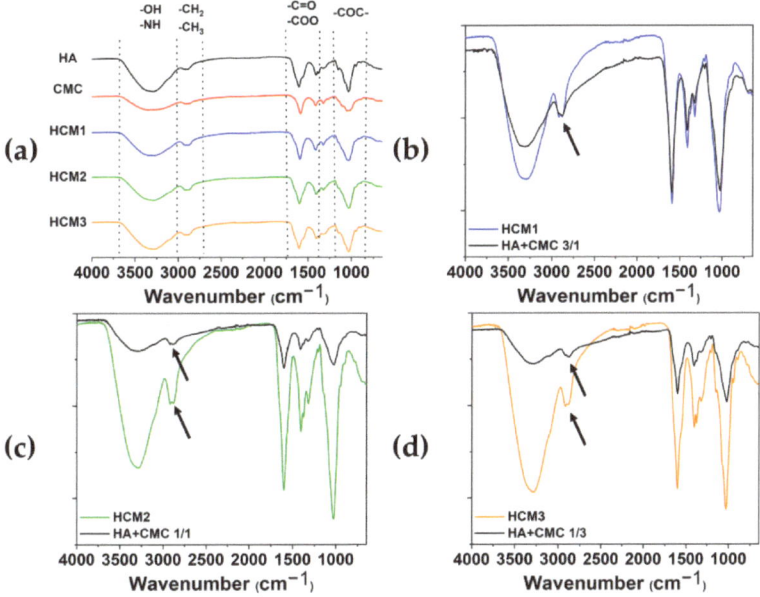

Figure 7. ATR-FTIR spectra of HA, CMC, HMC1, HMC2 and HMC3 (**a**); HCM1 and HA/CMC 3/1 before adding BDDE (**b**); HCM2 and HA/CMC 1/1 before adding BDDE (**c**); and HCM3 and HA/CMC 1/3 before adding BDDE (**d**). The arrows highlighted the changes in shape and intensity of the peak associated with the stretching of $-CH_2$ and $-CH_3$ after the crosslinking with BDDE.

### 2.4. Morphological SEM Evaluation

The morphology of the selected hydrogel HCM2, with higher mechanical performance, was qualitatively investigated by means of Scanning Electron Microscopy analysis (Figure 8). The images acquired show the dense crosslinking networks responsible for the rheological properties and for the thermal stability of the hydrogel. In particular, Figure 8c shows information about the cross section of the HCM2 gels, in which it is possible to observe the porosity and scaffold interconnection [27].

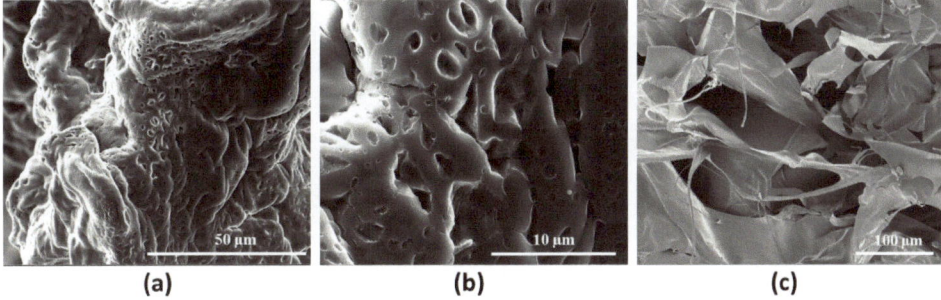

Figure 8. Representative SEM images of HCM2 at (**a**) 3000× and (**b**) 12,000× magnification and its cross section 800× (**c**) magnification.

## 2.5. Biological Response

In vitro biological response in terms of the cell viability percentage and morphological analysis of cells represents a key feature in evaluating the design of hydrogels useful as fillers in soft tissue augmentation applications. The first safety assessment has been investigated for the selected hydrogel HCM2, which possessed the best performance properties in terms of rheological properties and swelling. It has been widely established that the viscoelastic characteristics of the materials influence cell behavior, affecting the biocompatibility of the cells [28,29]. Biocompatibility results were first assessed via cell morphology (Figure 9a). Actin filaments, a constituent of the cytoskeleton of the cells, were stained with FITC phalloidin after 24 h of incubation with HCM2 hydrogel and in untreated control. Both treated and untreated L929 cells samples exhibited a typical non-cytotoxic fibroblast-like morphology [30]. The quantitative analysis of L929 cell viability percentage has been evaluated using an Alamar blue (AB) assay as reported in Figure 9b. The HCM2 hydrogels showed good safety after 24, 48, and 72 h of incubation with L929 cells, compared to the untreated cells control. In particular, after 24, 48, and 72 h of the incubation of HCM2, L929 cell viability percentage is around 90%, indicating, in accordance with the ISO 10993–5: 2009 standards [31], the good biocompatibility of the hydrogels. Indeed, these standards specify that cell viabilities greater than 70% indicate the non-cytotoxic behavior of tested biomaterials, thus, suggesting the absence of the toxic BDDE residues in the hydrogels [32]. Overall, these results indicated that the HCM2 hydrogels have good biocompatibility properties, indicating that the purification process of the HCM2 gel following crosslinking by means of BDDE was successfully achieved. In fact, all of the unreacted crosslinker, which is notoriously cytotoxic, appears to have been removed during the swelling process.

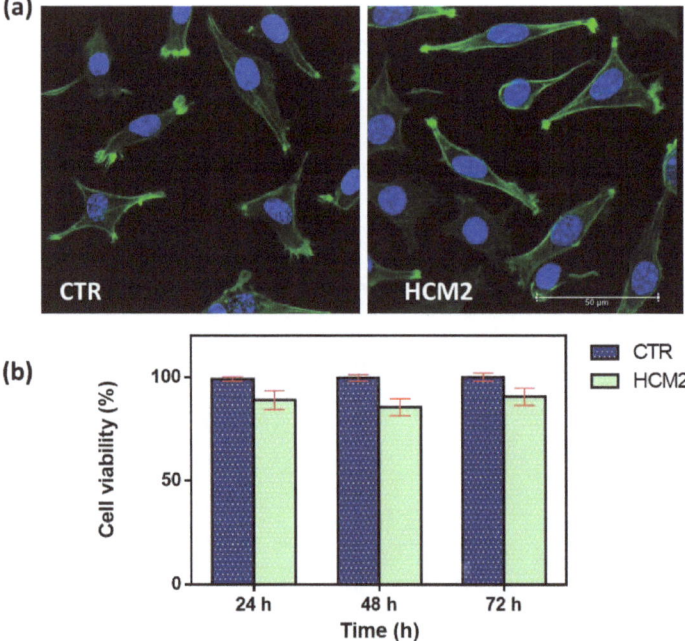

**Figure 9.** (a) Representative cell morphology of control L929 fibroblast cells and L929 after 24 h of incubation with the selected HCM2 hydrogels. Actin filaments, stained with phalloidin-FITC (green), and DAPI (blue)-stained nuclei cells. Images were acquired with a resolution of 1024 × 1024 pixels with a 63× oil immersion objective. (b) Cell viability percentage (%) of the L929 cells, after incubation at 24, 48, and 72 h compared to the control untreated cells. The data are representative of three repeated experiments in triplicate.

## 3. Conclusions

HA-based fillers are widely employed in soft tissue augmentation, both in cosmetic and in chirurgical practice, but the performance of HA-based devices is often affected by low mechanical properties and poor resistance to sterilization processes. In this work, for the first time, HA/CMC composite hydrogels crosslinked with BDDE were prepared. The optimal conditions in terms of operative temperature and reaction time to crosslink HA and CMC separately were evaluated with a rheological analysis of the hydrogels. The results suggested that the better conditions for reticulating HA and CMC were 25 °C for 24 h. Then, hydrogels with a different HA/CMC ratio were prepared, and their rheological properties were investigated before and after sterilization in an autoclave at 121 °C for 20 min. The hydrogel with an HA/CMC ratio of 1/1, labeled HCM2, exhibited the highest values of the viscoelastic moduli before and after the thermal treatment. The FTIR has given us information about the occurrence of chemical modification by means of BDDE in the composite hydrogels. The morphological analysis of HCM2 via SEM highlighted the densely reticulate structure of the hydrogel. Finally, the biocompatibility response, as shown by incubating HCM2 on L929 cell fibroblast, indicated a good cell viability percentage of around 90% at 24, 48, and 72 h and along with the morphological analysis, showed the overall success of the purification process by washing. However, significant research gaps remain, including a lack of long-term in vivo investigations and immune-toxicity evaluations. Further work is needed to understand the practical applications of these HA hydrogel composites. Nevertheless, the use of CMC resulted in improvements in the HA-based hydrogel properties, indicating that HA/CMC composite hydrogels cross-linked with BDDE represent a promising platform for the design of filler implants in soft tissue augmentation.

## 4. Materials and Methods

### 4.1. Hydrogel Preparation

Hyaluronic acid (803 KDa) and CMC (750 KDa) were kindly provided by Altergon s.r.l (Morra De Sanctis, Italy), BDDE was purchased from Sigma Aldrich. Two protocols for the crosslinking of HA with BDDE were evaluated as a starting point for the preparation of the hydrogels [19]. The conditions employed were reported in Tables 1 and 2, and were used to prepare HA/CMC-based hydrogels at different polymer ratios. According to the literature, the use of a high polymer concentration is necessary to obtain HA-based covalently crosslinked hydrogels [33,34]. HA and CMC hydrogels were also prepared as a reference. Briefly, for each protocol NaOH 1% $w/w$ solution was gradually added to dry polymer powder, and the mixture was gently stirred with a spatula, in order to promote polymer hydration and avoid the formation of bubbles. When the addition of NaOH solution was complete, the mixture was left for 24 h at room temperature, after that it appeared as homogeneous systems. Then, 25 µL of BDDE were added to the mixture, which was left reacting for the time required for each protocol (2 h at 50 °C or 24 h at RT). Once the time had passed, the mixture was neutralized by adding HCl, and the hydrogels were put in 200 mL of water for 3 days to remove unreacted BDDE.

### 4.2. Swelling Ratio

In order to evaluate the swelling ratio (SR) of each hydrogel prepared, during the purification the hydrogels were withdrawn periodically, dried of excess water, and weighed until the mass was stable over time. For all of the hydrogels the equilibrium was reached after 24 h. After purification was completed, the hydrogels were frozen at T = −80 °C and freeze dried. The dry mass was then weighed, and the swelling ratio was calculated according to Equation (1):

$$SR = \frac{W_h - W_d}{W_d} \qquad (1)$$

where $W_h$ is the mass of the hydrogel at the equilibrium and $W_d$ is the mass of the dry hydrogel.

## 4.3. Rheological Analysis

The viscoelastic properties of the hydrogels were evaluated by means of oscillatory regime tests using a rotational rheometer Haake Mars III (Thermo Fisher Scientific, Waltham, MA, USA) equipped with a parallel plate geometry, 35 mm plate diameter, 0.5 mm gap, and a thermostatic bath. Hydrogels are viscoelastic materials, and their mechanical response to a shear stress $\tau$ presents elements of an elastic solid (Hooke's law, Equation (2)) and of a Newtonian liquid (Newton's law, Equation (3)):

$$\tau = G\gamma \quad (2)$$

$$\tau = \eta \, d\gamma/dt \quad (3)$$

where G is the elastic modulus of the solid, $\eta$ is the dynamic viscosity of the liquid and $\gamma$ is the imposed deformation. When a sinusoidal strain with oscillatory frequency $\omega$ is exerted over time, the response of the elastic solid (Equation (3)) results shifted by 90° compared to that of the ideal fluid Equation (4). The frequency $\omega$ is the oscillation frequency in $s^{-1}$, which can also be reported as $2\pi f$, where f is in Hz.

$$\tau = G\gamma_0 \sin(\omega t) \quad (4)$$

The mechanical response of a viscoelastic material under a sinusoidal stress can be therefore written as in Equation (5):

$$\tau = G^* \gamma_0 \sin(\omega t + \delta) \quad (5)$$

where $G^*$ is the complex modulus of the material, $\gamma_0$ is the amplitude of the strain and $\delta$ is the shift with respect to the behavior of the ideal solid, and is comprised between 0° and 90°. Applying the sum sin identity, Equation (6) can be written as:

$$\tau = G^* \gamma_0 \sin(\omega t)\cos(\delta) + G^* \gamma_0 \cos(\omega t)\sin(\delta) \quad (6)$$

Equation (6) expresses the response of a viscoelastic material under sinusoidal strain as the sum of two contributions, one in phase with the strain, and one shifted by 90. Inside the equation it is possible to define two viscoelastic moduli (Equations (7) and (8)):

$$G' = G^* \cos(\delta) \quad (7)$$

$$G'' = G^* \sin(\delta) \quad (8)$$

where $G'$ is the storage modulus, and expresses the elastic response of the material, and $G''$ is the loss modulus, and expresses the viscous response of the material. Oscillatory tests allow, therefore, to break down into two contributions the mechanical response of a viscoelastic material.

In this work frequency sweep (FS) tests were performed to measure the viscoelastic moduli $G'$ and $G''$ as a function of frequency in the range 0.1–13 Hz at a fixed strain of 0.5%. The value of the strain was chosen to have a linear viscoelastic response, independent on the strain itself, and was determined by means of strain sweep tests. Form FS tests the ratio between $G''$ and $G'$, the loss factor tan $\delta$ (Equation (9)) was calculated as follows:

$$\tan\delta = \frac{G''}{G'} \quad (9)$$

tan $\delta$ expresses the ratio between the viscous and the solid ratio of the material [35].

Rheological analysis was also performed in order to evaluate the thermal stability of the hydrogels. The hydrogels were treated in an autoclave for 20 min at 121 °C, and FS tests were carried out on the heat-treated hydrogels. The reduction in mechanical properties was expressed for each material by dividing $G'$ at 1 Hz before the autoclave for $G'$ at 1 Hz after the autoclave.

*4.4. Fourier-Transformed Infrared (FT-IR) Analysis*

Portions from HA and CMC hydrogel and cross-linked composite hydrogels were obtained and characterized using Perkin Elmer Frontier Fourier Transform Infrared Spectroscopy FT-IR (Waltham, MA, USA), with a single-reflection, universal ATR-IR accessory. All spectra were recorded between 4000 and 650 cm$^{-1}$ with a resolution of 4 cm$^{-1}$ and the data were manipulated using OriginPro 2018 software.

*4.5. Morphological Analysis*

In order to obtain qualitative morphological information, scanning electron microscopy (Quanta 200 FEG, FEI Company, Hillsboro, OR, USA) was employed. The samples were lyophilized and platinum/palladium–sputtered to perform the analysis.

*4.6. Biological Resposnse*

4.6.1. Cell Culture

In order to evaluate the biological response of the HA/CMC composite hydrogels, mouse fibroblast L929 cells derived from mouse C34/An connective tissues (Sigma-Aldrich, Burlington, MA, USA) were grown in a T-75 cell culture flask (VWR, Radnor, PA, USA) at 37 °C and 5% $CO_2$. Cell culture medium Dulbecco's Modified Eagle's medium (DMEM) (Microgem, Naples, Italy) supplemented with 10% of fetal bovine serum and antibiotics (penicillin G sodium 100 U/mL, streptomycin 100 µg/mL) were used and changed every 3–4 days.

4.6.2. Cell Viability and Morphological Assay

To assess the cell morphology, L929 cells were seeded at a density of $1 \times 10^4$ cells/mL on fluorodish-35 mm (VWR, Radnor, PA, USA). The selected hydrogel HCM2 was sectioned and deposited in 3 wells of a 24 well plate and UV sterilization was carried out at 284 nm for 30 min. Subsequently, the DMEM was added until the samples were covered: these were left in an incubator at 37 °C for 24 h. The cells were incubated with the hydrogel eluate for 24 h. Then, the samples were washed three times with PBS and fixed with 10% formaldehyde for 1 h at 4 °C. Cells were permeabilized with Triton X-100 0.1% in PBS for 3–5 min. The actin filaments were stained with FITC phalloidin/PBS for 30 min at RT. Finally, after two washes with PBS to remove the unbound phalloidin, cell nuclei were stained with 4′,6-diamidino-2-phenylindole DAPI, (Sigma-Aldrich). The samples were observed using a confocal microscope system (Leica TCS SP8) with a 63× oil immersion objective. Images were acquired with a resolution of 1024 × 1024 pixels.

In order to study the cell viability, a density of $5 \times 10^3$ cells/mL of L929 cells were seeded on a 96-well plate (World Precision Instruments, Inc., Sarasota, FL, USA). The cells were then incubated for 24, 48, and 72 h with 200 µL of the HCM2 eluate. Then, the Alamar blue assay (AB) was performed by adding AB reagent, at 10% $v/v$ with respect to the medium to the samples and incubated at 37 °C for 4 h. The absorbance of the samples was measured using a spectrophotometer plate reader (Multilabel Counter, 1420 Victor, Perkin Elmer, Waltham, MA, USA) at 570 nm and 600 nm. The AB reagent dye indicates an oxidation-reduction by changing color in response to the chemical reduction in the growth medium, resulting from cell viability. Data are expressed as the percentage difference between treated and control to evaluate the percentage of reduction (Reduction %), which is calculated with the following formula (Equation (10)):

$$Reduction\ (\%) = \frac{(O_2 \times A_1) - (O_1 \times A_2)}{(O_2 \times P_1) - (O_1 \times P_2)} \times 100 \qquad (10)$$

where $O_1$ is the molar extinction coefficient ($E$) of oxidized AB at 570 nm; $O_2$ is the $E$ of oxidized AB at 600 nm; $A_1$ is the absorbance of test wells at 570 nm; $A_2$ is the absorbance of test wells at 600 nm; $P_1$ is the absorbance of the control well at 570 nm; and $P_2$ is the absorbance of the control well at 600 nm. The percentage of reduction for each sample was normalized to the percentage of reduction for the control to obtain the cell viability percentage.

**Author Contributions:** Conceptualization, A.B., F.D.S., M.d.G. and P.M.; methodology, F.D.S., M.d.G. and P.M.; validation, A.B.; investigation F.D.S., M.d.G. and P.M.; resources, A.B.; writing—original draft preparation, F.D.S., M.d.G.; writing—review and editing, F.D.S., M.d.G. and A.B.; visualization A.B.; supervision, A.B. All authors have read and agreed to the published version of the manuscript.

**Funding:** This research received no external funding.

**Institutional Review Board Statement:** Not applicable.

**Informed Consent Statement:** Not applicable.

**Data Availability Statement:** The data presented in this study are openly available in article.

**Acknowledgments:** The authors thank Fabio Borbone of the University of Naples "Federico II", for the acquisition of the SEM image present in Figure 8c.

**Conflicts of Interest:** The authors declare no conflicts of interest.

# References

1. Khalid, K.A.; Nawi, A.F.M.; Zulkifli, N.; Barkat, M.A.; Hadi, H. Aging and wound healing of the skin: A review of clinical and pathophysiological hallmarks. *Life* **2022**, *12*, 2142. [CrossRef] [PubMed]
2. Heitmiller, K.; Ring, C.; Saedi, N. Rheologic properties of soft tissue fillers and implications for clinical use. *J. Cosmet. Dermatol.* **2021**, *20*, 28–34. [CrossRef] [PubMed]
3. Della Sala, F.; Longobardo, G.; Lista, G.; Messina, F.; Borzacchiello, A. Effect of hyaluronic acid and mesenchymal stem cells secretome combination in promoting alveolar regeneration. *Int. J. Mol. Sci.* **2023**, *24*, 3642. [CrossRef] [PubMed]
4. Della Sala, F.; di Gennaro, M.; Lista, G.; Messina, F.; Ambrosio, L.; Borzacchiello, A. Effect of Hyaluronic Acid on the Differentiation of Mesenchymal Stem Cells into Mature Type II Pneumocytes. *Polymers* **2021**, *13*, 2928. [CrossRef] [PubMed]
5. Cassuto, D.; Bellia, G.; Schiraldi, C. An overview of soft tissue fillers for cosmetic dermatology: From filling to regenerative medicine. *Clin. Cosmet. Investig. Dermatol.* **2021**, *14*, 1857–1866. [CrossRef]
6. Petrie, K.; Cox, C.T.; Becker, B.C.; MacKay, B.J. Clinical applications of acellular dermal matrices: A review. *Scars Burn. Heal.* **2022**, *8*, 20595131211038313. [CrossRef] [PubMed]
7. Rohrich, R.J.; Ghavami, A.; Crosby, M.A. The role of hyaluronic acid fillers (Restylane) in facial cosmetic surgery: Review and technical considerations. *Plast. Reconstr. Surg.* **2007**, *120*, 41S–54S. [CrossRef]
8. Sionkowska, A.; Gadomska, M.; Musiał, K.; Piątek, J. Hyaluronic acid as a component of natural polymer blends for biomedical applications: A review. *Molecules* **2020**, *25*, 4035. [CrossRef]
9. Fallacara, A.; Manfredini, S.; Durini, E.; Vertuani, S. Hyaluronic acid fillers in soft tissue regeneration. *Facial Plast. Surg.* **2017**, *33*, 087–096. [CrossRef]
10. Damiano Monticelli, V.; Mocchi, R.; Rauso, R.; Zerbinati, U.; Cipolla, G.; Zerbinati, N. Chemical characterization of hydrogels crosslinked with polyethylene glycol for soft tissue augmentation. *Open Access Maced. J. Med. Sci.* **2019**, *7*, 1077.
11. Fundarò, S.P.; Salti, G.; Malgapo, D.M.H.; Innocenti, S. The rheology and physicochemical characteristics of hyaluronic acid fillers: Their clinical implications. *Int. J. Mol. Sci.* **2022**, *23*, 10518. [CrossRef] [PubMed]
12. Wende, F.J.; Gohil, S.; Mojarradi, H.; Gerfaud, T.; Nord, L.I.; Karlsson, A.; Boiteau, J.-G.; Kenne, A.H.; Sandström, C. Determination of substitution positions in hyaluronic acid hydrogels using NMR and MS based methods. *Carbohydr. Polym.* **2016**, *136*, 1348–1357. [CrossRef] [PubMed]
13. Faivre, J.; Pigweh, A.I.; Iehl, J.; Maffert, P.; Goekjian, P.; Bourdon, F. Crosslinking hyaluronic acid soft-tissue fillers: Current status and perspectives from an industrial point of view. *Expert Rev. Med. Devices* **2021**, *18*, 1175–1187. [CrossRef] [PubMed]
14. Daminato, E.; Bianchini, G.; Causin, V. New Directions in Aesthetic Medicine: A Novel and Hybrid Filler Based on Hyaluronic Acid and Lactose Modified Chitosan. *Gels* **2022**, *8*, 326. [CrossRef] [PubMed]
15. Tang, S.; Chi, K.; Xu, H.; Yong, Q.; Yang, J.; Catchmark, J.M. A covalently cross-linked hyaluronic acid/bacterial cellulose composite hydrogel for potential biological applications. *Carbohydr. Polym.* **2021**, *252*, 117123. [CrossRef] [PubMed]
16. Gopinath, V.; Kamath, S.M.; Priyadarshini, S.; Chik, Z.; Alarfaj, A.A.; Hirad, A.H. Multifunctional applications of natural polysaccharide starch and cellulose: An update on recent advances. *Biomed. Pharmacother.* **2022**, *146*, 112492. [CrossRef]
17. Varma, D.M.; Gold, G.T.; Taub, P.J.; Nicoll, S.B. Injectable carboxymethylcellulose hydrogels for soft tissue filler applications. *Acta Biomater.* **2014**, *10*, 4996–5004. [CrossRef]

18. Leonardis, M.; Palange, A.; Dornelles, R.F.; Hund, F. Use of cross-linked carboxymethyl cellulose for soft-tissue augmentation: Preliminary clinical studies. *Clin. Interv. Aging* **2010**, *5*, 317–322. [CrossRef]
19. Malson, T.; Lindqvist, B.L. Gel of Crosslinked Hyaluronic Acid for Use as a Vitreous Humor Substitute. US4716154A, 29 December 1987.
20. De Boulle, K.; Glogau, R.; Kono, T.; Nathan, M.; Tezel, A.; Roca-Martinez, J.-X.; Paliwal, S.; Stroumpoulis, D. A review of the metabolism of 1, 4-butanediol diglycidyl ether-crosslinked hyaluronic acid dermal fillers. *Dermatol. Surg.* **2013**, *39*, 1758–1766. [CrossRef]
21. Jeong, C.H.; Yune, J.H.; Kwon, H.C.; Shin, D.-M.; Sohn, H.; Lee, K.H.; Choi, B.; Kim, E.S.; Kang, J.H.; Kim, E.K. In vitro toxicity assessment of crosslinking agents used in hyaluronic acid dermal filler. *Toxicol. In Vitro* **2021**, *70*, 105034. [CrossRef]
22. Lawal, O.S.; Yoshimura, M.; Fukae, R.; Nishinari, K. Microporous hydrogels of cellulose ether cross-linked with di-or polyfunctional glycidyl ether made for the delivery of bioactive substances. *Colloid Polym. Sci.* **2011**, *289*, 1261–1272. [CrossRef]
23. Makvandi, P.; Della Sala, F.; di Gennaro, M.; Solimando, N.; Pagliuca, M.; Borzacchiello, A. A Hyaluronic Acid-Based Formulation with Simultaneous Local Drug Delivery and Antioxidant Ability for Active Viscosupplementation. *ACS Omega* **2022**, *7*, 10039–10048. [CrossRef] [PubMed]
24. Ghazali, S.R.; Kubulat, K.; Isa, M.; Samsudin, A.; Khairul, W.M. Contribution of methyl substituent on the conductivity properties and behaviour of CMC-alkoxy thiourea polymer electrolyte. *Mol. Cryst. Liq. Cryst.* **2014**, *604*, 126–141. [CrossRef]
25. Manju, S.; Sreenivasan, K. Conjugation of curcumin onto hyaluronic acid enhances its aqueous solubility and stability. *J. Colloid Interface Sci.* **2011**, *359*, 318–325. [CrossRef] [PubMed]
26. Li, X.; Ke, J.; Wang, J.; Kang, M.; Wang, F.; Zhao, Y.; Li, Q. Synthesis of a novel $CO_2$-based alcohol amine compound and its usage in obtaining a water-and solvent-resistant coating. *RSC Adv.* **2018**, *8*, 8615–8623. [CrossRef] [PubMed]
27. Chang, L.; Zhang, J.; Jiang, X. Comparative Properties of Hyaluronic Acid Hydrogel Cross-linked with 1, 4-Butanediol Diglycidyl Ether Assayed Using a Marine Hyaluronidase. *IOP Conf. Ser.: Mater. Sci. Eng.* **2019**, *493*, 012007. [CrossRef]
28. Guarino, V.; Altobelli, R.; della Sala, F.; Borzacchiello, A.; Ambrosio, L. Alginate processing routes to fabricate bioinspired platforms for tissue engineering and drug delivery. *Alginates Their Biomed. Appl.* **2018**, *11*, 101–120.
29. Borzacchiello, A.; Della Sala, F.; Ambrosio, L. Rheometry of polymeric biomaterials. In *Characterization of Polymeric Biomaterials*; Elsevier: Amsterdam, The Netherlands, 2017; pp. 233–253.
30. Bahcecioglu, G.; Hasirci, N.; Hasirci, V. Effects of microarchitecture and mechanical properties of 3D microporous PLLA-PLGA scaffolds on fibrochondrocyte and L929 fibroblast behavior. *Biomed. Mater.* **2018**, *13*, 035005. [CrossRef]
31. Iso, B.; STANDARD, B. Biological evaluation of medical devices. *Biomed. Saf. Stand* **1996**, *26*, 54.
32. Andrade del Olmo, J.; Pérez-Álvarez, L.; Sáez Martínez, V.; Benito Cid, S.; Pérez González, R.; Vilas-Vilela, J.L.; Alonso, J.M. Drug delivery from hyaluronic Acid–BDDE injectable hydrogels for antibacterial and anti-inflammatory applications. *Gels* **2022**, *8*, 223. [CrossRef]
33. Xue, Y.; Chen, H.; Xu, C.; Yu, D.; Xu, H.; Hu, Y. Synthesis of hyaluronic acid hydrogels by crosslinking the mixture of high-molecular-weight hyaluronic acid and low-molecular-weight hyaluronic acid with 1, 4-butanediol diglycidyl ether. *RSC Adv.* **2020**, *10*, 7206–7213. [CrossRef] [PubMed]
34. Yang, B.; Guo, X.; Zang, H.; Liu, J. Determination of modification degree in BDDE-modified hyaluronic acid hydrogel by SEC/MS. *Carbohydr. Polym.* **2015**, *131*, 233–239. [CrossRef] [PubMed]
35. Mezger, T. *The Rheology Handbook: For Users of Rotational and Oscillatory Rheometers*; Vincenz Network: Hanover, Germany, 2020.

**Disclaimer/Publisher's Note:** The statements, opinions and data contained in all publications are solely those of the individual author(s) and contributor(s) and not of MDPI and/or the editor(s). MDPI and/or the editor(s) disclaim responsibility for any injury to people or property resulting from any ideas, methods, instructions or products referred to in the content.

Article

# Regulated Self-Folding in Multi-Layered Hydrogels Considered with an Interfacial Layer

Jun Woo Lim [1,†], Sang Jin Kim [1,†], Jimin Jeong [1], Sung Gyu Shin [1], Chaewon Woo [1], Woonggyu Jung [2,*] and Jae Hyun Jeong [1,*]

1. Department of Chemical Engineering, Soongsil University, Seoul 06978, Republic of Korea; ljw9424@soongsil.ac.kr (J.W.L.); sj1229v@soongsil.ac.kr (S.J.K.); jamiej1123@soongsil.ac.kr (J.J.); whitegd45@ssu.ac.kr (S.G.S.); chaewon311@soongsil.ac.kr (C.W.)
2. Department of Biomedical Engineering, Ulsan National Institute of Science and Technology (UNIST), Ulsan 44919, Republic of Korea
* Correspondence: wgjung@unist.ac.kr (W.J.); nfejjh@ssu.ac.kr (J.H.J.); Tel.: +82-2-828-7043 (J.H.J.)
† These authors contributed equally to this work.

**Abstract:** Multi-layered hydrogels consisting of bi- or tri-layers with different swelling ratios are designed to soft hydrogel actuators by self-folding. The successful use of multi-layered hydrogels in this application greatly relies on the precise design and fabrication of the curvature of self-folding. In general, however, the self-folding often results in an undesired mismatch with the expecting value. To address this issue, this study introduces an interfacial layer formed between each layered hydrogel, and this layer is evaluated to enhance the design and fabrication precision. By considering the interfacial layer, which forms through diffusion, as an additional layer in the multi-layered hydrogel, the degree of mismatch in the self-folding is significantly reduced. Experimental results show that as the thickness of the interfacial layer increases, the multi-layered hydrogel exhibits a 3.5-fold increase in its radius of curvature during the self-folding. In addition, the diffusion layer is crucial for creating robust systems by preventing the separation of layers in the muti-layered hydrogel during actuation, thereby ensuring the integrity of the system in operation. This new strategy for designing multi-layered hydrogels including an interfacial layer would greatly serve to fabricate precise and robust soft hydrogel actuators.

**Keywords:** multi-layered hydrogel; interfacial layer; self-folding; soft hydrogel actuators

**Citation:** Lim, J.W.; Kim, S.J.; Jeong, J.; Shin, S.G.; Woo, C.; Jung, W.; Jeong, J.H. Regulated Self-Folding in Multi-Layered Hydrogels Considered with an Interfacial Layer. *Gels* **2024**, *10*, 48. https://doi.org/10.3390/gels10010048

Academic Editor: Shengshui Hu

Received: 17 November 2023
Revised: 20 December 2023
Accepted: 6 January 2024
Published: 10 January 2024

**Copyright:** © 2024 by the authors. Licensee MDPI, Basel, Switzerland. This article is an open access article distributed under the terms and conditions of the Creative Commons Attribution (CC BY) license (https://creativecommons.org/licenses/by/4.0/).

## 1. Introduction

Multi-layered hydrogels consisting of bi- or tri-layers with different swelling ratios are designed to serve as soft hydrogel actuators through self-folding, which can be induced by external stimuli. These hydrogels have been extensively studied for their potential use as bio-robots [1], active actuators [2–4], and drug delivery systems [5,6], with a focus on adjusting their mechanical properties to achieve various shapes [7–9]. For instance, recent studies have explored their potential in developing innovative soft actuators designed for live cell stimulation through self-folding, showcasing the adaptability and precision in controlling their mechanical properties [2]. This soft actuator exhibits the capability to stimulate live cell clusters through a dynamic interplay of compression and tension, triggered by changes in temperature. The self-folding mechanism allows for the precise manipulation of the curvature, ensuring adaptability and precision in live cell stimulation.

The successful utilization of multi-layered hydrogels in soft actuation critically depends on the precision with which the curvature of the self-folding is designed and fabricated [10–12]. Typically, the curvature of these multi-layered hydrogels has been predicted using the bimetallic strip equation, considering the mechanical properties of individual layers [13,14]. However, it is imperative to recognize the crucial role played by an overlooked component, the interfacial layer. The interfacial layer arises as a natural consequence of the manufacturing process, forming between adjacent layers of the multi-layered hydrogel through diffusion. Although not intentionally engineered, its presence exerts a substantial influence on the system's behavior. Specifically, the self-folding process, which stands as a critical element in the functionality of these hydrogel actuators, is intricately influenced by the properties of this interfacial layer. Neglecting the interfacial layer when designing and calculating the curvature of self-folding, often using approaches like the bimetallic strip equation, can lead to discrepancies between the anticipated and actual outcomes (Figure 1) [15–18]. By considering this layer, formed through diffusion, in the bimetallic strip equation for a tri-layered hydrogel, we observe a significant reduction in the mismatch between predicted and experimental curvature values. The interfacial layer introduces a previously unaccounted factor that shapes the hydrogel's response during self-folding.

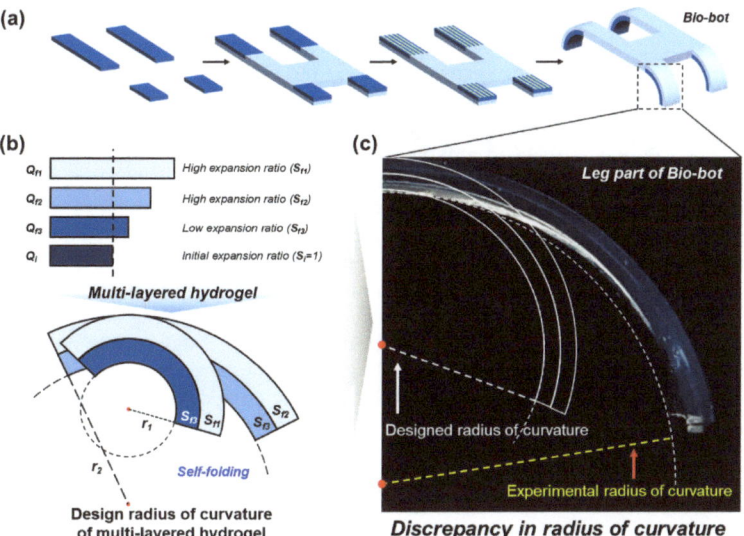

**Figure 1.** (**a**) Schematic description of the preparation of the biological machine (Bio-bot) by self-folding. (**b**) Schematic diagram illustrating the design of the radius of curvature for multi-layered hydrogels. The radius of curvature is predicted using the heat-induced bimetallic strip equation, with mechanical properties as the expansion ratio and elastic modulus of each layer. (**c**) Image depicting the radius of curvature of the multi-layered hydrogel, formed larger than the intended radius of curvature.

In this study, we examine the significance of this interfacial layer, emphasizing that it should not be disregarded when considering multi-layered hydrogels. By incorporating this naturally occurring layer as an essential component of the system, a more accurate representation of the self-folding behavior can be achieved, aligning the theoretical results with practical observations. Furthermore, we demonstrate that this interfacial layer plays a crucial role in maintaining the structural integrity of the multi-layered hydrogel during actuation. Its presence prevents the separation of layers and contributes to the robustness of the entire system. It acts as a diffusion layer that crosslinks with both layers, effectively binding them together and ensuring that the two distinct layers do not separate during actuation, which is vital for the structural integrity of the multi-layered hydrogel system. Thus, the consideration of the interfacial layer is essential for precision in self-folding as well as the creation of robust and dependable soft hydrogel actuators. In conclusion, this study highlights the underestimated role of the interfacial layer in multi-layered hydrogels. By recognizing and incorporating this layer into the design and analysis, we bridge the gap between theory and reality, ultimately leading to more accurate, reliable, and resilient soft hydrogel actuators [19–22].

## 2. Results and Discussion

### 2.1. Mechanical Properties of Individual Layers in Multi-Layered Hydrogels

In this section, we present a comprehensive overview of the molecular structure and synthetic pathway employed to fabricate the bi-layered hydrogel. The choice of acrylamide (AAm) as the monomer and N,N′-methylenebisacrylamide (MBA) as the cross-linker, along with the incorporation of the initiator Irgacure2959, is detailed. The concentration variations and the rationale behind the selection of MBA concentrations in the first and second layers are discussed. Furthermore, we elaborate on the photo-crosslinking process and conditions, emphasizing their role in the formation of a well-defined bi-layered hydrogel.

First, the investigation into the mechanical properties of hydrogels by varying the cross-linker concentration highlighted essential aspects of their behavior. The dependence of the expansion ratio (S) and elastic modulus (E) on the MBA concentration underscored the need to tailor these properties for specific applications, such as self-folding in multi-layered hydrogels. The expansion ratio and elastic modulus of the hydrogel, composed of AAm monomer at 20% ($w/v$) and varying concentrations of MBA (0.02–0.3% $w/v$), were measured. As the concentration of the cross-linker increased, the swelling ratio (S) and expansion ratio (E), calculated using the formulas described in the Materials and Methods section (Equations (1) and (2)), exhibited a decrease, while the elastic modulus (E) demonstrated an increase (see Figure 2 for details). The swelling ratio, calculated using Equation (1), represents the degree to which the hydrogel can absorb water depending on the mass ($W_d$) of the total polymer comprising the hydrogel. However, when inducing self-folding in multi-layered hydrogels, it is more important to consider how much each layer of the hydrogel actually expands compared to its initial size. While conventional approaches focus on adjusting the swelling ratio, our study emphasizes the importance of considering the expansion ratio, particularly in the context of self-folding applications. Evaluating how each layer of the hydrogel expands compared to its initial size provides a more accurate prediction of the radius of curvature for multi-layered hydrogels. This challenges the traditional reliance on the swelling ratio alone. Moreover, the trends observed in Figure 2 reveal specific concentration-dependent behaviors, further supporting the argument for tailoring mechanical properties to achieve optimal self-folding characteristics in multi-layered hydrogels. In summary, our approach shifts the focus from the conventional emphasis on adjusting the swelling ratio to a more nuanced consideration of the expansion ratio in multi-layered hydrogels. By evaluating how each layer expands relative to its initial size, we gain a more precise understanding of self-folding dynamics.

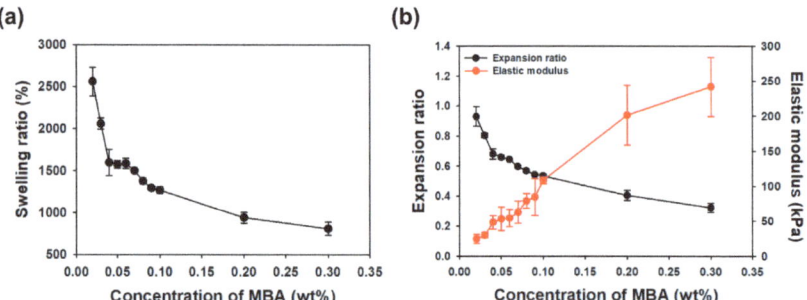

**Figure 2.** (**a**) The swelling ratio of the AAm hydrogels with varying MBA concentrations. (**b**) The expansion ratio and the elastic modulus of the AAm hydrogels at different MBA concentrations.

## 2.2. Design and Fabrication of Multi-Layered Hydrogels

A bi-layered hydrogel was formed using a photo-crosslinking process, consisting of two layers. The first layer was crosslinked using a fixed concentration of MBA (0.04% $w/v$), while the concentration of the second layer was adjusted to 0.08–0.3% ($w/v$). The radius of curvature (r) was calculated using the formulas described in the Materials and Methods section (Equation (4)) to predict the curvature of the multi-layered hydrogel upon self-folding. The calculated radius of curvature was then compared to the radius of curvature of the fully expanded bi-layered hydrogel (Figure 3a). However, the experimental radius of curvature of the bi-layered hydrogel did not match the calculated radius of curvature. For example, in the case of Bi-layered hydrogel (BH), such as BH-1, which had a difference in the expansion ratio of 0.11 between the two layers, it was predicted to be 1.44 mm. However, the experimental radius of curvature was 1.74 mm, representing a 1.2-fold difference from the designed radius of curvature. The curvature of BH-3 was also found to be significantly different from the designed radius of curvature, with an approximately 1.3-fold difference (see Figure 3b,c for details). These results indicate a significant discrepancy between the design value and the experimental value when predicting the radius of curvature using the bi-layered equation. Despite the observed disparities between the calculated and experimental values for the radius of curvature in the bi-layered hydrogel, our study emphasizes the significance of these findings in refining predictive models and underscores the need for a more comprehensive understanding of the underlying factors influencing self-folding behavior in multi-layered hydrogel systems.

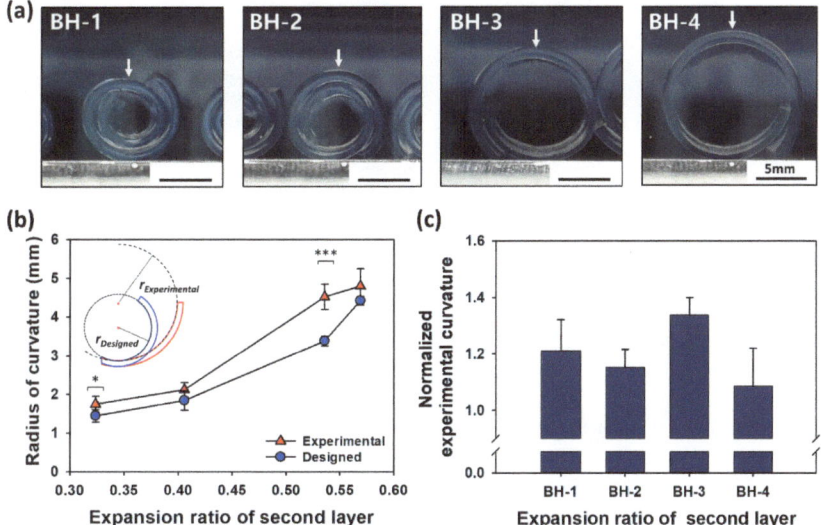

**Figure 3.** (a) Bright field images of bi-layered hydrogels with varying expansion ratios of the second layer. The abbreviations BH-1, BH-2, BH-3, and BH-4 correspond to different bi-layered hydrogel samples. (b) Calculated radius of curvature for the bi-layered hydrogel using a mathematical model developed for the curvature of a heat-induced bimetallic strip. Comparison of designed and experimental radius of curvatures based on differences in the expansion ratio of the two layers. (* $p < 0.05$, *** $p < 0.01$) (c) Normalized experimental curvature of BH-1, 2, 3, 4. Experimental curvatures for BH-1, 2, 3, 4 have been normalized to the designed radius of curvature.

## 2.3. Curvature Simulation of Multi-Layered Hydrogels

The experimental radius of curvature of the bi-layered hydrogel was observed to deviate significantly from the originally designed radius of curvature. This discrepancy was attributed to the formation of an interfacial layer at the interface between the two layers due to diffusion during the fabrication of the second layer of the multi-layered hydrogel (see Figure 4a for details). To examine this phenomenon in more detail, simulations were conducted, assuming that the physical properties of the interfacial layer fell between those of the first and second layers. Given that the interfacial layer naturally formed during the preparation of the multi-layered hydrogel, the bi-layered hydrogel was effectively treated as a tri-layered hydrogel for the purpose of simulations. The radius of curvature of the tri-layered hydrogel was then predicted using Equation (1) to account for the presence of the interfacial layer, employing MATLAB for the simulation.

$$r = \frac{2\left[\frac{\Delta\varepsilon_{12}(t_1+t_2)}{t_3 E_3} + \frac{\Delta\varepsilon_{23}(t_2+t_3)}{t_1 E_1} + \frac{\Delta\varepsilon_{13}(t_1+2t_2+t_3)}{t_2 E_2}\right]}{\frac{(t_1+t_2)^2}{t_3 E_3} + \frac{(t_2+t_3)^2}{t_1 E_1} + \frac{(t_1+2t_2+t_3)^2}{t_2 E_2} + \frac{4\left(\sum_1^3 E_n \frac{t_n^3}{12}\right)\left(\sum_1^3 t_n E_n\right)}{t_1 E_1 t_2 E_2 t_3 E_3}} \quad (1)$$

To corroborate the impact of the interfacial layer on the radius of curvature of the multi-layered hydrogel, BH-3 and BH-5 were prepared by altering the order of layer preparation. The resulting radius of curvature is displayed as Experimental values (1) and (2) in Figure 4b. The composition of the tri-layer hydrogel mirrored that of BH-3, but the thickness of the interfacial layer was varied from 10 to 80 µm, while the expansion ratio ranged from 0.484 to 0.660. The arrow in Figure 4b represents the experimental radius of curvature. These findings demonstrate that the radius of curvature of the bi-layered hydrogel can be precisely designed and predicted by taking into account the thickness of the interfacial layer. Furthermore, this meticulous exploration into the interfacial layer's impact on the radius of curvature provides valuable insights for enhancing the predictability and precision of multi-layered hydrogel designs. The comprehensive analysis sheds light on the interplay of the layer formation order, thickness variations, and expansion ratios, enabling more informed and tailored fabrication of hydrogels for diverse applications.

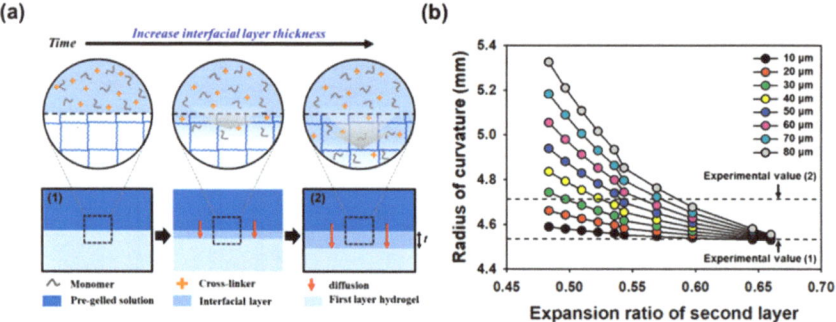

**Figure 4.** (**a**) Schematic representation of the formation of the interfacial layer through the diffusion of monomers and cross-linkers. (**a-1**) Designed curvature in the absence of the interfacial layer formation, (**a-2**) Experimental curvature observed after the interfacial layer formation. (**b**) Simulation of the radius of curvature was conducted by varying the expansion ratio and the thickness of the interfacial layer in the hydrogel. The arrow marks the experimental radius of curvature of the bi-layered hydrogel, which was prepared with the composition of BH-3.

In particular, the experimental value (1) of BH-3, prepared with the first layer having an MBA concentration of 0.04% ($w/v$), measured at 4.53 mm. This measurement matched the design value of the curvature radius of the tri-layered hydrogel, complete with a 10 µm interfacial layer and an expansion ratio of 0.660. Conversely, the experimental value (2) of BH-5, formed with the first layer having an MBA concentration of 0.08% ($w/v$), recorded a radius of curvature of 4.71 mm. The radius of curvature of the multi-layered hydrogel with experimental value (2) was estimated to have an interfacial layer thickness exceeding 30 µm. However, the interfacial layer, formed through diffusion, cannot expand beyond the first layer, with an expansion ratio lower than 0.536. Hence, it is reasonable to infer that the interfacial layer had a thickness ranging from 30 and 60 µm. These results affirm that the experimental and designed radius of curvature align when employing the radius of curvature equation for tri-layered hydrogels, which considers the interfacial layer as one layer in the bi-layered hydrogels. These findings highlight the nuanced impact of the MBA concentration on the experimental outcomes and emphasize the importance of accounting for interfacial layer dynamics in accurately predicting the radius of curvature in multi-layered hydrogel systems.

### 2.4. Evaluation of the Interfacial Layer in Multi-Layered Hydrogels

To investigate the impact of the interfacial layer's thickness on the radius of curvature of bi-layered hydrogels, the diffusion time was adjusted to various intervals, including 0, 0.2, 1, 5, and 10 min. Bi-layered hydrogels with BH-3 composition were created using a fixed concentration of 0.04% ($w/v$) MBA at the first layer. As shown in Figure 5(a-1–a-3), it becomes visually evident that the radius of curvature increases with longer diffusion times. The bi-layered hydrogel fabricated without any diffusion time and the one with 10 min of diffusion time exhibited 1.3-fold and 3.5-fold increases from the designed radius of curvature, respectively (Figure 5b). These results underscore the crucial role of the diffusion time in determining the interfacial layer's thickness, directly influencing the radius of curvature in bi-layered hydrogels. This insight offers a systematic approach to tailoring hydrogel designs for specific applications by controlling diffusion parameters during fabrication.

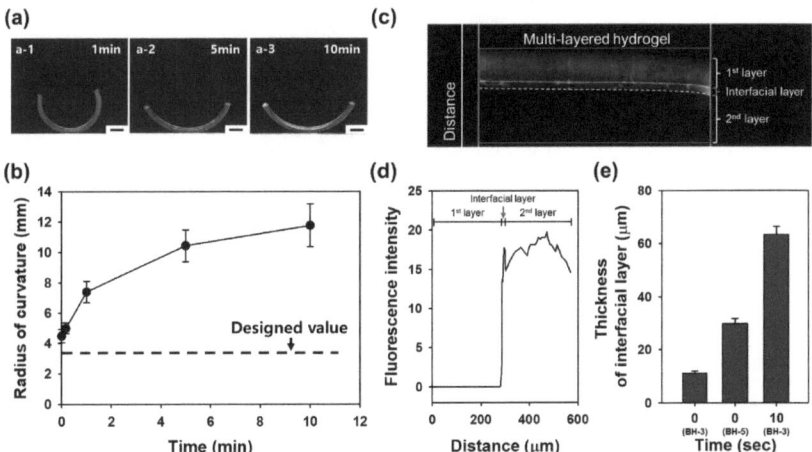

**Figure 5.** (**a**) Bright field images of bi-layered hydrogels with varying diffusion times for the second layer ((**a-1**): 1 min; (**a-2**): 5 min; (**a-3**): 10 min). (**b**) Radius of curvature of bi-layered hydrogels relative to the diffusion time of the second layer. (**c**) Cross-sectional fluorescence image of a multi-layered hydrogel with the first and second layers, including the interfacial layer. (**d**) Fluorescence intensity across the multi-layered hydrogel as a function of distance. (**e**) Thickness measurements of the interfacial layer in each bi-layered hydrogel (BH-3 and BH-5) at 0 and 10 s. (Scale bar: 2 mm).

Notably, the experimental radius of curvature in the bi-layered hydrogel with 10 s of diffusion time measured 5.01 mm, indicating that the thickness of the interfacial layer had grown to more than 60 µm, as simulated. These findings underscore the importance of the rapid manufacturing of multi-layered hydrogels to align with the designed radius curvature. Furthermore, to verify the actual thickness of the interfacial layer, a bi-layered hydrogel was prepared by introducing 0.0005% ($w/v$) of Fluorescein O,O'-dimethacrylate (Sigma, St. Louis, MO, USA) into the second layer. Immediately after fabrication, an image was captured using a fluorescence microscope (Figure 5c), and the thickness of the interfacial layer and each layer was measured based on the fluorescence intensity and distance within the image (Figure 5d). Bi-layered hydrogels were prepared with diffusion times of 0 and 10 s for BH-3 and 0 s for BH-5. As a result, the thickness of the interfacial layer in BH-3 and BH-5 at 0 s was measured as $11.10 \pm 0.82$ μm and $29.87 \pm 1.92$ μm, respectively. Additionally, the thickness of the interfacial layer of BH-3 at 10 s was measured as $63.42 \pm 2.91$ μm (Figure 5e). These results underscore the importance of considering the interfacial layer in the design of multi-layered hydrogel curvatures, aligning with previous simulation findings. In light of our findings on the interfacial layer's growth under varying diffusion times, we anticipate the need to precisely control the formation of interfacial layers and measure their properties for more accurate regulation.

On a different note, to assess the impact of diffusion of structural robustness, a multi-layered hydrogel was prepared using PEGDA. A bi-layered hydrogel was crafted using PEGDA575 at 20% ($w/v$) for the first layer and PEGDA3400 at 20% (w/v) for the second layer. Upon separating the prepared PEGDA bi-layered hydrogel into two layers, it was observed that the two layers separated due to inadequate diffusion at the interface (Figure 6b). This result corroborates our findings that the structural integrity of the multi-layered hydrogel is maintained only when proper diffusion occurs at the interface (Figure 6a). The interfacial layer acts as a diffusion layer that crosslinks with both layers, effectively binding them together and ensuring that the two distinct layers do not separate during actuation, which is vital for the structural integrity of the multi-layered hydrogel system.

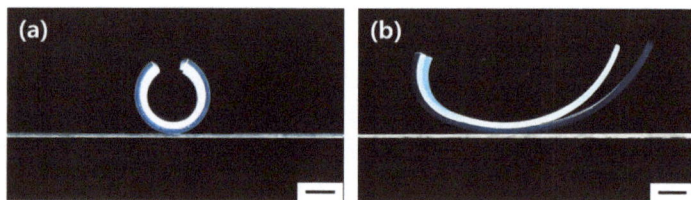

**Figure 6.** Bright field images of bi-layered hydrogels composed of PEGDA575 for the first layer and PEGDA3400 for the second layer, (**a**) with proper diffusion curves at the interface, and (**b**) with a minimal interfacial layer. (Scale bar: 2 mm).

Therefore, it is essential to consider diffusion at the interface of each layer during the design and fabrication of multi-layered hydrogels. The interfacial layer formed by diffusion must also be taken into account. This approach minimizes experimental errors by enabling accurate design and facilitating the prediction of the mechanical properties of the interfacial layer generated through diffusion. Furthermore, the diffusion layer is critical for ensuring the robustness of the system and preventing layer separation during actuation. In summary, the design of multi-layered hydrogels that accounts for the formation of the interfacial layer due to diffusion holds vast applications in fields such as bio-robotics, actuators, tissue engineering, and drug delivery [23–25].

## 3. Conclusions

In this study, we introduced an innovative strategy to enhance the precision and robustness of multi-layered hydrogels as soft hydrogel actuators by focusing on the interfacial layer, a natural byproduct of the manufacturing process formed by diffusion between the layers. Our experiments unequivocally reveal that as the interfacial layer's thickness increases, self-folding becomes significantly more predictable and precise, leading to a remarkable 3.5-fold increase in the radius of curvature. We also unveiled the interfacial layer's pivotal role in preserving structural integrity during the actuation. This groundbreaking approach not only reconciles theory with practice but also promises the development of highly accurate and reliable soft hydrogel actuators. The deliberate incorporation of the interfacial layer into the multi-layered hydrogel design represents a significant advancement with transformative potential in bio-robots, active actuators, tissue engineering, and drug delivery applications.

## 4. Materials and Methods

### 4.1. Hydrogel Preparation

Hydrogels were prepared using acrylamide (acrylamide, AAm, Sigma) and methylene bisacrylamide (N,N'-methylenebisacrylamide, MBA, Sigma) as monomers. The total concentration of AAm was fixed at 20% ($w/v$) and prepared by adjusting the concentration of MBA between 0.02% to 0.3% ($w/v$). An initiator, Irgacure2959 (2-hydroxy-40-(2-hydroxyethoxy)-2-methylpropiophenone, Sigma), was added to reach a final concentration of 0.2% ($w/v$). The hydrogels were formed by exposure to ultraviolet (UV) light (365 nm, VILBER LOURMAT, 4W) for 10 min. Subsequently, the gels were punched into 8 mm diameter disks and incubated in deionized (DI) water until fully swollen before characterization. The resulting hydrogels demonstrated tunable properties, showcasing the versatility of this synthesis approach for creating hydrogels with a range of mechanical and swelling characteristics suitable for diverse applications.

### 4.2. Characterizations of Hydrogels

The weight of hydrogel ($W_s$) was measured after 12 h of incubation in DI water. The weight of the dried hydrogel ($W_d$) was determined after drying at 60 °C over 24 h. The swelling ratio ($Q_m$) was calculated using Equation (2).

$$Q_m = \left(\frac{W_s - W_d}{W_d}\right) \times 100 \tag{2}$$

The expansion ratio, crucial for inducing self-folding in multi-layered hydrogels with equal dimensions for each layer, was characterized using the swelling ratio of hydrogels when prepared ($Q_i$) and after full incubation in DI water ($Q_f$). The expansion ratio, indicating one-dimensional expansion of the hydrogel, was calculated using Equation (3).

$$S = \left(\frac{Q_f}{Q_i}\right)^{\frac{1}{3}} - 1 \tag{3}$$

Here, $Q_i$ and $Q_f$ represent the swelling ratio ($Q_m$) before and after immersion in the aqueous solution. The mechanical properties of the hydrogel were evaluated using a universal testing machine (UTM, DrTech, Seongnam-si, Korea). Hydrogels, standardized to 8 mm diameter and 1 mm height, were compressed at a 10% strain and a constant rate of 1.0 mm/min, with a load range of 1.0 kg·f after incubation over 24 h.

### 4.3. Design and Fabrication of the Multi-Layered Hydrogels

The radius of curvature ($r$) of multi-layered hydrogels was determined using the bimetallic strip curvature Equation (4) [9,12]. In Equation (4), $E_1$ and $E_2$ represent the elastic modulus of each layer, $t_1$ and $t_2$ represent the thickness of each layer, and $\Delta\varepsilon$ represents the difference of expansion ratio of each layer.

$$r = \frac{E_1^2 t_1^4 + 4E_1 E_2 t_1^3 t_2 + 6E_1 E_2 t_1^2 t_2^2 + 4E_1 E_2 t_1 t_2^3 + E_2^2 t_2^4}{6E_1 E_2 (t_1 + t_2) t_1 t_2 \Delta\varepsilon} \tag{4}$$

The multi-layered hydrogels were composed of bi-layered hydrogels with different compositions (Table 1). The concentration of each layer of the bi-layered hydrogel was fixed at a final concentration of 20% ($w/v$) for AAm and 0.2% ($w/v$) for Irgacure2959. The MBA concentration in the first layer was fixed at 0.04% ($w/v$), while the second layer's concentration was adjusted to 0.08–0.3% ($w/v$) to vary the radius of curvature. The bi-layered hydrogel was assembled by first creating a hydrogel through UV exposure, followed by the preparation of second hydrogel layer on top of the first. The resulting multi-layered hydrogels were formed as strips measuring 30 mm in length and 5 mm in width, and 3 mm in thickness for each layer. The radius of curvature of the multi-layered hydrogels was measured after swelling in DI water for 24 h. The systematic control of the hydrogel composition and layer-specific adjustments allowed for the precise modulation of mechanical properties, enabling the tailoring of multi-layered hydrogels for applications demanding specific curvature characteristics and diverse functionalities.

Table 1. Preparation of bi-layered hydrogels (BH) with varying expansion ratio of second layer.

| Sample | First Layer | | | | Second Layer | | | |
|---|---|---|---|---|---|---|---|---|
| | Aam [a] | MBA [b] | Irgacure 2595 [c] | Expansion Ratio [d] | Aam [a] | MBA [b] | Irgacure 2595 [c] | Expansion Ratio [d] |
| BH-1 | 20 | 0.04 | 0.2 | 0.681 | 20 | 0.30 | 0.2 | 0.569 |
| BH-2 | 20 | 0.04 | 0.2 | 0.681 | 20 | 0.20 | 0.2 | 0.536 |
| BH-3 | 20 | 0.04 | 0.2 | 0.681 | 20 | 0.10 | 0.2 | 0.406 |
| BH-4 | 20 | 0.04 | 0.2 | 0.681 | 20 | 0.08 | 0.2 | 0.385 |
| BH-5 | 20 | 0.04 | 0.2 | 0.406 | 20 | 0.04 | 0.2 | 0.681 |

[a] Concentration of monomer %($w/v$), [b] Concentration of cross-linker %($w/v$), [c] Concentration of photo-initiator %($w/v$), [d] Expansion ratio of each layer.

### 4.4. Characterization and Simulation of Multi-Layered Hydrogels

The disparity between the designed and experimental radius of curvature ($r$) concerning the thickness of the interfacial layer was investigated through simulation. The radius of curvature was simulated using Equation (4) to calculate the radius of curvature of the tri-layered hydrogel, considering the presence of the interfacial layer in the bi-layered hydrogel. The thickness of the first and second layers of the bi-layered hydrogel was set at 300 µm for each, and the thickness of the interfacial layer was varied within the range of the first layer. The thickness of the interfacial layer was adjusted within the range of 10–60 um, and the composition of the first and second layers was the same as that of BH-3 in Table 1. The physical properties of the interfacial layer were determined within the mechanical properties range of the two layers. The simulations not only revealed insights into the correlation between the interfacial layer thickness and the observed disparity but also provided a basis for refining the design parameters for achieving precise control over the curvature in multi-layered hydrogel systems.

In addition, the diffusion time was adjusted to 0, 0.2, 1, 5 and 10 min to prepare a bi-layered hydrogel to determine the effect of the interfacial layer in multi-layered hydrogels. The multi-layered hydrogel was prepared in the same manner as in sample 3. Additionally, a multi-layered hydrogel was prepared using poly(ethyleneglycol) diacrylate (PEGDA) to confirm the effect of diffusion based on molecular weight. The concentration of the first layer was fixed at 20% ($w/v$) with PEGDA575 (Mn of 575 g/mol, Sigma), and concentration of the second layer was fixed at 20% ($w/v$) with PEGDA3400 (Mn of 3400 g/mol, Alfa Aesar). Irgacure2959 was added to reach a final concentration of 0.2% $w/v$ as an initiator. The variation in diffusion times and the utilization of poly(ethyleneglycol) di-acrylate (PEGDA) in multi-layered hydrogel synthesis further contribute to our understanding of interfacial layer dynamics, offering valuable insights for optimizing the design and performance of such hydrogel systems.

**Author Contributions:** Conceptualization, J.W.L., S.J.K. and J.H.J.; methodology, J.W.L., S.J.K., S.G.S. and J.H.J.; formal analysis, J.W.L., S.J.K., J.J., S.G.S. and J.H.J.; data curation, J.W.L., S.J.K., J.J. and C.W.; writing—original draft preparation, J.W.L., S.J.K. and J.H.J.; writing—review and editing, J.W.L. and J.H.J.; project administration, W.J. and J.H.J.; funding acquisition, W.J. and J.H.J. All authors have read and agreed to the published version of the manuscript.

**Funding:** This research was funded by the National Research Foundation of Korea (NRF-2019M2C8A 2058418) and by the Ministry of Education (NRF–2020R1A6A1A03044977).

**Data Availability Statement:** All data and materials are available on request from the corresponding author. The data are not publicly available due to ongoing researches using a part of the data.

**Conflicts of Interest:** The authors declare no conflict of interest.

## References

1. Cho, S.; Shin, S.G.; Kim, H.; Han, S.R.; Jeong, J.H. Self-folding of Multi-layered Hydrogel Designed for Biological Machine. *Polymer* **2017**, *41*, 346–351. [CrossRef]
2. Lim, J.W.; Kim, H.J.; Kim, Y.; Shin, S.G.; Cho, S.; Jung, W.G.; Jeong, J.H. An active and soft hydrogel actuator to stimulate live cell clusters by self-folding. *Polymers* **2020**, *12*, 583. [CrossRef] [PubMed]
3. Duan, J.; Liang, X.; Zhu, K.; Guo, J.; Zhang, L. Bilayer hydrogel actuators with tight interfacial adhesion fully constructed from natural polysaccharides. *Soft Matter* **2017**, *13*, 345–354. [CrossRef] [PubMed]
4. Bassik, N.; Abebe, B.; Laflin, K.; Gracias, D. Photolithographically patterned smart hydrogel based bilayer actuators. *Polymer* **2010**, *51*, 6093–6098. [CrossRef]
5. Baek, K.; Jeong, J.H.; Shkµmatov, A.; Bashir, R.; Kong, H. In Situ Self-Folding Assembly of a Multi-Walled Hydrogel Tube for Uniaxial Sustained Molecular Release. *Adv. Mater.* **2013**, *25*, 5568–5573. [CrossRef] [PubMed]
6. Chelu, M.; Musuc, A.M. Polymer Gels: Classification and Recent Developments in Biomedical Applications. *Gels* **2023**, *9*, 161. [CrossRef] [PubMed]
7. Guo, J.L.; Longaker, M. Bioprinted Hydrogels for Fibrosis and Wound Healing: Treatment and Modeling. *Gels* **2023**, *9*, 19. [CrossRef] [PubMed]
8. Antich, C.; JimUnez, G.; de Vicente, J.; Lopez-Ruiz, E.; Chocarro-Wrona, C.; Grinan-Lison, C.; Carrillo, E.; Montanez, E.; Marchal, J.A. Development of a Biomimetic Hydrogel Based on Predifferentiated Mesenchymal Stem-Cell-Derived ECM for Cartilage Tissue Engineering. *Adv. Healthc. Mater.* **2021**, *10*, 2001847. [CrossRef]
9. Mitragotri, S.; Lahann, J. Physical approaches to biomaterial design. *Nat. Mater.* **2009**, *8*, 15–23. [CrossRef]
10. Martínez-Sanz, M.; Mikkelsen, D.; Flanagan, B.; Gidley, M.J.; Gilbert, E.P. Multi-scale model for the hierarchical architecture of native cellulose hydrogels. *Carbohydr. Polym.* **2016**, *147*, 542–555. [CrossRef]
11. Xu, R.; Hua, M.; Wu, S.; Ma, S.; Zhang, Y.; Zhang, L.; Yu, B.; Cai, M.; He, X.; Zhou, F. Continuously growing multi-layered hydrogel structures with seamless interlocked interface. *Matter* **2022**, *5*, 634–653. [CrossRef]
12. Schmidt, B.V. Multicompartment hydrogels. *Macromol. Rapid Commun.* **2022**, *43*, 2100895. [CrossRef] [PubMed]
13. Lou, D.; Sun, Y.; Li, J.; Zheng, Y.; Zhou, Z.; Yang, J.; Pan, C.; Zheng, Z.; Chen, X.; Liu, W. Double lock label based on thermosensitive polymer hydrogels for information camouflage and multilevel encryption. *Angew. Chem.* **2022**, *134*, e202117066. [CrossRef]
14. Ma, C.; Lu, W.; Yang, X.; He, J.; Le, X.; Wang, L.; Zhang, J.; Serpe, M.J.; Huang, Y.; Chen, T. Bioinspired anisotropic hydrogel actuators with on–off switchable and color-tunable fluorescence behaviors. *Adv. Funct. Mater.* **2018**, *28*, 1704568. [CrossRef]
15. Jeong, J.H.; Schmidt, J.J.; Cha, C.; Kong, H. Tuning responsiveness and structural integrity of a pH responsive hydrogel using a poly (ethylene glycol) cross-linker. *Soft Matter* **2010**, *6*, 3930–3938. [CrossRef]
16. Xu, B.; Jiang, H.; Li, H.; Zhang, G.; Zhang, Q. High strength nanocomposite hydrogel bilayer with bidirectional bending and shape switching behaviors for soft actuators. *RSC Adv.* **2015**, *5*, 13167–13170. [CrossRef]
17. Lei, J.; Zhou, Z.; Liu, Z. Side Chains and the Insufficient Lubrication of Water in Polyacrylamide Hydrogel—A New Insight. *Polymers* **2019**, *11*, 1845. [CrossRef] [PubMed]
18. Tan, J.; Xie, S.; Wang, G.; Yu, C.W.; Zeng, T.; Cai, P.; Huang, H. Fabrication and Optimization of the Thermo-Sensitive Hydrogel Carboxymethyl Cellulose/Poly (N-isopropylacrylamide-co-acrylic acid) for U (VI) Removal from Aqueous Solution. *Polymers* **2020**, *12*, 151. [CrossRef]
19. He, X.; Sun, Y.; Wu, J.; Wang, Y.; Chen, F.; Fan, P.; Zhong, M.; Xiao, S.; Zhang, D.; Yang, J.; et al. Dual-stimulus bilayer hydrogel actuators with rapid, reversible, bidirectional bending behaviors. *J. Mater. Chem. C* **2019**, *7*, 4897–5210. [CrossRef]
20. Sackett, S.D.; Tremmel, D.M.; Ma, F.; Feeney, A.K.; Maguire, R.M.; Brown, M.E.; Zhou, Y.; Li, X.; O'Brien, C.; Li, L.; et al. Extracellular matrix scaffold and hydrogel derived from decellularized and delipidized human pancreas. *Sci. Rep.* **2018**, *8*, 10452. [CrossRef]
21. Ionov, L. Biomimetic Hydrogel-Based Actuating Systems. *Adv. Funct. Mater.* **2013**, *23*, 4555–4570. [CrossRef]
22. Hamley, I.; Castelletto, V. Biological soft materials. *Angew. Chem. Int. Ed.* **2007**, *46*, 4442–4455. [CrossRef] [PubMed]
23. Loi, G.; Stucchi, G.; Scocozza, F.; Cansolino, L.; Cadamuro, F.; Delgrosso, E.; Riva, F.; Ferrari, C.; Russo, L.; Conti, M. Characterization of a Bioink Combining Extracellular Matrix-like Hydrogel with Osteosarcoma Cells: Preliminary Results. *Gels* **2023**, *9*, 129. [CrossRef] [PubMed]
24. Wang, X.; Huang, H.; Liu, H.; Rehfeldt, F.; Wang, X.; Zhang, K. Multi-Responsive Bilayer Hydrogel Actuators with Programmable and Precisely Tunable Motions. *Macromol. Chem. Phys.* **2019**, *220*, 1800562. [CrossRef]
25. Vasudevan, M.; Johnson, W. Thermal Bending of a Tri-Metal Strip. *Aeronaut. J.* **1961**, *44*, 507–509. [CrossRef]

**Disclaimer/Publisher's Note:** The statements, opinions and data contained in all publications are solely those of the individual author(s) and contributor(s) and not of MDPI and/or the editor(s). MDPI and/or the editor(s) disclaim responsibility for any injury to people or property resulting from any ideas, methods, instructions or products referred to in the content.

Article

# Thermoresponsive Alginate-Graft-pNIPAM/Methyl Cellulose 3D-Printed Scaffolds Promote Osteogenesis In Vitro

Aikaterini Gialouri [1,†], Sofia Falia Saravanou [2,†], Konstantinos Loukelis [3], Maria Chatzinikolaidou [3,4,*], George Pasparakis [2,*] and Nikolaos Bouropoulos [1,5,*]

1. Department of Materials Science, University of Patras, 26504 Patras, Greece; katerinagialoyri@gmail.com
2. Department of Chemical Engineering, University of Patras, 26500 Patras, Greece; faliasaravanou@hotmail.com
3. Department of Materials Science and Technology, University of Crete, 70013 Heraklion, Greece; loukelisk@yahoo.com
4. Foundation for Research and Technology Hellas (FORTH), Institute of Electronic Structure and Laser (IESL), 70013 Heraklion, Greece
5. Foundation for Research and Technology Hellas, Institute of Chemical Engineering and High Temperature Chemical Processes, 26504 Patras, Greece
* Correspondence: mchatzin@materials.uoc.gr (M.C.); gpasp@chemeng.upatras.gr (G.P.); nbouro@upatras.gr (N.B.)
† These authors contributed equally to this work.

**Citation:** Gialouri, A.; Saravanou, S.F.; Loukelis, K.; Chatzinikolaidou, M.; Pasparakis, G.; Bouropoulos, N. Thermoresponsive Alginate-Graft-pNIPAM/Methyl Cellulose 3D-Printed Scaffolds Promote Osteogenesis In Vitro. *Gels* **2023**, *9*, 984. https://doi.org/10.3390/gels9120984

Academic Editors: María Vivero-Lopez, Ana Paula Serro and Diana Silva

Received: 10 November 2023
Revised: 5 December 2023
Accepted: 12 December 2023
Published: 15 December 2023

**Copyright:** © 2023 by the authors. Licensee MDPI, Basel, Switzerland. This article is an open access article distributed under the terms and conditions of the Creative Commons Attribution (CC BY) license (https://creativecommons.org/licenses/by/4.0/).

**Abstract:** In this work, a sodium alginate-based copolymer grafted by thermoresponsive poly(*N*-isopropylacrylamide) (PNIPAM) chains was used as gelator (Alg-g-PNIPAM) in combination with methylcellulose (MC). It was found that the mechanical properties of the resulting gel could be enhanced by the addition of MC and calcium ions ($Ca^{2+}$). The proposed network is formed via a dual crosslinking mechanism including ionic interactions among $Ca^{2+}$ and carboxyl groups and secondary hydrophobic associations of PNIPAM chains. MC was found to further reinforce the dynamic moduli of the resulting gels (i.e., a storage modulus of ca. 1500 Pa at physiological body and post-printing temperature), rendering them suitable for 3D printing in biomedical applications. The polymer networks were stable and retained their printed fidelity with minimum erosion as low as 6% for up to seven days. Furthermore, adhered pre-osteoblastic cells on Alg-g-PNIPAM/MC printed scaffolds presented 80% viability compared to tissue culture polystyrene control, and more importantly, they promoted the osteogenic potential, as indicated by the increased alkaline phosphatase activity, calcium, and collagen production relative to the Alg-g-PNIPAM control scaffolds. Specifically, ALP activity and collagen secreted by cells were significantly enhanced in Alg-g-PNIPAM/MC scaffolds compared to the Alg-g-PNIPAM counterparts, demonstrating their potential in bone tissue engineering.

**Keywords:** sodium alginate; PNIPAM; 3D printing; MC3T3-E1; bone tissue engineering

## 1. Introduction

Tissue engineering is a growing interdisciplinary field in biomedical sciences combining materials science, chemistry, biology, medicine, and engineering sciences. The basic aim of tissue engineering is to develop biological substitutes that maintain, restore, or improve tissue function [1]. Today, a diverse range of biomaterials of synthetic or biological origin are widely used in clinical practice for tissue engineering applications. Examples of synthetic biomaterials include metals, polymers, ceramics, and composite materials. Commonplace examples of natural biomaterials are protein or polysaccharide-based biomaterials such as collagen, gelatin, chitosan, and silk. In addition, autologous grafts or decellularized biomaterials are widely used for tissue regeneration [2]. One of the limitations of using the classical techniques to fabricate tissue engineering scaffolds is the resulting poor microstructural architecture and the restriction needed to control interconnections between the pores. Compared to traditional techniques, three-dimensional (3D) printing makes reproducible and customized structures with near-perfect micro-architecture and morphology [3]. Several 3D constructs with applications in regenerative medicine have been fabricated in recent years. The most common printing techniques used are laser based, jet and extrusion based printing, and fused deposition modeling [4].

Of particular interest are printing methods that allow for the direct deposition of viscous aqueous mixtures in arbitrary shapes, such as soft scaffolds that support the proliferation of mammalian cell populations and tissue growth. In this context, responsive polymers have been utilized to construct 3D matrices that can respond to external stimuli and exert shape/volume change in a reversible and dynamic manner and often recapitulate intrinsic properties of the extracellular matrix (ECM). Arguably, temperature-responsive polymers are widely studied due to their versatile responsive properties near physiological body temperature, as well as their ability to tune their stimuli response fully isothermally by combination with other co-monomers and/or polymers [5]. In addition, responsive polymer networks can undergo large changes in their dynamic moduli, rendering them ideal soft biomaterials for 3D printing in biomedical applications [6].

Poly(N-isopropylacrylamide) (PNIPAM) has been studied as a thermoresponsive polymer because of its rapid phase transition, biocompatibility and lower critical solution temperature (LCST) at approximately 32 °C, which is close to physiological body temperature. PNIPAM contains both hydrophilic amide groups (–CONH–) and hydrophobic isopropyl (–CH(CH$_3$)$_2$) side chains; in an aqueous environment, PNIPAM chains undergo a reversible sol-gel transition. Below the LCST, the chains are fully dissolved in water, and the polymer exhibits coil-like conformation due to hydrogen bonding and van der Waals forces. Above the LCST, the chains become hydrophobic, leading to a globule-like structure [7–9].

The printability and the biological evaluation of many thermosensitive hydrogels of synthetic or biological origin have been reported in the literature. For instance, bioengineered 3D-printed skin constructs based on thermosensitive PNIPAM hydrogels have been successfully evaluated for skin tissue engineering [10]. Pluronic F-127, a polaxamer co-polymer composed of polyethylene oxide (PEO) and polypropylene oxide (PPO), was used in combination with gelatin and hyaluronan to fabricate vascular channels [11]. Furthermore, the new synthetic biocompatible polymer PolyIsoCyanide was 3D-printed in a complex hydrogel construct and used as a fugitive material that could be removed after thermal stimulation [12]. Moreover, incorporation of particles into thermosensitive hydrogels has been reported to create hybrid hydrogels with improved rheological and mechanical properties, which in turn improve their printability [13].

Alginate is a major polysaccharide found in marine brown seaweed. Sodium alginate is a biopolymer broadly used in the food and beverage, pharmaceutical, cosmetics and medical industries, and has attracted significant interest due to its biocompatibility and biodegradability. It can be modified via covalent bonding of functional compounds, due to the abundant carboxylate and hydroxyl units, leading to new properties and applications in wound healing, controlled delivery of bioactive molecules, and cell encapsulation. The most common method to crosslink sodium alginate and form a hydrogel is the use of divalent cations such as calcium ions (Ca$^{2+}$), resulting in crosslinks ionically formed by the "egg-box" model [14].

Methylcellulose (MC) is derived from cellulose, a linear polysaccharide comprising glucose units held together by 1-4-β-glucosidic linkages, the most abundant renewable polymer in nature synthesized from plants, algae, fungi, and some bacterial species [15]. Cellulose is biocompatible and has significant mechanical strength; however, natural cellulose is insoluble in water, limiting its biomedical applications. To overcome this, the hydroxyl group of cellulose can be substituted with a methyl group. MC is a water-soluble biopolymer utilized in pharmaceutics, cosmetics and food industry as an emulsifier or as a thickening agent. In hydrogel 3D-printing processes, MC improves rheological properties by enhancing ink viscosity, which is necessary to produce high-quality printed structures.

In addition, several studies utilize MC-based hydrogels for cell engineering applications, emphasizing the in vitro biocompatibility of this polysaccharide [16].

The scope of the present work is to investigate the 3D printability of synthetic Alg-g-PNIPAM and Alg-g-PNIPAM/MC hydrogels. To this end, the thermoresponsive behavior of the corresponding polymers was studied, and the printing profile of the hydrogels was characterized by rheological analysis. Furthermore, the structural and morphological characterization of the 3D-printed scaffolds were investigated, while their in vitro erosion characteristics were evaluated. The cytocompatibility of the fabricated scaffolds was assessed in terms of cell viability, proliferation, adhesion, and morphology using the pre-osteoblastic cell line MC3T3-E1, whose osteogenic behavior has been previously reported as tunable in the presence of thermoresponsive polymers [17]. Alkaline phosphatase (ALP) activity, calcium, and total collagen production by cells cultured on scaffolds were evaluated as osteogenic markers to validate the potential of the developed scaffolds in bone tissue engineering applications.

## 2. Results and Discussion

### 2.1. Synthesis of Alg-g-PNIPAM

The proposed Alg-g-PNIPAM bioink was synthesized via free radical polymerization of the amino-terminated PNIPAM chains followed by a grafting procedure on the sodium alginate backbone by carbodiimide chemistry (Figure 1). The formation of the final Alg-g-PNIPAM product was confirmed by $^1$H NMR and FTIR characterization. The peaks at 3.5–4.6 ppm correspond to the four protons of the alginate ring [18]; the six methyl protons of the isopropyl group of NIPAM are depicted at ~1.1 ppm, the methylene and methine protons at 1.3–2.2 ppm, and the proton of the isopropyl group are linked to the amide group (N-C-H) at ~3.9 ppm (Figure S1). From the FTIR data, the 3305 cm$^{-1}$ band is representative of the aminoterminated groups of the PNIPAM-NH$_2$ precursor, which are absent in the spectrum of the Alg-g-PNIPAM due to their conversion to amide bonds (dashed area, Figure S2). Additionally, the characteristic amide C=O stretching and N-H bending of the PNIPAM are observed in the FTIR spectra of both the PNIPAM-NH$_2$ polymer and the Alg-g-PNIPAM at 1750–1500 cm$^{-1}$, further indicating the successful grafting reaction (Figure S2).

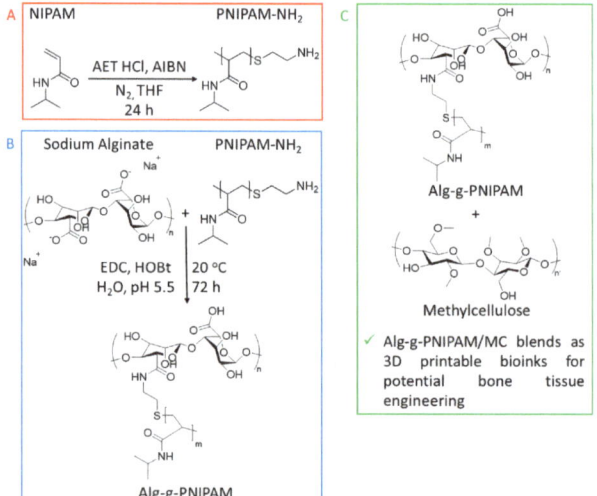

**Figure 1.** Synthetic route of the sodium alginate-based graft copolymer (**A**,**B**) and its combination with MC for the final ink material (**C**).

## 2.2. Thermosensitivity Measurements and Rheological Evaluation

The thermoresponsive behavior of PNIPAM precursor and Alg-g-PNIPAM was evaluated at a concentration of 4 mg/mL by measuring the absorption at 500 nm at different temperatures, i.e., 25–45 °C. The value of LCST was determined as the temperature onset point at which the solution turned cloudy. Both polymers exhibit sharp thermosensitive behavior. A LCST value at 35 °C was measured for PNIPAM slightly above 32 °C, which is attributed to the hydrophilic amino-end moieties that are known to shift the LCST at higher temperatures [19]. In the presence of alginate, the respective value for Alg-g-PNIPAM is again shifted at 37 °C, as presented in Figure 2a. This result is expected due to the hydrophilic nature of the alginate backbone, and interestingly, it is close to physiological body temperature, implying that the graft copolymer can be a good candidate for biomedical applications.

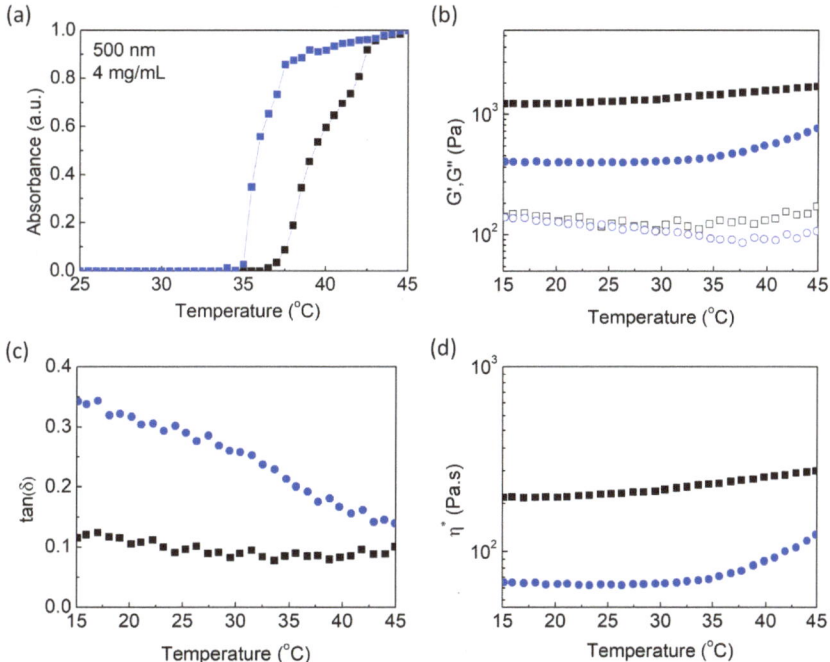

**Figure 2.** (**a**) Lower critical solution temperature measurements of PNIPAM (blue) and Alg-g-PNIPAM (black) polymers; (**b**) G′ (closed) and G″ (open) moduli, (**c**) tan (δ), and (**d**) complex viscosity as a function of temperature of Alg-g-PNIPAM (blue, circles) and of Alg-g-PNIPAM/MC (black, squares) aqueous solutions at a frequency of 6.28 rad/s, a strain amplitude of 0.1%, and during the heating cycle with a heating rate of 1 °C/min.

Alg-g-PNIPAM and Alg-g-PNIPAM/MC were used as main gelators to deconvolute the effect of each individual macromolecular component. The formulations were dissolved in 2 mm $Ca^{2+}$ aqueous solution. Oscillatory shear experiments reveal a thermo-induced gelation of the systems due to the hydrophobic associations of the PNIPAM pendants. Notably, the gel strengthening is observed at T > 35 °C (the LCST of the PNIPAM-$NH_2$ chains) upon a heating cycle at a ramp rate of 1 °C/min and a constant frequency of 6.28 rad/s; the storage modulus (G′) rapidly surpassed the loss one (G″), and the tangent δ is less than 0.2 for both systems. The rheological behavior of the 3D networks is enhanced in the entire temperature region, i.e., G′ constantly exceeds G″ by additional ionic interactions between the negatively ionized carboxylic groups of alginate or hydroxyl groups of MC and the positively charged $Ca^{2+}$ divalent cations, according to the "egg-box" mechanism, as presented in Figure 2b,c. The proposed f factor, i.e., f = $[Ca^{2+}]/[COO^-]$ or f = $[Ca^{2+}]/([COO^-]$ + $[OH^-])$ in molar ratio, is equal to f = 0.0095 in the Alg-g-PNIPAM hydrogel and f = 0.0056 in the Alg-g-PNIPAM /MC sample. Furthermore, the MC-thickening component creates a stable hydrogel adequate for 3D extrusion biomedical applications, as the G′ of the network is ~1280 Pa at room temperature and is equal to ~1500 Pa at physiological body temperature, i.e., a ~68% increase compared to the storage modulus of the MC-free network, while the tan (δ) of the MC-enriched hydrogels remains at ~0.1.

Moreover, in Figure 2d, the temperature dependance of the complex viscosity is demonstrated. The examined thermal region could be divided in two sections below and above ca. 35 °C. At T < 35 °C, the viscosity of the hydrogels remains constant as the 3D networks are formed due to ionic interactions, whereas at T > 35 °C, the secondary hydrophobic associations due to the PNIPAM pendant chains increase the viscosity values of the gelators. Indicatively, the viscous expansion between room and body temperature is 16% for the Alg-g-PNIPAM/MC system and 15% for the Alg-g-PNIPAM gel.

Additionally, the dependence of the storage and loss moduli was studied as a function of the angular frequency at room temperature (Figure 3a) close to the printer-bed temperature, i.e., 40 °C (Figure 3b). The G′ and G″ moduli of both gelators are almost independent of the frequency changes, denoting a solid-like behavior at both examined temperatures. At low temperature, the ionic interactions predominate in the network formation, whereas the MC augments the stability of the gel. Upon heating, the coexistence of the ionic and hydrophobic associations results in stronger hydrogel matrices.

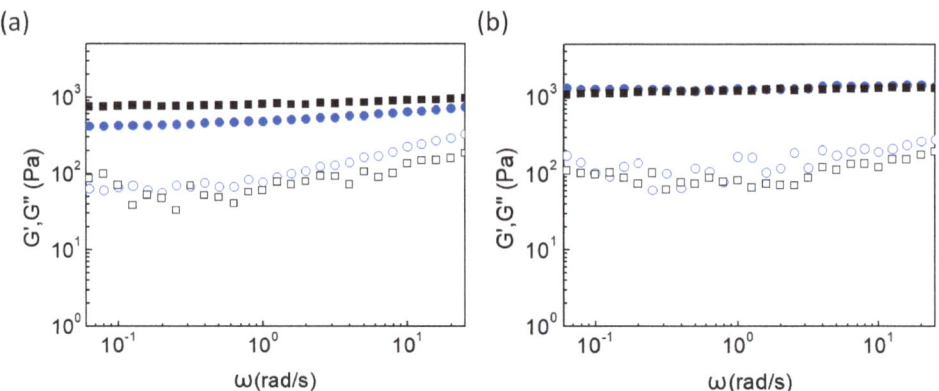

**Figure 3.** G′ (closed), G″ (open) versus angular frequency at (**a**) 25 °C and (**b**) 40 °C of Alg-g-PNIPAM (blue, circles) and of Alg-g-PNIPAM/MC (black, squares) hydrogels.

Considering the proposed materials as potential candidates for soft 3D printing, strain sweep and shear rate sweep tests were conducted in contemplation of bio-printability through extrusion. At low strain amplitude, the MC-rich system presents a stronger-gel behavior compared to the MC-free sample. As the strain amplitude is increased, the liquid-like threshold of the Alg-g-PNIPAM/MC is at a strain amplitude of approximately 10%, and the yield point of the Alg-g-PNIPAM is at 55%, denoting an easier-to-print material, as seen in Figure 4a. Besides, the shear-thinning effect is an important design criterion for efficient and accurate 3D printability, and both gelators meet this requirement, as shown in Figure 4b.

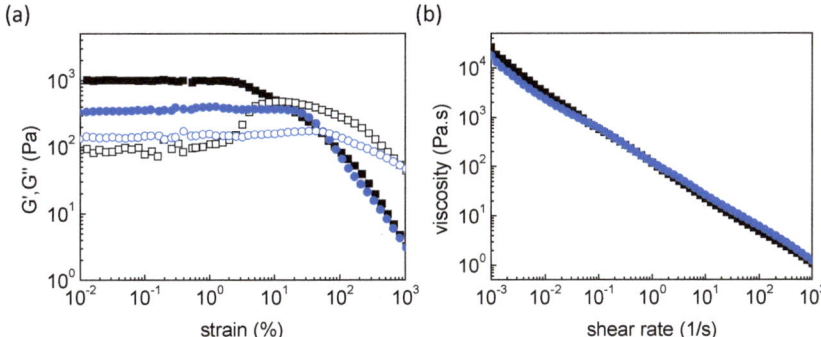

**Figure 4.** (**a**) G′ (closed) and G″ (open) upon strain sweep test and (**b**) viscosity as a function of shear rate at 25 °C (printing temperature) of Alg-g-PNIPAM (blue circles) and of Alg-g-PNIPAM/MC (black squares) hydrogels.

## 2.3. Characterization and Erosion Studies of 3D-Printed Scaffolds

### 2.3.1. Structural Characterization

Figure 5 shows the diffraction patterns of the dry hydrogel before printing and the lyophilized 3D-printed structures. All the diffractograms exhibit broad peaks, indicating the amorphous structure of all materials.

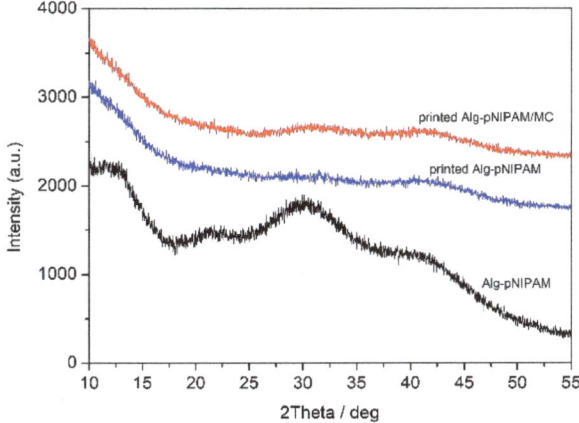

**Figure 5.** XRD patterns of Alg-g-PNIPAM and 3D-printed structures of Alg-g-PNIPAM and Alg-g-PNIPAM/MC.

### 2.3.2. Thermal Characterization

Thermogravimetric analysis was used to study the thermal behavior of raw materials and the printed samples. The results are depicted in Figure 6.

DSC thermogram of MC shows an endothermic peak at 77 °C, due to the loss of water (Figure 6A). The thermal behavior of sodium alginate is characterized by an endothermic peak at 79 °C, also attributed to water loss. A broad endothermic peak at 247 °C corresponds to degradation and depolymerization [20]. In the case of PNIPAM, except the water loss endotherm peak at 87 °C, the glass transition temperature Tg is observed at 148 °C, while the polymer is thermally stable at least until 300 °C [21]. Comparative assessment of both hydrogel inks Alg-g-PNIPAM and Alg-g-PNIPAM/MC revealed similar thermal behavior. Both samples are characterized by an alginate degradation peak at 253 and 257 °C, respectively.

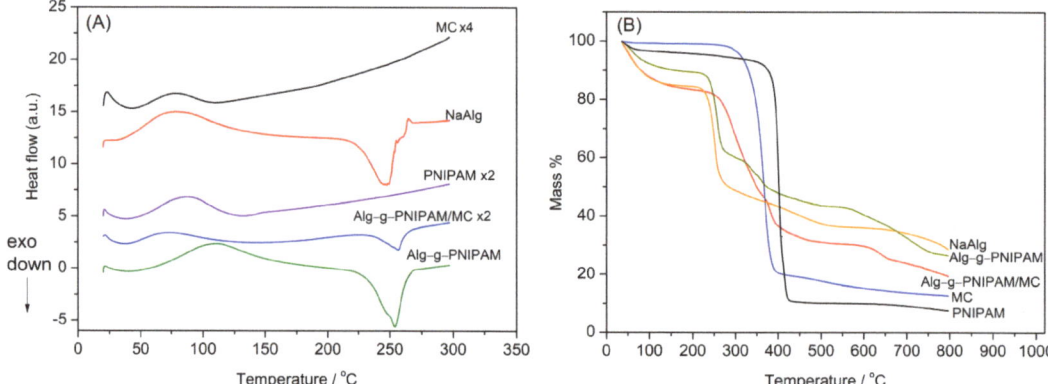

**Figure 6.** DSC (**A**) and TGA thermograms (**B**) of raw materials and printed objects. The DSC graphs have been shifted and multiplied by a factor for clarity.

In the TGA thermogram of MC, a decomposition peak is observed, which starts at 259 °C and finishes at 412 °C with a mass loss of 80%, while the total mass loss at 800 °C is 87% (Figure 6B). The sodium alginate decomposition occurs in three steps. The first step from 35 °C to 196 °C is due to dehydration, while decomposition starts at 196 °C and ends at 500 °C. Finally, the weight loss from 500 °C to 800 °C is attributed to $Na_2CO_3$ formation [20]. The thermogram of PNIPAM displays one distinctive degradation step which starts at 311 °C, due to degradation of the backbone of the polymer, and ends at 428 °C. Weight loss is continued until 800 °C due to main chain degradation [22,23]. Comparing the thermograms of sodium alginate with the grafted Alg-g-PNIPAM, it is shown that the grafted polymer exhibits higher thermal stability at least until 700 °C. After blending the Alg-g-PNIPAM copolymer with MC, the thermogram is more complex, and at 800 °C, the two copolymers Alg-g-PNIPAM and Alg-g-PNIPAM/MC show a remaining mass of 27 and 19%, respectively. The different thermal behavior of Alg-g-PNIPAM/MC can be attributed to the lower thermal stability of MC in comparison with sodium alginate and PNIPAM.

### 2.3.3. Morphological Characterization

The mean pore size of the wet Alg-g-PNIPAM sample was found equal to 2.04 ± 0.18 mm, and the dry 1.37 ± 0.14 mm, respectively. On the other hand, the average pore size in the presence of MC is 1.97 ± 0.17 mm for the wet scaffolds and 1.80 ± 0.11 mm for the dry ones. The examination of scaffolds' geometry revealed a decrease in pore size after freeze-drying, for both compositions, as seen in

Scanning electron microscopy (SEM) images from the lyophilized scaffolds are shown in Figure 8. Both compositions maintain their porous structure after lyophilization. It is interesting to mention that both samples show a rounded pore geometry, which is expected based on the soft nature of the gels. At a higher magnification, it can be observed that the specimen containing MC has a rougher surface texture. Figure 7.

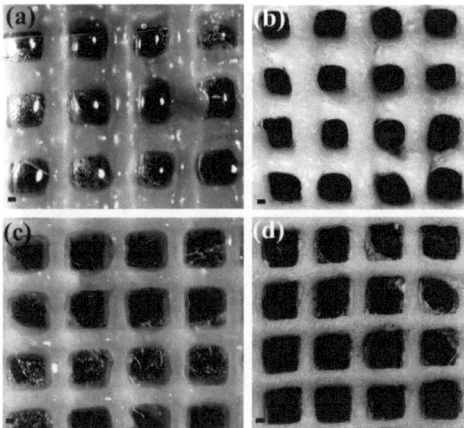

**Figure 7.** Digital microscope images of Alg-g-PNIPAM (**a**) before and (**b**) after freeze-drying and Alg-g-PNIPAM/MC (**c**) before and (**d**) after freeze-drying. Scale bar is equal to 500 μm.

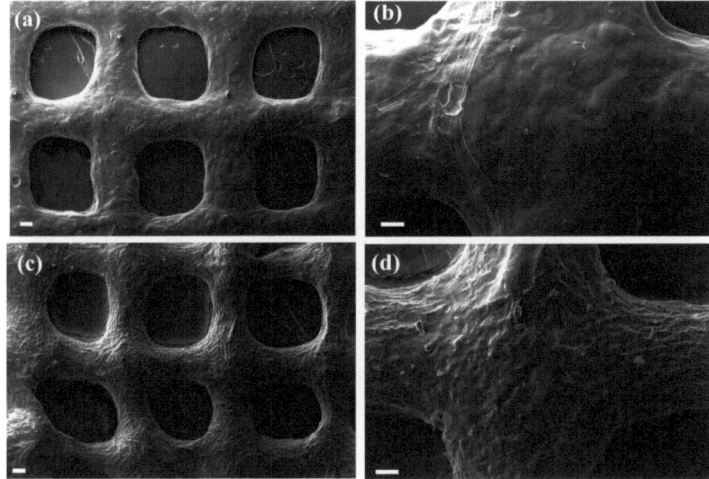

**Figure 8.** SEM images of 3D-printed scaffolds of Alg-g-PNIPAM (**a,b**) and Alg-g-PNIPAM/MC (**c,d**). Scale bar for (**a,c**) represents 200 μm and (**b,d**) is equal to 100 μm.

2.3.4. Erosion Studies

The erosion profile of 3D-printed Alg-g-PNIPAM/MC and Alg-g-PNIPAM domains have been examined, and the results are presented in Figure 9. The 3D-printed designs were freeze-dried, followed by hydration with distilled water at room and body temperature. The more stable material, Alg-g-PNIPAM/MC, was degraded only by 6% at 37 °C and 19% at 20 °C after 7 days (172 h). In contrast, Alg-g-PNIPAM was eroded by up to 26% at 20 °C after 5 days (120 h) and by up to 35% at 37 °C after 6 days (144 h). After these timeframes, the weaker bonds of Alg-g-PNIPAM domains were disintegrated. In Figure 7b, the moisturized MC-enriched and the MC-free samples are illustrated at day 7. After a week, the Alg-g-PNIPAM/MC seems almost intact compared to the Alg-g-PNIPAM. Hence, it was concluded that the MC-rich scaffolds could be promising candidates for a two-week pre-osteoblastic cell culture investigating cell viability, proliferation, and osteogenic differentiation, considering that MC is expected to enhance the cell proliferation.

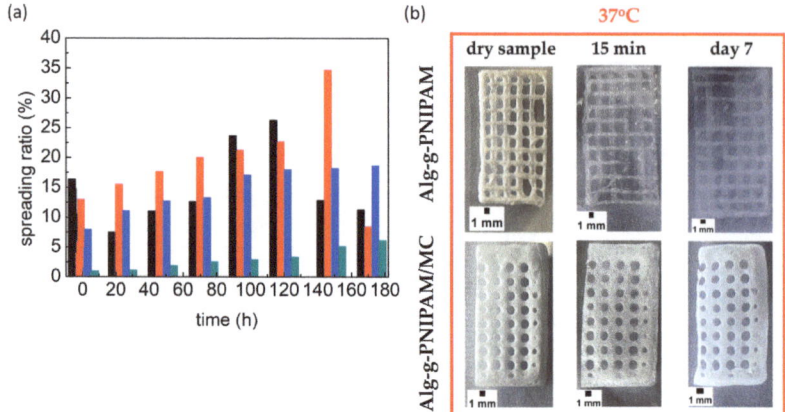

**Figure 9.** (**a**) Spreading ratio (%) defined as the percentage change in the pattern line width compared to the dry structure of Alg-g-PNIPAM at 20 °C (black) and at 37 °C (red) and of Alg-g-PNIPAM /MC at 20 °C (blue) and at 37 °C (cyan); (**b**) Photographs of the swelling samples up to day 7 at 37 °C.

## 2.4. Evaluation of Cytocompatibility, Cell Adhesion, Viability and Proliferation

### 2.4.1. Cell Viability and Proliferation

Cell viability and proliferation using pre-osteoblastic cells have been assessed at days 3, 5, and 7 (Figure 10a,b). The number of cells increased from day 3 to day 5 and up to day 7, demonstrating that the scaffold compositions promote cell proliferation. Both scaffold compositions, Alg-g-PNIPAM and Alg-g-PNIPAM/MC, showed similar cell viability at each time point, and these were found to be significantly lower compared to the tissue culture polystyrene (TCPS) control; however, they reached 80% viability of the control. The Alg-g-PNIPAM/MC scaffolds indicated a higher cell viability on day 5; however, this was not significant compared to the Alg-g-PNIPAM counterparts. These results show that the Alg-g-PNIPAM scaffolds are cytocompatible. Statistical analysis of each scaffold composition compared to the TCPS control revealed significant differences ($p < 0.0001$) at all time points. Another report on Alg-g-P(NIPAM)-based solutions of various concentrations showed 80% fibroblast cell viability after 24 h, indicating the cytocompatibility of this grafted co-polymer [20].

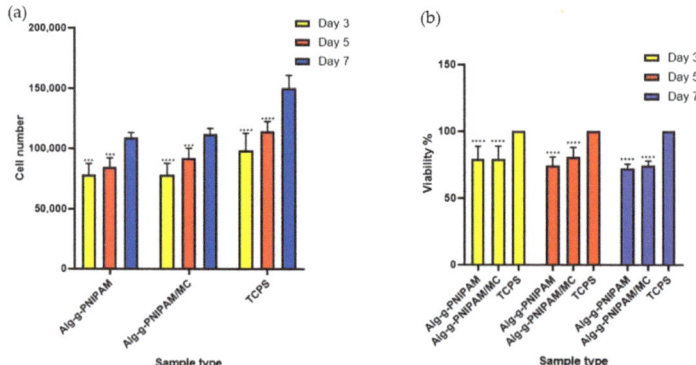

**Figure 10.** Cell viability and proliferation of pre-osteoblastic cells seeded on Alg-g-PNIPAM, Alg-g-PNIPAM/MC scaffolds and TCPS control at 3, 5 and 7 days expressed as OD values (**a**) and as cell viability percentage (**b**). Bars represent averages ± standard deviation of $n = 6$ (*** $p < 0.001$, **** $p < 0.0001$).

### 2.4.2. Cell Adhesion and Morphology Evaluation

We observed the cell adhesion on the surface of the scaffold struts and their morphology by means of SEM. Figure 11 (upper panel) shows the pre-osteoblastic cells adhered on both scaffold compositions after 7 days in culture. The cell nuclei of a dense cell layer covering both scaffold types are clearly visible in Figure 11 (lower panel). The morphology of adhered cells did not show any differences between the two compositions. The characteristic morphology of cell nuclei indicates that both scaffold types support cell adhesion. Although none of the scaffold compounds, alginate [24], PNIPAM [25] or MC [15], possess cell-specific binding sites to promote cell adhesion, both scaffold compositions showed adequate attachment of pre-osteoblasts. Similarly, other studies report on good cell adhesion on alginate [26] and PNIPAM [27]-based scaffolds combined with other biomaterials or coatings.

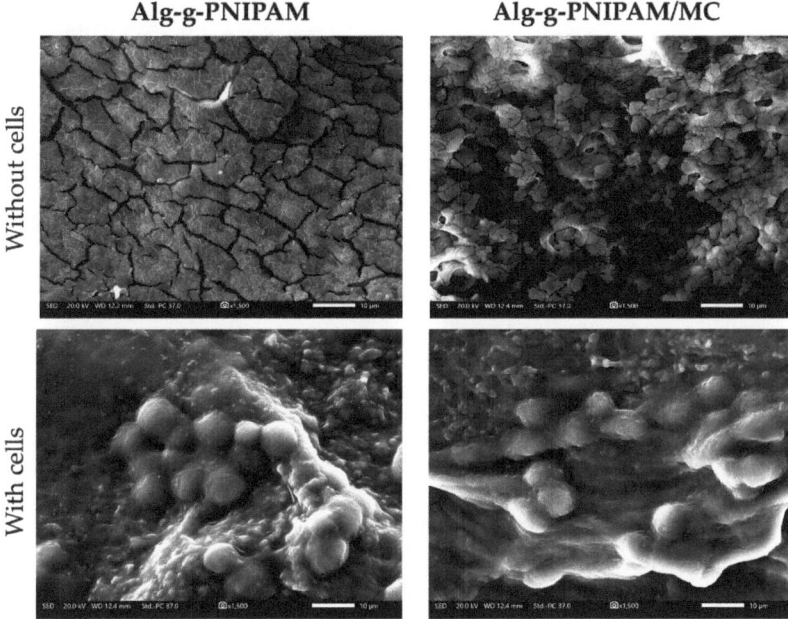

**Figure 11.** SEM images of scaffolds without (**upper panel**) and with cells (**lower panel**) at day 7. The surface of the Alg-g-PNIPAM scaffold is depicted in the upper left and Alg-g-PNIPAM/MC in the upper right images. The elongated morphology of adhered cells with visible cell nuclei is shown on the Alg-g-PNIPAM (**lower left**) and Alg-g-PNIPAM/MC (**lower right**) scaffolds. Magnification is ×1500, and scale bars represents 10 μm.

### 2.4.3. Evaluation of Osteogenic Differentiation Markers ALP Activity and Calcium and Collagen Production by Cells Cultured on Alg-g-PNIPAM and Alg-g-PNIPAM/MC Scaffolds

Bone formation is accompanied by specific enzymatic activity and expression characteristic of the osteoinduction process. Alkaline phosphatase (ALP) activity is elevated in areas of extracellular matrix mineralization. ALP activity was investigated at two early time points of culture, as it presents an early-phase osteogenesis marker. At day 3, both scaffold compositions demonstrate with a two-fold increase in enzyme activity, significantly higher than that of the TCPS control. At day 7, the ALP activity increased two-fold compared to day 3, and it is higher on both scaffold compositions compared to the control, with a significantly higher difference on the Alg-g-PNIPAM/MC scaffolds (Figure 12a).

The concentration of the calcium produced by osteoblasts was assessed on days 3, 7, 10 and 14 (Figure 12b) as a late marker of osteogenesis. The calcium concentration increased between the consecutive experimental time points. In particular, the production of calcium showed at least a 30% increase, with significantly higher levels in both scaffold types compared to the TCPS control at days 7, 10 and 14, with an approximate increase of 30–50% between the experimental timepoints. On days 7 and 10, the MC-containing scaffolds presented a significantly higher increase compared to the Alg-g-PNIPAM ones. Of note, the higher calcium content in both scaffold types may be attributed to the presence and release of $Ca^{2+}$ used for crosslinking the hydrogels before and after 3D printing.

Osteoblasts produce more collagen type I than any other cell. We have quantified the total collagen amount produced by cells cultured onto the scaffolds on days 7 and 14 (Figure 12c). Rich collagen synthesis with a concentration of 750 µg/mL was measured on day 7 for Alg-g-PNIPAM and 900 µg/mL for Alg-g-PNIPAM/MC scaffolds, while secreted collagen concentration increased by 10–25% on day 14. Both scaffold compositions depicted comparable collagen production with the TCPS control at day 7. However, on day 14, both scaffold compositions showed significantly higher collagen levels compared to the TCPS control, with the Alg-g-PNIPAM/MC scaffolds displaying significantly higher levels compared to the Alg-g-PNIPAM scaffolds.

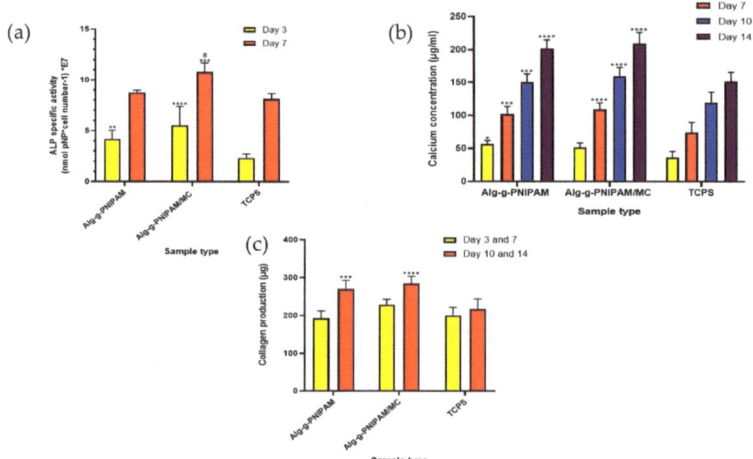

**Figure 12.** Normalized alkaline phosphatase activity of pre-osteoblasts cultured onto the two scaffold types and TCPS control on days 3 and 7 (**a**). Calcium concentration on days 3, 7, 10 and 14 (**b**). Collagen production by cells on days 7 and 14 (**c**). Bars represent averages ± standard deviation of $n = 6$ (** $p = 0.0098$, *** $p = 0.0004$, **** $p < 0.0001$). Asterisks (*) denote significant differences between each scaffold composition with the TCPS control, while hashtag (#) designates significant differences between Alg-g-PNIPAM and Alg-g-PNIPAM/MC scaffolds.

Alginate is one of the most prominently used biomaterials in bone tissue engineering due to its excellent gelling capacity and its ability to physically bind cations such as calcium and strontium, which are then available for release in aquatic environments, thus providing an osteogenic effect [28]. PNIPAM grafted onto gelatin has also been used as an injectable hydrogel for bone defect regeneration [29], making the mixing of these two biomaterials a rather promising combination for bone tissue engineering applications. The additional incorporation of MC into the alginate/PNIPAM blend was considered with regard to supporting the 3D-printing process, as well as due to its established osteogenic potential [30,31]. Alginate scaffolds have been previously reported to exhibit increased alkaline phosphatase activity, presenting elevated values between subsequent time points, compliant with the findings of this study [32]. Similarly, an injectable, thermo-responsive hyaluronic acid-g-chitosan-g-PNIPAM copolymer has been reported to show increased ALP activity and calcium deposition with progressing time in culture of bone marrow-derived mesenchymal stem cells [33]. In that particular work, calcium ions' concentration was determined through alizarin red staining over a period of 21 days, displaying a gradual increase, thus indicating a similar pattern to that evident in our measurements of calcium concentration in the scaffold supernatants.

## 3. Conclusions

In this study, we showed that it is possible to combine graft thermoresponsive alginates with MC to produce soft inks suitable for 3D printing in biomedical applications. It was possible to control the dynamic moduli of the resulting gels by the interplay of the different stimuli such as temperature and $Ca^{2+}$ ions. The presence of MC significantly improved the printing fidelity and the erosion profiles of the polymer networks post-printing. Cells cultured on both scaffold compositions, Alg-g-PNIPAM and Alg-g-PNIPAM/MC, displayed a significant increase in osteogenic markers including ALP activity, calcium, and collagen production compared to the control. Particularly, the MC-containing scaffolds indicated higher responses in these markers compared to the Alg-g-PNIPAM counterparts. These findings demonstrate the osteogenic potential of these gels and their excellent ability to act as 3D-printed scaffolds for bone tissue engineering.

## 4. Materials and Methods

### 4.1. Materials

*N*-isopropylacrylamide (NIPAM, Fluorochem), 2,2′-Azobis(2-methylpropionitrile) (AIBN, Sigma Aldrich, St. Louis, MO, USA), 2-aminoethanethiol hydro-chloride (AET HCl, Alfa Aesar), 1-hydroxybenzotriazole hydrate (HoBT, Fluka), 1-ethyl-3-(3-(dimethylamino) propyl) carbodiimide (EDC, Alfa Aesar), tetrahydrofuran (THF, Sigma Aldrich), calcium chloride dihydrate ($CaCl_2 \cdot 2H_2O$, Sigma Aldrich), sodium hydroxide (NaOH, Panreac), and acetone (Sigma-Aldrich) were used as purchased. Purified water (3D-$H_2O$) was provided by an ELGA Medica-R7/15 device. Sodium alginate (NaALG, Aldrich) with a molecular weight range of 120,000–190,000 g/mol and a mannuronic/guluronic ratio (M/G) of 1.53 (values are given by the supplier) was dissolved at 7 $w/v$% in 3D water, and was further purified against dialysis membrane (MWCO 12,000–14,000 Da) before being freeze-dried.

### 4.2. Synthesis of the Amino-Terminated PNIPAM-$NH_2$ Side

The polymeric side chains were synthesized by free radical polymerization. AIBN was used as an initiator and AET HCl as a chain transfer agent. Briefly, 4 g (0.035 mol) of NIPAM monomer units were dissolved in 40 mL THF. The mixture solution was degassed with nitrogen for 15 min. Then, the mixture was heated at 70 °C. Then, 0.02 g (0.177 mmol, 0.5% over the monomer concentration) AET and 0.058 g (0.354 mmol, 1% over the monomer concentration) AIBN were added to the mixture. The mixture was left under stirring for 24 h. The reaction was stopped by exposure to air at room temperature followed by purification against distilled water with a dialysis membrane (MWCO: 12,000–14,000 Da), in order to remove unwanted byproducts and impurities; finally, the product was freeze-dried and stored as white flakes. The number average molecular weight (Mn) was evaluated by acid–base titration of the amino-terminated groups of the polymer chains, as presented in Table 1.

**Table 1.** Molecular characteristics of the PNIPAM-$NH_2$.

| Grafting Chain | Mn (g/mol) [a] |
|---|---|
| PNIPAM-$NH_2$ | 13,740 |

[a] From acid–base titration.

### 4.3. Synthesis of the Alg-g-PNIPAM

The grafting reaction of the thermosensitive PNIPAM-$NH_2$ side chains onto the Alg backbone was accomplished by carbodiimide chemistry, forming an amide group by the carboxyl groups of the Alg monomer units' ring and the -$NH_2$ end-group of the PNIPAM chains [34,35]. EDC and HOBt were used as coupling agents. Then, 1.15 g of NaALG and 2.25 g of PNIPAM-NH2 were dissolved separately in 25 mL and 45 mL of 3D $H_2O$, respectively. The mixtures were left under stirring in room temperature for 24 h. Then, the PNIPAM solution was added in the Alg one. Subsequently, 0.111 g (0.89 mmol, 5% moles over the monomer moles) HOBt dissolved in 3 mL of 3D-water and 0.6279 g (0.003 mol, 20% moles over the monomer moles) of EDC dissolved in 5 mL of 3D $H_2O$ were added to the mixture and left under stirring at 20 °C for 48h. The pH of the Alg/PNIPAM mixture was adjusted at 5–6, and the EDC solution was added in two steps (half of the amount was added at the start of the reaction, and the rest after 24 h). The final mixture product was precipitated in acetone in order to remove the un-grafted PNIPAM-$NH_2$ chains, followed by filtration and dissolution of the precipitate in ~5 wt.% in distilled water. The pH of the mixture was set at 11 using NaOH (1M), and was further purified using a dialysis membrane (MWCO: 25,000 Da), before finally being freeze-dried. In Table 2 the molecular characteristics of the produced biopolymer are presented.

**Table 2.** Molecular characteristics of the polymer.

| Graft Copolymer | $M_w$ (g/mol) [a] | % Molar Composition Alg/Grafting Chain (mol/mol) [b] | % Weight Composition Alg/Grafting Chains (w/w) | Number of PNIPAM Side Chains Per NaALG Backbone [b] |
|---|---|---|---|---|
| Alg-g-PNIPAM | 167,470 | 74.4/25.6 | 83.6/16.4 | 2 |

[a] Calculated by the following equation: $M_{w,cop} = M_{w,Alg}/wt\%\ Alg$, $M_{w,Alg} = 140,000\ g/mol$; [b] Calculated by $^1H$ NMR.

*4.4. Proton Nuclear Magnetic Resonance ($^1H$ NMR)*

The molar composition (%) of the grafting PNIPAM chains onto the Alg was calculated by integrating the above characteristic areas, and the weight composition (%) of the graft copolymer was calculated by the molar composition using the $M_w$ building units of sodium alginate and the $M_w$ of the PNIPAM side chains. The samples were dissolved in $D_2O$ (peak ~4.8 ppm) and were analyzed using a Bruker Avance iii Hd Prodigy Ascend Tm 600 MHz spectrometer (Billerica, MA, USA).

*4.5. Hydrogel Preparation*

A solution of 5 wt.% Alg-g-PNIPAM containing 2 mm of $Ca^{2+}$ was prepared. Methylcellulose was added up to final mixture concentration 7.5 wt.%. The hydrogels were left under stirring at T = 10 °C up to full homogeneity.

*4.6. Rheological Studies*

The rheological evaluation of the Alg-g-PNIPAM and Alg-g-PNIPAM/MC hydrogel samples was conducted on a stress-controlled AR-2000ex rheometer (TA Instruments) with a cone and plate geometry (diameter 20 mm, angle 3°, truncation 111 μm). The hydrogels were loaded on a Peltier plate, which highly ensures the experimental temperature (±0.1 °C), and their thermo- and shear-responsiveness were measured. A solvent trap was located over the cone–plate geometry to prevent changes in the hydrogels' concentrations. All the measurements were operated in the linear viscoelastic regime (LVR), which has been verified for its sample by a strain sweep test and a constant angular frequency at 6.28 rad/s.

*4.7. Determination of the LCST*

The LCST temperature of PNIPAM and NaAlg-PNIPAM was measured by using a Shimadzu UV-1900i spectrophotometer (Shimadzu Co., Kyoto, Japan) at a wavelength of 500 nm. The instrument was equipped with a Peltier-controlled thermostated cell holder (TCC-100, Shimadzu Co., Kyoto, Japan)). Measurements were performed in the two polymer solutions at a concentration of 4 mg/mL.

*4.8. 3D Scaffold Design and Manufacturing*

The pattern of the 3D model designed using the online TinkerCAD™ design platform and exported to an .stl file. By loading the digital design on the open-source slicing software Ultimaker Cura 5.2.1, all printing parameters were set, and the final file exported in .gcode format. The generated gcode file was loaded to the 3D printer. The scaffolds were manufactured with a low-cost 3D FFF (fused filament fabrication) printer, which was modified to a hydrogel printer. In summary, the extruder of a Wanhao Duplicator i3 plus was removed, and next, a syringe support was designed and manufactured using a polylactic acid filament in a commercial 3D printer (Wanhao Duplicator i3 plus, Wanhao Ltd., Jinhua, China). A plastic 10 mL syringe was inserted in the extruder's position, and the head of the syringe was connected with a clamp to a pneumatic dispenser (DX-250, Metcal, Hampshire, UK). At the edge of the syringe, a plastic nozzle with a diameter of 400 μm was connected to a Luer lock fitting. The bed temperature was set at 40 °C, and a 60 mm glass Petri dish was adhered on the printing bed by means of tape. Afterwards, the hydrogel was placed in the syringe. An external air compressor (Mini 50, Airblock, Thessaloniki, Greece) was joined to the syringe, and by applying the appropriate pressure, the syringe plunger was forced to extrude the hydrogel through the nozzle. The 3D-printed objects were crosslinked with calcium ions directly by immersion after printing in a 0.5 M $CaCl_2 \cdot 2H_2O$ solution for 30 min. Finally, the scaffolds were washed thoroughly with distilled water and lyophilized (Telstar Cryodos, Terrassa, Spain) [30].

*4.9. Structural Characterization*

X-ray diffraction patterns were collected with a Bruker D8 Advance diffractometer (Bruker AXS GmbH, Karlsruhe, Germany) with monochromatic CuKα1 radiation (λ = 1.5406 nm) at a voltage of 40 kV and a current value of 40 mA. The angular scanning speed was 0.35 s/step. After lyophilization, the dried gels were placed on an XRD specimen holder and compacted with a glass slide to flatten them.

ATR-FTIR spectroscopy was performed on an FTIR spectrometer (IR Tracer-100, Schimadzu, Kyoto, Japan) using the ATR accessory MIRacle™ Single Reflection equipped with a ZnSe crystal. After the background signal collection, a small amount of the dried sample was placed in contact with the ATR crystal, and a pressure of 75 psi was applied. Spectra were recorded by averaging 25 scans at a resolution of 4 $cm^{-1}$ in the spectral range between 550 and 4000 $cm^{-1}$.

*4.10. Thermal Characterization*

For the differential scanning calorimetry (DSC) measurements, powdered samples (1–3 mg) were weighed and placed in aluminum pans, sealed hermetically, and assessed in a Q200 (TA Instruments, New Castle, DE, USA) differential scanning calorimeter. An empty aluminum pan used as the reference sample. The specimens were heated from 25 to 300 °C at a heating rate of 10 °C/min under nitrogen atmosphere at a flow rate of 50 mL/min.

Thermogravimetric analysis (TGA) measurements were performed through a TG Q500 thermogravimetric analyzer (TA Instruments, New Castle, DE, USA). About 5 mg of the samples were placed in a platinum pan and heated over a temperature range from 35 to 800 °C at a heating rate of 10 °C/min, under nitrogen with a flow rate of 60 mL/min.

*4.11. Morphological Characterization*

Scanning electron microscopy images (Zeiss EVO MA-10, Carl Zeiss, Oberkochen, Germany) determine the surface morphology of the 3D-printed scaffolds. The samples were placed on aluminum holders, fixed with a conductive silver paste, and then gold sputtered (BAL-TEC SCD-004). Also, images were taken using a digital microscope (Celestron LLC, Torrance, CA, USA) to measure the differences in pore diameters before and after freeze-drying. Pore sizes were quantified using ImageJ analysis software (Version 1.44p, National Institutes of Health: Bethesda, MD, USA).

*4.12. Cell Culture Maintenance*

MC3T3-E1 pre-osteoblastic cells (DSMZ Braunschweig, Germany) derived from mouse calvaria were cultured in a humidified incubator at 37 °C and 5% $CO_2$ (ThermoFisher, Waltham, MA, USA) in alpha-MEM supplemented with 10% ($v/v$) fetal bovine serum (FBS), 100 g/mL penicillin and streptomycin, 2 mm L-glutamine and 2.5 µg/mL amphotericin (all from PAN-Biotech, Aidenbach, Germany) (complete alpha-MEM). Cells were cultured to 90% confluence by medium change every two days, and detached using trypsin-0.25% ethylenediaminetetraacetic acid (EDTA) (Gibco, Thermo Fisher Scientific, Waltham, MA, USA) for passaging. All experiments were conducted with cell passages from 10 to 12. For the osteogenic potential assessment of the scaffolds, we applied osteogenic medium comprising complete alpha-MEM supplemented with 10 nM dexamethasone, 10 mm glycerophosphate, and 50 g/mL L-ascorbic acid 2-phosphate (Sigma-Aldrich, St. Louis, MO, USA).

Prior to cell seeding, all scaffolds were sterilized by immersion in 70% ethanol for 3 min, followed by 30 min of UV irradiation at 265 nm at both sides. Each scaffold was submerged for 10 min in culture medium. Then, the medium was removed and a 15 µL cell suspension with $3 \times 10^4$ cells was seeded onto each scaffold in a 96-well plate; 150 µL cell culture medium was added, and cell-loaded scaffolds were placed in the incubator.

*4.13. Cell Viability and Proliferation Assessment*

Cell viability assessment of cells cultured onto 3D-printed scaffolds is essential in any tissue engineering application. The cell viability of the scaffolds was investigated by employing a metabolic assay, the PrestoBlue™ (Invitrogen Life Technologies, Carlsbad, CA, USA) cell viability assay. The reagent contains resazurin, which changes color from blue to purple according to cell metabolism. On each scaffold, 10 µL of PrestoBlue™ reagent and 90 mL of fresh medium were added to a final volume of 100 mL for each well of a 96-well plate. The mixture was incubated for 1 h at 37 °C and then transferred to another 96-well plate to measure the absorbance in a spectrophotometer at 570 and 600 nm (Synergy HTX Multi-Mode Micro-plate Reader, BioTek, Bad Friedrichshall, Germany). The number of living cells was measured photometrically at 3, 5, and 7 days. For the determination of cell numbers from the absorbance values, we used a calibration curve of known pre-osteoblastic cell numbers in the same multi-well plate type.

*4.14. Cell Adhesion and Morphology*

The cell adhesion and morphology onto the scaffolds were monitored using SEM at 7 days. Prior to microscopy, the scaffolds were rinsed with phosphate-buffered saline (PBS) to remove the remaining culture medium, and were fixed using 4% paraformaldehyde for 20 min. The scaffolds were then dehydrated in increasing ethanol concentrations from 30, to 100% $v/v$. The scaffolds were then sputter-coated with a 20 nm gold layer (Baltec SCD 050, Baltec, Los Angeles, CA, USA) and observed by means of a scanning electron microscope (JEOL JSM-6390 LV, Tokyo, Japan) at an accelerating voltage of 20 kV.

*4.15. Osteogenic Potential Evaluation of Pre-Osteoblasts Seeded onto Scaffolds by Determination of the ALP Activity, Collagen and Calcium Secretion*

The ALP activity is indicative of the initial stages of osteogenesis. Briefly, the cell-seeded scaffolds remained in culture for 3 and 7 days using osteogenic medium. Each scaffold was rinsed with PBS and submerged in 200 µL lysis buffer (0.1% Triton X-100 in 50 mm Tris-HCl pH 10.5) to extract the cell lysate. The mixture of 100 µL lysate with 100 µL of 2 mg/mL p-nitrophenyl phosphate (pNPP, Sigma, St. Louis, MO, USA) solution was incubated at 37 °C for 1 h, and measured photometrically at 405 nm. The enzymatic activity was calculated using the following equation:

$$\text{Units} = \text{mmol p-nitrophenol/min}$$

and normalized to total protein in lysates determined using the Bradford protein concentration assay.

Calcium is one of the elemental components of bone tissue. Calcium mineralization is a crucial regulator for the formation of the bone matrix and a late marker of osteogenesis. Calcium secretion was determined by the O-cresol phthalein complexone (CPC) method. Culture supernatants were collected after 3, 7, 10 and 14 days, and 10 µL of each sample was mixed with 100 µL of calcium buffer and 100 µL of calcium dye CPC. The final solutions were transferred to a 96-well plate to measure the absorbance at 405 nm.

Collagen is a crucial element and the primary organic component of bone tissue, which plays a pivotal role in providing structural support in the formation of the extracellular matrix (ECM). Quantification of collagen secretion in culture supernatants was performed by the Sirius Red (Direct red 80, Sigma-Aldrich, St. Louis, MO, USA) staining method after 7 and 14 days. In brief, 25 µL of supernatants were diluted in 75 µL of ultrapure water at each time point. The solution was mixed with 1 mL of 0.1% Direct Red 80 and incubated at room temperature. After centrifugation of the samples, the pellets were rinsed with 0.5 M acetic acid for non-bound dye removal. Finally, 200 µL of NaOH 0.5 M was added to extract the collagen-bound dye complex. The absorbance of the solutions was measured with a spectrometer in a 96-well plate at 530 nm. A calibration curve correlates the quantity of collagen to mg/mL.

Statistical analysis was performed using GraphPad Prism 8.0.2 software (GraphPad Software, San Diego, CA, USA) and a two-way ANOVA followed by Tukey's multiple comparisons test. All values are expressed as average ± standard deviation (SD). The adjusted p value set * $p < 0.05$ compared each scaffold composition with the TCPS control at each time point. Statistical analysis was performed to compare the two scaffold compositions, Alg-g-PNIPAM and Alg-g-PNIPAM/MC.

**Supplementary Materials:** The following supporting information can be downloaded at: https://www.mdpi.com/article/10.3390/gels9120984/s1, Figure S1. $^1$H-NMR spectra of (a) Alg-g-PNIPAM and of (b) PNIPAM-NH$_2$. Figure S2. FTIR spectra of Alg-g-PNIPAM (a), Alg (b), and PNIPAM-NH2 (c).

**Author Contributions:** A.G. and S.F.S. performed experiments and analyzed data; K.L. performed in vitro cell biological studies, and analyzed data; M.C., G.P. and N.B. conceived of and overviewed the study; funding acquisition, N.B., G.P. and M.C. All authors contributed to the writing of the manuscript. All authors have read and agreed to the published version of the manuscript.

**Funding:** This research was supported by Grant (No. 81384) from the Research Committee of the University of Patras via the "C. CARATHEODORY" program (G.P.), and by the Hellenic Foundation for Research and Innovation (H.F.R.I.) under the "1st Call for H.F.R.I. Research Projects to support Faculty members and Re-searchers and the procurement of high-cost research equipment grant" (project number HFRI-FM17-1999).

**Institutional Review Board Statement:** Not applicable.

**Informed Consent Statement:** Not applicable.

**Data Availability Statement:** All data and materials are available on request from the corresponding author. The data are not publicly available due to ongoing research using a part of the data.

**Conflicts of Interest:** The authors declare no conflict of interest.

# References

1. Langer, R.; Vacanti, J.P. Tissue Engineering. *Science* **1993**, *260*, 920–926. [CrossRef]
2. Eldeeb, A.E.; Salah, S.; Elkasabgy, N.A. Biomaterials for Tissue Engineering Applications and Current Updates in the Field: A Comprehensive Review. *AAPS PharmSciTech* **2022**, *23*, 267. [CrossRef]
3. Tabriz, A.G.; Douroumis, D.; Boateng, J. 3D Printed Scaffolds for Wound Healing and Tissue Regeneration. In *Therapeutic Dressings and Wound Healing Applications*; Wiley: Hoboken, NJ, USA, 2020; pp. 385–398. [CrossRef]
4. Kontogianni, G.-I.; Bonatti, A.F.; De Maria, C.; Naseem, R.; Melo, P.; Coelho, C.; Vozzi, G.; Dalgarno, K.; Quadros, P.; Vitale-Brovarone, C.; et al. Promotion of In Vitro Osteogenic Activity by Melt Extrusion-Based PLLA/PCL/PHBV Scaffolds Enriched with Nano-Hydroxyapatite and Strontium Substituted Nano-Hydroxyapatite. *Polymers* **2023**, *15*, 1052. [CrossRef] [PubMed]

5. Pasparakis, G.; Tsitsilianis, C. LCST polymers: Thermoresponsive nanostructured assemblies towards bioapplications. *Polymer* **2020**, *211*, 123146. [CrossRef]
6. Suntornnond, R.; An, J.; Chua, C.K. Bioprinting of Thermoresponsive Hydrogels for Next Generation Tissue Engineering: A Review. *Macromol. Mater. Eng.* **2017**, *302*, 1600266. [CrossRef]
7. Singh, R.; Deshmukh, S.A.; Kamath, G.; Sankaranarayanan, S.K.R.S.; Balasubramanian, G. Controlling the aqueous solubility of PNIPAM with hydrophobic molecular units. *Comput. Mater. Sci.* **2017**, *126*, 191–203. [CrossRef]
8. Lencina, S.; Iatridi, Z.; Villar, M.; Tsitsilianis, C. Thermoresponsive hydrogels from alginate-based graft copolymers. *Eur. Polym. J.* **2014**, *61*, 33–44. [CrossRef]
9. Wang, Z.; Li, Y.; Wang, X.; Pi, M.; Yan, B.; Ran, R. A rapidly responsive, controllable, and reversible photo-thermal dual response hydrogel. *Polymer* **2021**, *237*, 124344. [CrossRef]
10. Zhang, J.; Yun, S.; Karami, A.; Jing, B.; Zannettino, A.; Du, Y.; Zhang, H. 3D printing of a thermosensitive hydrogel for skin tissue engineering: A proof of concept study. *Bioprinting* **2020**, *19*, e00089. [CrossRef]
11. Fitzsimmons, R.; Aquilino, M.; Quigley, J.; Chebotarev, O.; Tarlan, F.; Simmons, C. Generating vascular channels within hydrogel constructs using an economical open-source 3D bioprinter and thermoreversible gels. *Bioprinting* **2018**, *9*, 7–18. [CrossRef]
12. Celikkin, N.; Simó Padial, J.; Costantini, M.; Hendrikse, H.; Cohn, R.; Wilson, C.; Rowan, A.; Święszkowski, W. 3D Printing of Thermoresponsive Polyisocyanide (PIC) Hydrogels as Bioink and Fugitive Material for Tissue Engineering. *Polymers* **2018**, *10*, 555. [CrossRef] [PubMed]
13. Hu, C.; Ahmad, T.; Haider, M.S.; Hahn, L.; Stahlhut, P.; Groll, J.; Luxenhofer, R. A thermogelling organic-inorganic hybrid hydrogel with excellent printability, shape fidelity and cytocompatibility for 3D bioprinting. *Biofabrication* **2022**, *14*, 025005. [CrossRef] [PubMed]
14. Lee, K.Y.; Mooney, D.J. Alginate: Properties and biomedical applications. *Prog. Polym. Sci.* **2012**, *37*, 106–126. [CrossRef] [PubMed]
15. Mahboubian, A.; Vllasaliu, D.; Dorkoosh, F.A.; Stolnik, S. Temperature-Responsive Methylcellulose–Hyaluronic Hydrogel as a 3D Cell Culture Matrix. *Biomacromolecules* **2020**, *21*, 4737–4746. [CrossRef] [PubMed]
16. Contessi Negrini, N.; Bonetti, L.; Contili, L.; Farè, S. 3D printing of methylcellulose-based hydrogels. *Bioprinting* **2018**, *10*, e00024. [CrossRef]
17. Shymborska, Y.; Budkowski, A.; Raczkowska, J.; Donchak, V.; Melnyk, Y.; Vasiichuk, V.; Stetsyshyn, Y. Switching it up: The promise of stimuli-responsive polymer systems in biomedical science. *Chem. Rec.* **2023**, e202300217. [CrossRef]
18. Guo, H.; de Magalhaes Goncalves, M.; Ducouret, G.; Hourdet, D. Cold and Hot Gelling of Alginate-graft-PNIPAM: A Schizophrenic Behavior Induced by Potassium Salts. *Biomacromolecules* **2018**, *19*, 576–587. [CrossRef]
19. Halperin, A.; Kröger, M.; Winnik, F.M. Poly(N-isopropylacrylamide) Phase Diagrams: Fifty Years of Research. *Angew. Chem. Int. Ed.* **2015**, *54*, 15342–15367. [CrossRef]
20. Soares, J.; Santos, J.; Chierice, G.; Cavalheiro, É. Thermal behavior of alginic acid and its sodium salt. *Eclet. Quim.* **2004**, *29*, 57–64. [CrossRef]
21. Ciarleglio, G.; Toto, E.; Santonicola, M.G. Conductive and Thermo-Responsive Composite Hydrogels with Poly(N-isopropylacrylamide) and Carbon Nanotubes Fabricated by Two-Step Photopolymerization. *Polymers* **2023**, *15*, 1022. [CrossRef]
22. Shekhar, S.; Mukherjee, M.; Sen, A. Studies on thermal and swelling properties of Poly (NIPAM-co-2-HEA) based hydrogels. *Adv. Mater. Res.* **2012**, *1*, 269. [CrossRef]
23. Saeed, A.; Georget, D.M.R.; Mayes, A.G. Solid-state thermal stability and degradation of a family of poly(N-isopropylacrylamide-co-hydroxymethylacrylamide) copolymers. *J. Polym. Sci. Part A Polym. Chem.* **2010**, *48*, 5848–5855. [CrossRef]
24. Liu, S.; Kilian, D.; Ahlfeld, T.; Hu, Q.; Gelinsky, M. Egg white improves the biological properties of an alginate-methylcellulose bioink for 3D bioprinting of volumetric bone constructs. *Biofabrication* **2023**, *15*, 025013. [CrossRef]
25. Shimizu, T.; Yamato, M.; Kikuchi, A.; Okano, T. Cell sheet engineering for myocardial tissue reconstruction. *Biomaterials* **2003**, *24*, 2309–2316. [CrossRef] [PubMed]
26. Bousnaki, M.; Bakopoulou, A.; Papadogianni, D.; Barkoula, N.-M.; Alpantaki, K.; Kritis, A.; Chatzinikolaidou, M.; Koidis, P. Fibro/chondrogenic differentiation of dental stem cells into chitosan/alginate scaffolds towards temporomandibular joint disc regeneration. *J. Mater. Sci. Mater. Med.* **2018**, *29*, 97. [CrossRef]
27. Kim, H.; Witt, H.; Oswald, T.A.; Tarantola, M. Adhesion of Epithelial Cells to PNIPAm Treated Surfaces for Temperature-Controlled Cell-Sheet Harvesting. *ACS Appl. Mater. Interfaces* **2020**, *12*, 33516–33529. [CrossRef] [PubMed]
28. Hernández-González, A.C.; Téllez-Jurado, L.; Rodríguez-Lorenzo, L.M. Alginate hydrogels for bone tissue engineering, from injectables to bioprinting: A review. *Carbohydr. Polym.* **2020**, *229*, 115514. [CrossRef]
29. Ren, Z.; Wang, Y.; Ma, S.; Duan, S.; Yang, X.; Gao, P.; Zhang, X.; Cai, Q. Effective Bone Regeneration Using Thermosensitive Poly(N-Isopropylacrylamide) Grafted Gelatin as Injectable Carrier for Bone Mesenchymal Stem Cells. *ACS Appl. Mater. Interfaces* **2015**, *7*, 19006–19015. [CrossRef]
30. Fermani, M.; Platania, V.; Kavasi, R.-M.; Karavasili, C.; Zgouro, P.; Fatouros, D.; Chatzinikolaidou, M.; Bouropoulos, N. 3D-Printed Scaffolds from Alginate/Methyl Cellulose/Trimethyl Chitosan/Silicate Glasses for Bone Tissue Engineering. *Appl. Sci.* **2021**, *11*, 8677. [CrossRef]

31. von Strauwitz Né Ahlfeld, T.; Tiodorovic Neé Guduric, V.; Duin, S.; Akkineni, A.R.; Schütz, K.; Kilian, D.; Emmermacher, J.; Cubo Mateo, N.; Dani, S.; von Witzleben, M.; et al. Methylcellulose—A versatile printing material that enables biofabrication of tissue equivalents with high shape fidelity. *Biomater. Sci.* **2020**, *8*, 2102–2110. [CrossRef]
32. Luo, Y.; Chen, B.; Zhang, X.; Huang, S.; Wa, Q. 3D printed concentrated alginate/GelMA hollow-fibers-packed scaffolds with nano apatite coatings for bone tissue engineering. *Int. J. Biol. Macromol.* **2022**, *202*, 366–374. [CrossRef]
33. Liao, H.-T.; Chen, C.-T.; Chen, J.-P. Osteogenic Differentiation and Ectopic Bone Formation of Canine Bone Marrow-Derived Mesenchymal Stem Cells in Injectable Thermo-Responsive Polymer Hydrogel. *Tissue Eng. Part C Methods* **2011**, *17*, 1139–1149. [CrossRef] [PubMed]
34. Saravanou, S.F.; Ioannidis, K.; Dimopoulos, A.; Paxinou, A.; Kounelaki, F.; Varsami, S.M.; Tsitsilianis, C.; Papantoniou, I.; Pasparakis, G. Dually crosslinked injectable alginate-based graft copolymer thermoresponsive hydrogels as 3D printing bioinks for cell spheroid growth and release. *Carbohydr. Polym.* **2023**, *312*, 120790. [CrossRef] [PubMed]
35. Safakas, K.; Saravanou, S.-F.; Iatridi, Z.; Tsitsilianis, C. Alginate-g-PNIPAM-Based Thermo/Shear-Responsive Injectable Hydrogels: Tailoring the Rheological Properties by Adjusting the LCST of the Grafting Chains. *Int. J. Mol. Sci.* **2021**, *22*, 3824. [CrossRef] [PubMed]

**Disclaimer/Publisher's Note:** The statements, opinions and data contained in all publications are solely those of the individual author(s) and contributor(s) and not of MDPI and/or the editor(s). MDPI and/or the editor(s) disclaim responsibility for any injury to people or property resulting from any ideas, methods, instructions or products referred to in the content.

Article

# Carboxymethyl Chitosan Microgels for Sustained Delivery of Vancomycin and Long-Lasting Antibacterial Effects

Mehtap Sahiner [1,2], Aynur S. Yilmaz [2,3], Ramesh S. Ayyala [4] and Nurettin Sahiner [2,3,4,*]

[1] Department of Bioengineering, Faculty of Engineering, Canakkale, Onsekiz Mart University, Terzioglu Campus, Canakkale 17100, Turkey; sahinerm78@gmail.com
[2] Department of Chemical, Biological and Materials Engineering, University of South Florida, Tampa, FL 33620, USA; sanemyilmazz99@gmail.com
[3] Department of Chemistry, Faculty of Sciences, and Nanoscience and Technology Research and Application Center (NANORAC), Canakkale Onsekiz Mart University, Terzioglu Campus, Canakkale 17100, Turkey
[4] Department of Ophthalmology, Morsani College of Medicine, University of South Florida Eye Institute, 12901 Bruce B Down Blvd., MDC 21, Tampa, FL 33612, USA; rsayyala@gmail.com
* Correspondence: sahiner71@gmail.com or nsahiner@usf.edu

Citation: Sahiner, M.; Yilmaz, A.S.; Ayyala, R.S.; Sahiner, N. Carboxymethyl Chitosan Microgels for Sustained Delivery of Vancomycin and Long-Lasting Antibacterial Effects. *Gels* **2023**, *9*, 708. https://doi.org/10.3390/gels9090708

Academic Editor: Junfeng Shi

Received: 8 August 2023
Revised: 20 August 2023
Accepted: 28 August 2023
Published: 1 September 2023

Copyright: © 2023 by the authors. Licensee MDPI, Basel, Switzerland. This article is an open access article distributed under the terms and conditions of the Creative Commons Attribution (CC BY) license (https://creativecommons.org/licenses/by/4.0/).

**Abstract:** Carboxymethyl chitosan (CMCh) is a unique polysaccharide with functional groups that can develop positive and negative charges due to the abundant numbers of amine and carboxylic acid groups. CMCh is widely used in different areas due to its excellent biocompatibility, biodegradability, water solubility, and chelating ability. CMCh microgels were synthesized in a microemulsion environment using divinyl sulfone (DVS) as a crosslinking agent. CMCh microgel with tailored size and zeta potential values were obtained in a single stem by crosslinking CMCh in a water-in-oil environment. The spherical microgel structure is confirmed by SEM analysis. The sizes of CMCh microgels varied from one micrometer to tens of micrometers. The isoelectric point of CMCh microgels was determined as pH 4.4. Biocompatibility of CMCh microgels was verified on L929 fibroblasts with 96.5 ± 1.5% cell viability at 1 mg/mL concentration. The drug-carrying abilities of CMCh microgels were evaluated by loading Vancomycin (Van) antibiotic as a model drug. Furthermore, the antibacterial activity efficiency of Van-loaded CMCh microgels (Van@CMCh) was investigated. The MIC values of the released drug from Van@CMCh microgels were found to be 68.6 and 7.95 µg/mL against *E. coli* and *S. aureus*, respectively, at 24 h contact time. Disk diffusion tests confirmed that Van@CMCh microgels, especially for Gram-positive (*S. aureus*) bacteria, revealed long-lasting inhibitory effects on bacteria growth up to 72 h.

**Keywords:** carboxymethyl chitosan; microgel; sustained drug delivery; antibiotic; Fe (II) chelating activity; vancomycin; prolonged antibacterial effect

## 1. Introduction

Carboxymethyl chitosan (CMCh) is a chitosan (Ch) derivative composed of glucosamine units [1]. Ch is a biocompatible, biodegradable, non-toxic, antimicrobial, accessible natural polymer and has been used in biomedical applications due to its gene-carrying ability, bacteria growth-inhibitory effects, anti-oxidant and anti-inflammatory activities [1,2]. However, the low solubility of Ch in an aqueous environment restricts these biological properties [3]. Certain functional groups, such as amino and hydroxyl groups added onto Ch chains, allow Ch to be chemically modified so that the resultant structure possesses improved water solubility, allowing Ch derivatives multiple functions to be explored for new applications. Carboxymethylation is one of the methods for hydrophilic modification, and CMCh is proven to have numerous better biological properties and unique abilities, including wound healing capability, mucoadhesive, gelling, metal-ion chelating, moisture retention ability as well as injectability, bioimaging, gene and enzyme delivery aptitudes making CMCh as one of the most preferred materials in tissue engineering applications with its additional advantages such as degradability [1,4,5]. Depending on the carboxymethylation reaction, O-Carboxymethyl chitosan, N, O-Carboxymethyl chitosan, and N, N-Carboxymethyl chitosan with different biological activities can be obtained [6,7]. CMCh-based gels were found to be biocompatible with mesenchymal stem cells, osteoblasts, and fibroblasts [8–10]. Due to the presence of a higher number of chelating groups, supramolecular assemblies of CMCh, i.e., CMCh-based hydrogels, can increase the moisture retention time, improve water solubility, biodegradability, and enhance antibacterial activity [11,12]. Moreover, CMCh, as a drug carrier, is able to exhibit slower drug release and better adsorption ability compared to CH, which directly affects its bioavailability [13].

Novel gel-based drug carriers derived from natural polymers have been proposed for the treatment of a wide variety of diseases, including neurological disorders, diabetes, infections, psoriasis, glaucoma, and cancer [14–17], and the majority of them showed promising results. Considering the higher cost of manufacturing processes of synthetic drug molecules or drug delivery systems, the ease of production and modification of nature-based materials such as CH and CMCh are quite advantageous, especially in the biomedical, cosmetic, and food industries [1,18,19]. In literature, CMCh-based scaffolds, nanoparticles, and hydrogels showed controllable pore size, higher swelling ability, degradability, cell adhesion, and proliferation capabilities, e.g., Rao et al. reported that poly(vinyl alcohol) and CMC-containing wound dressing material and showed that these materials have high cell viability on connective tissue cells [20]. Another recently published study indicated that CMCh had to promote wound healing in burn wounds, such as wound contraction [18]. Local delivery of chemotherapeutics and their sustained release for up to 5 days was achieved using CMCh-based hydrogels [21]. In addition, CMCh showed pH-responsive swelling ability and high biocompatibility in normal cells. Recently, the controlled release of drugs from carriers via different factors such as pH, temperature, electrical field, and partition coefficient has become a highly preferred method [14,16,22]. Natural polymers such as CMCh can integrate and release drugs at a higher rate with their unique bio-beneficial properties with good swelling ability, entrapment, and degradation properties, maintaining the relevant drug concentration in the blood for an extended period [23,24]. CMCh-based polymeric structures were reported to be used for the delivery of a variety of antimicrobial agents such as ciprofloxacin [22], fluconazole [25], clindamycin [4], gentamycin sulfate [5], voriconazole [26]. To date, biocompatible Ch hydrogels crosslinked with different agents such as glutaraldehyde, formaldehyde, and oxalic acid for tissue engineering are reported [27,28]. Despite these gel-based polymeric structures that possess porous and interconnected structure, elasticity, flexibility, and moisturize-retention ability, the low mechanical strength, poor stability, and insolubility of hydrogels at certain pH conditions significantly restrict their use [27]. Likewise, the standardization of CS-based hydrogels, as well as their insufficient clinical utility, persist as substantial limitations [29]. On the other hand, nano- and microgels of chemically crosslinked Ch such as Ch:methacrylates, CMCh:sodium tripolyphosphate complex, CMCh:alginate complex, and CH:gelatin composites were proven to overcome most of these limitations and exhibit improved properties for drug delivery purposes including injectability [30–32].

Vancomycin (Van) is a first-generation glycopeptide antibiotic especially prescribed for hospital-acquired infections [33]. Van is suggested in many diseases such as colitis, skin and soft tissue infections, and endocarditis with FDA approval. The mechanism of action of Van is to stop the cell wall production of bacteria by binding to peptidoglycan precursors and inhibiting the protein–biosynthesis cycle. Hence, in addition to the main effects on Gram-positive bacteria, Van also has broad-spectrum antimicrobial effects. Van is administered parenterally or orally, but due to its hydrophilic nature, Van is released fast and has limited membrane permeability. Low systemic absorption (oral bioavailability of less than 10%) [34] and rapid clearance (elimination half-life is 5–11 h range) [35] of this drug are the important obstacles, and it was reported that Van should be used with a loading dose to ensure effective blood levels [36]. However, severe diarrhea, hearing loss, angioedema, neutropenia, nephrotoxicity, ototoxicity, and Van flushing syndrome (VFS, with quite varying incidences, were reported as non-negligible adverse effects of this antibiotic. These adverse effects strongly concern patients with diabetes, renal impairment, and hypersensitivity [35,37].

Several methods for Van loading into mesoporous materials or gel-based polymeric structures, including physical mixing and vacuum-assisted loading, were reported [38] to increase the bioavailability of Van. These reports confirmed that Van release from Van-loaded particles could last more than 5 days [38]. Poly(ethylene glycol) conjugated Van showed high drug loading content and stability [33]. Since Van is known as a critical situation medicine, it is important to provide optimum delivery conditions for this antibiotic. Rapid Van release, along with a prolonged local drug concentration for killing the bacteria in the initial stage, is preferred to obtain a better therapeutic response as well as to prevent continuous bacterial infections [39]. Van is a large antibiotic having approximately 1500 Da of molecular weight and a big molecular size ($3.2 \times 2.2$ nm) [40,41]. Therefore, the potential Van carriers should possess appropriate characteristics, e.g., size, pore, functional groups, etc., to encapsulate this drug as well as have controllable swelling or degradability capabilities. To date, Van@hydroxy propyl methyl cellulose microparticles [42], Van-encapsulated poly(epsilon-caprolactone) microparticles [43], Van@poly (lactic-co-glycolic acid)(PLGA) microspheres [44] with sustained release of Van up to 6.5, 7 and 28 days, respectively, are reported. In a recent study conducted by Yu et al., Van@PLGA microspheres with pH-responsive characteristics were proposed [45]. It was proven that Van-loaded microspheres were able to inhibit bacteria growth of *S. aureus* even on the 50th day of the drug release.

Here, we report CMCh microgels as a biocompatible drug delivery system designed for carrying active agents such as antibiotics, antifungals, antivirals, antineoplastic agents, and so on. Microgels of CMCh were synthesized in a microemulsion system. Functional and morphological characterization of CMCh microgels was elucidated by FTIR spectroscopy and scanning electron microscope (SEM) images. The hydrolytic degradation of the prepared microgels was determined by pH-dependent weight loss studies. In addition, the biocompatibility of CMCh microgels was assessed on connective tissue cells in vitro. Moreover, Van (Van HCl) as a model drug was loaded into CMCh microgels, and its release characteristics were studied because of the rapid elimination and low bioavailability nature of this antibiotic. The main aim of Van loading into CMCh microgels was to obtain a sustained drug release profile lasting more than 24 h. Furthermore, the antibacterial activity of CMCh microgels and its drug-loaded form were tested against both Gram-negative *Escherichia coli* and Gram-positive *Staphylococcus aureus*. Further, the prolonged drug release was planned to be confirmed by antimicrobial activity studies in vitro.

## 2. Results and Discussion

DVS forms ether bonds by the reaction of the primary hydroxyl groups of glucose units, e.g., in Hyaluronic acid) [46] or by reaction of amine groups, e.g., in the preparation of polyethyleneimine particles [47] under alkaline conditions. The reaction of DVS as a crosslinker in crosslinking is reaction is rapid, resulting in gel formation within minutes [48]. A schematic diagram of the synthesis of CMCh microgels with DVS is illustrated in Figure 1. It is apparent that spherical-shaped CMCh microgels with smooth surface morphology are successfully formed in two different reaction mediums. The SEM images of microgels synthesized in two different microenvironments are given in Figure 1. The dimensions of the microgels were determined using the ImageJ program. The microgels synthesized in 0.2 M AOT/isooctane medium (M1) were $5.8 \pm 4.8$ μm in size, while the microgels synthesized in 0.1 M lecithin/cyclohexane medium (M2) were found to be in $1.2 \pm 0.4$ μm. The yield was less than 25% in the M1 medium, whereas the yield was $89.4 + 3.3\%$ for the microgels prepared in M2. Therefore, the studied for the progress of this research were continued using CMCh particles prepared in M2 medium.

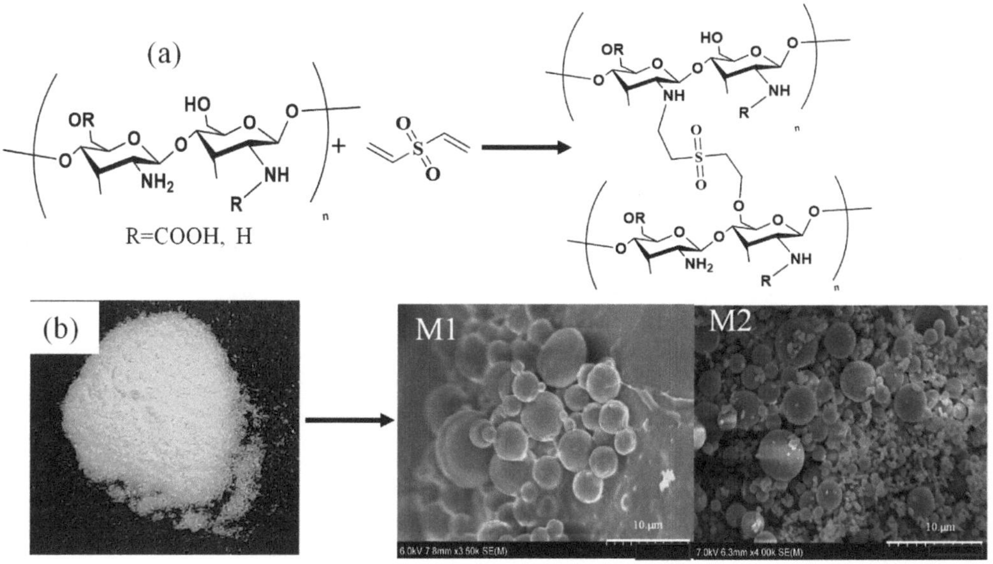

**Figure 1.** (a) Schematic presentation of CMCh microgels preparation and (b) photograph of powder CMCh molecules and the SEM images of CMCh microgels synthesized in M1 (0.2 M AOT/isooctane medium) and M2 (0.1 M lecithin/cyclohexane) medium, respectively.

The SEM images of Van@CMCh microgels are given in Figure S1. Although the CMCh microgels have a smooth spherical shape, the Van@CMCh microgels have reduced roundness and smoothness. The presence of drug within the network of CMCh upon drying can cause this kind of alteration of the smooth surface features.

From the FT-IR spectra of CMCh and CMCh microgels shown in Figure 2a, the wide-stretching peak in the 3200–3550 cm$^{-1}$ band is noticeable for both CMCh, i.e., for CMCh polymer and its microgel. It has been reported that DVS shows an FT-IR peak at 1312 cm$^{-1}$ (S=O asymmetric stretching vibrations) and at 1131 cm$^{-1}$ for S=O symmetric stretching vibrations [47]. Since the S=O (asymmetric stretching vibrations) peak of DVS seen in 1310 overlaps with the O-H peak of CHCh, and the S=O (symmetric stretching vibrations) peak in 1100 coincides with the ether peak, -CH$_2$-O- of CMCh, no new peak is visualized upon microgel formation by DVS crosslinking of CMCh. It is also evident that some of the stretching frequencies of CMCh and crosslinked CMCh microgel peaks overlapped, such as S=O stretching, C-O stretching, and O-H vibrations. On the other hand, the peak intensity belonging to R-O-R upon crosslinking of -OH groups of CHCh molecules with DVS at around 1100 cm$^{-1}$ is increased.

The thermal gravimetric analysis provides a quantitative measurement of weight change as the temperature is increased. Figure 2b shows a thermal gravimetric analysis of CMCh and CMCh microgels. In the thermal degradation graph of CMCh, about 1.2 wt% at 239 °C and about 29 wt% at 291 °C were degraded, and heating up to 600 °C, 49.8 wt% of CMCh remained. The CMCh microgels, on the other hand, retained 73.5 wt% of their initial weight at 249.1 °C. At 303.6 °C, 33.5% of its weight remained, and at about 600 °C, 3.4% of the weight of the CMCh microgels remained.

In a 1 mM KNO$_3$ solution, in varying pH (2.5–11.5) solutions, the zeta potential values of CMCh microgels' were investigated. The pH was found to be 10.22, and the zeta potential was $-16.9 \pm 1.2$ mV for 20 mg/mL microgel in 1 mM KNO$_3$. The zeta potentials against changing pH are given in Figure 3a. Based on these data, the isoelectric point (IEP) was calculated as pH 4.4. When the IEP is greater than 7, the surface is called a basic surface. When the IEP is less than 7, the surface is called an acidic surface. In this study, CMCh microgels IEP is less than 7. This indicates that the surface of the CMCh microgels has an acidic character. From this point of view, it can be deduced that the acidic characters due to -OH and -COOH groups on the surface are more than the basic characters such as -NH$_2$ groups.

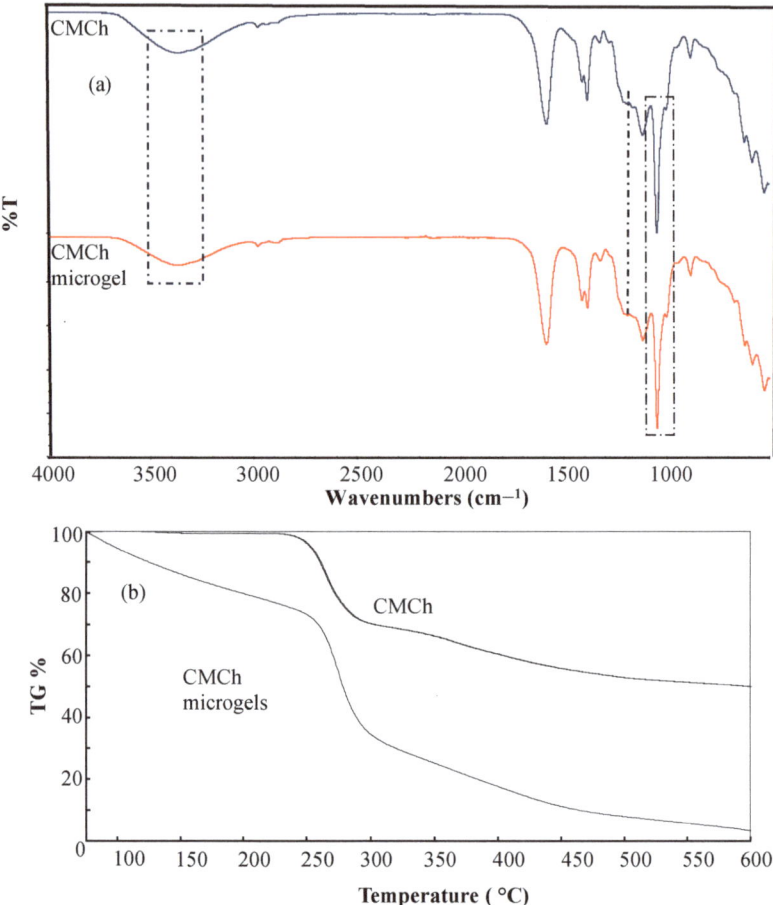

**Figure 2.** (**a**) FT-IR spectrum of CMCh and CMCh microgels and (**b**) and their thermal degradation (TG %) curves.

Fe(II) chelating activity of CMCh and CMCh microgels was investigated in the 125–2000 mg/mL concentration range. As seen in Figure 3b, Fe(II) chelating capability of both CMCh and CMCh microgels increases depending on the concentration of microgels. CMCh and CMCh microgel chelate Fe(II) ions at very high rates. At 250 mg/mL concentration, CMCh chelates 67.3 ± 9.3% of Fe(II) ion, while CMCh microgel chelates only 53.6 ± 10.3% of it. This can be understandable because of the crosslinked structure of CMCh microgels; some of the carboxylic acid amine groups from CMCh chains are not readily available for chelation with Fe(II) due to the microgel network.

The hydrolytic degradation of CMCh microgels was investigated by gravimetric analysis at pH 1, 7.4, and 9 as simulation environments for the stomach [49], physiological conditions [49], and duodenum [50] conditions, respectively. The incubation times of 24 h, 48 h, and 72 h were chosen, and their results are illustrated in Figure 4a as the weight loss (%) of the microgels.

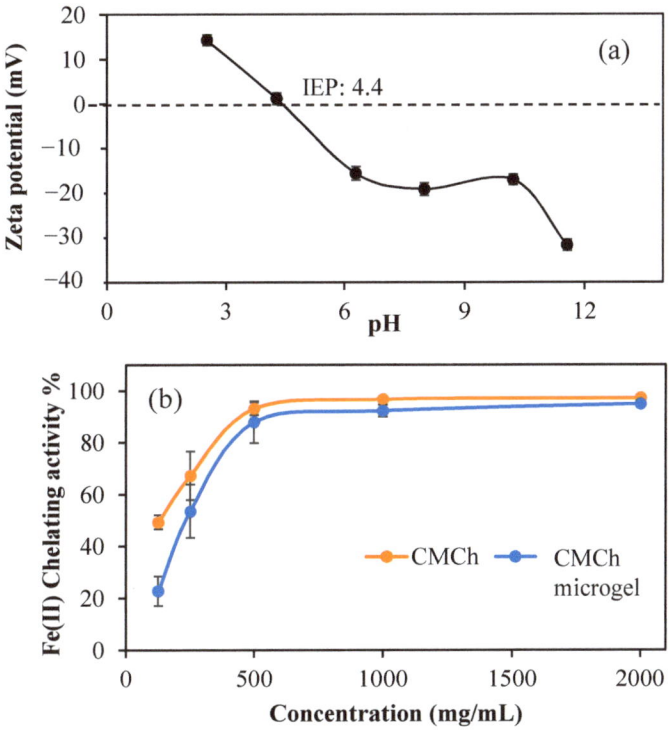

**Figure 3.** (**a**) Zeta potential measurements of CMCh microgels at various pHs in 0.01 M $KNO_3$ solutions and (**b**) Fe(II) chelating activity % of CMCh and CMCh microgels.

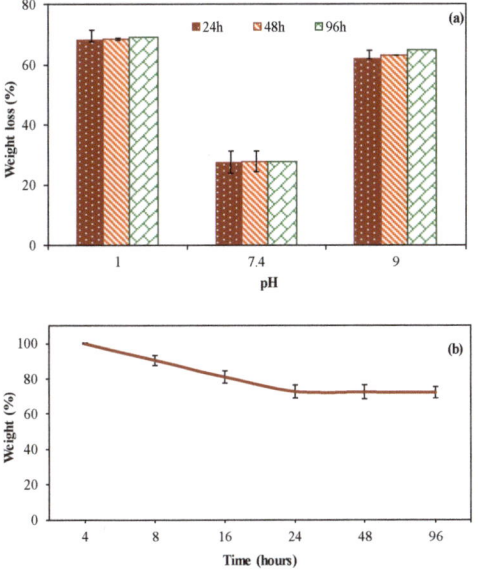

**Figure 4.** (**a**) Weight loss (%) of CMCh microgels at different pH conditions at 24 h, 48 h and 72 h incubation times and (**b**) gravimetric weight loss (%) of CMCh microgels with time at pH 7.4, 37 °C.

As seen in Figure 4a, the weight loss (%) of CMCh microgels was determined as 68.1% ± 3.7%, 27.3% ± 4.0%, 62.1% ± 2.3% at pH 1, 7.4, and 9, respectively, in 24 h. These results indicated that CMCh microgels are highly degradable under acidic (pH 1) and basic (pH 9) conditions. Further, it was found that CMCh microgels are highly stable under physiological conditions, pH 7.4. The hydrolytic degradation of CMCh microgels was also analyzed for longer incubation times, 48 h and 72 h. According to the degradation results, 68.4% ± 2.4%, 27.7% ± 4.1%, 63.2% ± 0.5% weight loss (%) at 48 h incubation, and 69.1% ± 0.5%, 27.8% ± 3.0%, 64.7% ± 2.9% were measured for 96 h incubation times at pH 1, 7.4, and 9, respectively. As seen, more than 50% of the microgels were degraded within the first 24 h at pH 1 and pH 9, indicating that hydrolytic degradation of CMCh microgels was in line with the desired degradation profile for certain applications. Considering the regional pH changes observed in the human body due to reasons such as inadequate blood perfusion, tumor formation, hypoxia, and inflammation [51,52], controllable drug release from CMCh microgels could be obtained by endogenous stimuli. For example, the tumor microenvironment is more acidic (pH 5.6 to 6.8) compared to physiological pH [53]. Therefore, it can be said that pH-dependent drug release for targeting certain body parts could be achieved by drug-loaded CMCh microgels with suitable drug loads. Furthermore, the degradation rate of the microgels did not change significantly between 24 h and 72 h, revealing that CMCh microgels can remain stable for up to 3 days after degradation at a certain rate. In addition, the hydrolytic degradation of CMCh microgels was performed at pH 7.4 and 37 °C at various times, as shown in Figure 4b. As seen at 4 h, no degradation of CMCh microgels was observed. It was observed that, at 8 h, 16 h, 24 h, 48 h, and 96 h incubation times, 9.57%, 19.1%, 27.3%, 27.78%, and 27.81% of the weight loss were observed, respectively, for the prepared CMCh microgels. Furthermore, by reducing the crosslinker ratio, microgels with higher degradability can be obtained [54]. The degradation profiles of 10% and 25% crosslinked CMCh microgels, which can be degraded at a higher rate, will be investigated in our future studies for different applications.

Fibroblasts are differentiated into many cell types, including adipogenic, chondrogenic, and osteogenic cells [55]. L929 fibroblasts are adhesive cells derived from Mouse C3/An connective tissue. L929 fibroblasts, with their reproducible biological responses and susceptibility to toxic effects, are considered correct cells for biocompatibility studies [56]. Therefore, CMCh and CMCh microgel samples up to 1000 µg/mL concentrations were used in biocompatibility studies on L929 fibroblasts at 24 h incubation time.

As illustrated in Figure 5, the cell viability (%) values of the natural CMCh polymer and CMCh microgels were found to be 97.2 ± 1.8% and 96.6 ± 1.7%, respectively, at the highest concentration, 1 mg/mL. According to the biocompatibility test data and the Anova statistical test results, there is no statistical difference between each (50–1000 ug/mL) concentration of CMCh and the control. For CMCh microgels, it was not statistically different from the control at all concentrations. Therefore, prepared microgels, with their excellent cell compatibility, allow them to be designed and safely used for biological, biomedical, and pharmaceutical applications.

Van is loaded into CMCh microgels for a short amount of time, 2 h, by adsorption method. Van loading amount of CMCh microgels was found to be 111.42 ± 7.08 mg/g. As previously stated in the previous sections, Van is a large active pharmaceutical compound antibiotic. Here, a high amount of drug loading for Van was reported. Van-loaded microgels (Van@CMCh) underwent SEM analysis to further examine the microgels in their drug-incorporated forms, and the corresponding images are given in Figure S1. The drug-loading process was carried out in an aqueous solution, and the loaded particles were washed with ethanol solution. In Figure S1, it is seen that the spherical shape of the prepared materials did not change after Van loading. It assumed that Van drug molecules (seen as smaller particles) are homogenously distributed into CMCh microgels. Van release studies from Van@CMCh microgels were performed at pH 7.4 and 37 °C in vitro, and the corresponding graphs are illustrated in Figure 6.

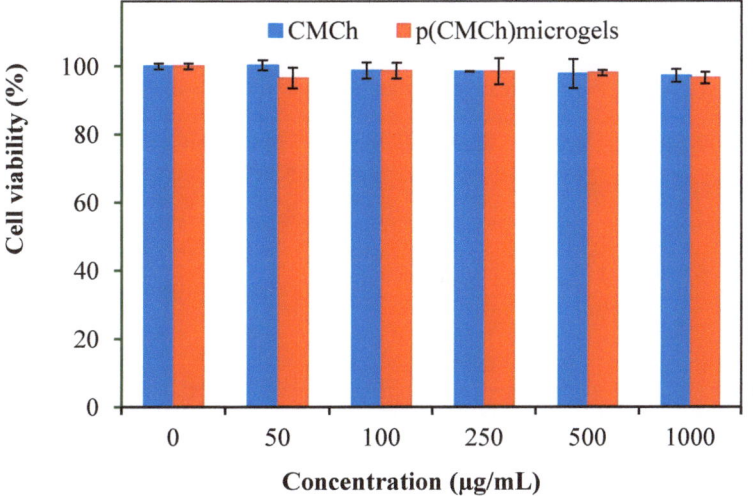

**Figure 5.** In vitro biocompatibility studies of CMCh molecules and microgels at 24 h incubation time.

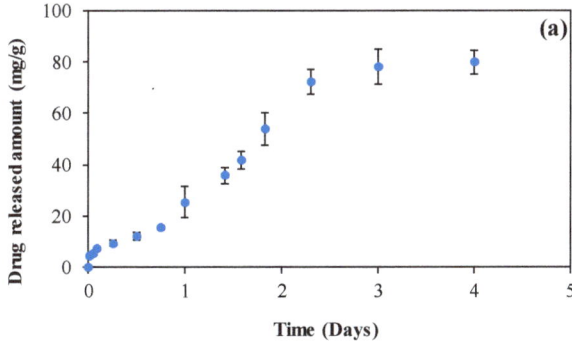

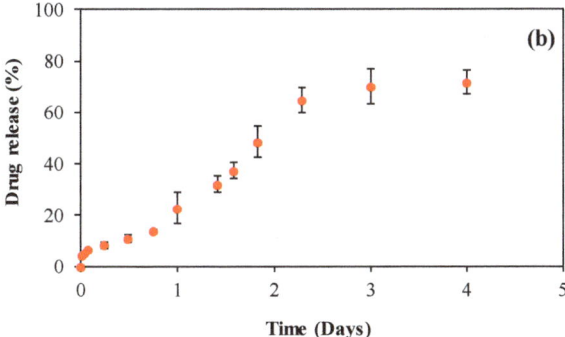

**Figure 6.** (a) Vancomycin (Van) release profile (mg/g) with time and (b) Van release (%) within 4 days.

As seen in Figure 6, within the first hour, the sixth hour, and twelfth hour of the drug release, 5.55 ± 0.1, 9.35 ± 1.1, and 12.14 ± 1.3 mg/g Van were released from Van@CMCh microgels which equal to 4.98%, 8.38% and 10.89% of the loaded Van, respectively. It can be said that even up to 30 h, 4.6 ± 0.1 mg/g (4.13% of the loaded antibiotic) was released, which indicates the rapid release of the antibiotic. Van release from Van@CMCh microgels was linear up to 50 h, whereas from 50 h to 96 h, the drug released was found at a considerably slower rate. Overall, 79.99 mg/g Van was released from Van@CMCh microgels, which shows that 71.8% of the loaded drug was released within 4 days, which also fits the desired amount. The calibration curve of Van in PBS solution at 280 nm is given in Figure S2. Moreover, different kinetic models such as zero order, first order, Higuchi, and Korsmeyer–Peppas models were applied to the drug release graphs to examine the Van release from CMCh microgels. The calculated values for each model were summarized in Table 1, and the corresponding release (%) vs time (day) plots for related models are given in Figure S3.

**Table 1.** The list of kinetic models and parameters calculated for the release of Van from Van@CMCh microgels.

| Kinetic Model | Parameters | Van@CMCh Microgels |
|---|---|---|
| Zero order | $k_0$ | 25.582 |
| | $R^2$ | 0.9652 |
| First order | $k_1$ | 0.327 |
| | $R^2$ | 0.9264 |
| Higuchi | $k_H$ | 30.667 |
| | $R^2$ | 0.8257 |
| Korsmeyer–Peppas | $k_{KP}$ | 22.437 |
| | n | 1.245 |
| | $R^2$ | 0.9763 |

Correlation coefficients ($R^2$) shown in Table 1 indicate that the Korsmeyer–Peppas kinetic model fitted Van release the best with a higher $R^2$ value, 0.9763. As seen, the Higuchi model was the one fitted with the lowest $R^2$ value of 0.8257. The $R^2$ values of the zero-order and first-order models for Van release are calculated as 0.9652 and 0.9264, respectively. The Korsmeyer–Peppas kinetic model is defined as a linear and non-linear regression model in which the n values determine the drug release mechanism [57]. Korsmeyer–Peppas model is reported to be applicable for the release of various drugs from the drug-loaded polymeric structures [58]. The n values for Van release from Van@CMCh microgels were calculated as 1.245, which exceeds the value of 0.89 and reveals that the release model fitted the super case II transport [57,58].

Antibacterial potencies of CMCh, CMCh microgels, Van@CMCh, and Van were tested against two common bacteria strains, *E. coli* and *S. aureus*, and the corresponding results were summarized in Table 2.

**Table 2.** MIC and MBC values of CMCh-based materials.

| Microorganisms | E. coli | | S. aureus | |
|---|---|---|---|---|
| Sample | MIC (µg/mL) | MBC (µg/mL) | MIC (µg/mL) | MBC (µg/mL) |
| CMCh | - | - | - | - |
| CMCh microgels | - | - | - | - |
| Van@CMCh microgels | 64 | - | 8 | 8 |
| Vancomycin * | 15 | 250 | 0.2 | 0.2 |

* Vancomycin aqueous solution was used as control.

As shown in Table 2, CMCh and CMCh microgels did not show any antibacterial effect on the studied microorganisms up to 2.5 mg/mL concentrations. On the other hand, on the well plate, transparent wells (an indicator of no visible growth) containing Van@CMCh microgels were detected, and MIC and MBC values were determined accordingly. MIC values of Van released from Van@CMCh microgels were found to be 64 µg/mL and 8 µg/mL for *E. coli* and *S. aureus*, respectively. Moreover, MBC values of

Van released from Van@CMCh microgels were found to be 8 µg/mL. Micro-titer assay results are also illustrated in Figure S4 as the reduction in the colony-forming units (expressed logarithmically). For *E. coli*, no concentration was detected that kills 99.9% of the microorganism within 24 h. These results are correlated with the literature because Van is mainly effective on Gram-positive bacteria and has mild antibacterial effects on Gram-negative bacteria [34]. Further, bare Van was tested as a control. MIC and MBC values of bare-Van were found to be 15 µg/mL and 250 µg/mL for *E. coli* and 0.2 µg/mL and 0.2 µg/mL for *S. aureus*, respectively.

Van@CMCh microgel suspension at 5 mg/mL concentration was freshly prepared, and 20 µL of the suspension was placed onto disks and incubated for 24 h, 48 h, and 72 h at 35 °C. The inhibition zone diameters and photographs of the disk diffusion assay are given in Table 3 and Figure 7, respectively.

Table 3. Inhibition zone diameters of Van@CMCh microgels (20 µL of 5 mg/mL).

| Inhibition Zone Diameter (mm) | 24 h | 48 h | 96 h |
|---|---|---|---|
| *E. coli* | - | - | 11.5 ± 1 |
| *S. aureus* | 11.6 ± 1.5 | 12.5 ± 1 | 14.5 ± 1.5 |

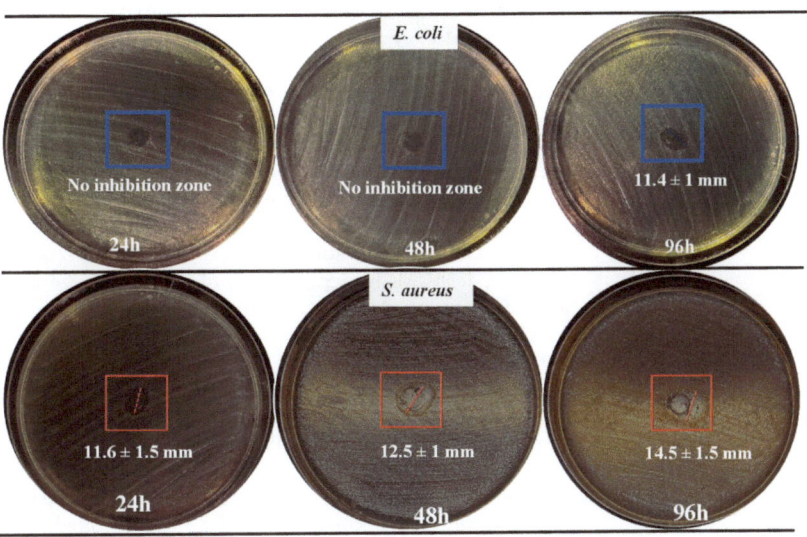

Figure 7. Photographs of inhibition zone diameters of Van@CMCh microgels (20 µL of 5 mg/mL) at 24 h, 48 h, and 96 h incubation times.

The inhibition zone for *E. coli* was detected only at 96 h as 11.4 ± 1 mm. Inhibition zone diameters for *S. aureus* were measured as 11.6 ± 1.5, 12.5 ± 1, and 14.5 ± 1.5 mm at 24 h, 48 h, and 72 h incubation times, respectively. It can be noticed that for *S. aureus*, especially at 48 h and 96 h, a relatively less symmetrical zone was detected due to the retention of the particle suspension on the disc, as can be seen in Figure 7. However, since the endpoint where bacteria could grow was seen, the zones were measured with average values. Disk diffusion test results are strongly correlated with the in vitro drug release studies of Van@CMCh microgels. The inhibition zones for *S. aureus* from 24 h to 72 h expanded significantly, revealing that the drug release continued, and the antibacterial effect lasted at least for 72 h. One of the areas where extended drug release is most needed is the administration of antimicrobial agents. Some of the antibiotics and antifungals currently used have become insufficiently effective due to unnecessary use, inadequate or ill-timed dosing, enzymatic inactivation of drugs, changes in drug targets, and excretion of drugs by active transport proteins in microorganisms [59]. Antimicrobial resistance, especially seen in bacterial species such as *Escherichia coli, Staphylococcus aureus, Klebsiella pneumoniae*, and *Streptococcus pneumonia*, reveals difficult cases to treat, such as nosocomial infections [60]. Under selective antibiotic pressure, drug-susceptible bacteria are destroyed or stop growing, while naturally resistant bacteria can survive [61]. Therefore, it is crucial that therapeutic agents are given at appropriate intervals and exhibit improved absorption and distribution profiles [62]. The extended release of certain antibiotics, such as Van, could improve their in vivo half-life and that bioavailability. In healthy adults, the terminal half-life of Van is reported as 4–6 h [34].

In the current study, Van release from CMCh microgels lasted for a significantly longer time (4 days), revealing that the prepared microgels were found quite successful as antibiotic carriers. Furthermore, wide-spectrum antibiotics, as well as anticancer, antifungal, antiviral, anti-inflammatory, antihistaminic drugs, and so on, can be loaded into CMCh microgels. Considering their degradability profile, controlled drug release for pH or temperature-sensitive drugs can also be readily achievable.

## 3. Conclusions

In this study, the single-step preparation of CMCh microgels via a microemulsion method by crosslinking with DVS was reported. The microsphere formation was confirmed via SEM and FT-IR spectroscopy analyses. The prepared CMCh microgels are spherical and $1.2 \pm 0.4$ μm size range. Despite the ease of their production, the possible residue from the organic solvent and chemicals can be identified as a limitation. Hence, the microemulsion polymerization technique requires a proper microgel/particle washing process to remove the surfactant from the environment, or the surfactant-free synthesis method needs to be considered for in vivo applications. The hydrolytic stability of CMCh microgels was confirmed that the microgel has a pH dependent on degradation profiles and can degrade up to about 70% at pH 1 and 9 while degrading about 30% at pH 7.4 up to 96 h contact times. The cell cytotoxicity results of CMCh microgels performed on L929 fibroblasts indicated that prepared CMCh microgels did not induce any significant toxicity even at 1 mg/mL concentration with cell viability values more than 95%. Moreover, the drug delivery efficiency of CMCh microgels was evaluated using Van, a large antibiotic with a rapid clearance profile and low bioavailability, as a model drug. Van-loaded CMCh microgels showed sustained drug release up to 96 h. Furthermore, the prolonged Van release ability of CMCh microgels was confirmed by antimicrobial activity studies on *E. coli* and *S. aureus* bacteria. The high drug loading capability of CMCh microgels ($111.42 \pm 7.08$ mg/g) suggests that other large drug molecules or drugs with stability and solubility issues could be delivered utilizing CMCh microgels as highly biocompatible drug delivery vehicles. Further, the drug release amount and kinetic can be readily controlled by a suitable amount of crosslinker used during particle preparation. Moreover, the higher surface area of microgels provides many advantages over common bulk hydrogel formulations. Therefore, CMCh microgels, with their adjustable degradability, pore characteristics, and controllable drug loading and release properties, have many advantages for various drugs with limited activities in the treatment of different diseases.

## 4. Materials and Methods

### 4.1. Materials

Carboxymethyl chitosan (Santa Cruz Biotechnology, Fischer Scientific, Deacetylation degree 90%, Hampton, New Hampshire) as a starting material and divinyl sulfone (DVS, >96%, TCI) as a chemical crosslinker was used in CMCh microgel preparation. L-alpha-Lecithin, granular, from soybean oil (Across), sodium bis(2-ethylhexyl) sulfosuccinate (AOT, 96%, Sigma Aldrich, St. Louis, MO, USA) as a surfactant, and Cyclohexane (Certified ACS, Fisher Chemical™, Pittsburgh, PA, USA, 99+%), 2,2,4-trimethylpentane (isooctane, Sigma) as a solvent were used as received in CMCh microgels preparation Vancomycin hydrochloride (Van HCl, Alfa Aesar, Thermo Fisher Scientific, Molecular Biology Grade, Waltham, MA, USA) was purchased and used as a model antibiotic for drug delivery studies. For the iron (II) chelating assay, 3-(2-pyridyl)-5,6-diphenyl-1,2,4-triazine-4′,4″-disulfonic acid sodium salt (ferrozine, $\geq$98%, from Santa Cruz Biotechnology, Dallas, TX, USA) and iron (II) sulfate heptahydrate ($FeSO_4 \cdot 7H_2O$, >99.5%, ACS reagent from Across Organics, Geel, Belgium) were used as received. For the cell viability tests, the L929 fibroblast cell line was obtained from the SAP Institute (Ankara, Turkey). Dulbecco's Modified Eagle's Medium (DMEM/F-12, 1:1) (L-Glutamine, 15 mM HEPES, 1.2 g/L $NaHCO_3$) as the cell culture medium was purchased from Pan Biontech GmbH, Aidenbach, Germany. Fetal bovine serum (FBS), antibiotic solution (penicillin–streptomycin), and trypsin-EDTA (0.25%) were used as received (Pan Biontech GmbH, Aidenbach, Germany). Trypan Blue (0.5% solution) was acquired from Biological Industries, and thiazolyl blue tetrazolium bromide (MTT) was obtained from BioFroxx (Einhausen, Germany). Dimethyl sulfoxide (DMSO, 99.9%, Carlo-Erba, Val-de-Reuil, France) was used as received. Molecular porous membrane tubing was obtained from Spectrum Laboratories (MWCO: 12-14 kD, Fischer Scientific, San Jose, CA, USA). For antibacterial activity tests, Gram-negative bacteria *Escherichia coli* ATCC 8739 and Gram-positive bacteria *Staphylococcus aureus* ATCC 6538 were obtained from KWIK-STIK™ Microbiologics (St. Cloud, MN, USA). Nutrient agar and nutrient broth as growth medium were purchased from BD Difco ™ (Becton, Dickinson and Company, Sparks, MD, USA) and used as received.

*4.2. Synthesis and Characterization of CMCh Microgels*

CMCh microgels were prepared by micro emission method in two different environments. Briefly, 0.05 g CMCh was dissolved in 1.5 mL 0.5 M NaOH solution. A total of 0.5 mL of this solution was placed in 0.1 M lecithin/cyclohexane medium. Then, 50 moles of crosslinking agent, DVS was put into the mixed solution. After an hour, the solution mixture was precipitated at 1000 rpm for 10 min. The CMCh microgels were washed 2 times with cyclohexane, 2 times with ethanol and 2 times with an ethanol: water (1:1) mixture. Finally, it was washed once with acetone.

Similarly, 0.5 mL of the CMCh solution was put into 0.2 M AOT/isooctane medium, and 50% DVS crosslinker was added and allowed to react for 1 h. The precipitation process was achieved by centrifuging twice at 1000 rpm with acetone.

Using the attenuated total reflection (ATR) technique, the spectra of CMCh and CMCh microgels were evaluated using Fourier transform infrared radiation (FT-IR, Nicolet iS10, Thermo, Boston, MA, USA). A thermogravimetric analyzer determined the percentage of CMCh microgels (exstar, SII TG/DTA6300, Seiko Ins. Corp, Tokyo, Japan). About 5 mg of CMCh sample was heated from 100 to 600 °C with a temperature increase of 10 °C/min under the influence of nitrogen gas flow of 200 mL/min for thermogravimetric analysis.

In 40 mL of 10 Mm $KNO_3$ solution, 20 mg of CMCh microgels were suspended. Zeta potential measurements performed in 40 mL of 1 mM $KNO_3$ solution to determine the surface charge of the microgels.

In accordance with the literature, chelating Fe (II) was performed [63]. For this purpose, 96 well plates were filled with 140 µL of CMCh and CMCh microgels. A 20 mL of 1 mM Fe(ll) aqueous solution was then added to the sample and measured at 562 nm with a Thermo Multiscan Go microplate reader. A second reading was performed after adding 40 µL of 2.5 mM ferrozine solution. The following formula was used to calculate Fe (II) chelating activity.

$$\text{Iron (II) chelating activity\%} = \left[1 - \frac{\Delta A_{562}^{sample}}{\Delta A_{562}^{control}}\right] \times 100$$

*4.3. Degradation Profile of CMCh Microgels*

The stability and degradation profiles of CMCh microgels were investigated by hydrolytic degradation studies at 24 h, 48 h, and 72 h incubation times at different solution pH values, pH 1, 7.4, and 9. For this goal, CMCh microgels weighing 20 mg were placed in 20 mL of buffered pH solutions of pH 1, 7.4, and 9, in centrifuge tubes and kept up to three days. At certain times, at 24 h, 48 h, and 72 h incubation times, samples were taken and centrifuged at 10,000 rpm for 15 min to precipitate the non-degraded CMCh microgels. Then, the supernatant was gently decanted, and the precipitated particles were dried in an oven at 50 °C overnight. Weight loss (%) was calculated by the difference between the initial microgel weight and the microgel weights at 24 h, 48 h, and 72 h incubation periods. Moreover, hydrolytic weight loss kinetics of CMCh microgels were studied in detail at pH 7.4 and 37 °C at 4 h, 8 h, 16 h, 24 h, 48 h and 96 h incubation. Hydrolytic degradation studies were performed three times, and the mean values are given with the standard deviations.

*4.4. In Vitro Cell Compatibility Studies of CMCh Microgels*

The cell viability of CMCh microgels was determined on the L929 fibroblast cell line via colorimetric MTT assay following the literature [64]. CMCh stock solution and CMCh microgel suspensions were prepared by weighing 10 mg of each CMCh-based material and suspending them in 10 mL of DMEM solution. Initial concentrations of CMCh-based samples were prepared at 1 mg/mL and diluted in DMEM to prepare different concentrations of samples. The fibroblasts were cultured in a DMEM medium containing 10% FBS and 1% antibiotic for 4 days. For the cytotoxicity analysis, 100 µL of cell suspension containing $1 \times 10^3$ cells/mL were seeded onto a 96-well plate and incubated at 37 °C in a 5% $CO_2$/95% air atmosphere. After 24 h incubation, cells were checked and interacted with 100 µL of CMCh-based samples at 0.05–1 mg/mL concentrations. After 24 h of the incubation period, the old culture media containing samples was discarded, and the cells were washed with phosphate buffer saline (PBS) solution two times. Then, 0.1 mL of fresh prepared MTT solution at 0.5 mg/mL was placed onto each well and incubated at 37 °C for 3 h in a dark condition. After this period, formazan crystals produced by the active mitochondria in viable cells were dissolved using 0.2 mL of DMSO and slowly mixed. After 20 min, the absorbance values of the wells were measured at 570 nm by using a plate reader (Thermo Scientific, Multiskan Sky, Waltham, MA, USA). GraphPad Prism 9 software was used for the statistical analysis of cytotoxicity analysis. Statistical differences between the control groups and samples were measured using Dunnett's multiple-comparison test and one-way ANOVA. A $p$-value $< 0.05$ was considered statistically significant.

*4.5. Drug Delivery Abilities of CMCh Microgels*

The drug-loading process of Vancomycin (Van) into CMCh microgels was completed by the soaking method (adsorption technique), as described in the literature [65]. Briefly, 0.1 g of CMCh microgels were immersed in 30 mL of 1 mg/mL Van aqueous solution and stirred at 5000 rpm for 2 h. The drugs can be loaded into polymeric particles for long periods, i.e., 12 h and 24 h [66], but the Van loading process was performed for a shorter amount of time (2 h) in order not to degrade 50% crosslinked microgels during the drug loading time. After Van loading, antibiotic-loaded microgels as Van@CMCh were precipitated in the same medium at 10,000 rpm for 5 min, then washed once with an ethanol solution to eliminate the drug molecules that adhered to the outer surface but did not penetrate the microgel structure. Finally, Van@CMCh microgels were dried at 50 °C oven overnight.

In vitro drug release studies of Van@CMCh, microgels proceeded at 37 °C and physiological pH condition (pH 7.4) to mimic the normal body temperature. First, Van@CMCh microgels of 50 mg were weighed and suspended in 1 mL of phosphate-buffered saline solution (PBS, sterilized) in dialysis tubing. Then, Van@CMCh containing dialysis membrane was placed in 20 mL of PBS solution in falcon tubes and kept in a shaking bath. Uv-Vis spectra were recorded at various times, i.e., each measurement was performed three times, and the results are given as the average values. The loaded and released amounts of antibiotic drug were calculated by using the calibration curve of Van at 280 nm in DI water and PBS solutions, respectively, via UV-Vis spectroscopy.

### 4.6. Antibacterial Activities of Van@CMCh Microgels

Bacteria growth inhibition and bactericidal effects of Van@CMCh were investigated against Gram-negative bacteria *E. coli* ATCC 8739 and gram-positive bacteria *S. aureus* ATCC 6538 by micro-titer dilution and inhibition zone assays as described in the literature [67,68]. CMCh, CMCh microgels, and Van@CMCh microgels weighing 50 mg were sterilized under UV irradiation at 355 nm and then suspended in 10 mL of PBS solution. Microgel-containing samples were sonicated for 30 s for homogenization and then immediately used.

#### 4.6.1. Broth Micro-Titer Dilution Assay

The bacterial suspensions were adjusted in nutrient broth (NB) to a McFarland standard of 0.5, which corresponds to $1.5 \times 10^8$ CFU/mL [69]. Then, 0.1 mL of NB was added to each well of the 96-well plate. Then, 0.1 mL of CMCh-based samples were placed in the first well of each column on the plate and diluted with the existing media. Lastly, 10 μL of bacteria inoculum at 0.5 McFarland was added to each well and gently mixed. A vancomycin aqueous solution of 1 mg/mL was used as a control. Bacteria containing well-plates were incubated at 35 °C for 24 h. After this time, MIC and MBC values of CMCh-based materials were determined as the concentration that showed no visible growth in the wells and killed 99.9% of the microorganisms, respectively.

#### 4.6.2. Zone of Inhibition Method

Following the micro-titer assay, zone of inhibition experiments were performed for Van@CMCh microgels against both bacteria strains at 24 h, 48 h, and 72 h incubation periods. For this, 100 μL of *E. coli* and *S. aureus* inoculums at 0.5 McFarland were poured onto nutrient agar solid growth medium on petri dishes. Then, sterilized 10 mm × 10 mm spherical-shaped filter papers were placed on the petri dishes. Immediately, 20 μL of Van@CMCh microgel suspensions at 5 mg/mL were gently placed onto filter papers and incubated at 35 °C for three days. After the incubation, disks were taken, and the inhibition zones observed around the disks were measured.

**Supplementary Materials:** The following supporting information can be downloaded at: https://www.mdpi.com/article/10.3390/gels9090708/s1, Figure S1: SEM images of Van@p(CMCh) microgels at different magnifications; Figure S2: The Calibration curve of Vancomycin (Van) in phosphate-buffered saline solution at 280 nm; Figure S3: The Van release% from Van@CMCh microgels vs. time plots for related kinetic models; Figure S4: Colony forming units (CFU/mL) of (a) *E. coli* and (b) *S. aureus* strains versus the concentration of the released amount of Van from Van@CMCh microgels at 24 h incubation time.

**Author Contributions:** Conceptualization, N.S.; methodology, M.S., A.S.Y. and N.S.; software, A.S.Y. and M.S.; validation, M.S.; formal analysis, A.S.Y. and M.S.; investigation, A.S.Y., M.S. and N.S.; resources, N.S. and R.S.A.; data curation, M.S. and A.S.Y.; writing—original draft preparation, M.S. and A.S.Y.; writing—review and editing, N.S. and R.S.A.; visualization, N.S. and R.S.A.; supervision, N.S.; project administration, N.S.; funding acquisition, N.S. and R.S.A. All authors have read and agreed to the published version of the manuscript.

**Funding:** Some parts of this work were supported by the startup funds (N. Sahiner) from the Ophthalmology Department at USF.

**Institutional Review Board Statement:** Not applicable.

**Informed Consent Statement:** Not applicable.

**Data Availability Statement:** The data presented in this study are available on request from the corresponding author.

**Conflicts of Interest:** The authors declare no conflict of interest.

## References

1. Aranaz, I.; Alcántara, A.R.; Civera, M.C.; Arias, C.; Elorza, B.; Caballero, A.H.; Acosta, N. Chitosan: An Overview of Its Properties and Applications. *Polymers* **2021**, *13*, 3256. [CrossRef] [PubMed]
2. Hayes, J.D.; Dinkova-Kostova, A.T.; Tew, K.D. Oxidative Stress in Cancer. *Cancer Cell* **2020**, *38*, 167–197. [CrossRef] [PubMed]
3. Aranaz, I.; Harris, R.; Heras, A. Chitosan Amphiphilic Derivatives. Chemistry and Applications. *Curr. Org. Chem.* **2010**, *14*, 308–330. [CrossRef]
4. Chaiwarit, T.; Sommano, S.R.; Rachtanapun, P.; Kantrong, N.; Ruksiriwanich, W.; Kumpugdee-Vollrath, M.; Jantrawut, P. Development of Carboxymethyl Chitosan Nanoparticles Prepared by Ultrasound-Assisted Technique for a Clindamycin HCl Carrier. *Polymers* **2022**, *14*, 1736. [CrossRef]
5. Wu, F.; Meng, G.; He, J.; Wu, Y.; Gu, Z. Antibiotic-Loaded Chitosan Hydrogel with Superior Dual Functions: Antibacterial Efficacy and Osteoblastic Cell Responses. *ACS Appl. Mater. Interfaces* **2014**, *6*, 10005–10013. [CrossRef]
6. Jaikumar, D.; Sajesh, K.; Soumya, S.; Nimal, T.; Chennazhi, K.; Nair, S.V.; Jayakumar, R. Injectable alginate-O-carboxymethyl chitosan/nano fibrin composite hydrogels for adipose tissue engineering. *Int. J. Biol. Macromol.* **2015**, *74*, 318–326. [CrossRef] [PubMed]
7. Kzk, A.-A.; Mhs, M. Hematology, bacteriology and antibiotic resistance in milk of water buffalo with subclinical mastitis. *OJVRTM Online J. Vet. Res.* **2019**, *23*, 1–8.
8. Tang, Y.; Sun, J.; Fan, H.; Zhang, X. An improved complex gel of modified gellan gum and carboxymethyl chitosan for chondrocytes encapsulation. *Carbohydr. Polym.* **2012**, *88*, 46–53. [CrossRef]
9. Zhao, X.; Li, P.; Guo, B.; Ma, P.X. Antibacterial and conductive injectable hydrogels based on quaternized chitosan-graft-polyaniline/oxidized dextran for tissue engineering. *Acta Biomater.* **2015**, *26*, 236–248. [CrossRef]
10. Zhou, Y.; Xu, L.; Zhang, X.; Zhao, Y.; Wei, S.; Zhai, M. Radiation synthesis of gelatin/CM-chitosan/β-tricalcium phosphate composite scaffold for bone tissue engineering. *Mater. Sci. Eng. C* **2012**, *32*, 994–1000. [CrossRef]
11. Phan, M.T.T.; Pham, L.N.; Nguyen, L.H.; To, L.P. Investigation on Synthesis of Hydrogel Starting from Vietnamese Pineapple Leaf Waste-Derived Carboxymethylcellulose. *J. Anal. Methods Chem.* **2021**, *2021*, 6639964. [CrossRef]
12. Liu, Y.; Chen, Y.; Zhao, Y.; Tong, Z.; Chen, S. Superabsorbent Sponge and Membrane Prepared by Polyelectrolyte Complexation of Carboxymethyl Cellulose/Hydroxyethyl Cellulose-$Al^{3+}$. *Bioresources* **2015**, *10*, 6479–6495. [CrossRef]
13. Das, S.S.; Kar, S.; Singh, S.K.; Hussain, A.; Verma, P.; Beg, S. Carboxymethyl chitosan in advanced drug-delivery applications. In *Chitosan in Drug Delivery*; Elsevier: Amsterdam, The Netherlands, 2022; pp. 323–360. [CrossRef]
14. Pakzad, Y.; Fathi, M.; Omidi, Y.; Mozafari, M.; Zamanian, A. Synthesis and characterization of timolol maleate-loaded quaternized chitosan-based thermosensitive hydrogel: A transparent topical ocular delivery system for the treatment of glaucoma. *Int. J. Biol. Macromol.* **2020**, *159*, 117–128. [CrossRef]
15. Zhao, D.; Song, H.; Zhou, X.; Chen, Y.; Liu, Q.; Gao, X.; Zhu, X.; Chen, D. Novel facile thermosensitive hydrogel as sustained and controllable gene release vehicle for breast cancer treatment. *Eur. J. Pharm. Sci.* **2019**, *134*, 145–152. [CrossRef] [PubMed]
16. Permana, A.D.; Utami, R.N.; Layadi, P.; Himawan, A.; Juniarti, N.; Anjani, Q.K.; Utomo, E.; Mardikasari, S.A.; Arjuna, A.; Donnelly, R.F. Thermosensitive and mucoadhesive in situ ocular gel for effective local delivery and antifungal activity of itraconazole nanocrystal in the treatment of fungal keratitis. *Int. J. Pharm.* **2021**, *602*, 120623. [CrossRef]
17. Iriventi, P.; Gupta, N.V.; Osmani, R.A.M.; Balamuralidhara, V. Design & development of nanosponge loaded topical gel of curcumin and caffeine mixture for augmented treatment of psoriasis. *DARU J. Pharm. Sci.* **2020**, *28*, 489–506. [CrossRef]
18. Gonçalves, R.C.; Signini, R.; Rosa, L.M.; Dias, Y.S.P.; Vinaud, M.C.; Junior, R.D.S.L. Carboxymethyl chitosan hydrogel formulations enhance the healing process in experimental partial-thickness (second-degree) burn wound healing. *Acta Cir. Bras.* **2021**, *36*, e360303. [CrossRef] [PubMed]
19. Shariatinia, Z. Carboxymethyl chitosan: Properties and biomedical applications. *Int. J. Biol. Macromol.* **2018**, *120*, 1406–1419. [CrossRef]
20. Rao, K.M.; Sudhakar, K.; Suneetha, M.; Won, S.Y.; Han, S.S. Fungal-derived carboxymethyl chitosan blended with polyvinyl alcohol as membranes for wound dressings. *Int. J. Biol. Macromol.* **2021**, *190*, 792–800. [CrossRef]
21. Pandit, A.H.; Nisar, S.; Imtiyaz, K.; Nadeem, M.; Mazumdar, N.; Alam Rizvi, M.M.; Ahmad, S. Injectable, Self-Healing, and Biocompatible N,O-Carboxymethyl Chitosan/Multialdehyde Guar Gum Hydrogels for Sustained Anticancer Drug Delivery. *Biomacromolecules* **2021**, *22*, 3731–3745. [CrossRef]
22. Zhao, L.; Zhu, B.; Jia, Y.; Hou, W.; Su, C. Preparation of Biocompatible Carboxymethyl Chitosan Nanoparticles for Delivery of Antibiotic Drug. *BioMed Res. Int.* **2013**, *2013*, 236469. [CrossRef]
23. Ullah, K.; Sohail, M.; Murtaza, G.; Khan, S.A. Natural and synthetic materials based CMCh/PVA hydrogels for oxaliplatin delivery: Fabrication, characterization, In-Vitro and In-Vivo safety profiling. *Int. J. Biol. Macromol.* **2018**, *122*, 538–548. [CrossRef]
24. Lyu, Y.; Azevedo, H.S. Supramolecular Hydrogels for Protein Delivery in Tissue Engineering. *Molecules* **2021**, *26*, 873. [CrossRef]
25. Lo, W.-H.; Deng, F.-S.; Chang, C.-J.; Lin, C.-H. Synergistic Antifungal Activity of Chitosan with Fluconazole against *Candida albicans*, *Candida tropicalis*, and Fluconazole-Resistant Strains. *Molecules* **2020**, *25*, 5114. [CrossRef] [PubMed]
26. Pardeshi, S.R.; More, M.P.; Patil, P.B.; Mujumdar, A.; Naik, J.B. Statistical optimization of voriconazole nanoparticles loaded carboxymethyl chitosan-poloxamer based in situ gel for ocular delivery: In vitro, ex vivo, and toxicity assesment. *Drug Deliv. Transl. Res.* **2022**, *12*, 3063–3082. [CrossRef] [PubMed]

27. Giri, T.K.; Thakur, A.; Alexander, A.; Ajazuddin; Badwaik, H.; Tripathi, D.K. Modified chitosan hydrogels as drug delivery and tissue engineering systems: Present status and applications. *Acta Pharm. Sin. B* **2012**, *2*, 439–449. [CrossRef]
28. Buranachai, T.; Praphairaksit, N.; Muangsin, N. Chitosan/Polyethylene Glycol Beads Crosslinked with Tripolyphosphate and Glutaraldehyde for Gastrointestinal Drug Delivery. *AAPS PharmSciTech* **2010**, *11*, 1128–1137. [CrossRef] [PubMed]
29. Tang, G.; Tan, Z.; Zeng, W.; Wang, X.; Shi, C.; Liu, Y.; He, H.; Chen, R.; Ye, X. Recent Advances of Chitosan-Based Injectable Hydrogels for Bone and Dental Tissue Regeneration. *Front. Bioeng. Biotechnol.* **2020**, *8*, 3063–3082. [CrossRef]
30. Liu, Q.; Zuo, Q.; Guo, R.; Hong, A.; Li, C.; Zhang, Y.; He, L.; Xue, W. Fabrication and characterization of carboxymethyl chitosan/poly(vinyl alcohol) hydrogels containing alginate microspheres for protein delivery. *J. Bioact. Compat. Polym.* **2015**, *30*, 397–411. [CrossRef]
31. Mathew, S.A.; Arumainathan, S. Crosslinked Chitosan–Gelatin Biocompatible Nanocomposite as a Neuro Drug Carrier. *ACS Omega* **2022**, *7*, 18732–18744. [CrossRef] [PubMed]
32. Quadrado, R.F.N.; Fajardo, A.R. Microparticles based on carboxymethyl starch/chitosan polyelectrolyte complex as vehicles for drug delivery systems. *Arab. J. Chem.* **2020**, *13*, 2183–2194. [CrossRef]
33. Zeng, D.; Debabov, D.; Hartsell, T.L.; Cano, R.J.; Adams, S.; Schuyler, J.A.; McMillan, R.; Pace, J.L. Approved Glycopeptide Antibacterial Drugs: Mechanism of Action and Resistance. *Cold Spring Harb. Perspect. Med.* **2016**, *6*, a026989. [CrossRef] [PubMed]
34. Patel, S.; Preuss, C.V.; Bernice, F. *Vancomycin*; StatPearls Publishing: Treasure Island, FL, USA, 2023.
35. Gomceli, U.; Vangala, S.; Zeana, C.; Kelly, P.J.; Singh, M. An Unusual Case of Ototoxicity with Use of Oral Vancomycin. *Case Rep. Infect. Dis.* **2018**, *2018*, 2980913. [CrossRef] [PubMed]
36. Beumier, M.; Roberts, J.A.; Kabtouri, H.; Hites, M.; Cotton, F.; Wolff, F.; Lipman, J.; Jacobs, F.; Vincent, J.-L.; Taccone, F.S. A new regimen for continuous infusion of vancomycin during continuous renal replacement therapy. *J. Antimicrob. Chemother.* **2013**, *68*, 2859–2865. [CrossRef]
37. Peng, Y.M.; Li, C.-Y.M.; Yang, Z.-L.M.; Shi, W.M. Adverse reactions of vancomycin in humans. *Medicine* **2020**, *99*, e22376. [CrossRef]
38. Ndayishimiye, J.; Cao, Y.; Kumeria, T.; Blaskovich, M.A.T.; Falconer, J.R.; Popat, A. Engineering mesoporous silica nanoparticles towards oral delivery of vancomycin. *J. Mater. Chem. B* **2021**, *9*, 7145–7166. [CrossRef]
39. Li, D.; Tang, G.; Yao, H.; Zhu, Y.; Shi, C.; Fu, Q.; Yang, F.; Wang, X. Formulation of pH-responsive PEGylated nanoparticles with high drug loading capacity and programmable drug release for enhanced antibacterial activity. *Bioact. Mater.* **2022**, *16*, 47–56. [CrossRef]
40. Rybak, M.J. The Pharmacokinetic and Pharmacodynamic Properties of Vancomycin. *Clin. Infect. Dis.* **2006**, *42*, S35–S39. [CrossRef]
41. Cauda, V.; Onida, B.; Platschek, B.; Mühlstein, L.; Bein, T. Large antibiotic molecule diffusion in confined mesoporous silica with controlled morphology. *J. Mater. Chem.* **2008**, *18*, 5888–5899. [CrossRef]
42. Mahmoudian, M.; Ganji, F. Vancomycin-loaded HPMC microparticles embedded within injectable thermosensitive chitosan hydrogels. *Prog. Biomater.* **2017**, *6*, 49–56. [CrossRef]
43. Le Ray, A.-M.; Chiffoleau, S.; Iooss, P.; Grimandi, G.; Gouyette, A.; Daculsi, G.; Merle, C. Vancomycin encapsulation in biodegradable poly(ε-caprolactone) microparticles for bone implantation. Influence of the formulation process on size, drug loading, in vitro release and cytocompatibility. *Biomaterials* **2003**, *24*, 443–449. [CrossRef]
44. Li, S.; Shi, X.; Xu, B.; Wang, J.; Li, P.; Wang, X.; Lou, J.; Li, Z.; Yang, C.; Li, S.; et al. In vitro drug release and antibacterial activity evaluation of silk fibroin coated vancomycin hydrochloride loaded poly (lactic-co-glycolic acid) (PLGA) sustained release microspheres. *J. Biomater. Appl.* **2022**, *36*, 1676–1688. [CrossRef] [PubMed]
45. Yu, X.; Pan, Q.; Zheng, Z.; Chen, Y.; Chen, Y.; Weng, S.; Huang, L. pH-responsive and porous vancomycin-loaded PLGA microspheres: Evidence of controlled and sustained release for localized inflammation inhibition in vitro. *RSC Adv.* **2018**, *8*, 37424–37432. [CrossRef]
46. Lai, J.-Y. Relationship between structure and cytocompatibility of divinyl sulfone cross-linked hyaluronic acid. *Carbohydr. Polym.* **2014**, *101*, 203–212. [CrossRef]
47. Sahiner, N.; Demirci, S.; Sahiner, M.; Al-Lohedan, H. The synthesis of desired functional groups on PEI microgel particles for biomedical and environmental applications. *Appl. Surf. Sci.* **2015**, *354*, 380–387. [CrossRef]
48. Collins, M.N.; Birkinshaw, C. Investigation of the swelling behavior of crosslinked hyaluronic acid films and hydrogels produced using homogeneous reactions. *J. Appl. Polym. Sci.* **2008**, *109*, 923–931. [CrossRef]
49. Patricia, J.J.; Dhamoon, A.S. *Physiology, Digestion*; StatPearls Publishing: Treasure Island, FL, USA, 2023.
50. Akiba, Y.; Mizumori, M.; Guth, P.H.; Engel, E.; Kaunitz, J.D. Duodenal brush border intestinal alkaline phosphatase activity affects bicarbonate secretion in rats. *Am. J. Physiol. Liver Physiol.* **2007**, *293*, G1223–G1233. [CrossRef]
51. Lin, B.; Chen, H.; Liang, D.; Lin, W.; Qi, X.; Liu, H.; Deng, X. Acidic pH and High-$H_2O_2$ Dual Tumor Microenvironment-Responsive Nanocatalytic Graphene Oxide for Cancer Selective Therapy and Recognition. *ACS Appl. Mater. Interfaces* **2019**, *11*, 11157–11166. [CrossRef]
52. Pugliese, S.C.; Poth, J.M.; Fini, M.A.; Olschewski, A.; El Kasmi, K.C.; Stenmark, K.R. The role of inflammation in hypoxic pulmonary hypertension: From cellular mechanisms to clinical phenotypes. *Am. J. Physiol. Cell. Mol. Physiol.* **2015**, *308*, L229–L252. [CrossRef]
53. Xiao, K.; Lin, T.-Y.; Lam, K.S.; Li, Y. A facile strategy for fine-tuning the stability and drug release of stimuli-responsive cross-linked micellar nanoparticles towards precision drug delivery. *Nanoscale* **2017**, *9*, 7765–7770. [CrossRef]

54. Foster, G.A.; Headen, D.M.; González-García, C.; Salmerón-Sánchez, M.; Shirwan, H.; García, A.J. Protease-degradable microgels for protein delivery for vascularization. *Biomaterials* **2017**, *113*, 170–175. [CrossRef] [PubMed]
55. Sudo, K.; Kanno, M.; Miharada, K.; Ogawa, S.; Hiroyama, T.; Saijo, K.; Nakamura, Y. Mesenchymal Progenitors Able to Differentiate into Osteogenic, Chondrogenic, and/or Adipogenic Cells In Vitro Are Present in Most Primary Fibroblast-Like Cell Populations. *Stem Cells* **2007**, *25*, 1610–1617. [CrossRef] [PubMed]
56. Wadajkar, A.S.; Ahn, C.; Nguyen, K.T.; Zhu, Q.; Komabayashi, T. In Vitro Cytotoxicity Evaluation of Four Vital Pulp Therapy Materials on L929 Fibroblasts. *ISRN Dent.* **2014**, *2014*, 191068. [CrossRef] [PubMed]
57. Wu, I.Y.; Bala, S.; Škalko-Basnet, N.; di Cagno, M.P. Interpreting non-linear drug diffusion data: Utilizing Korsmeyer-Peppas model to study drug release from liposomes. *Eur. J. Pharm. Sci.* **2019**, *138*, 105026. [CrossRef]
58. Unagolla, J.M.; Jayasuriya, A.C. Drug transport mechanisms and in vitro release kinetics of vancomycin encapsulated chitosan-alginate polyelectrolyte microparticles as a controlled drug delivery system. *Eur. J. Pharm. Sci.* **2018**, *114*, 199–209. [CrossRef]
59. Prestinaci, F.; Pezzotti, P.; Pantosti, A. Antimicrobial resistance: A global multifaceted phenomenon. *Pathog. Glob. Health* **2015**, *109*, 309–318. [CrossRef]
60. Mancuso, G.; Midiri, A.; Gerace, E.; Biondo, C. Bacterial Antibiotic Resistance: The Most Critical Pathogens. *Pathogens* **2021**, *10*, 1310. [CrossRef]
61. Aslam, B.; Wang, W.; Arshad, M.I.; Khurshid, M.; Muzammil, S.; Nisar, M.A.; Alvi, R.F.; Aslam, M.A.; Qamar, M.U.; Salamat, M.K.F.; et al. Antibiotic resistance: A rundown of a global crisis. *Infect. Drug Resist.* **2018**, *11*, 1645–1658. [CrossRef]
62. Ayukekbong, J.A.; Ntemgwa, M.; Atabe, A.N. The threat of antimicrobial resistance in developing countries: Causes and control strategies. *Antimicrob. Resist. Infect. Control.* **2017**, *6*, 47. [CrossRef]
63. Sahiner, M.; Demirci, S.; Sahiner, N. Enhanced Bioactive Properties of Halloysite Nanotubes via Polydopamine Coating. *Polymers* **2022**, *14*, 4346. [CrossRef]
64. Ghasemi, M.; Turnbull, T.; Sebastian, S.; Kempson, I. The MTT Assay: Utility, Limitations, Pitfalls, and Interpretation in Bulk and Single-Cell Analysis. *Int. J. Mol. Sci.* **2021**, *22*, 12827. [CrossRef] [PubMed]
65. ElShaer, A.; Ghatora, B.; Mustafa, S.; Alany, R.G. Contact lenses as drug reservoirs & delivery systems: The successes & challenges. *Ther. Deliv.* **2014**, *5*, 1085–1100. [CrossRef] [PubMed]
66. Bărăian, A.-I.; Iacob, B.-C.; Sorițău, O.; Tomuță, I.; Tefas, L.R.; Barbu-Tudoran, L.; Șușman, S.; Bodoki, E. Ruxolitinib-Loaded Imprinted Polymeric Drug Reservoir for the Local Management of Post-Surgical Residual Glioblastoma Cells. *Polymers* **2023**, *15*, 965. [CrossRef] [PubMed]
67. Sahiner, M.; Yilmaz, A.S.; Demirci, S.; Sahiner, N. Physically and Chemically Crosslinked, Tannic Acid Embedded Linear PEI-Based Hydrogels and Cryogels with Natural Antibacterial and Antioxidant Properties. *Biomedicines* **2023**, *11*, 706. [CrossRef] [PubMed]
68. Kowalska-Krochmal, B.; Dudek-Wicher, R. The Minimum Inhibitory Concentration of Antibiotics: Methods, Interpretation, Clinical Relevance. *Pathogens* **2021**, *10*, 165. [CrossRef]
69. Kralik, P.; Beran, V.; Pavlik, I. Enumeration of Mycobacterium avium subsp. paratuberculosis by quantitative real-time PCR, culture on solid media and optical densitometry. *BMC Res. Notes* **2012**, *5*, 114. [CrossRef]

**Disclaimer/Publisher's Note:** The statements, opinions and data contained in all publications are solely those of the individual author(s) and contributor(s) and not of MDPI and/or the editor(s). MDPI and/or the editor(s) disclaim responsibility for any injury to people or property resulting from any ideas, methods, instructions or products referred to in the content.

## Article

# Acceleration of Wound Healing in Rats by Modified Lignocellulose Based Sponge Containing Pentoxifylline Loaded Lecithin/Chitosan Nanoparticles

Pouya Dehghani [1], Aliakbar Akbari [1], Milad Saadatkish [1], Jaleh Varshosaz [2,\*], Monireh Kouhi [3] and Mahdi Bodaghi [4,\*]

1. Pharmacy Student's Research Committee, School of Pharmacy, Isfahan University of Medical Sciences, Isfahan 81746-73461, Iran
2. Novel Drug Delivery Systems Research Center, Department of Pharmaceutics, Faculty of Pharmacy, Isfahan University of Medical Sciences, Isfahan 81746-73461, Iran
3. Dental Materials Research Center, Dental Research Institute, School of Dentistry, Isfahan University of Medical Sciences, Isfahan 81746-73461, Iran
4. Department of Engineering, School of Science and Technology, Nottingham Trent University, Nottingham NG11 8NS, UK
* Correspondence: varshosaz@pharm.mui.ac.ir (J.V.); mahdi.bodaghi@ntu.ac.uk (M.B.)

**Abstract:** Dressing wounds accelerates the re-epithelialization process and changes the inflammatory environment towards healing. In the current study, a lignocellulose sponge containing pentoxifylline (PTX)-loaded lecithin/chitosan nanoparticles (LCNs) was developed to enhance the wound healing rate. Lecithin/chitosan nanoparticles were obtained by the solvent-injection method and characterized in terms of morphology, particle size distribution, and zeta potential. The lignocellulose hydrogels were functionalized through oxidation/amination and freeze-dried to obtain sponges. The prepared sponge was then loaded with LCNs/PTX to control drug release. The nanoparticle containing sponges were characterized using FTIR and SEM analysis. The drug release study from both nanoparticles and sponges was performed in PBS at 37 °C at different time points. The results demonstrated that PTX has sustained release from lignocellulose hydrogels. The wound healing was examined using a standard rat model. The results exhibited that PTX loaded hydrogels could achieve significantly accelerated and enhanced healing compared to the drug free hydrogels and the normal saline treatment. Histological examination of the healed skin confirmed the visual observations. Overall speaking, the in vivo assessment of the developed sponge asserts its suitability as wound dressing for treatment of chronic skin wounds.

**Keywords:** modified lignocellulose; pentoxifylline; hydrogels; nanoparticles; rat model; wound dressing

## 1. Introduction

Design and production of an effective wound dressing capable of improving the wound healing process is a major biomedical challenge [1–3]. An ideal wound dressing is required to maintain the wound moisture, protect the wound area from infection and injury, absorb exudate, and reduce wound pain. Moreover, it should carry wound healing factors such as growth factors or nitric oxide to enhance the healing process [1,4]. Conventional wound dressings such as cotton wool, bandage, lint, or gauze do not offer these properties sufficiently. Recently, modern wound dressings such as hydrogels, cellulose sponges, and nanofibers have been investigated to improve the wound healing process [5,6]. The selection of suitable materials is crucial in designing wound dressing since functions and properties of dressings are mainly determined by the dressing materials. Lignocellulose (LC), as the naturally most abundant and renewable resource with unique properties such as excellent mechanical performance, biocompatibility, biodegradability, and multiple functional group, is a proper material for the development of dressings [7,8]. However, due to its relatively poor water solubility, its application in the biomedical field is limited. The physical and chemical modification of LC is an effective way to improve its physicochemical and biological properties [9,10]. It was reported that surface modification of cellulose materials through oxidation and amination can affect the cell adhesion and cellular uptake due to the favorable electrostatic interaction [11,12]. Moreover, it can produce a hydrogel with ionic strength-responsive swelling properties capable of controlled drug release after changing the conditions [13]. Pentoxifylline (PTX), an anti-inflammatory drug, is a synthetic methyl xantine [14]. It was reported that PTX can increase the wound healing rate by modulating gene expression of MMP-1, MMP-3, and TIMP-1 in normoglycemic rats [15]. It can be applied orally or intravenously; however, there are some disadvantages, such as several side effects related to the gastrointestinal tract and in the central nervous system. The topical administration of PTX could be an alternative to heal skin disorders; however, the predominantly hydrophilic nature of the drug makes it difficult for it to penetrate the skin [16]. Thus, encapsulation of PTX into a colloidal delivery system would offer reduced side effects, frequency of administration, and increased bioavailability [17]. In the present study, to control PTX release, it was encapsulated in lecithin/chitosan nanoparticles (LCNs). It was shown that LCNs are highly effective at delivering therapeutic agents transdermally [18,19]. LCNs have the synergetic advantages of both lecithin and chitosan, resulting in increasing the drug retention time at the target site and enhancing penetration of the drug [20–24]. It was hypothesized that the inclusion of PTX loaded LCNs into lignocellulose hydrogels can produce a dressing with the ability to control the release capable of enhancing the wound healing process. In this study, the LC sponge was prepared by surface modification of LC hydrogels through carboxylation (oxidation) and amination following by freeze drying and was then filled with PTX loaded LCNs (Figure 1 shows the schematic illustration of experimental procedures). The physicochemical properties of the developed nanoparticles and sponge, as well as in vitro drug release and in vivo wound healing, were investigated to study the suitability of the developed sponge for wound healing applications. The results reveal that both the LCNs and LC sponge possess a controlled release of PTX. Moreover, the LC sponge containing PTX-loaded LCNs increased the rate of wound healing compared to the control group.

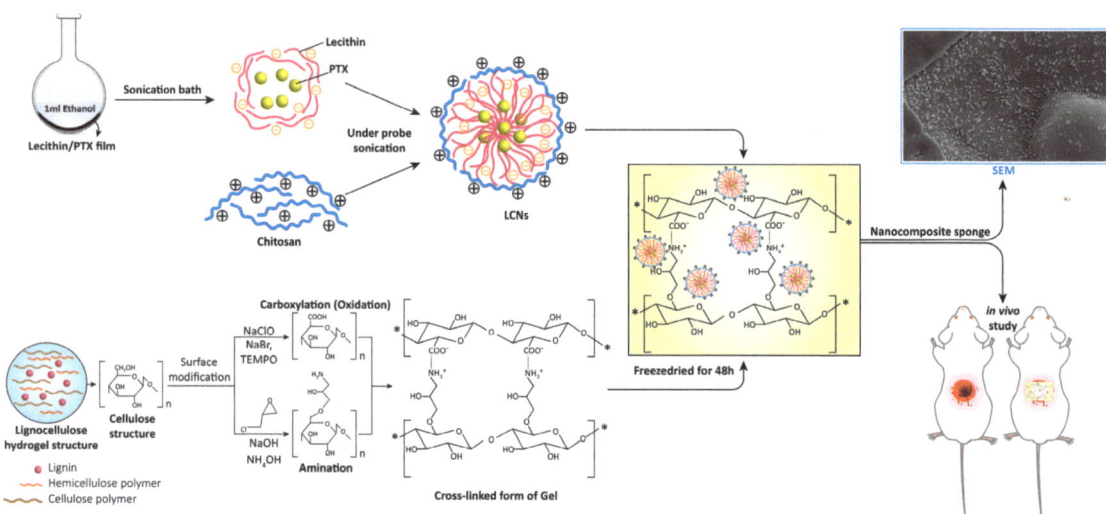

**Figure 1.** A schematic diagram representing the experimental design.

## 2. Results and Discussion

LCNs have great potential for the transdermal delivery of therapeutic agents due to their ability to encapsulate drugs and sustain release [25]. In the current study, the LCNs nanoparticles were obtained by the injection of methanolic lecithin dispersion into a chitosan solution. Positively charged chitosan interacted with negatively charged lecithin through self-assembly. The content of LCNs and PTX-loaded LCNs and their characteristics, including size, polydispersity index (PDI), zeta potential and drug entrapment efficiency (EE), are summarized in Table 1, in which the obtained results are in the range of 259–869 nm, 0.042–0.373, 7.43–43.1 mV, and 30–62%, respectively. The low PDI of the nanoparticles indicates the uniform size distribution in all formulations. The drug release behavior from LCNs with different chitosan, lecithin, and PTX content during 12 h is plotted in Figure 2. Results show that at constant PTX and lecithin, the higher concentration of chitosan causes a lower release rate with a more controlled manner. Formulations with low levels of chitosan and lecithin (S1, S2) show a considerable burst release revealing the disability of the nanoparticle wall to protect fast release of the highly water soluble PTX. The burst release of PTX may also be related to the accumulation of the drug on the surface region of the nanoparticles.

**Table 1.** The amount of chitosan, lecithin, and PTX in the developed nanoparticles, as well as properties of the developed nanoparticles, including size, polydispersity index, zeta potential, and drug entrapment efficiency.

| Sample No | Chitosan (mg/mL) | Lecithin (mg) | PTX (mg) | Size (nm) | PDI | Zeta Potential (mv) | EE (%) |
|---|---|---|---|---|---|---|---|
| S1 | 1 | 2.5 | 1 | 321.4 ± 21.6 | 0.275 | 13.1 ± 0.8 | 35 ± 7.3 |
| S2 | 1 | 2.5 | 2 | 259.3 ± 11 | 0.042 | 7.43 ± 0.35 | 47 ± 6.8 |
| S3 | 1 | 5 | 1 | 329.3 ± 25.9 | 0.188 | 18.3 ± 2.1 | 30 ± 0.3 |
| S4 | 1 | 5 | 2 | 508.2 ± 32 | 0.373 | 19 ± 2.2 | 45 ± 1.2 |
| S5 | 2 | 2.5 | 1 | 864.6 ± 33.4 | 0.121 | 14.6 ± 0.5 | 43 ± 0.4 |
| S6 | 2 | 2.5 | 2 | 306.1 ± 18.2 | 0.14 | 15.5 ± 1.1 | 41 ± 6.5 |
| S7 | 2 | 5 | 1 | 365.2 ± 28.1 | 0.059 | 17.8 ± 1.8 | 62 ± 0.3 |
| S8 | 2 | 5 | 2 | 429.4 ± 25.3 | 0.199 | 18.3 ± 0.09 | 45 ± 3.2 |

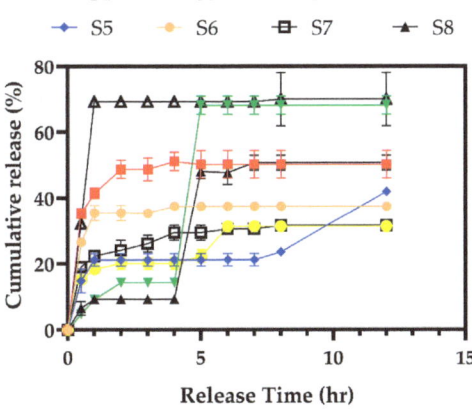

**Figure 2.** In vitro release profile of the PTX from different formulations of nanoparticles (the properties of nanoparticles and their formulations are summarized in Table 1).

A full factorial design was used to optimize the process parameters for encapsulation of PTX in LCNs. The design parameters are the concentration of chitosan, lecithin, and PTX as their values are mentioned in Table 1. According to the results of Design Expert Software, the optimum formulation contained 1.9 mg/mL chitosan, 3.89 mg lecithin, and 1.89 PTX, which was predicted to have a particle size of 422.9 nm and drug entrapment of 43.5%. The proposed formulation was synthesized for the preparation of LC sponges containing LCNs/PTX. The morphologies of the optimized LCNs/PTX and prepared LC sponge are demonstrated in Figure 3. As can be seen in higher magnification images, the nanoparticles are entrapped successfully in the sponge pores. They have a uniform shape with obvious agglomeration.

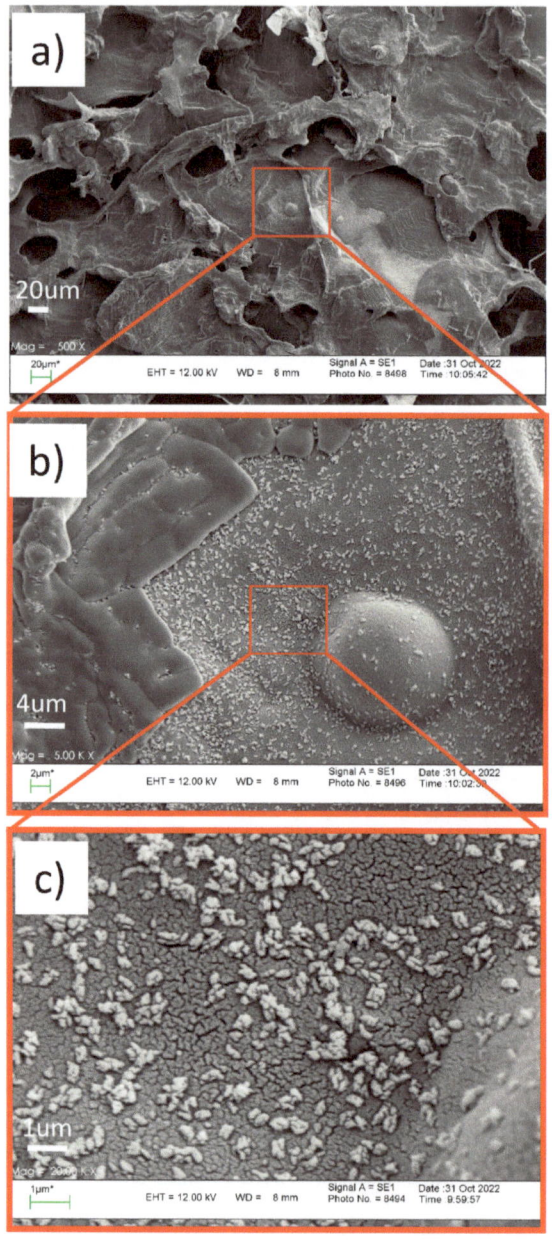

**Figure 3.** SEM micrograph of the LC sponge containing LCNs/PTX at (**a**) 500× and (**b**) 5000× and (**c**) 20,000X, (LCNs/PTX nanoparticles on the surface of sponge are more visible in the 20,000× image).

FTIR spectroscopy analysis was performed to study the possible interaction of LC hydrogels with nanoparticles and the drug, as well as the effect of modification on the lignocellulose chemical bonds. Figure 4 illustrates the FTIR spectra of PTX, LCNs, unmodified hydrogels, TEMPO-modified hydrogels, epichlorohydrin (EPI)/TEMPO-modified hydrogels, and EPI/TEMPO modified hydrogels containing LCNs/PTX. In the FTIR spectra of PTX, peaks at 1700 cm$^{-1}$, 1718 cm$^{-1}$, 1548 cm$^{-1}$, 756 cm$^{-1}$, and 1412 cm$^{-1}$ contribute to the C-O stretching vibration of the amide bond, the C-O stretching vibration of ketone, the N-H bending vibration of the amide, the C-N out of plane wagging of the amide, and the C-N stretching of the amide, respectively [26]. All TEMPO-modified hydrogels, EPI/TEMPO-modified hydrogels, and EPI/TEMPO modified hydrogels containing LCNs/PTX exhibit a broad absorption peak at 3000–3600 cm$^{-1}$, 2900 cm$^{-1}$, 1300–1400 cm$^{-1}$, and 1000–1200 cm$^{-1}$, which are attributed to O-H single bond stretching, C-H single bond stretching, O-H single bond and C-H single bond bending, and C–OH single bond and C-O single bond asymmetric stretching, respectively [27]. The peak at 1600 cm$^{-1}$ is related to the C-O double bond stretching of carboxylate groups (COONa and COOCa) [28]. The O-H single bond peak intensity at 3000−3600 cm$^{-1}$ reduces in EPI/TEMPO modified hydrogels, while the C-H single bond peak intensity at 1425 cm$^{-1}$ increases in EPI/TEMPO modified hydrogels compared to TEMPO-modified hydrogels and unmodified hydrogels. These results clearly reveal the formation of covalent bonds between EPI and the hydroxyl groups [29].

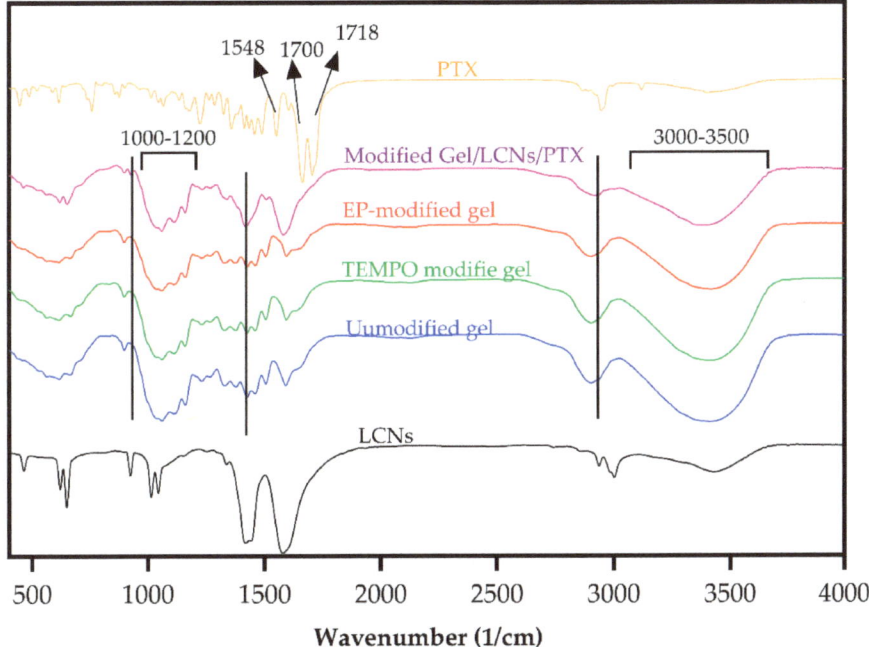

**Figure 4.** FTIR spectra of PTX, LCNs, unmodified hyhdrogels, TEMPO-modified hydrogels, EPI/TEMPO-modified hydrogels, and EPI/TEMPO-modified hydrogels containing LCNs/PTX.

Figure 5 depicts the PTX release curves from unmodified hydrogels containing LCNs/PTX and modified hydrogels containing LCNs/PTX with different LC/nanoparticle ratios. At a higher concentration of LCNs/PTX, the hydrogels show burst release of approximately 75% in the first 3 h. Among different samples, the LC:LCNs/PTX (1:1) sample with a lower nanoparticle concentration shows a more sustained release rate such that, in the first 9 h, 75% of the drug was released.

In numerous studies, PTX was found to facilitate the wound healing process in a wide variety of pathological conditions, including ulcers in a venous leg, the syndrome of diabetes, wounds caused by physicochemical factors, and radiation-related injuries [30,31]. Similarly, hydrogel materials exhibited superiority over traditional wound treatment (ointment, etc.) due to their well-known properties of high biocompatibility, biodegradability, very low immunogenicity, excellent drug delivery (e.g., antibiotics), and ease of use. More importantly, treatment with the hydrogels led to a preservation of the breathability and good moistening of the tissue, which is due to the galenics of the gels consisting of water [32]. Furthermore, the promotion of re-vascularization, a significant lower infection rate due to moist wound management, less scarring, and aesthetically better healing results can be achieved by applying hydrogels in wound treatments [33]. Figure 5 demonstrates the wound healing process in different groups of NS treatment, PTX solution, LC sponge, and LC sponge containing LCNs/PTX, on days 3, 7, 14, and 21 after surgery. In all groups, the wound size reduces because of the body's natural response and the biological proceeding of wound healing. Still, on the third and seventh days of the test, the size of the wound in the LC containing LCNs/PTX nanoparticles is obviously smaller than other groups. The results of the wound contraction measurement summarized in Figure 6 reveal that the size contraction from the PTX containing sponge was significantly larger than the other group on day 7 ($p \leq 0.05$). On the 14th and 21st days, wound healing proceeded in the LCNs/PTX-loaded sponge group with almost the same initial speed; however, the healing rate of the wound in other groups decreased significantly ($p < 0.05$). In general, inner-group statistical comparison between wound size revealed that the wounds in the sponge group containing PTX significantly increased on days 14, 7, 3, and 21, respectively, 27.83 ± 2.83, 42.39 ± 2.99, 83.56 ± 5.03, and 97.44 ± 0.85, which was faster than other groups. By comparing the treatments with the developed hydrogels and normal saline, it was concluded that all the treatments developed in this study increased the wound healing rate compared to normal saline. Additionally, by comparing the two groups of PTX solution and LC sponge (without PTX), it can be seen that despite the better effect of PTX, there is no significant difference between the two groups.

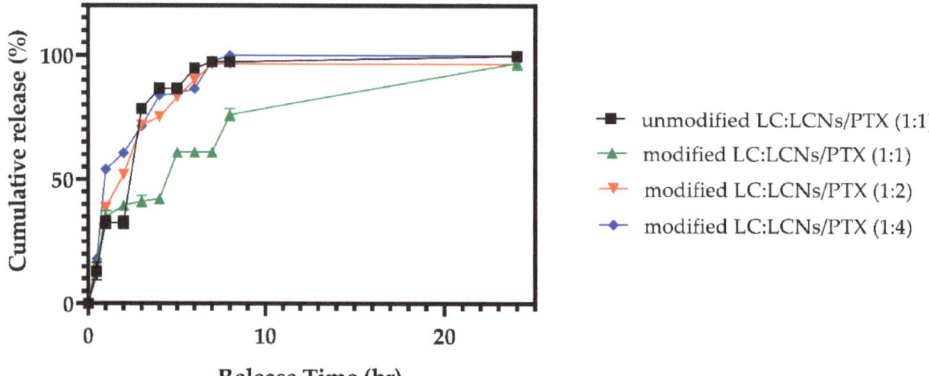

**Figure 5.** In vitro release profile of the PTX from unmodified hydrogels containing LCNs/PTX and modified hydrogels containing LCNs/PTX with different LC:LCNs/PTX ratios. Modified hydrogels containing lower concentrations of LCNs/PTX revealed a lower release rate compared to other samples.

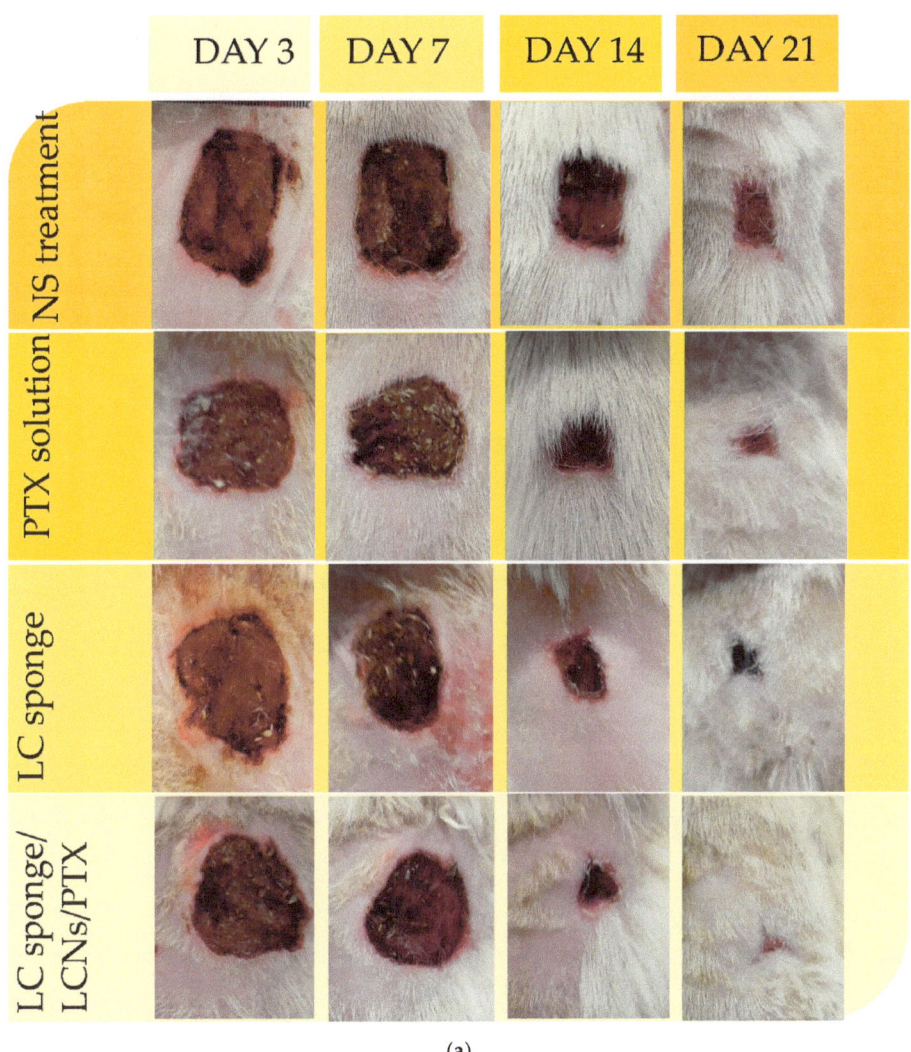

(a)

**Figure 6.** *Cont.*

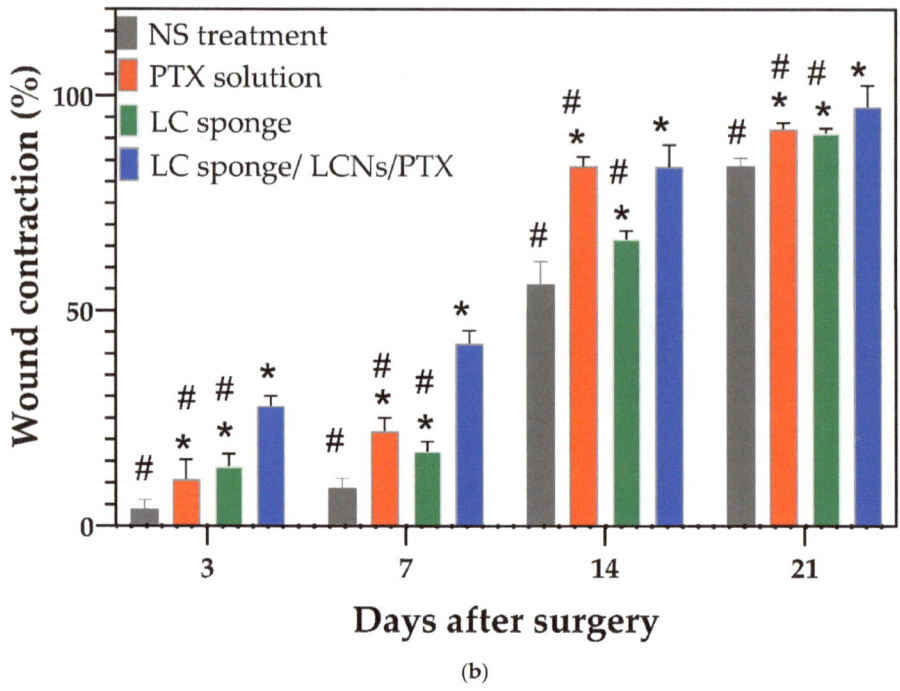

**Figure 6.** Wound healing process of treated rats with normal saline, free PTX, LC sponge, and LC sponge containing LCNs/PTX, (**a**) Photographical images of wound contraction on different days of control and treated animals. (**b**) Representation of percentage wound contraction of control and treated wounds. Values are expressed as mean ± SD for twenty-four animals. *: indicates the significant differences with NS treatment group and #: indicates the significant differences with LC sponge containing LCNs/PTX group.

According to the microscopic observations, aseptic conditions were confirmed to be in all the wounds. As shown in Figure 7, in wounds treated with normal saline, the inflammatory response was higher than the other treated wounds. Necrotic debris on the surface of the tissue was almost removed in the wound treated with the LC sponge containing LCNs/PTX. As observed in H&E staining images, the areas associated with neovascularization (open lumen vascular structures and endothelial cell clusters) were greater in the blank group than other groups treated with wound dressings. The reduction of these areas in LC sponge containing LCNs/PTX wound dressings groups indicates the completion of the healing process. Other signs of wound healing, in addition to neovascularization areas visible in this group, include the thickness of the epidermis layer (E) and the formation of the keratin layer (K) on the epidermis layer, as well as the presence of skin appendages such as new follicles (N.F) and sebaceous glands (Sc. G) in the dermis layer. These can be attributed to the role of PTX in facilitating tissue repair and the wound healing process. PTX activity can be divided into four main categories, including: (1) PTX can increase the deformation ability of red blood cells (RBCs), while decreasing the RBCs aggregation and vasoconstriction and has reducing effects on blood viscosity [34]. (2) PTX showed immunological activities that are effective by various mechanisms such as: inhibiting the activation of T and B lymphocytes, reducing the release of TNF-α from monocytes, releasing of peroxides from neutrophils, reducing the neutrophil degranulation, decreasing the leukocytes aggregation/ adhesion, and increasing their deformability and chemotaxis. PTX also modulates or blocks the inflammatory actions of IL-1 and TNF-α on neutrophils and has some effects on other cytokines, such as IL-6, IL-8, VEGF, and TGF-β1 [15,35,36] (3). Effects on platelet aggregation and adhesion [37], and finally (4), effects on properties of connective tissue and direct wound healing such as decreasing the levels of fibroblast collagen, fibronectin, and glycosaminoglycans, as well as reducing the response of fibroblast to IL-1 and TNF-α [38].

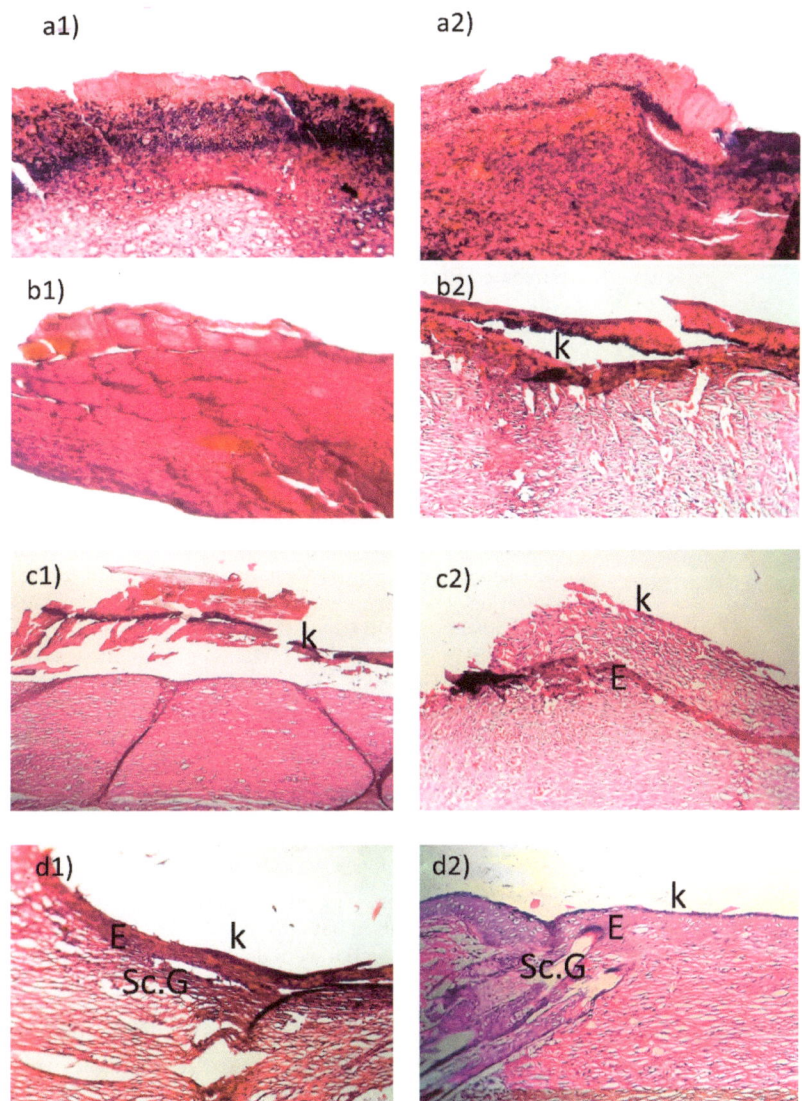

Figure 7. Histological changes during the wound-healing process. Histology of the wound tissue of experimental groups, (a1,a2) NS treatment, (b1,b2) PTX solution, (c1,c2) LC sponge and (d1,d2) LC sponge containing LCNs/PTX. N.F, follicles; Sc. G, sebaceous glands; K, keratin layer; E, epidermis layer.

## 3. Conclusions

In this study, the EPI/TEMPO modified lignocellulose hydrogels were developed and filled with LCNs containing PTX to control drug release for wound healing applications.

Using full factorial design software, an optimized formulation of nanoparticles prepared from different concentrations of ingredients was suggested based on the results of particle size, zeta potential, and drug load/release. The prepared nanoparticles were entrapped successfully in the sponge pores and attached to its pore wall, as is evident from SEM observation. The drug release study from hydrogels revealed that a lower concentration of PTX in the hydrogels resulted in more sustained drug release. The results of the wound healing evaluation in the animal model indicated the higher healing rate by treatment with sponges containing PTX-loaded LCNs. The developed sponge with control release behavior has shown the potential to be used in wound treatment applications.

## 4. Materials and Methods

### 4.1. Materials

PTX was kindly provided by Amin Pharmaceutical Company (Isfahan, Iran), soy lecithin (Degussa, GmbH, Freising, Germany), low molecular weight chitosan from Sigma Company (Saint Louis, MO, US), glacial acetic acid, and all other reagents were purchased from Merck Chemical Company (Darmstadt, Germany), LC nanofibrils hydrogel was purchased from Nano Novin Polymer (Gorgan, Iran).

### 4.2. Preparation of Chitosan Coated Lecithin Nanoparticles

In order to prepare drug-loaded lecithin nanoparticles, the desired amount of the drug was dissolved in methanol, followed by the dissolution of lecithin in the prepared drug solution using sonication. The methanol was evaporated using a rotary evaporator to obtain the PTX-loaded lecithin film. The prepared film was dispersed in 1 mL methanol using a sonication bath to re-disperse the drug-loaded lecithin films. The chitosan solution was prepared by dissolving defined amounts of chitosan (Table 1) in 5 mL distilled water containing 1% acetic acid for 24 h, and the pH was increased to 4.7 by adding NaOH 1M solution. For the preparation of chitosan-coated lecithin nanoparticles (LCNs/PTX), the methanolic lecithin–PTX solution was added to the chitosan solution under probe sonication. The blank nanoparticles were prepared using a similar procedure without the PTX addition. A full factorial design suggested by the Design Expert software (version 10.0.7, US) was used for optimization of the chitosan concentration (A), lecithin amount (B), and PTX amount (C) as main independent variables. NPs size, zeta potential, loading efficiency, and release efficiency were studied as dependent variables. The minimum and maximum levels of each parameter were determined based on preliminary experiments. The composition of prepared NPs and observed responses are presented in Table 1. Design Expert software was used to analyze the statistical significance and optimization of the NPs formulation.

### 4.3. Particle Size and Zeta Potential Measurement

First, 0.1 mg of prepared nanoparticles was dispersed in 10 mL of deionized water for 30 min in a sonication bath (85 W, 42 kHz), and their particle size, polydispersity index (PDI), and zeta potential were read using Zetasizer (Zetasizer 3600, Malvern Instrument Ltd., Worchestershire, UK).

### 4.4. Entrapment Efficiency

To determine the entrapment efficiency of the nanoparticles, the unentrapped PTX was measured by centrifuging a suspension of LCNs-PTX nanoparticles in deionized water at 12,000 rpm. The UV absorbance of the supernatant was read spectrophotometrically at 274 nm. Then, the entrapment efficiency (EE%) was calculated as:

$$EE\% = (\text{Initial drug quantity} - \text{Unentrapped quantity of drug})/(\text{Total drug}) \times 100\%.$$

*4.5. Surface Modification of LC Hydrogels*

The surface modification of LC hydrogels was performed through carboxylation (oxidation) and amination according to the previously reported method with some minor modification [13]. For the oxidation reaction, 3.145 g of LC hydrogels (equal to 0.1 g of dry weight) was suspended in an 80 mL solution containing 15 mg 2,2,6,6-tetramethylpiperidine-1-oxy (TEMPO) and 98.87 mg of sodium bromide. Then, 101 µL of sodium hypochlorite was added to start the oxidation reaction. The pH was kept at 10.5 by adding an appropriate amount of 1 M sodium hydroxide. The reaction was continued on the stirrer at room temperature and 600 rpm for 4 h. Finally, 2 mL of ethanol was added to the reaction mixture. The obtained hydrogel was washed with distilled water by centrifugation at 7800 rpm for 10 min and repeated until the oxidized hydrogel reached the pH of approximately 7. For the amination reaction, a 3.145 g sample of hydrogel was added to the 40 mL of 1 M sodium hydroxide solution and heated with constant stirring to 60 °C. Then, 140 µL of epichlorohydrin was added and allowed to react with stirring at 600 rpm and 60 °C for 2 h. The solution was washed with distilled water by centrifugation at 7800 rpm for approximately 10 min until the pH reached below 12. The precipitated hydrogel was re-suspended in 40 mL of 0.01 M sodium hydroxide, and the proper amount of ammonium hydroxide (29.4% *w/v*) was added to adjust the pH of the solution to 12. The mixture was then left to stir at 600 rpm and 60 °C for an additional 2 h. The resultant hydrogel was washed and centrifuged until the pH was reduced to 7 and stored at 4 °C.

*4.6. Preparation of LC Sponge Containing Drug Loaded Nanoparticles*

The cationic and anionic forms of the surface modified nanofibers were added together in equal amounts to obtain the cross-linked form hydrogel. To load nanoparticles in the prepared hydrogel, 2 mg of nanoparticles were dispersed in 1 mL of deionized water and mixed evenly with different amounts of the prepared hydrogels (0.5, 1, and 2 g). The mixtures were then frozen overnight and freeze dried for 48 h to obtain nanocomposite sponges. For comparison purposes, LCNs and LCNs/PTX nanoparticles were also added to 2 g of non-modified hydrogel and freeze dried in a similar manner to assess the effect of cross-linking on the release profile of drug.

*4.7. Fourier Transform Infrared Spectroscopy*

All samples including LCNs/PTX nanoparticles and sponges were ground well with potassium bromide and made into pellets. The spectra of samples were recorded using a FTIR spectrometer (JASCO model FT/IR 6300FV, Tokyo, Japan) in the 350–4000-cm$^{-1}$ region.

*4.8. SEM Observation*

The morphologies of LCNs/PTX nanoparticles and dried hydrogels were studied using field emission scanning electron microscopy (FESEM-SEM Hitachi F41100, Japan) at an acceleration voltage of 20 kV.

*4.9. In Vitro Drug Release Studies*

Drug release studies were carried out using the dialysis technique, in which 0.5 mL of the drug/nanoparticles dispersion or 1.5–2 g sponge were placed inside a dialysis bag, sealed, and submerged in release media (19.5 mL of phosphate buffer saline solution and Tween 80 at pH 7.4) while stirred at 200 rpm. Through the dialysis membrane, the released drug diffused to the outer release media, from where the samples were taken for further analysis. To measure the amount of drug released, 1 mL of the release medium was centrifuged for 5 min at 12,000 rpm and UV-spectrophotometrically examined at 274 nm, and returned to the test media. The studies were repeated three times, and the results were reported as mean ± standard deviation.

*4.10. In Vivo Wound Healing Studies*

All experimental procedures were in accordance with approved guidelines of the Animal Committee of Isfahan University of Medical sciences, Iran (ethical code No.: IR.MUI.RESEARCH.REC.1397.267). A full-thickness excisional wound model was used to evaluate the wound healing ability of the prepared hydrogels. Twenty-four healthy adult male Wistar rats (8–10 months old, weighing 220–250 g) were obtained from the animal lab of the Department of Pharmaceutics, Isfahan University of Medical Sciences. The animals were kept at a controlled ad libitum condition of $22 \pm 2\ °C$ and 50–60% humidity. Then, the rats were randomly divided into four groups (6 rats in each group), including blank group, PTX group, PTX free sponge, PTX loaded sponge. The animals were anesthetized by intraperitoneal injection of 10 mg/kg xylazine and 80 mg/kg ketamine, then their backs were shaved, and an approximate 3 cm midline incision was made in the skin and subcutaneous tissue. The wound in blank group was treated with 1 mL of normal saline. In other groups the content of PTX was set to be 1.5 mg. The wound was covered with a gauze dressing and the dressing was changed every day.

4.10.1. Macroscopic Observation of the Wound-Healing Process

After 3, 7, 14, and 21 days of wound creation, the healing progress was recorded using a digital camera. The rate of wound closure and reduction in the wound size were determined by measuring wound area using an image analyzing program. The wound closure was calculated as:

$$\text{Wound closure (\%)} = [1 - (\text{open wound area}/\text{initial wound area})] \times 100$$

After the study period, the animals were sacrificed and skin tissue samples were removed surgically for histological evaluation.

4.10.2. Histological Examination

On day 21 post-treatment, the treated excision wounds were immediately fixed in 10% formaldehyde, dehydrated with ethanol (70%), and then were embedded in paraffin blocks. The blocks were sectioned into 5 μm slides and stained with hematoxylin-eosin (H&E) to study the best stage of healing. Stained samples were examined under a light microscope (Olympus CX 21, Tokyo, Japan), and a digital camera was used to take the photo image from stained slides.

*4.11. Statistical Analysis*

The data were presented as mean and standard deviation. The results were statistically analyzed by IBM SPSS Statistics 26 Software using a one-way ANOVA test. In all of the evaluations, $p < 0.05$ was considered statistically significant.

**Author Contributions:** Conceptualization, J.V.; methodology, P.D., and A.A. and M.S.; software, M.K. and M.S.; validation, J.V. and M.K.; formal analysis, M.K. and M.B.; investigation, M.K. and M.B.; resources, M.B.; data curation, P.D. and A.A.; writing—original draft preparation, P.D.; writing—review and editing, M.K. and M.B.; visualization, M.K.; supervision, J.V.; project administration, P.D.; funding acquisition, P.D. All authors have read and agreed to the published version of the manuscript.

**Funding:** Pharmacy Student Research Committee of Isfahan University of Medical Sciences with the project No 197084.

**Institutional Review Board Statement:** All experimental procedures were in accordance with approved guidelines of the Animal Committee of Isfahan University of Medical sciences, Iran (ethical code No.: IR.MUI.RESEARCH.REC.1397.267-28 Oct-2018).

**Informed Consent Statement:** Not applicable.

**Data Availability Statement:** All data are available on request.

**Acknowledgments:** Authors are grateful to any support from the Pharmacy Student Research Committee of Isfahan University of Medical Sciences with the project No 197084.

**Conflicts of Interest:** The authors declare no conflict of interest.

## References

1. Boateng, J.S.; Matthews, K.H.; Stevens, H.N.; Eccleston, G.M. Wound Healing Dressings and Drug Delivery Systems: A Review. *J. Pharm. Sci.* **2008**, *97*, 2892–2923. [CrossRef] [PubMed]
2. Fonder, M.A.; Lazarus, G.S.; Cowan, D.A.; Aronson-Cook, B.; Kohli, A.R.; Mamelak, A.J. Treating the chronic wound: A practical approach to the care of nonhealing wounds and wound care dressings. *J. Am. Acad. Dermatol.* **2008**, *58*, 185–206. [CrossRef] [PubMed]
3. Wangsawangrung, N.; Choipang, C.; Chaiarwut, S.; Ekabutr, P.; Suwantong, O.; Chuysinuan, P.; Techasakul, S.; Supaphol, P. Quercetin/Hydroxypropyl-β-Cyclodextrin Inclusion Complex-Loaded Hydrogels for Accelerated Wound Healing. *Gels* **2022**, *8*, 573. [CrossRef] [PubMed]
4. Reesi, F.; Minaiyan, M.; Taheri, A. A novel lignin-based nanofibrous dressing containing arginine for wound-healing applications. *Drug Deliv. Transl. Res.* **2018**, *8*, 111–122. [CrossRef]
5. Zhang, K.; Bai, X.; Yuan, Z.; Cao, X.; Jiao, X.; Li, Y.; Qin, Y.; Wen, Y.; Zhang, X. Layered nanofiber sponge with an improved capacity for promoting blood coagulation and wound healing. *Biomaterials* **2019**, *204*, 70–79. [CrossRef] [PubMed]
6. Raju, N.R.; Silina, E.; Stupin, V.; Manturova, N.; Chidambaram, S.B.; Achar, R.R. Multifunctional and Smart Wound Dressings—A Review on Recent Research Advancements in Skin Regenerative Medicine. *Pharmaceutics* **2022**, *14*, 1574. [CrossRef]
7. Kalinoski, R.; Shi, J. Hydrogels derived from lignocellulosic compounds: Evaluation of the compositional, structural, mechanical and antimicrobial properties. *Ind. Crop. Prod.* **2019**, *128*, 323–330. [CrossRef]
8. Nasution, H.; Harahap, H.; Dalimunthe, N.F.; Ginting, M.H.S.; Jaafar, M.; Tan, O.O.H.; Aruan, H.K.; Herfananda, A.L. Hydrogel and Effects of Crosslinking Agent on Cellulose-Based Hydrogels: A Review. *Gels* **2022**, *8*, 568. [CrossRef]
9. Liu, K.; Dai, L.; Li, C. A lignocellulose-based nanocomposite hydrogel with pH-sensitive and potent antibacterial activity for wound healing. *Int. J. Biol. Macromol.* **2021**, *191*, 1249–1254. [CrossRef]
10. Cui, N.; Xu, Z.; Zhao, X.; Yuan, M.; Pan, L.; Lu, T.; Du, A.; Qin, L. In Vivo Effect of Resveratrol-Cellulose Aerogel Drug Delivery System to Relieve Inflammation on Sports Osteoarthritis. *Gels* **2022**, *8*, 544. [CrossRef]
11. Haldar, D.; Purkait, M.K. Micro and nanocrystalline cellulose derivatives of lignocellulosic biomass: A review on synthesis, applications and advancements. *Carbohydr. Polym.* **2020**, *250*, 116937. [CrossRef] [PubMed]
12. Riva, L.; Lotito, A.D.; Punta, C.; Sacchetti, A. Zinc- and Copper-Loaded Nanosponges from Cellulose Nanofibers Hydrogels: New Heterogeneous Catalysts for the Synthesis of Aromatic Acetals. *Gels* **2022**, *8*, 54. [CrossRef]
13. Spaic, M.; Small, D.P.; Cook, J.R.; Wan, W. Characterization of anionic and cationic functionalized bacterial cellulose nanofibres for controlled release applications. *Cellulose* **2014**, *21*, 1529–1540. [CrossRef]
14. Bessler, H.; Gilgal, R.; Djaldetti, M.; Zahavi, I. Effect of Pentoxifylline on the Phagocytic Activity, cAMP Levels, and Superoxide Anion Production by Monocytes and Polymorphonuclear Cells. *J. Leukoc. Biol.* **1986**, *40*, 747–754. [CrossRef] [PubMed]
15. Babaei, S.; Bayat, M. Pentoxifylline Accelerates Wound Healing Process by Modulating Gene Expression of MMP-1, MMP-3, and TIMP-1 in Normoglycemic Rats. *J. Investig. Surg.* **2015**, *28*, 196–201. [CrossRef] [PubMed]
16. Lin, H.-Y.; Yeh, C.-T. Controlled release of pentoxifylline from porous chitosan-pectin scaffolds. *Drug Deliv.* **2010**, *17*, 313–321. [CrossRef] [PubMed]
17. Cavalcanti, A.L.M.; Reis, M.Y.F.A.; Silva, G.C.L.; Ramalho, Í.M.M.; Guimarães, G.P.; Silva, J.A.; Saraiva, K.L.A.; Damasceno, B.P.G.L. Microemulsion for topical application of pentoxifylline: In vitro release and in vivo evaluation. *Int. J. Pharm.* **2016**, *506*, 351–360. [CrossRef]
18. Dong, W.; Ye, J.; Wang, W.; Yang, Y.; Wang, H.; Sun, T.; Gao, L.; Liu, Y. Self-Assembled Lecithin/Chitosan Nanoparticles Based on Phospholipid Complex: A Feasible Strategy to Improve Entrapment Efficiency and Transdermal Delivery of Poorly Lipophilic Drug. *Int. J. Nanomed.* **2020**, *15*, 5629–5643. [CrossRef]
19. Liu, Y.; Liu, L.; Zhou, C.; Xia, X. Self-assembled lecithin/chitosan nanoparticles for oral insulin delivery: Preparation and functional evaluation. *Int. J. Nanomed.* **2016**, *11*, 761–769. [CrossRef]
20. Chaves, L.L.; Silveri, A.; Vieira, A.C.C.; Ferreira, D.; Cristiano, M.C.; Paolino, D.; Di Marzio, L.; Lima, S.C. pH-responsive chitosan based hydrogels affect the release of dapsone: Design, set-up, and physicochemical characterization. *Int. J. Biol. Macromol.* **2019**, *133*, 1268–1279. [CrossRef]
21. Cosco, D.; Failla, P.; Costa, N.; Pullano, S.; Fiorillo, A.; Mollace, V.; Fresta, M.; Paolino, D. Rutin-loaded chitosan microspheres: Characterization and evaluation of the anti-inflammatory activity. *Carbohydr. Polym.* **2016**, *152*, 583–591. [CrossRef] [PubMed]
22. Di Francesco, M.; Primavera, R.; Fiorito, S.; Cristiano, M.C.; Taddeo, V.A.; Epifano, F.; Di Marzio, L.; Genovese, S.; Celia, C. Acronychiabaueri Analogue Derivative-Loaded Ultradeformable Vesicles: Physicochemical Characterization and Potential Applications. *Planta Medica* **2016**, *83*, 482–491. [CrossRef]
23. Ma, Q.; Gao, Y.; Sun, W.; Cao, J.; Liang, Y.; Han, S.; Wang, X.; Sun, Y. Self-Assembled chitosan/phospholipid nanoparticles: From fundamentals to preparation for advanced drug delivery. *Drug Deliv.* **2020**, *27*, 200–215. [CrossRef] [PubMed]

24. Sonvico, F.; Cagnani, A.; Rossi, A.; Motta, S.; Di Bari, M.; Cavatorta, F.; Alonso, M.J.; Deriu, A.; Colombo, P. Formation of self-organized nanoparticles by lecithin/chitosan ionic interaction. *Int. J. Pharm.* **2006**, *324*, 67–73. [CrossRef]
25. Saha, M.; Saha, D.R.; Ulhosna, T.; Sharker, S.M.; Shohag, H.; Islam, M.S.; Ray, S.K.; Rahman, G.S.; Reza, H.M. QbD based development of resveratrol-loaded mucoadhesive lecithin/chitosan nanoparticles for prolonged ocular drug delivery. *J. Drug Deliv. Sci. Technol.* **2021**, *63*, 102480. [CrossRef]
26. Shuwaili, A.H.A.L.; Rasool, B.K.A.; Abdulrasool, A.A. Optimization of elastic transfersomes formulations for transdermal delivery of pentoxifylline. *Eur. J. Pharm. Biopharm.* **2016**, *102*, 101–114. [CrossRef]
27. Shi, S.; Zhu, K.; Chen, X.; Hu, J.; Zhang, L. Cross-Linked Cellulose Membranes with Robust Mechanical Property, Self-Adaptive Breathability, and Excellent Biocompatibility. *ACS Sustain. Chem. Eng.* **2019**, *7*, 19799–19806. [CrossRef]
28. Shimizu, M.; Saito, T.; Isogai, A. Water-resistant and high oxygen-barrier nanocellulose films with interfibrillar cross-linkages formed through multivalent metal ions. *J. Membr. Sci.* **2015**, *500*, 1–7. [CrossRef]
29. Lee, K.; Jeon, Y.; Kim, D.; Kwon, G.; Kim, U.-J.; Hong, C.; Choung, J.W.; You, J. Double-crosslinked cellulose nanofiber based bioplastic films for practical applications. *Carbohydr. Polym.* **2021**, *260*, 117817. [CrossRef]
30. Aghajani, A.; Kazemi, T.; Enayatifard, R.; Amiri, F.T.; Narenji, M. Investigating the skin penetration and wound healing properties of niosomal pentoxifylline cream. *Eur. J. Pharm. Sci.* **2020**, *151*, 105434. [CrossRef]
31. Ahmadi, M.; Khalili, H. Potential benefits of pentoxifylline on wound healing. *Expert Rev. Clin. Pharmacol.* **2016**, *9*, 129–142. [CrossRef]
32. Rüther, L.; Voss, W. Hydrogel or ointment? Comparison of five different galenics regarding tissue breathability and transepidermal water loss. *Heliyon* **2021**, *7*, e06071. [CrossRef] [PubMed]
33. Hoeksema, H.; De Vos, M.; Verbelen, J.; Pirayesh, A.; Monstrey, S. Scar management by means of occlusion and hydration: A comparative study of silicones versus a hydrating gel-cream. *Burns* **2013**, *39*, 1437–1448. [CrossRef] [PubMed]
34. Lim, A.A.T.; Washington, A.P.; Greinwald, J.H.; Lassem, L.F.; Holtel, M.R. Effect of pentoxifylline on the healing of guinea pig tympanic membrane. *Annals of Otology, Rhinology & Laryngology* **2000**, *109*, 262–266.
35. Falanga, V.; Fujitani, R.M.; Diaz, C.; Hunter, G.; Jorizzo, J.; Lawrence, P.F.; Lee, B.Y.; O Menzoian, J.; Tretbar, L.L.; Holloway, G.A.; et al. Systemic treatment of venous leg ulcers with high doses of pentoxifylline: Efficacy in a randomized, placebo-controlled trial. *Wound Repair Regen.* **1999**, *7*, 208–213. [CrossRef] [PubMed]
36. Babaei, S.; Bayat, M.; Nouruzian, M.; Bayat, M. Pentoxifylline improves cutaneous wound healing in streptozotocin-induced diabetic rats. *Eur. J. Pharmacol.* **2013**, *700*, 165–172. [CrossRef]
37. Natarajan, S.; Williamson, D.; Stiltz, A.J.; Harding, K. Advances in Wound Care and Healing Technology. *Am. J. Clin. Dermatol.* **2000**, *1*, 269–275. [CrossRef]
38. Siang, R.; Teo, S.Y.; Lee, S.Y.; Basavaraj, A.K.; Koh, R.Y.; Rathbone, M.J. Formulation and evaluation of topical pentoxifylline-hydroxypropyl methylcellulose gels for wound healing application. *Int. J. Pharm. Pharm. Sci.* **2014**, *6*, 535–539.

*Communication*

# Hydrotropic Hydrogels Prepared from Polyglycerol Dendrimers: Enhanced Solubilization and Release of Paclitaxel

Tooru Ooya [1,2,*] and Jaehwi Lee [3]

1. Department of Chemical Science and Engineering, Graduate School of Engineering, Kobe University, 1-1 Rokkodai-cho, Nada-ku, Kobe 657-8501, Japan
2. Center for Advanced Medical Engineering Research & Development (CAMED), Kobe University, 1-5-1 Minatojima-Minamimachi, Kobe 657-0047, Japan
3. College of Pharmacy, Chung-Ang University, 84 Heukseok-ro, Dongjak-gu, Seoul 06974, Korea
* Correspondence: ooya@tiger.kobe-u.ac.jp; Tel.: +81-78-803-6255

**Abstract:** Polyglycerol dendrimers (PGD) exhibit unique properties such as drug delivery, drug solubilization, bioimaging, and diagnostics. In this study, PGD hydrogels were prepared and evaluated as devices for controlled drug release with good solubilization properties. The PGD hydrogels were prepared by crosslinking using ethylene glycol diglycidylether (EGDGE). The concentrations of EGDGE and PGDs were varied. The hydrogels were swellable in ethanol for loading paclitaxel (PTX). The amount of PTX in the hydrogels increased with the swelling ratio, which is proportional to EGDGE/OH ratio, meaning that heterogeneous crosslinking of PGD made high dense region of PGD molecules in the matrix. The hydrogels remained transparent after loading PTX and standing in water for one day, indicating that PTX was dispersed in the hydrogels without any crystallization in water. The results of FTIR imaging of the PTX-loaded PGD hydrogels revealed good dispersion of PTX in the hydrogel matrix. Sixty percent of the loaded PTX was released in a sink condition within 90 min, suggesting that the solubilized PTX would be useful for controlled release without any precipitation. Polyglycerol dendrimer hydrogels are expected to be applicable for rapid release of poorly water-soluble drugs, e.g., for oral administration.

**Keywords:** polyglycerols; dendrimers; hydrogels; paclitaxel; release; solubilization; FTIR imaging; hydrotrope

Citation: Ooya, T.; Lee, J. Hydrotropic Hydrogels Prepared from Polyglycerol Dendrimers: Enhanced Solubilization and Release of Paclitaxel. *Gels* **2022**, *8*, 614. https://doi.org/10.3390/gels8100614

Academic Editors: Diana Silva, Ana Paula Serro and María Vivero-Lopez

Received: 1 September 2022
Accepted: 22 September 2022
Published: 26 September 2022

**Publisher's Note:** MDPI stays neutral with regard to jurisdictional claims in published maps and institutional affiliations.

**Copyright:** © 2022 by the authors. Licensee MDPI, Basel, Switzerland. This article is an open access article distributed under the terms and conditions of the Creative Commons Attribution (CC BY) license (https://creativecommons.org/licenses/by/4.0/).

## 1. Introduction

Dendritic glycerol is a glycerol molecule with a branched chemical structure similar to those of polyglycerol dendrimer (PGD) (Figure 1a) and hyperbranched polyglycerol (HPG) (Figure 1b). Dendritic glycerol consists of a polyether structure, similar to the highly biocompatible polyethylene glycol (PEG), as the backbone with a branched structure, and has many hydroxyl groups at the ends [1–3]. In addition to its high water solubility and biocompatibility, it has advantages such as higher thermal stability compared to PEG [4]. One reason for using PGDs in bio-applications is that the single molecular weight and monodisperse nature of PGDs with perfect degree of branching (DB) make them suitable for targeted modification of molecular ends. In addition, the ability to precisely control the molecular weight allows for a detailed examination of the effects of molecular weight, molecular size, and number of end groups on the system. Hyperbranched polyglycerol is easy to synthesize and can generate high-molecular-weight polymers in a single step, which is difficult with the stepwise organic synthesis that is required for PGD. To date, PGDs and HPGs have been used as nanocapsules for drug delivery [5,6], dispersions of metal ions using host–guest interactions [7], protein adsorption inhibitory surfaces [8–12], hydrogels [13–17], human serum albumin substitutes [18], organ preservation solutions [19], and solvents for poorly water soluble drugs as hydrotropes [20–26].

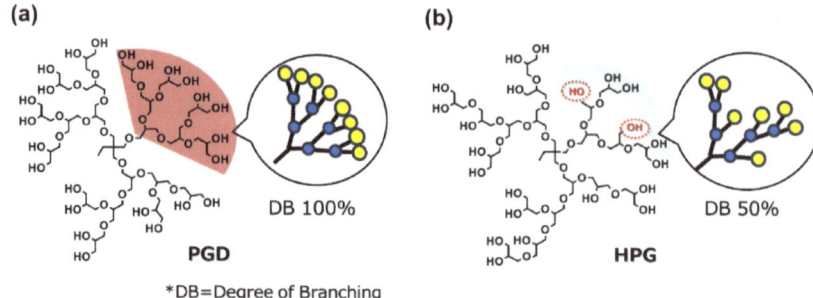

**Figure 1.** Representative structure of (**a**) PGD and (**b**) HPG; OH groups in the dotted circles constructs liner parts, resulting in decreasing DB.

Hydrogels are composed of a three-dimensional polymer network formed by crosslinked hydrophilic polymer chains. They are insoluble in water and can hold a large amount of internal water [27,28]. The preparation of hydrogels using HPG has been reported by Oudshoorn et al. The gelation is carried out using potassium persulfate as a radical initiator [13]. Other crosslinking methods include enzymatic catalysis [14] and biomimetic mineralization [15]. The expected advantages of HPG hydrogels are that the low viscosity of HPG in water enables the preparation of hydrogels with high polymer concentrations and controlled mechanical properties [29]. Hyperbranched polyglycerol hydrogels have reportedly been used as bioinks for microfabrication [30], scaffolds for living cells [14,31,32], and drug delivery systems for poorly-water-soluble drugs and proteins [33]. However, PGD-crosslinked hydrogels have not been reported so far.

We have shown that PGDs are a hydrotrope for paclitaxel (PTX; 5β,20-Epoxy-1,2α,4,7β, 13α-hexahydroxytax-11-en-9-one4,10-diacetate2-benzoate13-ester with (2R,3S)-N-benzoyl-3-phenyllisoserine, molecular formula $C_{47}H_{51}NO_{14}$, corresponding to molecular weight of 853.91 Da). Paclitaxel is a well-known anti-tumor agent with poor water solubility (0.6 ± 0.08 µg/mL [34]), so that it is clinically formulated in a mixture composed of 1:1 blend of Cremophor EL (polyethoxylated castor oil) and ethanol [35]. Paclitaxel promotes the assembly of microtubles, resulting in cancer cells death via protecting from the disassembly of microtubules induced by cold or calcium treatment [36]. Polyethylene glycol 400 (PEG400), which is known as a co-solvent for dissolving PTX at high concentration, requires about 50 wt% to dissolve 0.1 mg/mL of PTX, while about 10 wt% is sufficient for PGD [20]. Furthermore, precipitation occurs upon dilution, suggesting that PGD does not incorporate PTX but functions as a "hydrotropic" molecule that dissolves PTX by interacting with the surrounding PTX molecules. In order to apply this dissolution property, the dissolved PTX release must be controlled. In the present study, PGD-crosslinked hydrogels were prepared by using ethylene glycol diglycidylether (EGDGE). If the local concentration of PGDs can be increased by chemical crosslinking, the solubility of PTX is expected to increase. Crosslinking conditions such as the solvent and concentration were varied, and the obtained hydrogels were evaluated in terms of swelling, PTX loading, dissolution state of PTX, and release of PTX.

## 2. Results and Discussion

Polyglycerol dendrimer of generation 3 (PGD-G3) was crosslinked by reaction with EGDGE (Figure 2). It is well-known that hydroxyl groups in water-soluble polysaccharides can be modified by glycidyl ethers in NaOH aqueous solution [37] and in DMSO in the presence of appropriate catalysts such as DMAP [38]. Since PGD-G3 has many hydroxyl groups, these methods are applicable to the preparation of crosslinked PGD-G3 hydrogels. As shown in Table 1, the concentrations of EGDGE and PGD-G3 were varied. Since the solubility of EGDGE in 1 M NaOH solution is limited, additional 1 M NaOH was added to the reaction mixture when the ratio of EGDGE and hydroxyl groups of PGD-G3 was increased. This results in decreasing the final concentration of PGD-G3. The swelling ratio in water increased with increasing concentration of EGDGE and decreasing concentration of PGD-G3. All the hydrogels prepared in 1 M NaOH reached their equilibrium swelling at around 10 h (Figure 3). When both DMSO and DMAP were used for gel preparation, the concentrations of EGDGE and PGD-G3 both increased, and the swelling ratio for G3-EG(DMSO)0.75 was the lowest among all samples. All the hydrogels were stiff, except for G3-EG(NaOH)0.75, which could not maintain a disc shape. Taking the largest swelling ratio of G3-EG(NaOH)0.75 into account, the crosslinking condition for G3-EG(NaOH)0.75 induced partially intramolecular crosslinking of PGD-G3 molecules. The hydrogels were also swellable in ethanol, which allows loading of PTX.

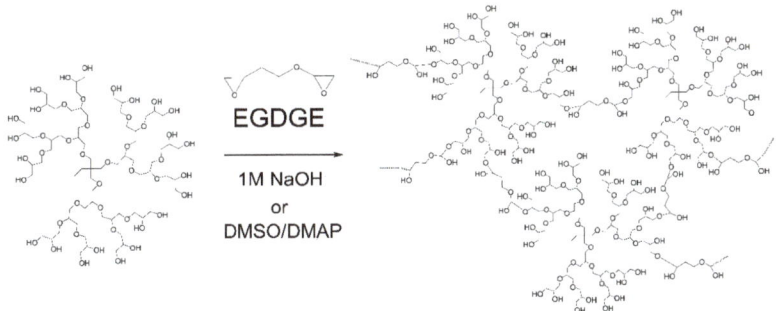

**Figure 2.** Synthetic scheme of PGD-G3 hydrogels using EGDGE as crosslinking agent (G3-EG hydrogels).

**Table 1.** Preparation conditions and swelling properties of PGD-G3 hydrogels.

| Sample Code | Conc. of EGDGE (mmol/mL) | Conc. of PGD-G3 (wt. %) | EGDGE/OH Groups | Swelling Ratio (q) | |
|---|---|---|---|---|---|
| | | | | In Water | In Ethanol |
| G3-EG(NaOH)0.15 | 0.682 | 32 | 0.15 | 5.39 | 1.16 |
| G3-EG(NaOH)0.5 | 1.261 | 18 | 0.5 | 9.12 | – |
| G3-EG(NaOH)0.75 | 1.215 | 11 | 0.75 | 36.20 | 4.28 |
| G3-EG(DMSO)0.75 | 3.035 | 30 | 0.75 | 1.94 | – |
| G3-EG(NaOH)1.0 | 1.621 | 11 | 1.0 | 11.39 | 3.83 |

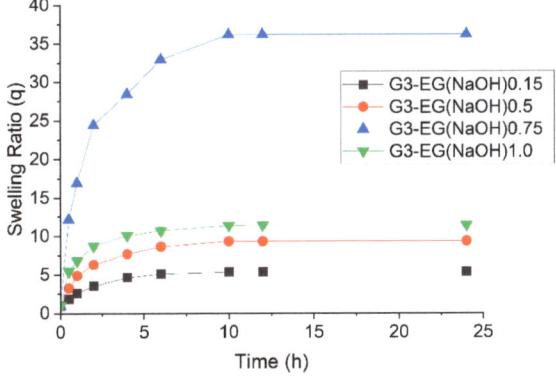

**Figure 3.** Swelling behavior of G3-EG hydrogels in water at 25 °C: black: G3-EG(NaOH)0.15, red: G3-RG(NaOH)0.15, blue: G3-EG(NaOH)0.75, green: G3-EG(NaOH)1.0. Detailed preparation conditions of each hydrogel are summarized in Table 1 (mean ± S.E.M., n = 3).

In order to compare the influence of the generation of PGD, PGD-G4 hydrogels were prepared in a similar manner in NaOH (Table 2). The swelling ratio for each hydrogel was slightly smaller than that for each G3-EG hydrogel, even with similar concentrations of EGDGE and PGDs. This may be due to the denser network of hydroxyl groups that can act as a more hyperbranched structure on the nanoscale than those in PGD-G3. However, the tendency of swelling for PGD-G3 and PGD-G4 was the same.

Table 2. Preparation conditions and swelling properties of PGD-G4 hydrogels.

| Sample Code | Conc. of EGDGE (mmol/mL) | Conc. of PGD-G4 (wt. %) | EGDGE/OH Groups | Swelling Ratio (q) | |
|---|---|---|---|---|---|
| | | | | In Water | In Ethanol |
| G4-EG(NaOH)0.15 | 0.658 | 32 | 0.15 | 4.68 | 1.36 |
| G4-EG(NaOH)0.5 | 1.217 | 18 | 0.5 | 7.41 | 2.05 |
| G4-EG(NaOH)0.75 | 1.171 | 11 | 0.75 | 37.44 | 2.45 |
| G4-EG(NaOH)1.0 | 1.564 | 11 | 1.0 | 9.95 | 2.14 |

Figure 4 shows the PTX amount loaded into hydrogels as a function of the swelling ratio in water. The PTX amount is proportional to the swelling ratio (except for G3-EG(NaOH)0.75 and G4-EG(NaOH)0.75 in Tables 1 and 2), indicating that the dissolved PTX was entrapped in the spaces between the crosslinks in the swollen hydrogel. In other words, heterogeneous crosslinking of PGD made a high-density region of PGD molecules in the matrix, and this region is likely to act as a hydrotrope. To confirm the PTX solubility, a small amount of water was added to the G4-EG(NaOH)1.0 hydrogel after loading PTX to reach maximum swelling, and the gel was left to stand for one day. The hydrogel remained transparent although its shape collapsed (Figure 5), indicating that PTX was dispersed in the hydrogels without any crystallization in water.

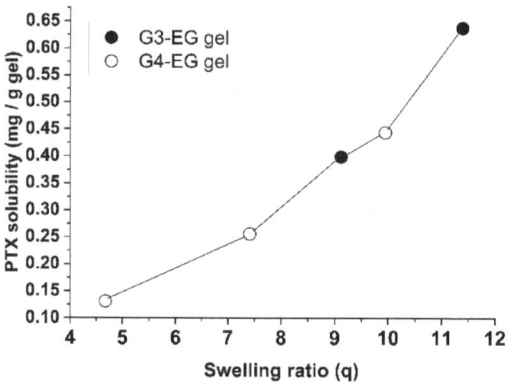

Figure 4. PTX solubility, determined by HPLC, per 1 g dried hydrogels (G3-EG(NaOH)0.5, G3-EG(NaOH)1.0, G4-EG(NaOH)0.15, G4-EG(NaOH)0.5, and G4-EG(NaOH)1.0) as a function of swelling ratio in water.

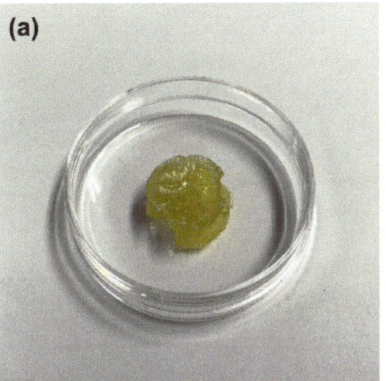 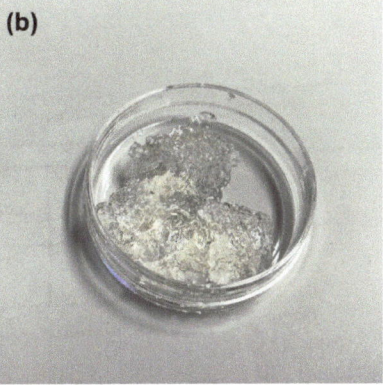

**Figure 5.** Photographs of (**a**) PTX-loaded dried G4-EG(NaOH)1.0 hydrogel and (**b**) swollen PTX-loaded G4-EG(NaOH)1.0 hydrogel. A small amount of water was added to G4-EG(NaOH)1.0 hydrogel after loading PTX, and the hydrogel was left to stand for one day.

In order to check the degree of PTX dispersion, FTIR imaging was performed using the PTX-loaded dried G4-EG(NaOH)1.0 hydrogel. Typical FTIR spectra of PTX-rich and PTX-poor regions are shown in Figure 6a. Bands due to hydroxyl groups were observed from 3000 to 3500 cm$^{-1}$ in both cases with the same absorbance (data not shown), which is consistent with a previous report on polyglycerol-based hydrogels [39]. However, the absorbance at 1700 to 1730 cm$^{-1}$ increased in the PTX-rich region (see red circle in Figure 6a), suggesting the presence of carbonyl groups of PTX in the PTX-rich region. Based on these results, the absorbance values around 3000–3500 cm$^{-1}$ and 1700–1730 cm$^{-1}$ were adopted for 2D imaging of the PGD matrix and PTX distribution for 0.1 mm-thick gels (Figure 6b). As shown in Figure 6c, hydroxyl groups of PGD were not homogeneously distributed in the measured area, suggesting heterogeneous crosslinking by EGDGE. Interestingly, carbonyl groups of PTX were likely to be located in the PGD-rich region (Figure 6d). These results indicated that PTX was dispersed in the gel matrix, and it was assumed that PGD and PTX molecules interacted in the hydrogel and remained in a dissolved state.

Finally, PTX release from the PTX-loaded G4-EG(NaOH)1.0 hydrogel was evaluated in vitro. Approximately 60% of PTX was released in 90–150 min (Figure 7). Since PTX normally takes 1–2 days to be released when distributed and retained in the hydrophobic domain of micelles such as PEG-PLA block copolymers [40], these results suggest that the rapid release was governed by hydrotropic dissolution based on intermolecular interactions between PTX and the crosslinked PGD molecules. Under the release experimental conditions, PTX precipitation was not observed, and the hydrogel remained transparent even after 700 min. From this result, it is suggested that the remained PTX in the hydrogel (approximately 40%) was still solubilized in the hydrogel and entrapped in the G4-crosslinked matrix. The calculated amount of the released PTX at 150 min was 5.5 µg/mL, the concentration of which is reported to decrease cell viability to less than 40% using PTX-loaded nanoparticles against MCF-7 cells [41]. From these results and reports, the release of PTX can be achieved at the therapeutic level in vitro. We think that the PTX-loaded G4-EG or G3-EG hydrogels can be fabricated to nanogels by further chemical modification of OH-groups in combination with "click" chemistry [42] or the miniemulsion technique [43]. Such nanogels could be applicable for effective oral chemotherapy [41].

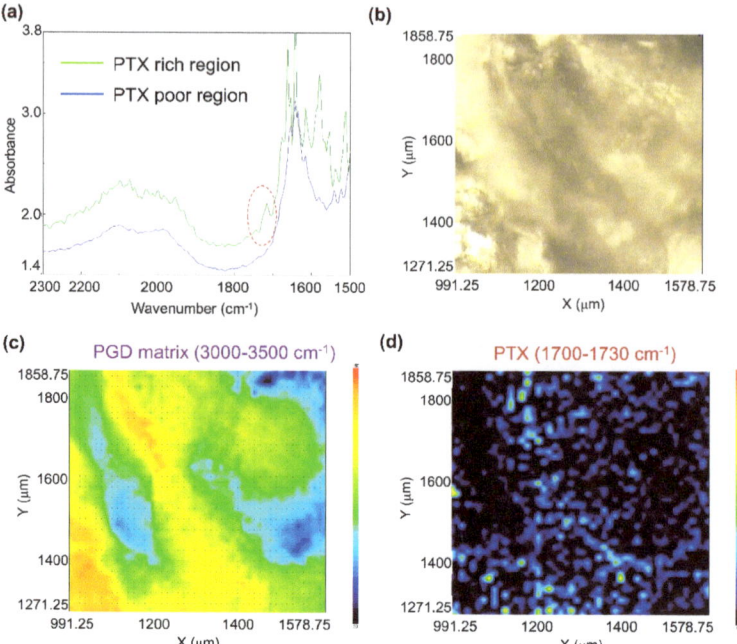

**Figure 6.** (**a**) Typical FTIR spectra of PTX-loaded dried G4-EG(NaOH)1.0 hydrogel (green line: PTX-rich region; blue line: PTX-poor region), (**b**) micrograph of PTX-loaded dried G4-EG(NaOH)1.0 hydrogel, (**c**) FTIR image of PTX-loaded dried G4-EG(NaOH)1.0 hydrogel using absorbance around 3000–3500 cm$^{-1}$ (hydroxyl groups of PGD matrix), (**d**) FTIR image of PTX-loaded dried G4-EG(NaOH)1.0 hydrogel using absorbance around 1700–1730 cm$^{-1}$ (carbonyl groups of PTX). Measurement area: 600 × 600 μm.

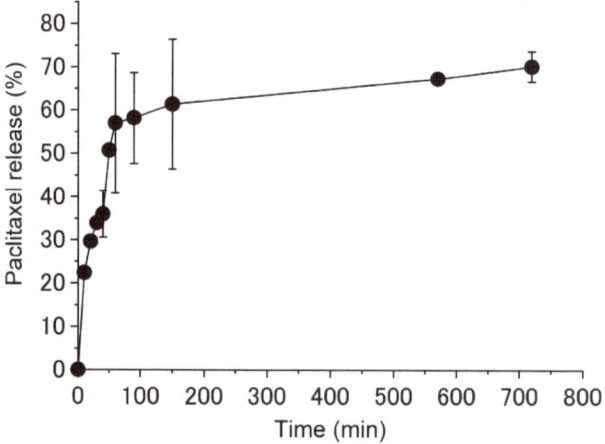

**Figure 7.** Cumulative release profiles for PTX loaded in G4-EG(NaOH)0.75 hydrogel. *N,N*-Diethylnicotinamide (1.5 M in PBS, pH 7.4) solution was used as the release medium to maintain an infinite sink condition. Samples were taken at predetermined time intervals and assayed for PTX by isocratic reverse-phase HPLC (mean ± S.E.M., n = 3).

## 3. Conclusions

Polyglycerol dendrimers crosslinked by reacting with EGDGE exhibited an increased local concentration in the gel matrix, and PTX was successfully loaded into the hydrogels. The amount of loaded PTX was proportional to the swelling ratio of the hydrogels, regardless of the generation of PGD (G3 and G4). Fourier-transform infrared spectroscopic imaging of the PTX-loaded PGD hydrogels proved that PTX was retained and distributed in the hydrogel matrix after loading based on hydrotropic solubilization. Sixty percent of PTX was released in a few hours without recrystallization in the sink state, which is expected to be applicable for rapid release of PTX, e.g., for oral administration.

## 4. Materials and Methods

### 4.1. Materials

1,1,1-Tris(hydroxymethyl)propane (TMP) and dimethylaminopyridine (DMAP) were purchased from FUJIFILM Wako Pure Chemical Corporation (Osaka, Japan). Allyl chloride, N-methylmorpholine N-oxide (NMO), 50 wt.% sodium hydroxide solution, and paclitaxel (PTX) were purchased from Sigma-Aldrich (Tokyo, Japan). Tetrabutylammonium bromide, $OsO_4$ (4% in water), and ethylene glycol diglycidylether (EGDGE) were purchased from Tokyo Chemical Industry Co., Ltd. (Tokyo, Japan). Generation 3 and 4 PGDs (PGD-G3: $M_n$ = 1689, m/z 1712 $[M + Na]^+$, calculated by MALDI-TOF-MS spectrometry (Voyager 2000, AB SCIEX); PGD-G4: $M_n$ = 3508, m/z 3491 $[M-H_2O]$) were prepared as described in previous papers [25,44].

### 4.2. Preparation of PGD Hydrogels Crosslinked by Ethylene Glycol Diglycidyl Ether (EGDGE)

PGD-G3 and G4 hydrogels using EGDGE as a crosslinking agent (G3-EG and G4-EG hydrogels) were prepared by the following methods.

#### 4.2.1. PGD-G3 Hydrogels Prepared in NaOH Aqueous Solution [G3-EG(NaOH)]

PGD-G3 hydrogels were prepared by crosslinking using EGDGE in 1 M NaOH aqueous solutions at 60 °C overnight in a sample bottle (molding size: 14 × 3.5 mm). The detailed conditions and swelling properties are summarized in Table 1 and Figure 3. The number following (NaOH) in the sample code refers to the ratio of EGDGE and hydroxyl groups in one PGD-G3 molecule.

The swelling ratio was calculated by the following equation:

$$\text{Swelling ratio (q)} = W_s / W_d$$

where $W_s$ is the weight of swollen hydrogel and $W_d$ is the weight of the dried hydrogel.

#### 4.2.2. PGD-G3 Hydrogels Prepared in DMSO [G3-EG(DMSO)]

PGD-G3 hydrogels were prepared by crosslinking using EGDGE in DMSO in the presence of dimethylaminopyridine (DMAP) as a catalyst at 60 °C overnight. The detailed conditions and swelling properties are summarized in Table 1. The number following (DMSO) in the sample code refers to the ratio of EGDGE and hydroxyl groups in one PGD-G3 molecule. The swelling ratio was calculated by the method described above.

#### 4.2.3. PGD-G4 Hydrogels Prepared in NaOH Aqueous Solution [G4-EG(NaOH)]

In a similar manner to G3-EG hydrogels, PGD-G4 hydrogels using EGDGE as a crosslinking agent (G4-EG hydrogels) were prepared. The detailed conditions and swelling properties are summarized in Table 2.

### 4.3. PTX Loading of G3-EG and G4 Hydrogels

Each dried hydrogel was weighed to obtain the initial weight of the gel. PTX was dissolved in EtOH (1 mg/mL), and the hydrogels were placed in an EtOH solution of PTX for 2 days to allow the gels to reach swelling equilibrium. After two days, the gels were withdrawn from the PTX solution and stored in an oven (37 °C) until dry. After drying, acetonitrile was added to the gel to extract trapped PTX. The extraction % of PTX was more than 98% (9.0 μg/mL), because acetonitrile is a good solvent for PTX [45]. The PTX concentration was then measured using high-performance liquid chromatography (HPLC) (Agilent 1100 series) using a Symmetry column (Waters, Milford, MA, USA) at 25 °C. The mobile phase consisted of acetonitrile–water (45:55, v/v) with a flow rate of 1.0 mL/min. A diode array detector was used with a detection wavelength of 227 nm. The PTX concentrations in the samples were obtained from a calibration curve.

### 4.4. FTIR Imaging

The dried G4-EG(NaOH)1.0 hydrogel was sliced into samples with a thickness of about 0.1 mm. The sliced sample was placed on the sample holder of a multichannel infrared microscope system (FTIR-6200 with IMV-4000, JASCO, Tokyo, Japan). FTIR imaging was performed under the following conditions:

Objective focusing mirror magnification: ×16

Number of measurement points: 48 × 48

Measurement region: 1 point 12.5 × 12.5 μm (48 × 12.5 = 600 × 600 μm)

Number of integrations: 16

Resolution: 8 cm$^{-1}$

### 4.5. PTX Release from G4-EG Hydrogels

The PTX-loaded G4-EG(NaOH)1.0 hydrogel was immersed in an aqueous solution. N,N-Diethylnicotinamide (1.5 M in PBS, pH 7.4) solution was used as the release medium to maintain an infinite sink without requiring simulated flow conditions [46]. Samples were taken at predetermined time intervals and assayed for PTX by isocratic reverse-phase HPLC (see Section 4.3).

**Author Contributions:** Conceptualization, T.O.; methodology, T.O. and J.L.; validation, T.O. and J.L.; formal analysis, T.O. and J.L.; investigation, T.O. and J.L.; data curation, T.O. and J.L.; writing—original draft preparation, T.O.; writing—review and editing, T.O.; visualization, J.L.; supervision, T.O. and J.L.; project administration, T.O. and J.L.; funding acquisition, T.O. All authors have read and agreed to the published version of the manuscript.

**Funding:** This study was financially supported by the JSPS KAKENHI (Grant No. JP22H04545) and Eno Scientific Foundation (under the auspices of The Ministry of Education, Culture, Sports and Technology).

**Data Availability Statement:** The dataset generated during the current study are not publicly available but are available from the corresponding author on reasonable request.

**Acknowledgments:** We thank Kinam Park (Purdue University) for valuable support of this research. We thank Moemi Matsuda (Kobe University) for help with PGD synthesis.

**Conflicts of Interest:** The authors declare no conflict of interest.

## References

1. Pouyan, P.; Cherri, M.; Haag, R. Polyglycerols as Multi-Functional Platforms: Synthesis and Biomedical Applications. *Polymers* **2022**, *14*, 2684. [CrossRef] [PubMed]
2. Daniel, W.; Stiriba, S.E.; Holger, F. Hyperbranched Polyglycerols: From the Controlled Synthesis of Biocompatible Polyether Polyols to Multipurpose Applications. *Acc. Chem. Res.* **2010**, *43*, 129–141. [CrossRef]
3. Bochenek, M.; Oleszko-Torbus, N.; Wałach, W.; Lipowska-Kur, D.; Dworak, A.; Utrata-Wesołek, A. Polyglycidol of Linear or Branched Architecture Immobilized on a Solid Support for Biomedical Applications. *Polym. Rev.* **2020**, *60*, 717–767. [CrossRef]
4. Calderón, M.; Quadir, M.A.; Sharma, S.K.; Haag, R. Dendritic Polyglycerols for Biomedical Applications. *Adv. Mater.* **2010**, *22*, 190–218. [CrossRef] [PubMed]

5. Türk, H.; Shukla, A.; Alves Rodrigues, P.C.; Rehage, H.; Haag, R. Water-Soluble Dendritic Core–Shell-Type Architectures Based on Polyglycerol for Solubilization of Hydrophobic Drugs. *Chem. A Eur. J.* **2007**, *13*, 4187–4196. [CrossRef] [PubMed]
6. Kim, T.H.L.; Yu, J.H.; Jun, H.; Yang, M.Y.; Yang, M.J.; Cho, J.W.; Kim, J.W.; Kim, J.S.; Nam, Y.S. Polyglycerolated Nanocarriers with Increased Ligand Multivalency for Enhanced in Vivo Therapeutic Efficacy of Paclitaxel. *Biomaterials* **2017**, *145*, 223–232. [CrossRef] [PubMed]
7. Zhou, L.; Gao, C.; Hu, X.; Xu, W. General Avenue to Multifunctional Aqueous Nanocrystals Stabilized by Hyperbranched Polyglycerol. *Chem. Mater.* **2011**, *23*, 1461–1470. [CrossRef]
8. Yeh, P.-Y.J.; Kainthan, R.K.; Zou, Y.; Chiao, M.; Kizhakkedathu, J.N. Self-Assembled Monothiol-Terminated Hyperbranched Polyglycerols on a Gold Surface: A Comparative Study on the Structure, Morphology, and Protein Adsorption Characteristics with Linear Poly(Ethylene Glycol)s. *Langmuir* **2008**, *24*, 4907–4916. [CrossRef]
9. Wang, X.; Gan, H.; Zhang, M.; Sun, T. Modulating Cell Behaviors on Chiral Polymer Brush Films with Different Hydrophobic Side Groups. *Langmuir* **2012**, *28*, 2791–2798. [CrossRef]
10. Wang, S.; Zhou, Y.; Yang, S.; Ding, B. Growing Hyperbranched Polyglycerols on Magnetic Nanoparticles to Resist Nonspecific Adsorption of Proteins. *Colloids Surf. B Biointerfaces* **2008**, *67*, 122–126. [CrossRef]
11. Wyszogrodzka, M.; Haag, R. Synthesis and Characterization of Glycerol Dendrons, Self-Assembled Monolayers on Gold: A Detailed Study of Their Protein Resistance. *Biomacromolecules* **2009**, *10*, 1043–1054. [CrossRef] [PubMed]
12. Yamazaki, M.; Sugimoto, Y.; Murakami, D.; Tanaka, M.; Ooya, T. Effect of Branching Degree of Dendritic Polyglycerols on Plasma Protein Adsorption: Relationship between Hydration States and Surface Morphology. *Langmuir* **2021**, *37*, 8534–8543. [CrossRef] [PubMed]
13. Oudshoorn, M.H.M.; Rissmann, R.; Bouwstra, J.A.; Hennink, W.E. Synthesis and Characterization of Hyperbranched Polyglycerol Hydrogels. *Biomaterials* **2006**, *27*, 5471–5479. [CrossRef]
14. Wu, C.; Strehmel, C.; Achazi, K.; Chiappisi, L.; Dernedde, J.; Lensen, M.C.; Gradzielski, M.; Ansorge-Schumacher, M.B.; Haag, R. Enzymatically Cross-Linked Hyperbranched Polyglycerol Hydrogels as Scaffolds for Living Cells. *Biomacromolecules* **2014**, *15*, 3881–3890. [CrossRef] [PubMed]
15. Postnova, I.; Silant'ev, V.; Kim, M.H.; Song, G.Y.; Kim, I.; Ha, C.S.; Shchipunov, Y. Hyperbranched Polyglycerol Hydrogels Prepared through Biomimetic Mineralization. *Colloids Surf. B Biointerfaces* **2013**, *103*, 31–37. [CrossRef] [PubMed]
16. Steinhilber, D.; Haag, R.; Sisson, A.L. Multivalent, Biodegradable Polyglycerol Hydrogels. *Int. Artif. Organs* **2011**, *34*, 118–122. [CrossRef] [PubMed]
17. Ying, H.; He, G.; Zhang, L.; Lei, Q.; Guo, Y.; Fang, W. Hyperbranched Polyglycerol/Poly(Acrylic Acid) Hydrogel for the Efficient Removal of Methyl Violet from Aqueous Solutions. *J. Appl. Polym. Sci.* **2016**, *133*, 1–11. [CrossRef]
18. Kainthan, R.K.; Janzen, J.; Kizhakkedathu, J.N.; Devine, D.V.; Brooks, D.E. Hydrophobically Derivatized Hyperbranched Polyglycerol as a Human Serum Albumin Substitute. *Biomaterials* **2008**, *29*, 1693–1704. [CrossRef]
19. Gao, S.; Guan, Q.; Chafeeva, I.; Brooks, D.E.; Nguan, C.Y.C.; Kizhakkedathu, J.N.; Du, C. Hyperbranched Polyglycerol as a Colloid in Cold Organ Preservation Solutions. *PLoS ONE* **2015**, *10*, e0116595. [CrossRef]
20. Ooya, T.; Lee, J.; Park, K. Effects of Ethylene Glycol-Based Graft, Star-Shaped, and Dendritic Polymers on Solubilization and Controlled Release of Paclitaxel. *J. Control Release* **2003**, *93*, 121–127. [CrossRef]
21. Ooya, T.; Ogawa, T.; Takeuchi, T. Temperature-Induced Recovery of a Bioactive Enzyme Using Polyglycerol Dendrimers: Correlation between Bound Water and Protein Interaction. *J. Biomater. Sci. Polym. Ed.* **2018**, *29*, 701–715. [CrossRef] [PubMed]
22. Kimura, M.; Ooya, T. Enhanced Solubilization of α-Tocopherol by Hyperbranched Polyglycerol-Modified β-Cyclodextin. *J. Drug Deliv. Sci. Technol.* **2016**, *35*, 30–33. [CrossRef]
23. Park, J.H.; Huh, K.M.; Lee, S.C.; Lee, W.K.; Ooya, T.; Park, K. Nanoparticulate Drug Delivery Systems Based on Hydrotropic Polymers, Dendrimers, and Polymer Complexes. In Proceedings of the 2005 NSTI Nanotechnology Conference and Trade Show—NSTI Nanotech 2005, Anaheim, CA, USA, 8–12 May 2005; pp. 124–127.
24. Ooya, T.; Huh, K.M.; Saitoh, M.; Tamiya, E.; Park, K. Self-Assembly of Cholesterol-Hydrotropic Dendrimer Conjugates into Micelle-like Structure: Preparation and Hydrotropic Solubilization of Paclitaxel. *Sci. Technol. Adv. Mater.* **2005**, *6*, 452–456. [CrossRef]
25. Ooya, T.; Lee, J.; Park, K. Hydrotropic Dendrimers of Generations 4 and 5: Synthesis, Characterization, and Hydrotropic Solubilization of Paclitaxel. *Bioconjug. Chem.* **2004**, *15*, 1221–1229. [CrossRef] [PubMed]
26. Ooya, T.; Lee, S.C.; Huh, K.M.; Park, K. Hydrotropic Nanocarriers for Poorly Soluble Drugs. In *Nanocarrier Technologies: Frontiers of Nanotherapy*; Springer Nature: Cham, Switzerland, 2006; Volume 9781402050, pp. 51–73. ISBN 9781402050411.
27. Appel, E.A.; Forster, R.A.; Rowland, M.J.; Scherman, O.A. The Control of Cargo Release from Physically Crosslinked Hydrogels by Crosslink Dynamics. *Biomaterials* **2014**, *35*, 9897–9903. [CrossRef]
28. Pramanik, B.; Ahmed, S. Peptide-Based Low Molecular Weight Photosensitive Supramolecular Gelators. *Gels* **2022**, *8*, 533. [CrossRef]
29. Pedron, S.; Pritchard, A.M.; Vincil, G.A.; Andrade, B.; Zimmerman, S.C.; Harley, B.A.C. Patterning Three-Dimensional Hydrogel Microenvironments Using Hyperbranched Polyglycerols for Independent Control of Mesh Size and Stiffness. *Biomacromolecules* **2017**, *18*, 1393–1400. [CrossRef]
30. Hong, J.; Shin, Y.; Kim, S.; Lee, J.; Cha, C. Complex Tuning of Physical Properties of Hyperbranched Polyglycerol-Based Bioink for Microfabrication of Cell-Laden Hydrogels. *Adv. Funct. Mater.* **2019**, *29*, 16–19. [CrossRef]

31. Kapourani, E.; Neumann, F.; Achazi, K.; Dernedde, J.; Haag, R. Droplet-Based Microfluidic Templating of Polyglycerol-Based Microgels for the Encapsulation of Cells: A Comparative Study. *Macromol. Biosci.* **2018**, *18*, 1800116. [CrossRef]
32. Randriantsilefisoa, R.; Hou, Y.; Pan, Y.; Camacho, J.L.C.; Kulka, M.W.; Zhang, J.; Haag, R. Interaction of Human Mesenchymal Stem Cells with Soft Nanocomposite Hydrogels Based on Polyethylene Glycol and Dendritic Polyglycerol. *Adv. Funct. Mater.* **2020**, *30*, 1905200. [CrossRef]
33. Park, H.; Choi, Y.; Jeena, M.T.; Ahn, E.; Choi, Y.; Kang, M.G.; Lee, C.G.; Kwon, T.H.; Rhee, H.W.; Ryu, J.H.; et al. Reduction-Triggered Self-Cross-Linked Hyperbranched Polyglycerol Nanogels for Intracellular Delivery of Drugs and Proteins. *Macromol. Biosci.* **2018**, *18*, 1700356. [CrossRef] [PubMed]
34. Guo, D.D.; Xu, C.X.; Quan, J.S.; Song, C.K.; Jin, H.; Kim, D.D.; Choi, Y.J.; Cho, M.H.; Cho, C.S. Synergistic Anti-Tumor Activity of Paclitaxel-Incorporated Conjugated Linoleic Acid-Coupled Poloxamer Thermosensitive Hydrogel in Vitro and in Vivo. *Biomaterials* **2009**, *30*, 4777–4785. [CrossRef] [PubMed]
35. Singla, A.K.; Garg, A.; Aggarwal, D. Paclitaxel and Its Formulations. *Int. J. Pharm.* **2002**, *235*, 179–192. [CrossRef]
36. Weaver, B.A. How Taxol/Paclitaxel Kills Cancer Cells. *Mol. Biol. Cell* **2014**, *25*, 2677–2681. [CrossRef]
37. Jensen, M.; Birch Hansen, P.; Murdan, S.; Frokjaer, S.; Florence, A.T. Loading into and Electro-Stimulated Release of Peptides and Proteins from Chondroitin 4-Sulphate Hydrogels. *Eur. J. Pharm. Sci.* **2002**, *15*, 139–148. [CrossRef]
38. Van Dijk-Wolthuis, W.N.E.; Franssen, O.; Talsma, H.; van Steenbergen, M.J.; Kettenes-van den Bosch, J.J.; Hennink, W.E. Synthesis, Characterization, and Polymerization of Glycidyl Methacrylate Derivatized Dextran. *Macromolecules* **1995**, *28*, 6317–6322. [CrossRef]
39. Salehpour, S.; Zuliani, C.J.; Dube, M.A. Synthesis of Novel Stimuli-Responsive Polyglycerol-Based Hydrogels. *Eur. J. Lipid Sci. Technol.* **2012**, *114*, 92–99. [CrossRef]
40. Huh, K.M.; Lee, S.C.; Cho, Y.W.; Lee, J.; Jeong, J.H.; Park, K. Hydrotropic Polymer Micelle System for Delivery of Paclitaxel. *J. Control Release* **2005**, *101*, 59–68. [CrossRef]
41. Le, Z.; Chen, Y.; Han, H.; Tian, H.; Zhao, P.; Yang, C.; He, Z.; Liu, L.; Leong, K.W.; Mao, H.-Q.; et al. Hydrogen-Bonded Tannic Acid-Based Anticancer Nanoparticle for Enhancement of Oral Chemotherapy. *ACS Appl. Mater. Interfaces* **2018**, *10*, 42186–42197. [CrossRef]
42. Chen, W.; Achazi, K.; Schade, B.; Haag, R. Charge-Conversional and Reduction-Sensitive Poly(Vinyl Alcohol) Nanogels for Enhanced Cell Uptake and Efficient Intracellular Doxorubicin Release. *J. Control Release* **2015**, *205*, 15–24. [CrossRef]
43. Steinhilber, D.; Sisson, A.L.; Mangoldt, D.; Welker, P.; Licha, K.; Haag, R. Synthesis, Reductive Cleavage, and Cellular Interaction Studies of Biodegradable, Polyglycerol Nanogels. *Adv. Funct. Mater.* **2010**, *20*, 4133–4138. [CrossRef]
44. Lee, H.; Ooya, T. Generation-Dependent Host-Guest Interactions: Solution States of Polyglycerol Dendrimers of Generations 3 and 4 Modulate the Localization of a Guest Molecule. *Chem. A Eur. J.* **2012**, *18*, 10624–10629. [CrossRef] [PubMed]
45. Lee, S.C.; Huh, K.M.; Lee, J.; Cho, Y.W.; Galinsky, R.E.; Park, K. Hydrotropic Polymeric Micelles for Enhanced Paclitaxel Solubility: In Vitro and In Vivo Characterization. *Biomacromolecules* **2007**, *8*, 202–208. [CrossRef] [PubMed]
46. Finkelstein, A.; McClean, D.; Kar, S.; Takizawa, K.; Varghese, K.; Baek, N.; Park, K.; Fishbein, M.C.; Makkar, R.; Litvack, F.; et al. Local Drug Delivery via a Coronary Stent with Programmable Release Pharmacokinetics. *Circulation* **2003**, *107*, 777–784. [CrossRef] [PubMed]

*Review*

# Novel Therapeutic Hybrid Systems Using Hydrogels and Nanotechnology: A Focus on Nanoemulgels for the Treatment of Skin Diseases

Kamil Sghier [1], Maja Mur [2], Francisco Veiga [3,4], Ana Cláudia Paiva-Santos [3,4,*] and Patrícia C. Pires [3,4,5,*]

1. Faculty of Pharmacy, Masaryk University, Palackého tř. 1946, Brno-Královo Pole, 612 00 Brno, Czech Republic
2. Faculty of Pharmacy, University of Ljubljana, Aškerčeva c. 7, 1000 Ljubljana, Slovenia
3. Faculty of Pharmacy, University of Coimbra, Azinhaga de Santa Comba, 3000-548 Coimbra, Portugal; fveiga@ci.uc.pt
4. REQUIMTE/LAQV, Group of Pharmaceutical Technology, Faculty of Pharmacy, University of Coimbra, 3000-548 Coimbra, Portugal
5. CICS-UBI—Health Sciences Research Centre, University of Beira Interior, 6201-001 Covilhã, Portugal
* Correspondence: acsantos@ff.uc.pt (A.C.P.-S.); patriciapires@ff.uc.pt (P.C.P.)

**Citation:** Sghier, K.; Mur, M.; Veiga, F.; Paiva-Santos, A.C.; Pires, P.C. Novel Therapeutic Hybrid Systems Using Hydrogels and Nanotechnology: A Focus on Nanoemulgels for the Treatment of Skin Diseases. *Gels* **2024**, *10*, 45. https://doi.org/10.3390/gels10010045

Academic Editors: Junfeng Shi and Esmaiel Jabbari

Received: 19 November 2023
Revised: 4 January 2024
Accepted: 4 January 2024
Published: 6 January 2024

**Copyright:** © 2024 by the authors. Licensee MDPI, Basel, Switzerland. This article is an open access article distributed under the terms and conditions of the Creative Commons Attribution (CC BY) license (https://creativecommons.org/licenses/by/4.0/).

**Abstract:** Topical and transdermal drug delivery are advantageous administration routes, especially when treating diseases and conditions with a skin etiology. Nevertheless, conventional dosage forms often lead to low therapeutic efficacy, safety issues, and patient noncompliance. To tackle these issues, novel topical and transdermal platforms involving nanotechnology have been developed. This review focuses on the latest advances regarding the development of nanoemulgels for skin application, encapsulating a wide variety of molecules, including already marketed drugs (miconazole, ketoconazole, fusidic acid, imiquimod, meloxicam), repurposed marketed drugs (atorvastatin, omeprazole, leflunomide), natural-derived compounds (eucalyptol, naringenin, thymoquinone, curcumin, chrysin, brucine, capsaicin), and other synthetic molecules (ebselen, tocotrienols, retinyl palmitate), for wound healing, skin and skin appendage infections, skin inflammatory diseases, skin cancer, neuropathy, or anti-aging purposes. Developed formulations revealed adequate droplet size, PDI, viscosity, spreadability, pH, stability, drug release, and drug permeation and/or retention capacity, having more advantageous characteristics than current marketed formulations. In vitro and/or in vivo studies established the safety and efficacy of the developed formulations, confirming their therapeutic potential, and making them promising platforms for the replacement of current therapies, or as possible adjuvant treatments, which might someday effectively reach the market to help fight highly incident skin or systemic diseases and conditions.

**Keywords:** anti-aging; nanoemulgels; nanoemulsions; neuropathy; skin cancer; skin infection; skin inflammation; topical administration; transdermal administration; wound healing

## 1. Introduction

### 1.1. The Skin: Properties and Advantages and Limitations as a Drug Delivery Route

The skin consists of three primary layers: the epidermis, the dermis, and the hypodermis [1–3]. The epidermis, the outermost layer, is divided into several sublayers, with the deepest layer being known as the *stratum basale* (also called the basal layer). This layer contains rapidly dividing basal cells that continually undergo mitosis to replace the cells lost from the skin's surface. As these basal cells divide and mature, they gradually move upwards toward the skin's surface [4,5]. Above the *stratum basale* lies the *stratum spinosum*, which provides strength and support to the epidermis. The cells in this layer have spiny projections that interlock with neighboring cells, enhancing tissue integrity [6,7]. Further up is the *stratum granulosum*, where the cells begin to produce large amounts of keratin and other proteins. As these cells mature, they form flattened, granular layers, preparing to become the outermost protective barrier of the skin [7,8]. Finally, we have the *stratum corneum*, the outermost layer of the epidermis. This layer is composed of tough, flattened, and fully keratinized cells known as corneocytes. These corneocytes are continuously shed and replaced by new cells from the lower layers. The *stratum corneum* acts as a formidable barrier, preventing the entry of pathogens and chemicals, and excessive water loss [9–11]. Beneath the epidermis lies the dermis, a thicker and more complex layer of the skin, primarily composed of connective tissue, which includes collagen and elastic fibers, providing structural support and elasticity to the skin. The dermis houses blood vessels, nerves, hair follicles, sebaceous glands, and sweat glands. It also contains sensory receptors and specialized cells like Merkel cells, responsible for detecting touch and pressure [12–14]. Lastly, the hypodermis, or subcutaneous tissue, forms the deepest layer of the skin. It is mainly composed of adipose tissue (fat) that acts as an insulator and cushion, regulating body temperature and providing protection to underlying organs and structures [15,16].

Several factors can affect the properties of the skin. Environmental factors such as ultraviolet (UV) radiation from the sun can cause skin damage, premature aging, and increase the risk of skin cancer. Pollution, chemicals, and harsh weather conditions can also impact the skin's health and appearance [17–19]. Additionally, lifestyle choices, such as diet, smoking, and alcohol consumption, can also influence the skin's elasticity and overall health. Hormonal changes, stress, and certain medical conditions may also lead to skin issues, which can consequently evolve into pathological conditions, such as acne, eczema, and psoriasis, among others [20–23]. Understanding the intricate organization of the skin's layers and their functions is essential for maintaining healthy skin. Proper care, protection from harmful environmental factors, and a balanced lifestyle can contribute to the overall health and well-being of this remarkable organ [24,25].

Furthermore, the fact that the skin, as the body's largest external organ, serves as a protective barrier against various external factors, makes drug delivery on or through the skin a significant challenge, since the skin exhibits a very low or even nonexistent permeability to most drug molecules [26–28]. Factors such as *stratum corneum* composition, hydration, anatomical site, and individual variations contribute to the overall high complexity of skin drug permeation, and the understanding of the skin's characteristics is essential for optimum drug delivery, in both topical and transdermal administration [24,29,30].

Transdermal and topical drug administration are both methods of delivering medications to the body through the skin. However, there are some key differences between the two [31,32]. Transdermal application refers to the delivery of medications through the skin, and into the bloodstream, allowing active ingredients to have systemic effects since they are meant to be distributed throughout the body. Hence, the formulation is designed to penetrate the skin's surface and reach the bloodstream [33,34]. It is commonly used for medications that require slow, continuous release into the bloodstream over an extended period of time, and often used for systemic conditions such as hormone replacement therapy, pain management, or nicotine replacement therapy, among other applications [35–37]. Drugs are usually delivered through patches, gels, or creams, specifically designed to facilitate absorption through the skin and controlled release into the bloodstream [38–41]. On the other hand, topical administration involves applying medications directly to the skin's surface to exert local effects on the area of application. In this case, the drug is meant to stay primarily at the site of application and will probably not be designed to penetrate deeply into the skin's layers and reach the bloodstream [42–44]. Hence, this administration route is typically used for localized conditions, such as skin infections, rashes, inflammation, and other skin-related issues [45–47]. Formulations are designed to target specific areas without affecting the entire body, and come in various forms, including creams, ointments, lotions, sprays, and foams, depending on the intended application [48–50].

Nevertheless, although topical and transdermal drug delivery have gained significant interest due to their numerous advantages, including noninvasiveness, easy administration, and possibility of a localized therapy, conventional pharmaceutical forms often lack an answer to the many challenges that drug delivery on or through the skin presents, such as drug permeation and/or retention, drug metabolism by the skin's enzymes, or effective solubilization of hydrophobic drugs [33,51–53]. Here, novel approaches using nanotechnology for drug encapsulation and delivery can be the answer.

*1.2. Nanotechnology for Efficient Skin Drug Delivery: A Special Focus on Nanoemulsions and Nanoemulgels*

Nanosystems are generally characterized as being structures in the nanosize range which are capable of encapsulating drug molecules for improved drug delivery. Several different types of nanosystems have been developed over the years, each with their own specific composition and characteristics (Figure 1) [54–57].

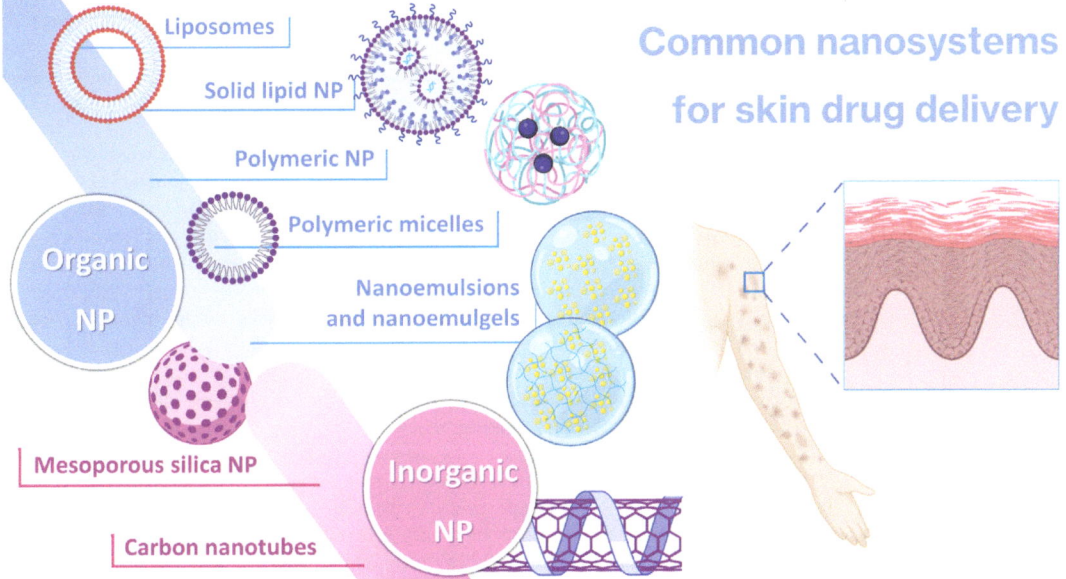

**Figure 1.** Schematic representation of common nanosystem categories for skin drug delivery (produced with Biorender).

Nanoparticles (NPs) are the most common general type of nanosystems, with the term typically referring to particles within the size range of 1 to 100 nm [58,59]. Their adjustable physicochemical properties, such as size, shape, and surface characteristics, contribute to tailored and improved physiological performance, when compared to conventional pharmaceutical forms, generally leading to more effective treatment and minimization of the side effects [60–63]. NPs also hold significant potential for targeted delivery of active ingredients to specific locations, in contrast to conventional formulations which usually require the administration of larger quantities of active ingredients in order to reach a therapeutically effective response, and often fail to target the desired area of interest, resulting in an extensive and potentially harmful penetration of the drugs into healthy tissues, hence causing systemic side effects [60,64,65]. Key considerations include these particles' size and structure, which is essential to optimize their performance and minimize potential harm, with surface modification by different types of substances, such as polymers and other molecules, being important in what concerns stability, specificity, and therapeutic effectiveness [66–68]. NPs can be divided into various subcategories, including organic and inorganic nanocarriers [69–71].

Inorganic NPs include mesoporous silica nanoparticles (MSNs), which are nanoscale structures composed of silica, with large surface areas and well-defined pores. Their ordered porous structure, with uniform pore sizes ranging from 2 to 50 nm, allows for efficient encapsulation and controlled release of active substances [72,73]. The large surface area of MSNs provides a space for active ingredient loading, leading to high drug-loading capacity and enhanced therapeutic efficacy, but despite these advantages, MSN synthesis requires precise control over particle size, pore site, and surface chemistry, which can be complex and time-consuming [74–76]. Carbon nanotubes (CNTs) are another type of inorganic nanocarrier, being cylindrical structures composed of carbon atoms arranged in hexagonal lattice, producing a large surface area that offers significant loading capacity for efficient encapsulation of therapeutic agents [77,78]. They possess unique advantages, such as high mechanical strength, exceptional electrical conductivity, thermal stability, protection of incorporated substances from degradation, controlled release capacity, and targeting ability [79–81]. These properties, combined with their nanoscale dimensions, make CNTs attractive candidates for therapeutic applications. Nevertheless, they have quite significant disadvantages, especially their poor water solubility and potential for toxicity [82–84]. Given the potential toxicity of inorganic nanoparticles, among other disadvantages, organic nanocarriers have been preferred for various applications [85,86].

There are many types of organic nanocarriers, all having in common high biocompatibility with the human body, an essential criterion for drug delivery in order to minimize toxicity and side effects [87,88]. Liposomes, the first to be developed, and also the first to effectively reach the pharmaceutical market, are spherical nanocarriers with single (small or large unilamellar vesicles) or multiple (multilamellar vesicles) bilayered membranes, formed of natural or synthetic lipids, which enclose an aqueous core [89–92]. They have showed a number of advantages compared to conventional systems, including enhanced drug delivery, drug protection from degradation, improved bioavailability, and even prolonged half-life in the blood circulation when functionalized with specific polymers (such as polyethylene glycol (PEG)) on their surface (stealth or PEGylated liposomes) [82,93,94]. On the other hand, solid lipid nanoparticles (SLNs) are formed by dispersing melted solid lipids in water, while emulsifiers are employed to stabilize the dispersion. These nanocarriers provide a lipophilic lipid matrix that facilitates the dispersion or dissolution of lipophilic drugs, providing controlled drug delivery, biocompatibility, high drug payload, and improved bioavailability of poorly water-soluble drugs [82,95,96]. Nevertheless, some important disadvantages of SLNs reside in their inability to encapsulate hydrophilic drug molecules, and stability issues have been reported, including burst release of the drugs during storage [97,98]. Another relevant nanocarrier subcategory is the polymeric micelles (PMs), self-assembled nanostructures formed by amphiphilic block copolymers in aqueous solutions. These nanosystems consist of a hydrophobic core, inside which hydrophobic drugs can be encapsulated, and a hydrophilic shell, stabilizing the micelle, which is made possible by the unique structure of block copolymers. These structures can also be surface-functionalized with targeting ligands to achieve specific delivery to target cells or tissues [99–101]. On the other hand, also composed of polymers, polymeric nanoparticles (PNPs) are nanocarriers that are fabricated using biocompatible polymers, typically poly (lactic-co-glycolic acid) (PLGA), PEG, chitosan, or polycaprolactone (PCL), amongst others. PNPs have the ability to encapsulate and protect various types of drugs including small molecules, proteins, peptides, and nucleic acids. The encapsulation also improves their stability, and controls their release kinetics, which allows for sustained or targeted drug delivery [102–104]. However, both PMs and PNPs have been reported to exhibit toxicity, slow clearance, formulation instability issues due to aggregation, and low drug-loading capacity [102,105,106].

Additionally, all mentioned nanocarrier types have complex preparation methods, which are time-consuming, costly, and often lead to scale-up issues, while also many times not being environment-friendly due to the use of organic solvents and the need for high energy amounts during production [107–109].

Given the overall stability, toxicity, low drug-loading, and problematic preparation issues that are seen with other nanocarrier types, nanoemulsions and nanoemulgels have arisen as advantageous solutions with high potential. Nanoemulsions are kinetically stable biphasic dispersions of two immiscible phases, an oil phase and a water phase, which are combined with surfactants and/or cosurfactants to increase their stability. Two main types of nanoemulsions exist, water-in-oil (W/O) or oil-in-water (O/W), with the latter being the most common [110–112]. Usually, O/W nanoemulsions contain from 5% to 20% oils/lipids, with this amount increasing up to 70% when the nanoemulsion is W/O. The type of the oil used in the formulation of nanoemulsions depends on the active substances that are intended to be solubilized, with the selection often being performed based on a solubility criterion (oils that solubilize the active substance the best) [111,113]. On the other hand, surfactants are amphiphilic molecules, which are used to stabilize the nanoemulsions, to reduce the interfacial tension and prevent the aggregation of the internal phase droplets. They usually have an ability to absorb quickly at the oil–water interface, and provide steric, electrostatic, or dual electro-steric stabilization. Cosurfactants may be used as surfactant complements to strengthen the interfacial film, if they fit suitably in areas which are structurally weaker [114–116]. There is an ongoing effort in both industrial preparations and scientific works to ensure that the components which are used in the development of nanoemulsions be strictly nontoxic, generally-regarded-as-safe (GRAS) excipients, in order to ensure maximum biocompatibility and decrease the propensity for side effects potentiated by the formulation [117,118]. Nanoemulsions can be produced by high-energy, low-energy, or low- and high-energy combination methods [117,119]. High-energy emulsification methods depend on mechanical devices, which use energy for creating powerful disruptive forces to reduce the size of the formed droplets. Ultrasonicators, microfluidizers, and high-pressure homogenizers can be used, including on an industrial level. These methods have the advantage of being able to nanoemulsify almost any oil, but the dependence on instrumental techniques is associated with high costs and the generation of high temperatures, which might not be feasible for all formulation components (for example, thermolabile drugs) [120,121]. Hence, low-energy methods, such as phase inversion or spontaneous emulsification, are the most advantageous, since it is the energy stored in the system that is used to produce the ultrafine droplets, leading to lower production costs and an easy application [122,123]. One of the most used low-energy methods is the phase inversion temperature method, to take advantage of changes in the aqueous/oil solubility of surfactants in response to temperature fluctuation. This will include the conversion of a W/O to an O/W nanoemulsion, or the reverse, via an intermediary bicontinuous phase. The change of temperature from low to high causes the opening and reversal of interfacial structure, which leads to phase inversion. However, this method has a substantial disadvantage, which is that it cannot be used for thermosensitive drugs [124–126]. Hence, the most beneficial low-energy method ends up being spontaneous emulsification, in which the components are usually just mixed with each another, resorting to manual mixing or low-energy mixers (such as mechanical stirring or vortex stirring). This, of course, only happens when the right components are mixed in the right proportions, with the oils, surfactants and cosurfactants having to be miscible with one another, and with reasonably high amounts of surfactants/cosurfactants being valuable for maximum formulation stabilization, homogeneity, and low droplet size [127–129]. Regardless of the used preparation method, nanoemulsions' droplet size usually varies between 20 and 500 nm, hence being small and responsible for these formulations' clear or hazy appearance [117,130]. This type of nanosystem can be used in many different dosage forms, such as creams, sprays, gels, aerosols, or foams, and various administration routes, such as intravenous, oral, intranasal, pulmonary, ocular, topical, or transdermal drug delivery [111,131,132]. They have several relevant advantages, such as a high solubilization capacity, great long-term physical stability, which reduces the propensity for conventional destabilization mechanisms to occur (such as creaming, coalescence, or Ostwald ripening), and a very large surface area available for drug absorption to occur [133–135]. Nevertheless, for certain applications, nanoemulsions might not be the ideal dosage form, since they tend to be fluid and do not usually have bioadhesion capability, which can lead to a short retention time of the preparation at the site of administration, an important parameter in transdermal and topical administration. In order to tackle these issues, nanoemulsions can be transformed into nanoemulgels [136,137].

Nanoemulgels are hybrid colloidal systems composed of nanosized oil droplets dispersed in an aqueous gel matrix, combining the properties of both nanoemulsions and hydrogels, which solve important issues such as spreadability and skin retention of the preparations by increasing their viscosity [136,138]. The methodology for preparing a nanoemulgel entails the creation of a gel-based formulation using common and already described emulsification techniques, similar to those that are used in the preparation of nanoemulsions [139,140]. As topical or transdermal administration systems, they function as drug reservoirs, facilitating the controlled release of the drug from the inner phase to the outer phase, and subsequently onto the skin. Aside from controlled drug delivery, nanoemulgels exhibit several advantages over alternative topical formulations, which include compatibility with the skin, high viscosity, adhesiveness, good spreadability, and long-lasting therapeutic effects [139,140]. In what concerns common excipients that are used as part of a nanoemulgel's composition, oil selection is usually dependent on the intended hydrophobicity, viscosity, permeability, and stability in the formulated nanoemulsion, but oils from natural sources, such as oleic acid, are commonly used, or chemically modified oils with medium chain mono-, di-, or tri-glycerides, such as Capryol® 90, Miglyol® 812, or Labrafac™ [141–143]. Additionally, there are several categories of surfactants, and among the most used are cationic surfactants (such as amines and quaternary ammonium compounds, or lecithin), anionic surfactants (such as sodium bis-2-ethylhexylsulfosuccinate, or sodium dodecyl sulfate), zwitterionic surfactants (such as phospholipids), and non-ionic surfactants (such as Tweens, Lauroglycol® 90, Cremophor® EL, or Cremophor® RH 40) [137,144,145]. Frequently used cosurfactants include propylene glycol, PEG 400, ethyl alcohol, or Transcutol®, with alcohol-based cosurfactants hence being the most preferred due to their capability to split between oil and water phases, thereby improving their miscibility [139,146,147]. Lastly, there are different types of gelling agents used in nanoemulgel formation, including natural gelling agents such as bio-polysaccharides (such as pectin, carrageenan, alginic acid, locust bean gum, or gelatine), derivates of bio-polysaccharides (such as xanthan gum, starch, dextran, or acacia gum), as well as semisynthetic (such as hydroxypropyl cellulose, ethyl cellulose, and sodium alginate) and synthetic polymers (such as carbomers or poloxamers) [136,137,148].

When developing a novel nanoemulgel, several properties have to be evaluated, in order to assess whether the preparation has optimum characteristics for the intended application, including droplet size, polydispersity index (PDI), zeta potential, pH, rheological properties, stability, bioadhesion, spreadability, in vitro drug release, ex vivo drug permeation, toxicity potential, and in vitro and/or in vivo assessment of the therapeutical potential of the developed formulation for the intended purpose. This review focuses on the latest advances regarding the development of novel nanoemulgels for transdermal or topical administration, for the treatment of several highly impactful diseases and conditions (Figure 2), such as skin wound healing, skin and appendage infections, skin inflammatory diseases, skin cancer, neuropathy, and skin aging. A critical analysis is performed regarding formulation composition and preparation, and all relevant characterization parameters, in order to assess the true potential of these formulations as novel functional platforms for drug delivery onto or through the skin (summary in Table 1).

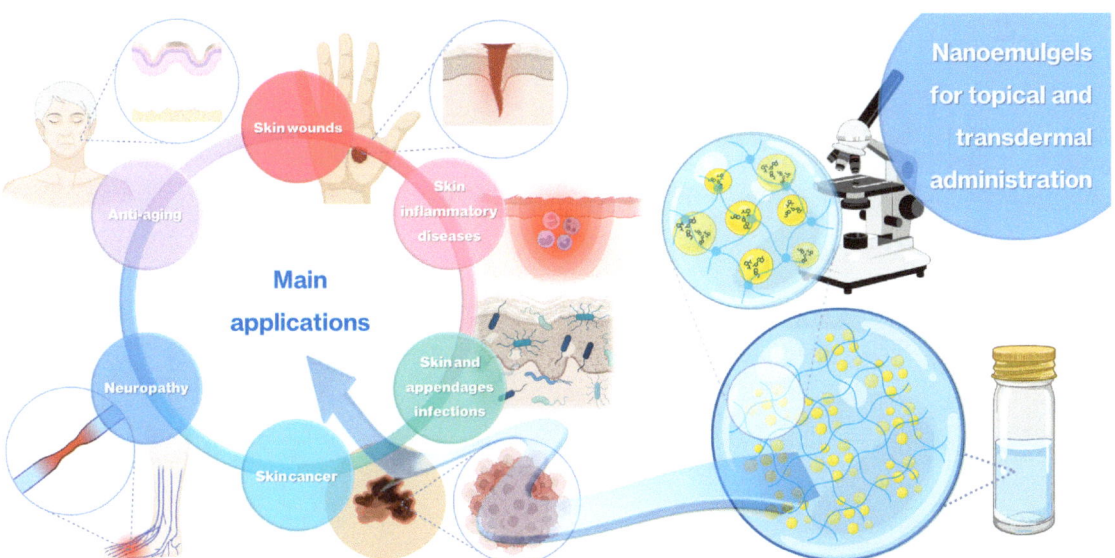

**Figure 2.** Schematic representation of nanoemulgel structure, and main applications of nanoemulgels in highly incident diseases, for topical and transdermal administration (produced with Biorender).

**Table 1.** Summary of the most relevant analyzed information of each included study, namely, disease intended to treat, encapsulated molecule(s), main formulation composition, droplet size, PDI, ZP, pH, main in vitro and/or in vivo therapeutic efficacy-related results, and corresponding reference.

| Disease Intended to Treat | Encapsulated Molecule(s) | Main Formulation Composition | Droplet Size (nm) | PDI | ZP (mV) | pH | Main In Vitro and/or In Vivo Therapeutic Efficacy-Related Results | Reference |
|---|---|---|---|---|---|---|---|---|
| Skin cancer | Chrysin | Capryol® 90, Tween® 80, Transcutol® HP, Pluronic® F127, water | <300 | 0.26 | −15 | NR | Strong antiproliferative effect in human and murine melanoma and human epidermoid carcinoma cell lines | [142] |
| Skin cancer and psoriasis | Leflunomide | Capryol® 90, Cremophor® EL, Transcutol® HP, Pluronic® F-127, water | 98.7 to 280.92 | 0.2 to 0.3 | −7.8 | NR | Significant antimelanoma activity by inducing apoptosis and inhibiting tumor cell proliferation in melanoma cells; potent antipsoriatic activity by inhibiting human keratinocyte proliferation and reducing proinflammatory cytokine levels | [146] |
| Skin wound healing | Atorvastatin | Liquid paraffin, Tween® 80, propylene glycol, carboxymethyl cellulose, water | 100 to 200 | <0.300 | −20 to −30 | 7.6 to 7.8 | Positive effect on wound healing, showing reduced inflammation, increased angiogenesis, and marked improvement in the skin's histological architecture, after topical application on rat skin for 21 days | [149] |
| Skin wound healing | Eucalyptus oil | Black seed oil, Tween® 80, Span® 60, propylene glycol, Carbopol® 940, water | 139 ± 5.8 | <0.450 | −28.05 | 5 to 6 | Significant improvement in wound healing in a rabbit model, with almost complete wound contraction after a 15-day period | [150] |
| Skin wound healing | Tocotrienols and naringenin | Capryol® 90, Solutol® HS15, Transcutol® P, Carbopol® 934 or Carbopol® 940, water | 145.6 ± 12.5 | 0.452 ± 0.03 | −21.1 ± 3.32 | 4.9 to 5.3 | NR | [151] |
| Skin wound healing | Thymoquinone | Black seed oil, Kolliphor® EL, Transcutol® HP, Carbopol® 940, water | 40.02 to 99.66 | 0.052 to 0.542 | −26.7 to −30.6 | 5.53 ± 0.04 | Accelerated wound closure in an in vivo rat wound model, evidenced by reduced wound size, enhanced re-epithelization, and increased collagen deposition | [152] |

Table 1. Cont.

| Disease Intended to Treat | Encapsulated Molecule(s) | Main Formulation Composition | Droplet Size (nm) | PDI | ZP (mV) | pH | Main In Vitro and/or In Vivo Therapeutic Efficacy-Related Results | Reference |
|---|---|---|---|---|---|---|---|---|
| Skin wound healing | Curcumin | Labrafac™ PG, Tween® 80, PEG 400, Carbopol® 940, water | 49.61 to 84.23 | 0.10 to 0.23 | −15.96 ± 0.55 to −20.26 ± 0.65 | 5.53 ± 0.03 | Significant wound-healing activity in Wistar rats, with almost complete wound healing after 20 days, with reduced inflammatory cells and extensive collagen fiber production | [153] |
| Skin infections | Miconazole nitrate | Olive oil, almond oil, Tween® 80, Span® 80, Carbopol® 940, water | 170 | 0.193 | <−30 | NR | Significant antifungal activity against selected fungal strains (*Candida albicans*) | [154] |
| Skin infections | Omeprazole | Olive oil, Span® 80, Tween® 80, chitosan, water | 369.7 ± 8.77 | 0.316 | −15.3 ± 6.7 | 6.21 ± 0.21 | Substantial antibacterial effects against both Gram-negative bacteria (*Escherichia coli*, *Klebsiella pneumoniae*, *Pseudomonas aeruginosa*) and Gram-positive bacteria (*Staphylococcus aureus*) | [155] |
| Skin infections | Ebselen | Captex® 300 EP/NF, Kolliphor® ELP, dimethylacetamide, Soluplus®, Aquaphor, water | 54.82 ± 1.26 | NR | −1.69 | NR | Potent antifungal activity against multi-drug-resistant *Candida albicans* and *Candida tropicalis* | [156] |
| Skin appendage infections (nails) | Ketoconazole | Labrafac™ Lipophile WL1349, Polysorbate 80, PEG 400, Carbopol® Ultrez 21, glycerin, methylparaben, thioglycolic acid, aminomethyl propanol, water | 77.52 ± 0.92 | 0.128 ± 0.035 | −5.44 ± 0.67 | 6.4 ± 0.24 | Significant antifungal activity against clinical isolates of dermatophytes, namely, *Trichophyton rubrum* and *Candida albicans* | [157] |

Table 1. Cont.

| Disease Intended to Treat | Encapsulated Molecule(s) | Main Formulation Composition | Droplet Size (nm) | PDI | ZP (mV) | pH | Main In Vitro and/or In Vivo Therapeutic Efficacy-Related Results | Reference |
|---|---|---|---|---|---|---|---|---|
| Skin infections | Fusidic acid | Myrrh oil, Tween® 80, Transcutol® P, CMC, water | 116 to 226 | NR | NR | 6.61 ± 0.23 | Substantial antibacterial activity, with significant inhibition zones against Staphylococcus Aureus, Bacillus subtilis, Enterococcus faecalis, Candida albicans, Shigella, and Escherichia coli | [158] |
| Skin cancer | Imiquimod and curcumin | Oleic acid, Tween® 20, Transcutol® HP, Carbopol® 934, water | 78.39 | 0.254 | −18.7 | 5.5 | Did not lead to the appearance of psoriasis-like symptoms after topical application to mice | [159] |
| Skin inflammatory diseases | Meloxicam | Eucalyptus oil, Tween® 80, PEG 400, Transcutol® P, distilled water | 139 ± 2.31 to 257 ± 3.61 | NR | NR | 6.58 ± 0.21 | Confirmed anti-inflammatory effects with reduced inflammation percentage in in vivo study | [160] |
| Skin inflammatory diseases | Brucine | Myrrh oil, Tween® 80, PEG 400, ethyl alcohol, carboxymethyl cellulose, water | 151 ± 12 | 0.243 | NR | 6.2 ± 0.2 | Anti-inflammatory effects (significant decrease in inflammation) and antinociceptive effects (reduction in writhing movements) in an animal model, after topical application | [161] |
| Diabetic neuropathy | Capsaicin | Eucalyptus oil, Tween® 80, propylene glycol, ethanol, isopropyl alcohol, Carbopol® 940, water | 28.15 ± 0.24 | 0.27 ± 0.05 | NR | NR | Significant in vivo efficacy in alleviating mechanical allodynia | [162] |
| Skin aging | Retinyl palmitate | Capryol® 90, Captex® 355, Kolliphor® EL, Transcutol® HP, Carbopol® 940, glycerin, water | 16.71 | 0.015 | −20.6 | 5.53 ± 0.06 | NR | [163] |

CMC—carboxymethyl cellulose; NR—not reported; PDI—polydispersity index; PEG—polyethylene glycol; ZP—zeta potential.

## 2. Topical and Transdermal Nanoemulgels for the Treatment of Skin Diseases and Other Applications

### 2.1. Skin Wound Healing

Skin wounds are physical injuries of the skin tissue, which lead it to break and open. Wound healing is a complex process, which requires a coordinated series of cellular and molecular events, which aim to restore the integrity and functionality of damaged tissues [164–166]. It includes inflammation, cell migration, tissue formation, and remodeling, and due to lack of effectiveness or slow therapeutic action, new strategies are needed for efficient and fast wound healing [167–169]. In this context, Morsy et al. [149] developed a novel nanoemulgel formulation containing atorvastatin for wound healing application. Despite the fact that atorvastatin (ATR) is primarily prescribed as a lipid-lowering medication, to manage cholesterol levels, high cholesterol levels have been associated with impaired wound healing, which means that atorvastatin may indirectly contribute to improved wound healing outcomes [170–172]. Furthermore, this drug also shows anti-inflammatory properties, and angiogenic effects, which promote the formation of new blood vessels, leading to an adequate blood supply that is crucial for tissue regeneration [173–175]. Additionally, it has also been described as targeting and inhibiting the growth of microorganisms, including common wound pathogens, and has been linked to other pleiotropic effects, such as antioxidant properties, modulation of cellular signaling pathways, and promotion of cell proliferation and migration, which make this drug a quite relevant candidate due to multiple wound-healing-related beneficial effects [173,176,177]. The incorporation of ATR into a nanoemulgel formulation intended to allow a controlled and sustained release at the wound site, maximizes its potential therapeutic effects. Hence, the developed nanoemulgel was prepared by using a combination of high-pressure homogenization and ultrasonication techniques. First, the gel was prepared by adding sodium carboxymethyl cellulose (CMC) to water and stirring continuously until gel formation. After that, a primary O/W emulsion was made, containing ATR solubilized in a mixture of liquid paraffin, Tween® 80, and propylene glycol, to which water was slowly added, and vortexed. Then, to this drug-loaded emulsion, the polymeric gel base was added and mixed for 5 min. Afterward, this primary emulgel was sonicated, for 10 min, in order to reduce its droplet size and finally obtain the required nanoemulgel. The developed formulations were characterized for particle size, PDI, zeta potential, viscosity, spreadability, in vitro drug release, stability, ex vivo permeation, and in vivo wound healing properties. The physicochemical characterization showed a small particle size of approximately 100 to 200 nm (Figure 3A,B), a good homogeneity with a PDI value of less than 0.3, and the zeta potential was found to be within the range of $-20$ to $-30$ mV, indicating good stability and preventing particle aggregation. Additionally, the viscosity and spreadability of the nanoemulgel was determined to be in the range suitable for topical application, ensuring ease of spreading and adherence to the wound area. The in vitro drug release profile (Figure 3C), determined across semipermeable cellulose membranes for 6 h, demonstrated a sustained release of the drug from the developed nanoemulgel over time, with a reduced initial burst release when compared to a CMC gel, and with a higher overall release when compared to an emulgel. Moreover, stability studies, which it underwent for a duration of 6 months, under storage conditions of 60% relative humidity and a temperature of 4 °C, indicated no noticeable alterations in several evaluated properties, such as color, appearance, spreading, or viscosity. In addition, an ex vivo permeation study (Figure 3D), through excised rat skin, revealed that the ATR nanoemulgel had a higher permeation, both in what concerns cumulative amount and velocity, than the drug-loaded emulgel, gel, or solution, after 2 h, also exhibiting the shortest lag time. Furthermore, a histopathological analysis of rats' skin, after nanoemulgel topical application in an in vivo study, supported its positive effect on wound healing, showing reduced inflammation and increased angiogenesis, with a marked improvement in the skin histological architecture, and considerable healing after 21 days of ATR nanoemulgel treatment (Figure 3E–G). Additionally, although the developed gel-based formulations may encounter drawbacks such as limited residence time at the application site, the nanoemulgel formulation can address this concern by exhibiting enhanced retention in the affected area, and therefore prolonged retention, enabling a steady release of the drug, and facilitating extended contact with the skin surface.

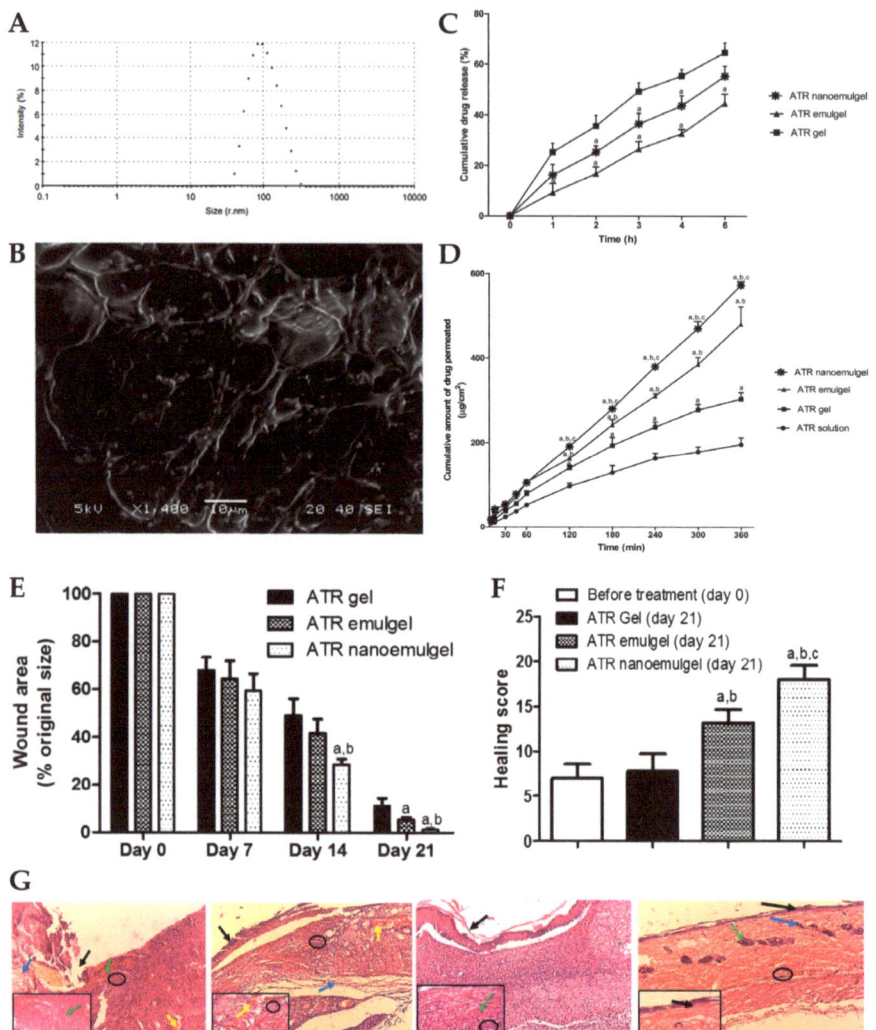

Figure 3. (A) Droplet size distribution of the developed ATR nanoemulgel; (B) surface morphology of the developed ATR nanoemulgel; (C) in vitro drug release profiles of the developed ATR nanoemulgel, compared to an emulgel and gel; (D) ex vivo drug permeation profiles of the developed ATR nanoemulgel, compared to an emulgel, a gel, and a solution; (E) wound area variation of rat skin after topical administration of the developed ATR formulations, after 0, 7, 14, and 21 treatment days; (F,G)—healing score (F) and photomicrographs (G) of rat skin before treatment (a), and after 21 days of topical administration of an ATR gel (b), an ATR emulgel (c), or an ATR nanoemulgel (d), where black arrows represent the absence of epidermal layer epithelization, blue arrows represent loss of collagen fibers normal arrangement in the dermal layer, green arrows represent severe congestion, yellow arrows represent hemorrhage, and black circles represent inflammatory cell infiltrations; ATR—atorvastatin; adapted from Morsy et al. [149], reproduced with permission from MDPI (Creative Commons CC BY 4.0 470 license).

Wounds are strongly connected to health disorders such as immune system diseases, diabetes, chronic peripheral vascular disorders, and various infectious and inflammatory diseases. In this context, chronic wounds pose a significant healthcare challenge due to their slow healing and susceptibility to infections [178,179]. Eucalyptol has been reported to function as a good penetrant in transdermal and topical drug delivery systems and is also claimed to possess antibacterial properties against human and food-borne pathogens [180,181]. Hence, Rehman et al. [150] developed a nanoemulgel for wound healing incorporating eucalyptus oil, obtained from *Eucalyptus globulus*, into a nanoemulgel (Figure 4A), developing an effective platform designed to enhance the stability, permeation, and controlled release of eucalyptol, one of the main constituents of eucalyptus oil, thereby promoting its therapeutic efficacy in wound healing. The preparation of the nanoemulgel was divided into two steps, with the first including the preparation of different primary O/W nanoemulsions by solvent emulsification diffusion method. The nanoemulsions were made of an aqueous phase containing the hydrophilic surfactant Tween® 80 and distilled water, and an oil phase containing black seed oil, the hydrophobic surfactant Span® 60, and the cosurfactant/cosolvent propylene glycol. These two phases were mixed together using a magnetic stirrer, and then the nanoemulsion was produced by droplet size reduction using a high-speed homogenizer. From the selected primary nanoemulsion, nanoemulgels were created, where Carbopol® 940 was used as the gelling agent. A Carbopol gel was produced by adding it to distilled water and mixing until a clear solution was formed, and then the pre-prepared nanoemulsion was added to the gel, in ratio of 1:1, with the pH being adjusted to a value of 5–6 using triethanolamine. These nanoemulgels were subsequently subjected to characterization. For the stability studies, temperature tests and centrifugation were used, with all formulations being subjected to storage at different conditions, namely, 8 °C, 25 °C, 40 °C, and 40 °C, with 40% relative humidity, for 28 days. Results showed that all the formulations were stable under the studied conditions, with no phase separation being observed after subsequent centrifugation. A Fourier-transform infrared spectrophotometer analysis was also employed to investigate the chemical interactions and compatibility between the components. By analyzing the produced spectra, it was possible to identify specific functional groups and molecular vibrations, confirming the absence of any major chemical changes or incompatibilities that could potentially affect the stability or therapeutic properties of the formulations. It was also shown that the pH had a major effect on the stability of the systems, as triethanolamine, used to adjust the formulation's pH to simulate the pH of the skin, affected the transparency and disrupted the internal structure of the formulations. Furthermore, organoleptic homogeneity tests were performed, where changes in color, phase separation, consistency, and liquefaction were observed. All the formulations were observed to be off-white in color, and smooth in terms of consistency, and showed also reasonable to good spreadability. Additionally, the drug content analysis showed that the drug was uniformly distributed throughout the nanoemulgels. In the study of in vitro drug release (Figure 4B), the nanoemulgel which released the highest amount of drug was selected. The selected nanoemulgel's particle size, PDI, and zeta potential were also determined. The droplet size was found to be around $139 \pm 5.8$ nm, the PDI was less than 0.45, and the zeta potential was measured to be $-28.05$ mV. After formulation physicochemical characterization, an in vivo study evaluated the wound-healing activity of the nanoemulgel in rabbits. The percentage of wound contraction was measured over a 15-day period, and they compared a negative control group, a nanoemulgel group, and a standard commercial product group. The results showed that the percentage of wound contraction for the nanoemulgel group on day 15 was 98.17%, indicating a significant improvement in wound healing compared to the negative control group (70.84%). Additionally, a statistical analysis using one-way ANOVA confirmed that the developed nanoemulgel had a wound-healing activity similar to that of the commercial cream, confirming its effectiveness in promoting wound healing. Thus, the performed comprehensive characterization studies, including physicochemical analysis and in vivo wound evaluations, provide valuable insights into the developed formulation's potential efficacy. Overall, based on the provided information, we can conclude that the topically applied nanoemulgel formulation containing eucalyptus oil as an active ingredient demonstrated significant wound-healing activity and stability in the tested conditions, hence being a potentially novel and effective strategy for skin wound healing.

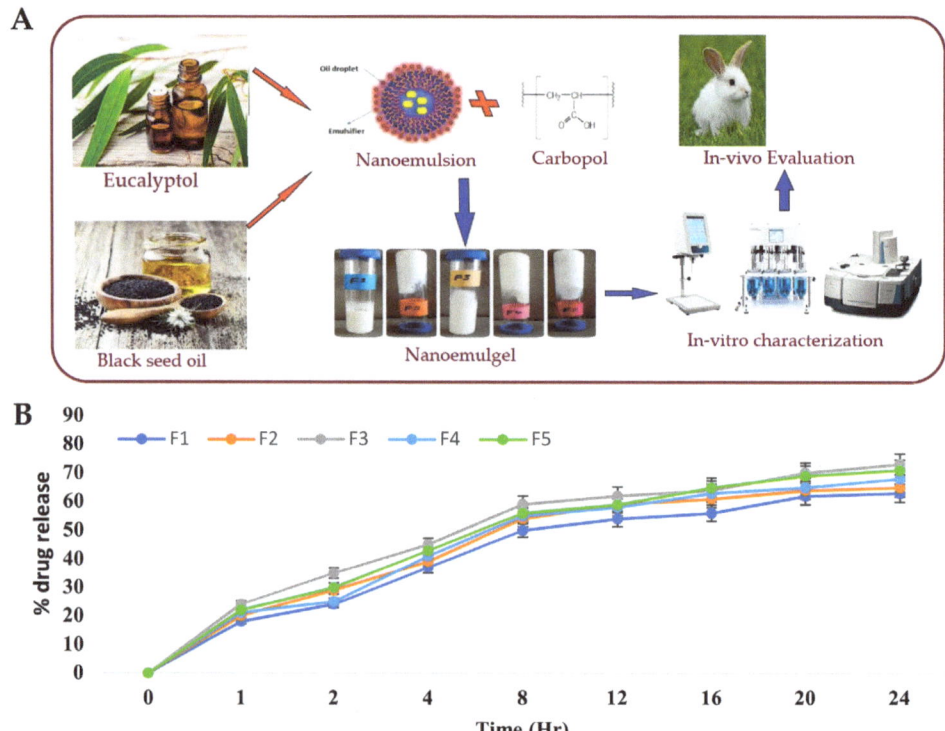

Figure 4. (A) Schematic representation of the developed eucalyptus oil nanoemulgel, including partial composition and general indication of performed studies; (B) in vitro eucalyptol release profiles of different nanoemulgels; adapted from Rehman et al. [150], reproduced with permission from MDPI (Creative Commons CC BY 4.0 470 license).

Diabetes mellitus remains a significant global health concern, affecting millions of individuals worldwide. One of the most debilitating complications of diabetes is the development of chronic, nonhealing wounds. These wounds pose substantial challenges for patients and healthcare providers, leading to increased morbidity, including the risk of limb amputation, and significant healthcare costs [182–184]. To address this critical issue, researchers are continuously exploring innovative approaches to improve wound healing and management. Hence, a study conducted by Yeo et al. [151] also focused on the fabrication and characterization of a topical nanoemulgel, containing tocotrienols and naringenin, for the management of diabetic wounds. In that study, researchers aimed to develop a nanoemulgel formulation taking advantage of the therapeutic potential of two bioactive compounds: tocotrienols, which are members of the vitamin E family with potent antioxidant and anti-inflammatory properties, and naringenin, a flavonoid known for its wound healing and tissue regeneration capabilities [185–188]. The study aimed to explore the synergistic effects of these compounds, with the goal of creating a multifunctional nanoemulgel that could accelerate wound closure, improve tissue repair, and alleviate the underlying inflammation often associated with diabetic wounds. With the aim of entrapping naringenin within the oil droplets of the o/w nanoemulgel, the oil phase was chosen according to the highest achievable solubilization for this compound. Among the tested oils, Capryol® 90 demonstrated the highest solubilization of naringenin, and hence was selected. Furthermore, the addition of tocotrienols to Capryol® 90 did not significantly impact the solubility of naringenin, suggesting successful and stable encapsulation of the drug with the chosen oil base. Additionally, in order to ensure the safety and biocompatibility of the formulation, the selection of appropriate surfactants is vital. Nonionic surfactants were prioritized due to their GRAS status and their ability to withstand pH or ionic strength changes. Solutol® HS15, with an HLB value of 15.2, was chosen as the primary surfactant. It offers several advantages, including low toxicity, good biocompatibility, and permeation-enhancement ability, making it a suitable choice for this formulation. To further stabilize the nanoemulsion and achieve a flexible interfacial film, Transcutol® P, a cosurfactant and cosolvent, was selected to work in combination with Solutol® HS15. Transcutol® P has been extensively studied for its skin-permeation-enhancing properties, without significantly affecting the diffusion of permeants across the skin. Moreover, it has the capability of creating an intracutaneous depot, increasing the reservoir capacity in the *stratum corneum* for incorporated therapeutics. These properties make Transcutol® P an ideal choice for the development of a nanoemulgel, especially for topical application. The preliminary nanoemulsions were prepared using the spontaneous emulsification method, followed by sonication and vortexing for increased homogenization. For the drug-loaded nanoemulsions, naringenin was solubilized in the mixture of Capryol 90 and tocotrienols, and then the surfactant and cosurfactant were added. Distilled water was added dropwise with continuous stirring, to achieve a final naringenin concentration of 2 mg/mL. To formulate and optimize a stable primary nanoemulsion, several critical characteristics were considered, including droplet size, PDI, and zeta potential, as they significantly impact the in vivo stability and overall performance of the formulations. The mean droplet size for most blank nanoemulsions was found to be less than 200 nm, with a trend for increased droplet size with higher oil concentrations, which aligns with previous findings in the literature, where an increment in oil content usually leads to larger droplet sizes due to a reduction in specific surface area. The zeta potential of the preliminary nanoemulsions ranged from $-4$ mV to $+11$ mV. Although nanoemulsions with low absolute zeta potential values can theoretically be more prone to being unstable, due to low electrostatic repulsion, several studies have shown that stable nanoemulsions can be formulated using nonionic surfactants with low surface charge values. This observation was confirmed in the study, where the formulated nanoemulsions, containing nonionic surfactants, exhibited good thermodynamic stability, in thermodynamic stability studies involving centrifugation, heating–cooling cycles, and freeze–thaw cycles, with no signs of phase separation, cracking, or creaming, hence ensuring their potential as stable delivery systems. The stable and optimized naringenin nanoemulsion was chosen as the starting point to formulate the nanoemulgels, being mixed with 1%, 1.5%, or 2% $w/v$ Carbopol® 934 or Carbopol® 940 gel bases. The ratio of naringenin nanoemulsion to gel base was 1:1, and the mixture was stirred for 10 min at 500 rpm. Triethanolamine was used to adjust the pH to 4.9–5.3, ensuring a slightly acidic environment beneficial for wound healing. The nanoemulgels were allowed to stand for 24 h to remove trapped air. The addition of the gelling agent led to a significant reduction in the size of the dispersed droplets. Furthermore, the zeta potential of the nanoemulgels decreased when compared to the optimized nanoemulsion, indicating a more negative surface charge, attributed to the presence of carboxylate ions on the Carbopol molecules. The incorporation of the gel matrix also further stabilized the dispersed oil droplets, indirectly supporting the stability of the nanoemulgel formulations. The optimized nanoemulgel had nanometric globules ($145.6 \pm 12.5$ nm) with a PDI of $0.452 \pm 0.03$ and a zeta potential of $-21.1 \pm 3.32$ mV. The spreadability of the formulations decreased with an increase in gelling agent concentration, which is typical for fluid gels that are easily spread over the affected area. The optimized nanoemulgel showed good spreadability and a high viscosity of 297600 cP. The nanoemulgels also showed promising mucoadhesive properties, with better adhesion being proportional to the concentration and grade of Carbopol that was used. In vitro drug release studies, in phosphate buffer saline (PBS), showed that Carbopol® 934-containing nanoemulgels exhibited higher drug release compared to those with Carbopol® 940, possibly due to their lower viscosity. Additionally, an initial burst release was observed in the first 2 h of the assay, which has been deemed typical of polymer-matrix-based formulations. The in vitro drug release of naringenin from the nanoemulgel was revealed to have controlled and sustained characteristics, adding up to 74.62 ± 4.54% within a time period of 24 h. Hence, overall, the findings from that study provide valuable insights into the formulation of stable naringenin nanoemulsions and nanoemulgels, paving the way for potential applications in wound healing treatments and other topical delivery approaches. Further studies and clinical evaluations may be needed to fully explore the therapeutic potential of the developed formulations in wound healing or other dermatological conditions.

Algahtani et al. [152] also aimed to develop a novel topical nanoemulgel formulation, loaded with thymoquinone (TMQ), and evaluate its effectiveness in wound healing. TMQ is a bioactive compound derived from *Nigella sativa*, and it has been widely studied for its potential therapeutic effects, including antioxidant, antimicrobial, anti-inflammatory, and wound healing properties. It has shown promise in promoting wound closure and tissue regeneration. However, its effectiveness in wound healing is limited by poor water solubility and low skin permeation [189–191]. Hence, the study aimed to develop a nanosystem formulation, encapsulating TMQ within the internal phase's oil droplets, using black seed oil as a natural carrier, and stabilized by a surfactant and cosurfactant mixture, with the addition of the aqueous phase leading to a primary O/W nanoemulsion. To prepare the TMQ-loaded nanoemulsion, a high-energy method was used, employing an ultrasonication technique. Initially, the researchers prepared a coarse emulsion by combining 5% $w/w$ (50 mg/g) of TQM with the mixture of the oil phase and a surfactant/cosurfactant mixture (Kolliphor® EL/Transcutol® HP). The components were mixed by using a vortex mixer, and then added to the aqueous phase, while continuously vortexing for 1 min. Then, in order to improve the emulsion's properties, the coarse emulsion was subjected to ultrasonication, using an ultrasonic homogenizer in a water bath. This process helped to break down larger droplets into smaller ones, and enhance the stability of emulsion, transforming it into a nanoemulsion. The increase in surfactant mixture concentration decreased the mean droplet size, and the ultrasonication time also significantly influenced the mean droplet size and PDI of the nanoemulsions. When the ultrasonication time increased from 3 to 5 min, the mean droplet size decreased. Nevertheless, an ultrasonication time higher than that, leading to an excessive exposure to ultrasonication energy known as overprocessing, caused intense turbulence, leading to collisions between the nanoemulsion's droplets and their subsequent coalescence, resulting in larger droplet sizes. The average globule sizes of selected nanoemulsions varied between 40.02 and 99.66 nm, and the PDI value between 0.052 and 0.542. Nanoemulsion viscosity was also measured, being between 71.04 mPas and 88.82 mPas, and drug content varied between 98.74% and 99.32%. The zeta potential of these primary nanoemulsions was measured to be in the range of $-26.7$ to $-30.6$ mV, which is expected to contribute to their stability, since high repulsive forces between the nanoemulsion droplets might help prevent their coalescence. Furthermore, the preliminary formulations' stability was effectively assessed, and all were found to be stable when subjected to various tests, including heating–cooling cycles, centrifugation, and freeze–thaw cycles. The formulations that demonstrated thermodynamic stability were selected for the in vitro drug release studies (Figure 5A) using the dialysis bag technique. The bags were filled with 1 mL of TMQ nanoemulsion formulation, and throughout the study, at specific time intervals, aliquots were withdrawn and replaced with PBS, up to 24 h. The aliquots were then analyzed by using UV-spectroscopy to quantify the amount of TMQ released from each formulation. After 12 h, more than 80% of the drug was released from all screened nanoemulsions, with the maximum cumulative drug release (at 24 h) varying between 84.3% and 87.1%. Additionally, comparing the TMQ nanoemulsions to a TMQ aqueous suspension, a significantly higher drug release was observed in the nanoemulsions. A selected drug-loaded nanoemulsion was then incorporated into a hydrogel system, creating the intended semisolid dosage form, a nanoemulgel. To form the nanoemulgel, the selected nanoemulsion was uniformly dispersed into a Carbopol® 940 gel matrix. This step aimed to create a final concentration of 0.5% TQM in nanoemulgel form, with the desired consistency, making it more suitable for topical application and ensuring a patient-friendly experience. The pH of the developed nanoemulgel was found to be within the range of skin's acid mantle (5.53), making it suitable for topical use, ensuring compatibility with the skin, and minimizing potential irritation. Moreover, the prepared nanoemulgel exhibited a similar rheological behavior to a placebo gel, demonstrating pseudoplastic behavior, with thixotropic properties, which is desirable for topical application. It also demonstrated excellent spreadability, making it suitable for topical application on wounded skin, and with the spreading area increasing proportionally with the applied force. Drug skin permeability and deposition investigation was conducted in an ex vivo study, using a Franz diffusion cell, on excised skin from Wistar rats, comparing the developed nanoemulgel to a conventional gel formulation. For that study, a shaved and excised dorsal skin sample from a Wistar rat was placed between the donor and receptor compartments of the Franz diffusion cell. Then, 500 mg of the test formulation was placed in the donor compartment, and the receptor compartment was filled with phosphate buffer. At various intervals, 1 mL aliquots were withdrawn and replaced with fresh media. Subsequently, these aliquots were diluted and quantified using a UV-spectrophotometer to estimate drug permeation through time. Results confirmed the expected enhancement in drug permeation, with the TMQ nanoemulgel showing a significantly enhanced cumulative drug permeation (549.16 $\mu g/cm^2$) when compared to the conventional gel form (120.75 $\mu g/cm^2$). Also, the percutaneous drug flux of TMQ from the nanoemulgel was approximately five times higher (23.14 $\mu g/cm^2 \cdot h$) than from the conventional gel form (4.78 $\mu g/cm^2 \cdot h$), as was the permeability coefficient (9.26 $K \times 10^{-3}$ vs. 1.91 $K \times 10^{-3}$). This enhanced permeation may be attributed to the presence of surfactant/cosurfactant mixture in the developed formulation, and to the small droplet size characteristic of the developed nanosystem. In addition, in order to estimate drug deposition in rat skin, the tape stripping technique was employed. After the 12 h ex vivo skin permeability study, the skin sample was removed from the assembly and washed with buffer. The first strips were discarded, and the subsequent 15 strips were used to remove the subcutaneous layer. The treated skin sample and the stripped tape were then chopped and incubated in ethanol to fully extract the drug. The incubated sample was sonicated and centrifugated before analyzing the extracted drug by using a UV-spectrophotometer, to determine the amount of drug deposited in the skin. The skin deposition of TMQ from the nanoemulgel form was significantly higher, measured to be 965.65 $\mu g/cm^2$, compared to 150.93 $\mu g/cm^2$ from the gel formulation. Moreover, the local accumulation efficiency of the nanoemulgel was higher by a factor of 1.4 when compared to the conventional gel, indicating a greater drug accumulation in the skin for localized and prolonged therapeutic action. Moreover, according to the in vivo studies performed in a Wistar rat wound model, the application of the nanoemulgel on the wounds resulted in accelerated wound closure, evidenced by reduced wound size, enhanced re-epithelization, and increased collagen deposition, with higher efficacy than control groups (no treatment, standard 1% $w/w$ silver sulfadiazine cream, and TMQ conventional gel) (Figure 5B). On the fourth day post-wounding, untreated rats displayed a hard thrombus swelling and exudates at the wound area. In contrast, animals from other groups exhibited a comparatively softer thrombus with reduced inflammation and no discharge. By the eighth day, reddish connective tissue, or granulation tissue, started forming in all groups. The complete epithelization time for the untreated group was 16.6 days. The groups with 1% silver sulfadiazine cream, TMQ conventional gel, and TMQ nanoemulgel had significantly shorter complete epithelization periods of 11.6, 14.33, and 10.33 days. A histopathological analysis (Figure 5C), on day 20 after treatment, revealed that the TMQ-nanoemulgel-treated group displayed larger amounts of granulation tissue and fewer mononuclear inflammatory cells compared to animals treated with 1% silver sulfadiazine and TMQ conventional gel. These findings suggest that the developed TMQ nanoemulgel has a significant wound healing potential, comparable to the 1% silver sulfadiazine cream, making it a promising formulation for topical wound healing applications. Thus, a topical TMQ nanoemulgel formulation was developed, by combining biocompatible polymers, oils, surfactants, and cosurfactants, resulting in a stable nanoemulgel with enhanced drug delivery potential and, hence, improved therapeutic efficacy, making it a promising formulation for potential use in dermatological applications, and specifically for improved skin wound healing.

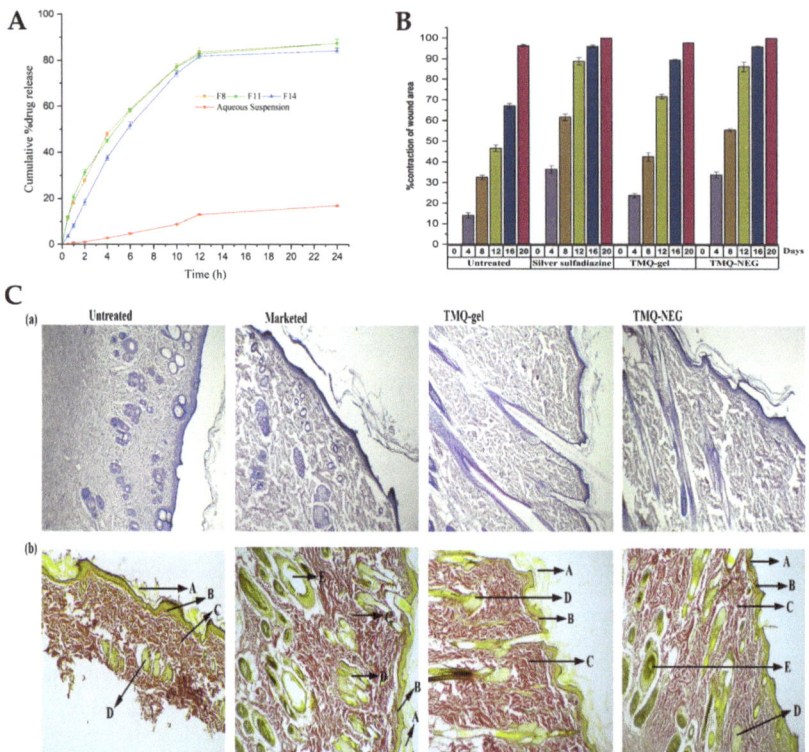

Figure 5. (A) In vitro drug release profiles of the TMQ-loaded nanoemulsions, compared to a drug aqueous suspension; (B) percentage of contraction of wound area variation, in the in vivo study, assessed for 20 days, with topical application of the developed TMQ nanoemulgel (TMQ-NEG), a conventional TMQ gel (TMQ-gel), a silver sulfadiazine formulation (Silver sulfadiazine), or no treatment (Untreated); (C) histopathology analysis of the rat's skin at day 20, newly healed, after topical application of the developed TMQ nanoemulgel (TMQ-NEG), a conventional TMQ gel (TMQ-gel), a silver sulfadiazine formulation (Marketed), or no treatment (Untreated), stained with hematoxylin-eosin (a) or Van Gieson (b), with arrows indicating the stratum corneum (A), the papillary dermis (B), collagen fibers (C), sebaceous glands (D), or hair follicles (E); adapted from Algahtani et al. [152], reproduced with permission from MDPI (Creative Commons CC BY 4.0 470 license).

Recently, natural compounds like curcumin have gained attention for their potential therapeutic effects on wound healing due to their anti-inflammatory, antioxidant, and antimicrobial properties. However, the clinical application of curcumin has been limited by its low solubility in aqueous media and poor skin permeability [192–194]. To tackle these issues, Algahtani et al. [153] developed a novel approach to enhance the wound healing potential of curcumin, through its formulation into a nanoemulgel, using the high-energy emulsification method of ultrasonication, known for its efficiency in producing nanosized droplets, in order to ensure uniform dispersion and stability of the nanoformulation. The first step involved the preparation of a preliminary O/W nanoemulsion, where curcumin was encapsulated within the oil droplets using Labrafac™ PG as the selected oil, and using a surfactant–cosurfactant system (Tween® 80 and PEG 400). The droplet size of the curcumin-loaded nanoemulsion system was significantly influenced by the ratio of the Smix phase in the formulation components. Specifically, when the nanoemulsion system had a Smix ratio of 2:1, the droplet size was notably reduced, compared to the nanoemulsion system with a Smix ratio of 1:1. Additionally, the droplet size was found to be significantly influenced by the ultrasonication time. Specifically, 5 min of ultrasonication resulted in a notable reduction in droplet size, compared to only 3 min of ultrasonication, at constant Smix concentration and ratio. Additionally, the effect was more pronounced at a lower Smix concentration, rather than at higher ones. Nevertheless, the Smix ratio and ultrasonication time did not have a remarkable effect on the PDI. Selected preliminary nanoemulsions achieved a droplet size of less than 100 nm, with mean droplet sizes in a range from 49.61 to 84.23 nm, and PDI values from 0.10 to 0.23. Thermodynamic stress testing was also conducted on curcumin-loaded nanoemulsions, to assess their stability. The formulations exhibited high stability under heat–cooling cycles, freeze–thaw cycles, and centrifugation. This stability was attributed to the reasonably high zeta potential values, ranging from $-15.96 \pm 0.55$ mV to $-20.26 \pm 0.65$ mV, which minimized droplet coalescence and physical instability. The viscosity of the selected preliminary nanoemulsion systems was also evaluated, at room temperature, by rotational viscosimeter, with the results ranging from 83.74 mPas to 89.82 mPas. The difference in viscosity values was attributed to the increased concentration of surfactant between formulations (a reasonably viscous component). A UV-visible spectrophotometric analysis confirmed that the selected nanoemulsion formulations had high curcumin content, ranging from 98.86% to 99.23%. Thus, these formulations showed high drug encapsulation. With droplet sizes around 50 nm and desirable surface charges, the selected nanoemulsions were chosen for further investigation, as they were proven ideal for topical application, enabling improved skin permeability and deeper penetration. Hence, in vitro drug release was assessed, with all nanoemulsions being able to release up to 85% of curcumin within the first 12 h, while the release from aqueous suspension (control) was only around 10% at the same time. The small droplet size of the nanoemulsions could have been a critical factor in positively influencing the in vitro drug release, since it produces a large surface area for drug diffusion and, potentially, absorption to occur. The next step involved incorporating the curcumin nanoemulsion into a Carbopol® 940 gel matrix, in order to form the curcumin nanoemulgel. The purpose was that this combination could take advantage of the benefits of both nanoemulsion and gel systems, both the high drug-loading capacity and improved skin permeation of nanoemulsions, and the enhanced stability and ease of application of gels. The drug concentration achieved in the final nanoemulgel preparation was 0.5% $w/w$ of curcumin, and the formulation exhibited a favorable physicochemical profile, with a measured gel strength of 46.33 s, while the placebo gel system had a strength of 44.66 s. Additionally, the drug content uniformity of the curcumin nanoemulgel was calculated, showing a uniform dispersion of curcumin within the hydrogel, with a uniformity of 98.93%. Furthermore, the curcumin nanoemulgel exhibited a similar rheological profile to the placebo gel, with the incorporation of curcumin not affecting its rheological behavior, and exhibiting a thixotropic behavior, which is desirable for topical pharmaceutical dosage forms. Hence, the developed nanoemulgel proved to have the desired consistency for patient-friendly topical use, which aligned with small droplet size, appropriate zeta potential, a pH within the acceptable range for skin application, and optimal drug release properties, suggested its safe and effective application for wound healing. This adequacy for skin application was confirmed by the ex vivo skin permeation study results, conducted on excised rat skin, on Franz diffusion cells, to assess the ability of the curcumin nanoemulgel to penetrate the skin barrier. The results showed improved skin permeation of curcumin from the nanoemulgel formulation, further confirming its potential as an effective topical delivery system for wound healing applications. The cumulative amount of curcumin which permeated through the skin from the nanoemulgel was 773.82 μg/cm$^2$, versus only 156.90 μg/cm$^2$ from the conventional gel formulation. In addition, the curcumin nanoemulgel showed a six-fold increase in percutaneous drug flux compared to the conventional curcumin gel. Similarly, the permeability coefficient from the curcumin nanoemulgel also increased approximately six-fold when compared to the conventional gel. Furthermore, the permeation enhancement ratio and local accumulation efficiency of curcumin from the nanoemulgel were significantly higher than from the gel, with a shorter lag time (0.75 h versus 2.37 h). Moreover, the in vivo wound-healing activity (Figure 6A,B) from the nanoemulgel and gel formulations were evaluated and compared to a commonly used silver sulfadiazine formulation. The topical application of these formulations was performed and monitored, in Wistar rats, for 20 days. The results showed that all treated groups exhibited significant wound-healing activity compared to the untreated group. The curcumin nanoemulgel demonstrated almost equivalent wound-healing activity to the silver sulfadiazine gel, with both formulations leading to almost complete wound healing at day 20. Moreover, the histopathological evaluation (Figure 6C) confirmed the enhanced wound-healing activity of curcumin from the nanoemulgel formulation, showing reduced inflammatory cells and extensive collagen fiber production.

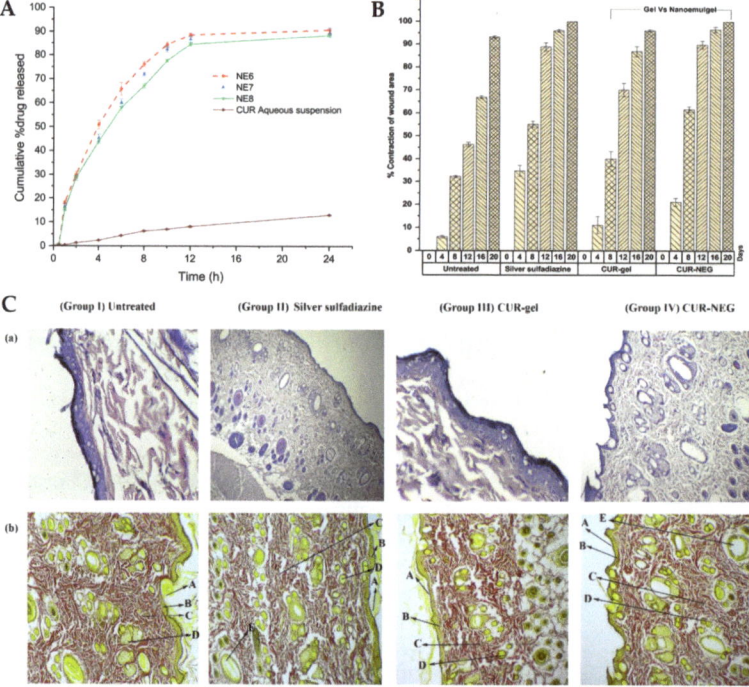

**Figure 6.** (**A**) In vitro cumulative drug release from the developed preliminary curcumin nanoemulsions, compared to the control (drug suspension); (**B**) in vivo wound-healing activity, in a rat model, of the developed curcumin nanoemulgel (CUR-NEG), compared to a curcumin conventional gel (CUR-gel), a marketed control formulation (Silver sulfadiazine), or no treatment (Untreated), including contraction of wound area percentage; (**C**) histopathology analysis of the rat's skin tissue at day 20 after treatment, including indications for the *stratum corneum* (A), the papillary dermis (B), collagen fibers (C), sebaceous glands (D), and hair follicles (E), (a) Stained with hematoxylin-eosin; (b) stained with vangeison to observe collagen formation (at 10× magnification); adapted from Algahtani et al. [153], reproduced with permission from MDPI (Creative Commons CC BY 4.0 470 license).

*2.2. Skin and Skin Appendage Infections*

The escalating global health threat posed by drug-resistant microbial and fungal infections necessitates the development of innovative and effective therapeutic strategies [195,196]. In this context, miconazole nitrate is a broad-spectrum antifungal medication commonly used to treat various fungal infections, especially on the skin. However, its therapeutic efficacy can be limited by factors such as poor drug permeation and low bioavailability [197,198]. Hence, a study by Tayah et al. [154] aimed to address these limitations by formulating miconazole into a nanoemulgel for topical application. Formulation composition was determined by selecting the excipients in which miconazole was most soluble, by dissolving it in various oils and surfactants. It was observed that olive oil and almond oil had the highest solubility, and among the surfactants, Tween® 80 and Span® 80 demonstrated the greatest ability to solubilize the drug. These oils and surfactants were selected as the drug vehicle, and hence preliminary O/W nanoemulsions were produced, using the self-emulsification technique. Optimized nanoemulsion formulations were chosen based on a ternary phase diagram (Figure 7A), according to which it was evident that the almond-oil-based formulations displayed the smallest particle size (170 nm) and PDI values (0.193). Therefore, the formulation containing almond oil was selected for subsequent experiments. Miconazole nanoemulgel formulations were then prepared by incorporating the preliminary miconazole nanoemulsions into a Carbopol® 940 hydrogel, with Carbopol at different concentrations. The particle size and PDI remained consistent when the transformation of the nanoemulsion into nanoemulgels occurred (Figure 7B), with the three Carbopol concentrations (0.4%, 0.6%, and 0.8%) showing similar behavior, with particle sizes ranging from 170 to 180 nm. The zeta potential results confirmed the potential stability of the nanoemulgel formulations, with values just below −30 mV. Regarding the rheological properties of the nanoemulgel, its viscosity decreased with an increase in the shear rate, which indicated pseudoplastic behavior. The release of the drug was evaluated and compared to a market product, using the dialysis method. The results showed an inverse relationship between Carbopol concentration and release profile, which is in accordance with an increased viscosity. Hence, the formulation with the lowest Carbopol concentration (0.4%) exhibited the highest cumulative drug release. Furthermore, another in vitro drug release assay was also performed, this time using a Franz cell diffusion test (Figure 7C), to measure the cumulative drug release from the selected miconazole nanoemulgel (0.4% Carbopol) and compare it to a conventional marketed Daktazol® cream (same drug molecule). After 6 h, the developed miconazole nanoemulgel exhibited a cumulative drug release of 29.67%, while the conventional Daktazol® cream achieved a release of 23.79%. Hence, the developed nanoemulgel achieved a higher drug release within the studied timeframe, while still retaining a controlled release profile. Then, the antifungal activity of the developed miconazole nitrate-loaded nanoemulgel was evaluated against selected fungal strains, and its performance was also compared to that of the conventional gel formulation. The antifungal activity was assessed by conducting an agar-based test on *Candida albicans*, and the size of the inhibition zone was measured as an indicator of effectiveness. The miconazole nanoemulgel demonstrated the highest activity, with an inhibition zone of 40.9 ± 2.3 mm, showing significant improvements in antifungal activity when compared to the marketed formulation, hence suggesting a promising approach to overcome the limitations of current conventional gel formulations. These findings support the conclusion that the developed novel miconazole nanoemulgel formulation exhibited improved antifungal activity, as well as increased but controlled drug release, and other desirable characteristics, compared to the conventional cream. Nevertheless, further research will be required to evaluate the safety and in vivo efficacy of this novel formulation, and to address regulatory considerations for its potential clinical use. Yet, the development of this miconazole nitrate-loaded nanoemulgel formulation shows promising results in enhancing the antifungal activity of miconazole, with improved drug delivery properties, presenting a potential solution to enhance the therapeutic efficacy of miconazole nitrate in the treatment of skin fungal infections.

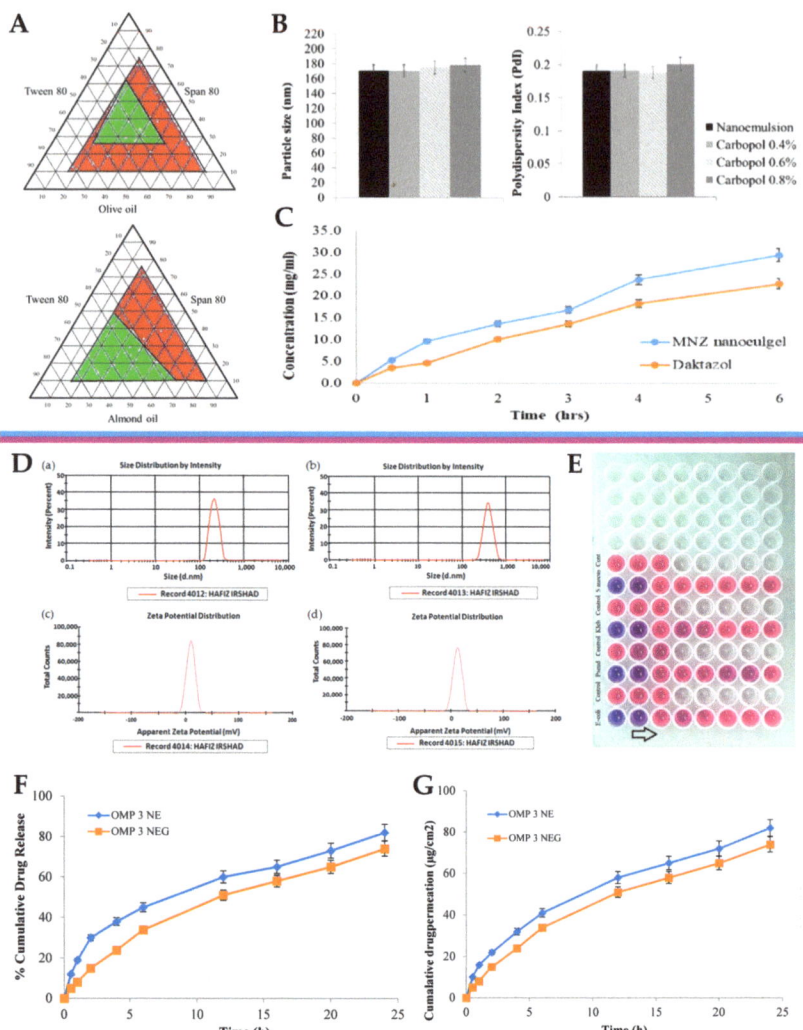

**Figure 7.** (**A**) Ternary phase diagrams of the preliminary nanoemulsions containing either olive oil, Tween® 80, and Span® 80, or almond oil, Tween® 80, and Span® 80; (**B**) droplet size and polydispersity index of the developed miconazole nitrate preliminary nanoemulsion and nanoemulgel formulations; (**C**) in vitro drug release profiles, in Franz diffusion cells, of the developed miconazole nitrate nanoemulgel, compared to the marketed product, Daktazol® cream; adapted from Tayah et al. [154], reproduced with permission from Elsevier (license number 5671991093263); (**D**) droplet size (a) and zeta potential (c) of the developed omeprazole-loaded nanoemulsion, and droplet size (b) and zeta potential (d) of the developed omeprazole-loaded nanoemulgel; (**E**) minimum inhibitory concentration determination assay of the developed omeprazole-loaded nanoemulgel, against selected bacterial strains, using a 96-well microplate (arrow shows decrescent antimicrobial activity); (**F**) cumulative drug release percentage from the developed omeprazole-loaded nanoemulsion and nanoemulgel formulations; (**G**) cumulative drug permeation from the developed omeprazole-loaded nanoemulsion and nanoemulgel formulations; adapted from Ullah et al. [155], reproduced with permission from MDPI (Creative Commons CC BY 4.0 470 license).

In another study, Ullah et al. [155] aimed to exploit the intrinsic antimicrobial properties of omeprazole, while utilizing chitosan's unique properties to enhance drug delivery and targeting capabilities. In this context, the combination of omeprazole, a widely used proton pump inhibitor with known antimicrobial properties, and chitosan, a natural biopolymer with remarkable biocompatibility and mucoadhesive characteristics, holds significant promise [199,200]. Hence, the fabrication and comprehensive characterization of a novel omeprazole-based chitosan nanoemulgel formulation, intended for potential antimicrobial application, was performed. A preliminary O/W nanoemulsion was previously produced, with an oil phase consisting of olive oil and Span® 80, and where omeprazole was solubilized. The aqueous phase was prepared by dissolving Tween® 80 in distilled water. Then, both phases were mixed, with the oil phase being gradually added dropwise to the aqueous phase, and with both phases being heated and stirred. The mixture was then gradually cooled, still with continuous stirring, to ensure thorough homogenization and formation of a homogenous nanoemulsion. Different nanoemulsions were prepared, with varying concentrations of constituents, for optimization purposes. Then, the optimized nanoemulsion was transformed into a nanoemulgel by replacing the aqueous phase with a chitosan solution (0.1% $w/w$) prepared in 1% acetic acid. The optimized preliminary nanoemulsion showed a high entrapment efficiency of 81.36 ± 1.98%, while the resulting nanoemulgel exhibited similar values, with an entrapment efficiency of 78.23 ± 3.76%. The optimized nanoemulgel exhibited a droplet size of 369.7 ± 8.77 nm, and a PDI of 0.316, with the zeta potential values of the optimized drug-loaded nanoemulsion and nanoemulgel formulations being −11.2 ± 5.4 mV and −15.3 ± 6.7 mV, respectively, indicating a potentially relevant physical stabilization of the system due to the electrostatic repulsion between the droplets (Figure 7D). In the rheological analysis, the nanoemulgel exhibited higher viscosity compared to the nanoemulsion, which was to be expected, being attributed to the presence of the highly viscous Carbopol gel base, making it beneficial for the application of the formulation on the skin. Additionally, both the nanoemulsion and nanoemulgel formulations demonstrated excellent spreadability and extrudability, ensuring convenient application. Moreover, the pH values of the optimized nanoemulsion and nanoemulgel formulations remained consistently within the acceptable range of human skin pH, throughout an evaluation period of 38 days, highlighting their suitability for effective transdermal drug delivery. The overall stability of the optimized drug-loaded nanoemulgel formulation and the changes in particle size during storage were also assessed. The formulation's appearance and clarity remained constant throughout the stability testing, and there was no change in the nanoemulgel's particle size during storage at room temperature. However, under accelerated storage conditions, the particle size progressively increased from 369.7 ± 8.77 nm to 405 ± 9.65 nm, 480 ± 8.87 nm, and 529 ± 9.41 nm, at the end of the first, second, and third months, respectively. This observation indicates that higher storage temperatures led to a rise in particle size. Hence, while the optimized nanoemulgel formulation demonstrated stability under standard storage conditions, it experienced particle clustering when exposed to elevated temperatures. During the in vitro drug release analysis (Figure 7F), the nanoemulsion demonstrated a rapid initial release of omeprazole, with approximately 85.28% of the drug being released after 24 h. Conversely, the nanoemulgel exhibited a slightly lower initial burst release, with approximately 82.16% of the drug being released after the same time period. Hence, both formulations displayed a controlled drug release profile, but with high cumulative value, ideal for topical application. The study also aimed to evaluate the skin permeation capabilities of both the nanoemulsion and nanoemulgel formulations, a factor that is closely related to the preparations' potential efficacy for transdermal drug administration. To assess this parameter, an ex vivo permeation assay was conducted (Figure 7G), using rabbit skin, providing insights into the drug permeation profiles of these formulations. The results revealed that the nanoemulsion demonstrated a higher permeation (82.18 ± 1.66 µg/cm$^2$) when compared to the nanoemulgel (72.21 ± 1.71 µg/cm$^2$), but this fact could simply be related to the higher viscosity of the nanoemulgel, which can be beneficial where formulation retention at the application site is concerned. Moreover, the success of transdermal drug permeation is influenced by several physicochemical properties, including particle size, zeta potential, and surface area. These factors play a pivotal role in determining the diffusion rate and the formulation's ability to penetrate the skin effectively. Since both formulations have similar composition, in this context the surfactant Tween® 80 likely played a vital role, by inducing lipid packing fluidization and optimizing the aqueous content in the *stratum corneum*, through a skin lipid extraction method. Additionally, Span® 80, as a cosolvent and cosurfactant, also exerted an impact on drug permeation, contributing to these formulations' favorable permeation profile. Furthermore, the combined effect of these formulations' components probably reduced the epidermis' barrier functions, providing the formulations with an advantage in promoting transdermal drug delivery. Additionally, the optimized nanoemulgel formulation displayed superior antibacterial effects compared to nanoemulsion (Figure 7E). The nanoemulgel exhibited reduced minimum inhibitory concentration (MIC) values against both Gram-negative bacteria (*Escherichia coli, Klebsiella pneumoniae, Pseudomonas aeruginosa*) and Gram-positive bacteria (*Staphylococcus aureus*). The formulation's excipients and unique characteristics likely also contributed to its enhanced antibacterial activity. This improved effect was attributed to the nanoemulgel's ability to transiently open tight junctions on the bacterial membrane, thereby increasing its antibacterial efficacy. Additionally, the presence of unsaturated fatty acids such as lactic acid in the formulation, further contributed to its antibacterial properties, by causing bacterial cell membrane rupture and eventual lysis. Therefore, overall, the nanoemulgel demonstrated promising potential as an effective strategy for controlling bacterial growth and promoting rapid healing. This is particularly critical in preventing the persistence of bacterial strains in injured skin tissues. In conclusion, the developed omeprazole-loaded chitosan nanoemulgel formulation represents a promising advancement in the field of antimicrobial drug delivery, offering new opportunities to combat microbial infections effectively and contribute to improved patient outcomes, while holding significant potential for therapeutic benefits in targeted drug delivery through the skin.

Fungal infections pose a substantial public health concern, affecting a significant number of individuals globally. Conventional antifungal therapies often encounter challenges such as drug resistance and limited efficacy, and hence the need for the exploration of alternative treatment strategies arises [201–203]. In this context, another study, by Vartak et al. [156], focused on the development and characterization of a novel Ebselen nanoemulgel, intended for the effective treatment of topical fungal infections. Ebselen (EB) is a well-established antioxidant and anti-inflammatory synthetic compound; nevertheless, it has limited solubility in commonly used solvents [204–206]. But since it has been proven to have reasonable solubility in dimethylacetamide (DMA), this cosolvent was selected to be part of the formulation's composition. The rest of the excipients were also selected on a highest drug solubility basis, with Kolliphor® ELP being selected as surfactant, and medium chain triglycerides (Captex® 300 EP/NF) as oil phase. Additionally, in order to prevent drug precipitation and enhance formulation stability, a gelling polymer mixture was added to the external phase, hence producing an O/W nanoemulgel. Different polymers and polymer combinations were evaluated. This was achieved by mixing the components in a 5:7 ratio of oil to surfactant, and then mixing the resulting nanoemulsion in a 1:1 ratio with the gel bases, to create the final nanoemulgels. Hydroxypropyl methylcellulose (HPMC) K4M and Aquaphor, both present at a concentration of 0.5% $w/w$, formed a clear system when combined with 1 mg of EB in DMA. However, an increase in EB loading to approximately 2 mg caused immediate precipitation of EB. Similarly, when EB was added to a Carbopol® 974P and Poloxamer 407 gel, at different gelling concentrations and loading levels, precipitation of the drug occurred. In contrast, the nanoemulgel prepared using Soluplus® showed a delayed precipitation effect. This polymer was also proven to be an effective solubilizer for EB, hence playing a crucial role as both a solubility and drug-loading enhancer, and a formulation viscosity enhancer and stabilizer. Optical microscopy (Figure 8A) and scanning electron microscopy (SEM) (Figure 8B) images of

EB-loaded nanoemulgels supported the stability of the Soluplus gel, when compared to all other gel compositions, not showing any drug precipitation. In contrast, the HPMC K4M nanoemulgels displayed distinct precipitation of EB throughout the system, manifesting as large, irregularly shaped crystals with small oil globules. In the optimized Soluplus® spontaneously formed nanoemulgel, EB was present at 1% $w/w$, and the nanosystem revealed a droplet size of 54.82 ± 1.26 nm and zeta potential of −1.69 mV. Furthermore, the findings of rheological studies showed that the optimized EB-loaded nanoemulgel displayed a non-Newtonian fluidic behavior. The observed decrease in viscosity with increasing rotational speed confirms the pseudoplastic behavior of the nanoemulgel, rendering it an ideal choice for topical application. In the in vitro drug release study (Figure 8C), HPMC K4M and Aquaphor nanoemulgels initially showed similar drug release, but after 24 h, the HMPC K4M gel exhibited a three-times higher drug release. Interestingly, the release profile of EB in a DMA solution was similar to that of the Soluplus® nanoemulgel, which demonstrated approximately two- and four-times higher release than the HPMC and Aquaphor formulations, respectively, after 24 h. Additionally, the optimized nanoemulgel exhibited a controlled release profile. Furthermore, the membrane deposition study (Figure 8D), which quantified the drug entrapped within the used membrane, showed that the nanoemulgel prepared using Soluplus® displayed a significantly higher drug deposition, compared to all other formulations, from 2.3- to 5-fold higher. Relating to the formulations' antifungal activity, resazurin, a redox indicator, was employed to assess the viability of fungal organisms, more specifically, multi-drug-resistant *Candida albicans* and *Candida tropicalis*, in culture plates. In a 48 h study, conducted in RPMI media, the EB-loaded optimized nanoemulgel demonstrated potent antifungal activity, exhibiting an MIC of 20 µM, exhibiting higher efficacy than the control, terbinafine hydrochloride (100 µM), a known antifungal drug. In conclusion, the developed novel topical EB nanoemulgel revealed improved drug solubility, membrane permeability, and deposition, and even displayed substantial antifungal activity against *Candida* infections (with higher potency than terbinafine hydrochloride, a commonly used antifungal agent), making it a potentially more effective alternative than existing conventional treatments.

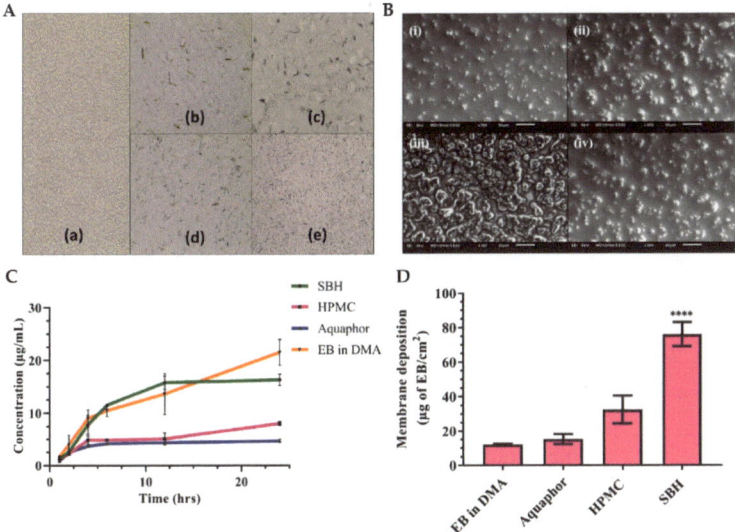

**Figure 8.** (**A**) Optical microscopy images of the EB-loaded preliminary nanoemulgels, containing either Soluplus® (a), HPMC (b), Poloxamer 407 (c), Carbopol® 974P (d), or Aquaphor (e); (**B**) scanning electron microscopy images of the optimized nanoemulgels, either drug-loaded Soluplus® formulation (i), Soluplus® vehicle (ii), drug-loaded HPMC formulation (iii), or HPMC vehicle (iv); (**C**,**D**) in vitro

cumulative drug release (**C**) and membrane drug deposition (**D**) of the different EB-loaded nanoemulgels; **** $p < 0.0001$; DMA—dimethylacetamide; EB—Ebselen; HPMC—hydroxypropyl methylcellulose; SBH—Soluplus®; adapted from Vartak et al. [156], reproduced with permission from Elsevier (license number 5672000024708).

Mahtab et al. [157] also developed a novel formulation for fungal infection treatment, namely, a transungual delivery system utilizing a nanoemulgel with ketoconazole for the efficient management of onychomycosis, a prevalent fungal infection that affects the nails and presents significant challenges in terms of treatment effectiveness and patient adherence [207,208]. At first, different primary nanoemulsions were produced, using GRAS excipients, with Labrafac ™ Lipophile WL1349 (medium-chain triglycerides of caprylic (C8) and capric (C10) acids) as the oil phase (highest ketoconazole solubility capacity), Polysorbate 80 as the surfactant, and PEG 400 as the cosurfactant. The surfactant and cosurfactant were combined in varying weight ratios (1:0, 1:1, 1:2, 2:1, 3:1, 1:3, and 4:1), with increasing concentrations of either the surfactant or cosurfactant in relation to each other. After solubilization of the drug within it, the mixture of oil, surfactant, and cosurfactant was then added dropwise to the aqueous phase, under moderate agitation on a vortex mixer, to form the nanoemulsions. Subsequently, high-pressure homogenization was used to achieve a desirable droplet size, and the resulting products were visually evaluated for clarity, transparency, and flowability. The selected primary nanoemulsion, with a surfactant-to-cosurfactant ratio of 3:1, revealed a mean droplet size of $77.52 \pm 0.92$ nm, a PDI of $0.128 \pm 0.035$, a zeta potential of—$5.44 \pm 0.67$ mV, a pH of $6.2 \pm 0.34$, and a viscosity of $20.00 \pm 1.24$ cP. Additionally, stability studies confirmed the physicochemical stability of the system for 3 months, with the optimized nanoemulsion only showing very negligible changes in the previously evaluated parameters, with no phase separation of flocculation being observed. From the optimized nanoemulsion, nanoemulgels were prepared by adding Carbopol® Ultrez 21 as gelling agent, glycerin as humectant, methylparaben playing the role of antimicrobial agent, thioglycolic acid as a penetration enhancer, and aminomethyl propanol as a pH-adjusting agent, to achieve a pH in the range of 6 to 6.5. Different nanoemulgels were prepared, using a combination of high-pressure homogenization and ultrasonication techniques, with different ratios of Carbopol-to-thioglycolic acid: 0.5 to 1.0, 1.0 to 1.5, and 1.5 to 1.0. Among these variations, the 1.0 to 1.5 proportion was found to have the best visual appearance, being glossy, creamy, viscous, and having a smooth and homogeneous texture, with no signs of phase separation, and was therefore selected for further studies. The physicochemical properties of the optimized nanoemulgel were extensively characterized, including pH ($6.4 \pm 0.24$), viscosity ($1142 \pm 10.33$ cP), spreadability ($3.5 \pm 0.22$ g cm/s), extrudability ($1.4 \pm 0.56$ g/cm$^2$), firmness ($13450 \pm 231$ g), consistency ($6133 \pm 19$ g s), cohesiveness ($169 \pm 2.23$ g), and adhesiveness ($20.4 \pm 0.81$ g). Additionally, in vitro drug release studies showed that the optimized nanoemulgel exhibited sustained release of ketoconazole from the nanoemulgel over time, suggesting its potential for prolonged therapeutic effectiveness. The observed sustained release pattern indicates that the formulation was able to release the drug gradually over time, which can be potentially beneficial for treatment outcomes, as it might ensure a continuous and prolonged presence of the antifungal agent at the site of infection. Moreover, ex vivo transungual permeation results, in a goat hooves model, showed that the optimized nanoemulgel demonstrated higher drug permeation ($77.54 \pm 2.88$%) when compared to the primary optimized nanoemulsion ($62.49 \pm 2.98$%) and an aqueous drug suspension ($38.54 \pm 2.54$%) during a period of 24 h. Additionally, the nanoemulgel's antifungal activity was assessed against clinical isolates of dermatophytes, namely, *Trichophyton rubrum* and *Candida albicans*, using the agar diffusion method. The ketoconazole-loaded nanoemulgel displayed superior antifungal activity when compared to a conventional gel, with a larger zone of inhibition, highlighting its potential for better therapeutic outcomes. The developed formulation's skin irritation potential was also assessed on rats, with the nanoemulgel showing minimal skin irritation, with no significant signs of erythema or edema, compared to the positive control. In addition, the histopathological analysis showed no significant pathological changes, indicating its safety for topical application. Therefore, overall, the results suggest that the optimized nanoemulgel holds promise as an effective delivery system for nail fungal infection treatment.

Another study, by Almostafa et al. [158], also focused on the development of a novel approach to combat skin bacterial infections, using a nanoemulgel formulation containing fusidic acid (FA) and myrrh oil. FA is a potent antibiotic, effective against a panoply of Gram-positive bacteria, but it has poor water solubility, which poses a significant challenge for its effective formulation [209–211]. Myrrh oil, a traditional herbal extract, was used to modify and enhance the transdermal delivery of FA. In addition to these components, Tween® 80 was used as a surfactant, Transcutol® P as a cosurfactant, and CMC as a viscosifying and gelling agent. In initial studies, it was observed that increasing oil concentrations led to a relative increase in particle size for all preliminary nanoemulsions. Contrarily, a decrease in nanoemulsion particle size was observed with an increase in surfactant concentration, while keeping the oil concentration constant, due to a reduction in surface tension. The particle size of all developed preliminary nanoemulsions ranged from 116 to 226 nm. Moreover, the in vitro release of FA from the fabricated nanoemulsion formulations was evaluated, and results showed that the percentage of FA released from all nanoemulsion formulations ranged from 40.1 to 75.6%, exhibiting a controlled release profile. Increasing oil concentration resulted in a decrease in the percentage of FA released, due to the resulting larger particle size, since it provides a smaller surface area. On the other hand, increasing Tween 80 and Transcutol P concentrations led to a substantial increase in the percentage of FA release, which can be attributed not only to these excipients' strong solubilizing ability, but also to the resulting smaller particle size. Hence, overall, the particle size of the formulations was revealed to have a crucial role in the drug release process, as systems with smaller particle sizes achieved maximum drug release. The preliminary nanoemulsion formulation was hence transformed into a nanoemulgel by adding CMC to the external phase, and this formulation was also characterized for relevant parameters. The pH value for the FA nanoemulgel formulation was found to be within an acceptable range at 6.61, making it potentially safe for topical application and minimizing the risk of skin irritation. The viscosity was also measured to assess the formulations' rheological behavior, as this influences drug diffusion and in vitro release. The viscosity of the developed nanoemulgel was found to be 25265.0 cP, higher than that of the corresponding CMC gel formulation, 15245.0 cP, which was also measured for comparison purposes. Hence, both fell within an appropriate range for topical application. The spreadability, which determines whether the uniform application of the formulation on the skin is possible or not, was also determined. Results showed that the FA gel exhibited a spreadability of 40.5 mm, while the nanoemulgel formulation had a spreadability of 33.6 mm, both indicating excellent spreadability despite the observed difference between the two formulations. The formulations were also studied for stability during storage at 4 °C and 25 °C, for 1 and 3 months, and both showed nonsignificant variation in physical properties under all studied conditions, compared to their freshly prepared counterparts. The in vitro drug release (Figure 9A) of FA from the developed gel and nanoemulgel formulations was also evaluated and compared to an FA suspension. The results revealed that the drug was completely released from the suspension after 120 min, reaching 99.5% of cumulative release. In contrast, the FA gel and nanoemulgel formulations released 80.3% and 59.3% of FA after 180 min, exhibiting a more controlled release profile, as is intended for topically applied formulations. These results can be explained by the substantially increased viscosity of the gel and, especially, the nanoemulgel formulations, which slowed down the diffusion of the drug from the formulations and into and through the membrane. Additionally, the nanoemulgel formulation could have acted as a drug reservoir, where the drug passed from the inner phase to the outer one, further slowing the release rate. Furthermore, FA release from both preparations remained consistent during storage (Figure 9C,D), when compared to freshly prepared formulations. The results confirm the stability of the formulations and demonstrate the efficacy of nanoemulgels as nanocarriers. Skin permeability studies (Figure 9B) were also conducted, with the permeability of FA when incorporated into different formulations being evaluated using excised animal skin. The results further supported the potential of the developed formulations, since they showed that the FA nanoemulgel exhibited the highest permeability, followed by the FA gel, and only then the FA suspension, which exhibited the lowest skin permeability. This proven superiority of the nanoemulgel and gel formulations when compared to the drug suspension could be due to the incorporation of Transcutol P, a known effective permeation enhancer, especially effective in increasing permeation through the skin (more specifically, the *stratum corneum*), which certainly contributed to the significantly improved drug permeation profile through rat skin. The safety of the developed formulations was also tested, with the animals' back skin being treated with the test formulations, and then undergoing a thorough examination to check for any sensitivity reactions. Fortunately, no signs of inflammation, irritation, erythema, or edema were observed on the inspected area during the entire 7-day study period. These results indicate that the formulations are potentially safe and well tolerated by the skin, without causing any adverse reactions or sensitivity issues. Moreover, the antibacterial activity of the developed FA nanoemulgel was evaluated (Figure 9E) against various microorganisms and compared to a placebo nanoemulgel and a common marketed cream. The FA nanoemulgel showed significant antibacterial activity against *Staphylococcus Aureus*, *Bacillus subtilis*, and *Enterococcus faecalis*, with a larger inhibition zone than the placebo or the marketed cream. Moreover, both the FA-loaded nanoemulgel and the placebo nanoemulgel exhibited significant inhibition zones against *Candida albicans*, *Shigella*, and *Escherichia coli*, while the marketed cream showed a negative effect against these bacteria. The combination of FA and myrrh oil in the developed nanoemulgel likely contributed to this enhanced antibacterial activity. In conclusion, FA was successfully incorporated into a nanoemulgel prepared with myrrh essential oil, showing good physical properties for topical application, enhanced skin permeation, no skin irritation, and potent antibacterial and antifungal activity, against several different types of microorganisms, with the combination of FA and myrrh essential oil showing synergistic effects. Hence, the results demonstrated that the developed nanoemulgel showed promise as a potential topical or transdermal drug delivery system for the treatment of skin bacterial or fungal infections, offering a new platform for innovative and effective topical treatments, and providing a basis for future research and clinical applications in dermatology.

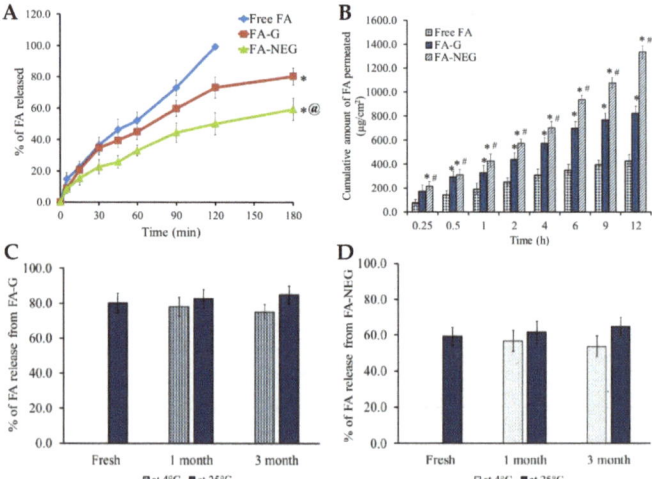

**Figure 9.** *Cont.*

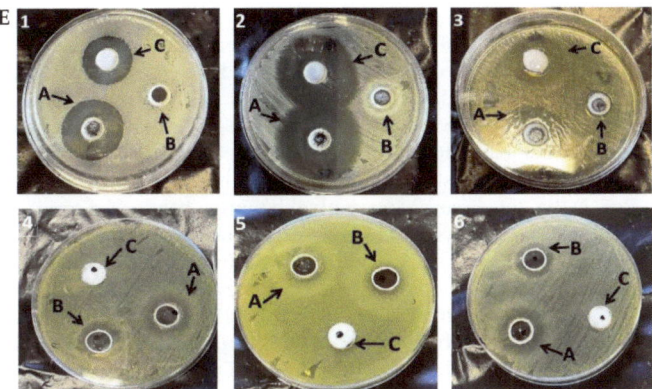

**Figure 9.** (**A**) In vitro FA release profiles from the developed nanoemulgel (FA-NEG), compared to a conventional gel (FA-G) and a drug suspension (Free FA), where * $p < 0.05$ compared to the drug suspension, and @ $p < 0.05$ compared to the conventional gel; (**B**) ex vivo FA permeation profiles, across rat skin, from the developed nanoemulgel (FA-NEG), compared to a conventional gel (FA-G) and a drug suspension (Free FA), where * $p < 0.05$ compared to the drug suspension, and # $p < 0.05$ compared to the conventional gel; (**C,D**) variation of the in vitro drug release during stability studies, under storage at 4 °C and 25 °C for 1 and 3 months, for the developed nanoemulgel (C) and conventional gel (D) formulations; (**E**) inhibition zone diameter photographs after treatment with the developed FA-loaded nanoemulgel (A), placebo nanoemulgel (B), or marketed FA formulation (C), on *Bacillus subtilis* (1), *Staphylococcus aureus* (2), *Enterococcus faecalis* (3), *Candida albicans* (4), *Shigella* (5), and *Escherichia coli* (6); FA—fusidic acid; adapted from Almostafa et al. [158], reproduced with permission from MDPI (Creative Commons CC BY 4.0 470 license).

## 2.3. Skin Cancer

Skin cancer is a prevalent form of cancer with increasing incidence rates worldwide. Conventional treatment options often come with limitations, such as systemic side effects and inadequate skin permeation [212–214]. To fight these limitations, Nagaraja et al. [142] also developed a novel topical nanoemulgel formulation, incorporating nanosized droplets encapsulating the drug, chrysin, a flavonoid with proven anticancer potential, into a gel matrix for improved drug delivery and extended release, aiming for it to be an effective skin cancer treatment [215–217]. Stable lipid-based nanoemulsifying preconcentrates containing chrysin, made of Capryol® 90 (oil and hydrophobic surfactant), Tween® 80 (hydrophilic surfactant), and Transcutol® HP (cosurfactant and cosolvent), were designed to be easily reconstituted in a gel base made of Pluronic® F127 (gelling agent). The formulation components were selected in order to achieve the highest chrysin solubilization and, consequently, drug strength in the final formulation. Additionally, a surfactant mixture of 2:1 of Tween® 80 to Transcutol® HP was proven to be the best combination for highest solubilization. The physicochemical characterization of the nanoemulsifying drug delivery system showed its small particle size (Figure 10A), of less than 300 nm, a good homogeneity, with a PDI value around 0.26, indicating narrow droplet size distribution, and a negative zeta potential value, of about −15 mV, indicative of reasonable formulation stabilization potential. Moreover, the formulation was further stabilized by the incorporation of the oil droplets in a semisolid gel base. The chrysin-nanoemulsifying preconcentrate was added to water containing the gelling agent, Pluronic® F-127, at 10 °C, for easier mixture and overall nanoemulgel preparation, due to this polymer's thermosensitive properties. In addition, the droplet size and PDI values of the system did not change significantly after a 3-month storage period, hence ensuring its good stability under the studied conditions. Regarding the results of the performed ex vivo permeation studies (rat abdominal skin), while a chrysin conventional Pluronic® gel base, used as control, showed poor penetration through the skin, the developed nanoemulgel exhibited a superior performance due to the presence of nanosized droplets, which resulted in better and faster skin permeation. In vitro cytotoxicity studies of the developed chrysin nanoemulgel were performed on various skin cancer cell lines, including A375 (human melanoma, Figure 10B), A375.S2 (human melanoma), SK-MEL-2 (human melanoma, Figure 10C), B16-F1 (murine melanoma), and A431 (human epidermoid carcinoma). Results showed that chrysin had matrix metalloproteinase-2 inhibitory activity, leading to a strong antiproliferative effect, with cell mobility and migration inhibition, with the developed nanoemulgel having a better performance than the control formulation, leading to deeper changes in cancer cell morphology. Furthermore, biocompatibility tests were also carried out, on L929 cells (noncancerous murine cells), which showed no changes in cell growth, therefore making the developed formulation safe for topical application. In conclusion, the incorporation of chrysin into a topical nanoemulgel formulation resulted in a significant enhancement of its therapeutic response in cytotoxicity studies, with the developed drug delivery platform technology exhibiting substantial advantages when compared to conventional formulations, such as its versatility, extended skin permeation, and retention, and the ability to potentially reduce systemic drug absorption.

A different study, by Algahtani et al. [159], aimed to develop a novel formulation that combined the immunomodulatory effects of imiquimod (IMQ) and the anti-inflammatory properties of curcumin (CUR), for skin cancer treatment, using a nanoemulgel delivery system (Figure 11A). IMQ is a commonly used chemotherapeutic agent for skin cancer [218,219]. On the other hand, the combination with CUR has been shown to improve the therapeutic effectiveness of various chemotherapeutics, with this molecule also having intrinsic anticancer properties [220,221]. Nevertheless, the topical delivery of IMQ and CUR can be difficult due to their poor solubility and low skin penetration properties, and IMQ has been shown to lead to psoriasis-like lesions when applied topically [222,223]. Hence, the purpose of the incorporation of both IMQ and CUR into a topical nanoemulgel was to not only enhance drug permeation and provide sustained release of these drugs, but also reduce IMQ's topical side effects, leading to improved therapeutic outcomes for skin cancer patients. For the selection of the most adequate excipients, IMQ solubility studies were performed at room temperature, and results showed that this drug's solubility was maximized in oleic acid, in the oil category, which best mixed with Tween® 20, in the surfactant category, and Transcutol® HP, in the cosurfactant/cosolvent category, to form a stable nanoemulsion system through spontaneous emulsification, an advantageous low-energy method. Different ratio mixtures of oil and surfactant/cosurfactant were tested, and the different nanoemulsions were formed by addition of the aqueous phase. The preparation procedure included the accurate weighting of the necessary quantity of CUR and dissolving it completely in a homogeneous mixture of oil, surfactant, and cosurfactant, with the help of vortex mixing. Then, the aqueous phase was added, and the mixture was vortexed again. Clear, easily flowable, and transparent formulations were selected for further studies and characterized for relevant parameters. The analysis showed that the optically clear nanoemulsion formulations were composed of fine dispersed droplets in nanosized dimensions, ranging between 91.07 and 98.88 nm (Figure 11B). An increase in oil droplet size was correlated with an increase in oleic acid concentration and a decrease in surfactant/cosurfactant concentration. The zeta potential of the droplet surface was negative, ranging from $-10.9$ to $-35.8$ mV (Figure 11B), due to the presence of anionic groups in the fatty acids and glycols present in the nanoemulsions' composition. This could be a potential advantage, as nanoemulsion droplets with a negative zeta potential tend to be more able to penetrate deeper in the skin. For the thermodynamic stability studies, the nanoemulsion formulations were subjected to stress conditions, namely, centrifugation, and heating–cooling and freeze–thaw cycles. Results confirmed their stability, since they did not show any physical changes such as phase separation, creaming, cracking, or coalescence. The final selected preliminary nanoemulsion formula, with a drug strength of 15 mg/mL, had a high percentage of drug content of 99.26%, a narrow droplet size distribution (10.57 nm) with a PDI of 0.094, a negative zeta potential of $-18.7$ mV, and a viscosity of 125.48 cP. The in vitro drug release for the optimized nanoemulsion was also performed (Figure 11C) using the dialysis bag technique. The release of IMQ from an aqueous suspension (control) was just 11% at 24 h, while the release from the preliminary nanoemulsion, containing IMQ only, was quite high, being around 92%. The incorporation of CUR into the nanoemulsion did not affect the release of IMQ, since the drug release from the formulation combining both drugs was 92.84% for IMQ and 83.94% for CUR. The selected nanoemulsion was then transformed into a nanoemulgel by incorporating Carbopol® 934 as the gelling agent. The nanoemulgel had a mean droplet size of 78.39 nm and PDI of 0.254, which were still considered adequate for system stability and topical application. In spreadability studies, the spreading factors of the placebo nanoemulgel, the IMQ-loaded nanoemulgel, and the IMQ-CUR-loaded nanoemulgel were found to be equivalent, being equal to 0.82 $cm^2/m$, 0.85 $cm^2/m$, and 0.87 $cm^2/m$, respectively. Hence, all formulations showed good extrudability potential from the container tube, adequate for patient-friendly applications. The overall drug content in the final nanoemulgel formulation was quite high, being around 99%, and the measured pH was 5.5, adequate in order to minimize skin irritation, since the ideal value is between 5 and 6. Ex vivo skin permeation and deposition studies were then performed in rat skin using Franz diffusion cells. Again, different formulations were compared, namely, the placebo nanoemulgel, the IMQ-loaded nanoemulgel, and the IMQ-CUR-loaded nanoemulgel. Formulations were introduced in the donor compartment, and the aliquots were taken at specified periods of time, filtered, and analyzed using UV-visible spectrophotometry. The tape stripping technique was used to determine the amount of drug deposited on the skin layer. The skin deposition of IMQ from the IMQ-nanoemulgel was 1205.2 $\mu g/cm^2$, and 1367 $\mu g/cm^2$ from the IMQ-CUR-nanoemulgel, which was 5 times higher than that obtained from a conventional gel formulation (control). CUR skin deposition was also around 9 times higher (5178 $\mu g/cm^2$) in the developed IMQ-CUR-nanoemulgel than in the conventional gel formulation (570 $\mu g/cm^2$). The percutaneous IMQ drug flux was also determined, being equal to 0.042 $\mu g/cm^2$.h for the IMQ-nanoemulgel, and 0.071 $\mu g/cm^2$.h for the IMQ-CUR-nanoemulgel, which was around 18 times higher than for the conventional gel formulation (0.004 $\mu g/cm^2$.h). Hence, the developed nanoemulgel showed an improved permeation profile when compared to a conventional gel formulation, with the nanoemulgel containing both IMQ and CUR displaying apparent synergistic effects. In vivo studies were also performed, on mice, and skin pathological changes after 10 days of topical application of different formulations were monitored. Psoriasis-like symptoms started to appear on mice treated with the IMQ conventional gel formulation from the 2nd day of application and worsened until the 10th day. On the other hand, mice treated with the IMQ-nanoemulgel exhibited a delayed appearance of these symptoms, which was possibly correlated with a more controlled drug release from the formulation. Fortunately, the application of the IMQ-CUR-nanoemulgel did not lead to the appearance of psoriasis-like symptoms, which was connected with the antipsoriatic activity of CUR. Additionally, on the 11th day after the application of the formulations on the skin, the skin was collected for histopathology analysis (Figure 11D). While untreated skin showed regular epidermis and dermis, skin treated with the conventional gel showed hyperkeratosis, parakeratosis, acanthosis, and epidermal infiltrates. The IMQ-nanoemulgel treated skin showed comparable results, but less thickening of the epidermis layer, but the skin treated with the IMQ-CUR-nanoemulgel showed similar characteristics to untreated normal skin, with only a reduced number of infiltrates being observed. Hence, a stable nanoemulgel with optimal rheological properties, allowing easy spreadability on the skin, exhibiting high encapsulation efficiency for both IMQ and CUR, with sustained release profiles and enhanced permeation across the skin barrier, was successfully developed, while reducing psoriasis-like lesions in an animal model.

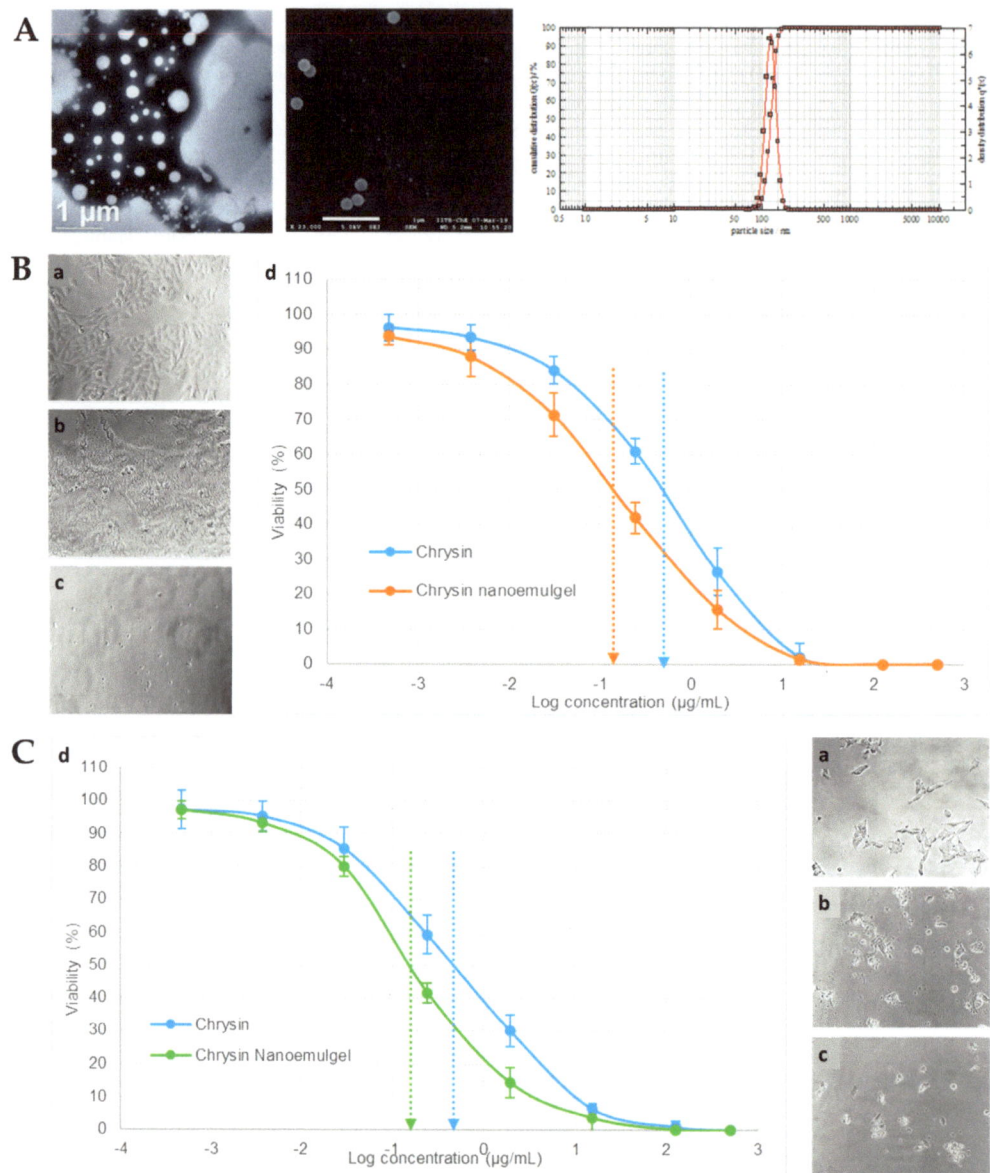

**Figure 10.** (**A**) Developed chrysin nanoformulation's droplet size and size distribution, with transmission electron microscopy image (left), scanning electron microscopy image (middle), and photon cross-correlation spectroscopy results (right); (**B**) A375 cells' morphological observation and growth inhibition after no treatment (control cells, a), treatment with chrysin solution (b), or treatment with the developed chrysin nanoemulgel (c), and respective in vitro cytotoxicity profile (cell viability %) (d); (**C**) SK-MEL-2 cells' morphological observation and growth inhibition after no treatment (control cells, a), treatment with chrysin solution (b), or treatment with the developed chrysin nanoemulgel (c), and respective in vitro cytotoxicity profile (cell viability %) (d); adapted from Nagaraja et al. [142], reproduced with permission from MDPI (Creative Commons CC BY 4.0 470 license).

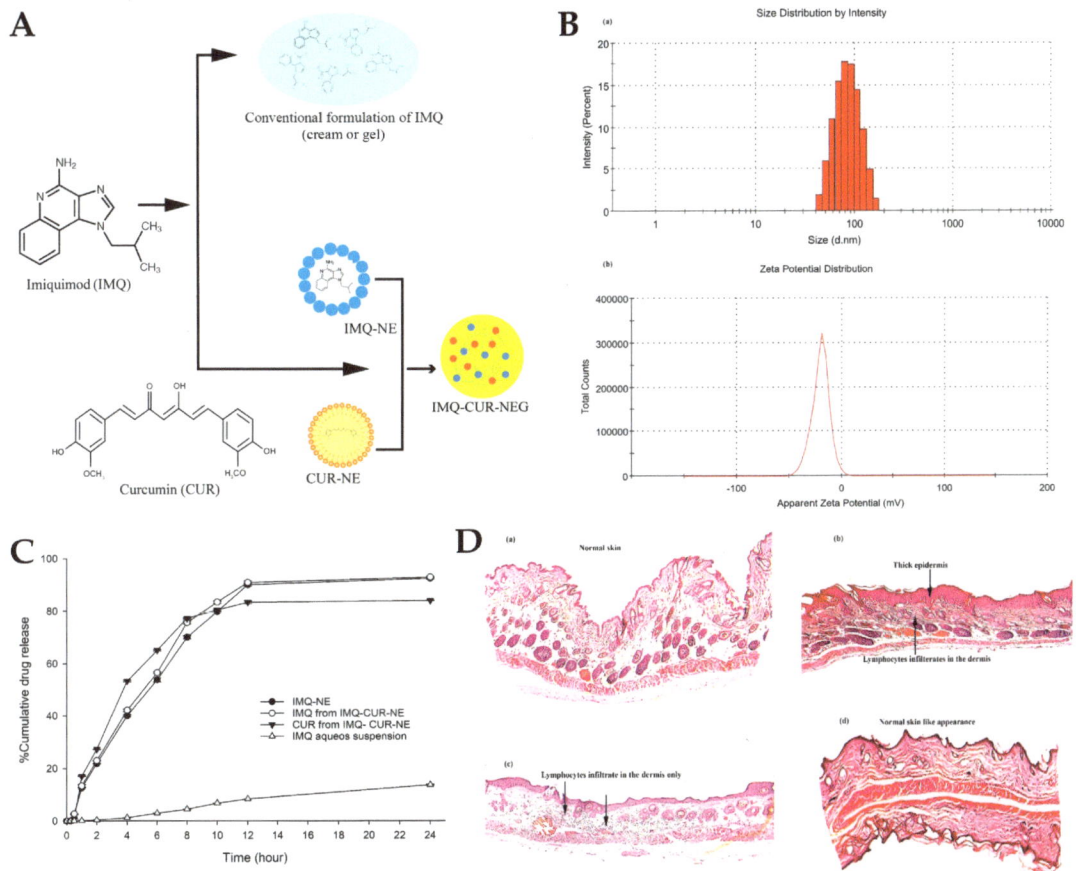

**Figure 11.** (**A**) Schematic representation of the developed IMQ-CUR-nanoemulgel; (**B**) droplet size (a) and zeta potential (b) distribution of the developed preliminary nanoemulsion; (**C**) in vitro IMQ and CUR drug release profiles from different formulations, including an IMQ-nanoemulsion (IMQ-NE, with no CUR), IMQ-CUR-nanoemulsion (with both IMQ and CUR), and an IMQ aqueous suspension (control); (**D**) histopathology images of the mice's skin after the ten days of topical treatment, with either the IMQ gel (b), the IMQ-nanoemulgel (c), the IMQ-CUR-nanoemulgel (d), or no treatment (a); CUR—curcumin; IMQ—imiquimod; adapted from Algahtani et al. [159], reproduced with permission from MDPI (Creative Commons CC BY 4.0 470 license).

## 2.4. Skin Inflammatory Diseases

Psoriasis is a chronic inflammatory skin disorder affecting millions of individuals worldwide, being characterized by abnormal keratinocyte proliferation and inflammation. Current treatment options have limitations; hence, there arises the need for the development of novel and effective therapies [224–226]. In this context, another study, by Pund et al. [146], investigated the transcutaneous delivery of leflunomide, an immunomodulatory drug, using a nanoemulgel formulation for the treatment of both melanoma and psoriasis [227–229]. Based on leflunomide solubility studies, the excipients selected to form the preliminary O/W nanoemulsion were Capryol® 90, Cremophor® EL, and Transcutol® HP, as oil base, surfactant, and cosolvent, respectively. These components were found to have good miscibility with each other, and the formed preliminary nanoemulsion was transformed into a nanoemulgel by incorporating it into a gel matrix made of Pluronic®

F-127. The preliminary nanoemulsions were evaluated for droplet size, which was found to be in the range of 98.7 to 280.92 nm, with a PDI between 0.2 and 0.3, indicative of narrow size distribution, and a slightly negative zeta potential of $-7.8$ mV. These characterization parameters were also measured in the nanoemulgel, with no significant changes being observed. Regarding the nanoemulgel's viscosity, the shear thinning nature of the poloxamer gel was demonstrated by the decrease in viscosity at higher shear rates, which allows for effortless dispensing of the product from the container and smooth application on the skin. Additionally, the nanoemulgel demonstrated potent antipsoriatic activity by inhibiting the proliferation of human keratinocytes (HaCaT cell line) and reducing proinflammatory cytokine levels, such as interleukin-6 and tumor necrosis factor-alpha. The suppression of these proinflammatory cytokines is crucial in mitigating the inflammatory response associated with psoriasis. Moreover, the developed formulations also exhibited antipsoriatic activity by effect on leukocyte infiltration and keratinocyte proliferation. Furthermore, the nanoemulgel displayed significant antimelanoma activity by inducing apoptosis in A375 and SK-MEL-2 melanoma cells and inhibiting tumor cell proliferation. These effects were attributed to the cytotoxicity of leflunomide, which can target multiple signaling pathways involved in cancer growth and survival. Safety assays involved systemic biocompatibility assessment, by hemolytic toxicity evaluation, with results showing that the nanoemulgel had a minimal 0.25% hemolytic toxicity compared to the positive control (Triton-X), suggesting its compatibility with blood cells and potential for reducing the risk of adverse effects. In conclusion, the development and characterization of a novel leflunomide-loaded nanoemulgel was successful, with the results indicating that the developed formulation possesses favorable physicomechanical characteristics for transcutaneous delivery and exhibiting promising in vitro antipsoriatic and antimelanoma activity, suggesting its potential as a novel therapeutic approach for the treatment of these diseases.

In another study, by Shehata et al. [160], the authors presented the development, characterization, and optimization of a novel eucalyptus-oil-based nanoemulgel loaded with meloxicam (Figure 12A) aimed at enhancing the anti-inflammatory efficacy of the drug for topical application. Nonsteroidal anti-inflammatory drugs, such as meloxicam (MX), have been widely used to alleviate inflammatory conditions; however, their application is often limited by issues like systemic toxicity and poor drug delivery to the targeted site [230,231]. To overcome these limitations, a nanoemulgel formulation was designed to facilitate controlled drug release and improve local drug concentration, as chronic inflammation is a critical pathophysiological process associated with numerous dermatological disorders with a strong inflammatory basis, hence necessitating effective and targeted therapeutic interventions. Therefore, in their research, several preliminary nanoemulsion formulations were developed by carefully blending specific ingredients. To create the oily phase, 1% ($w/w$) of MX was combined with a precise amount of eucalyptus oil and a cosolvent/cosurfactant, Transcutol® P. For the aqueous phase, varying quantities of Tween® 80 (as a surfactant) and PEG 400 (as a cosurfactant) were mixed with distilled water. The two phases were then meticulously merged, but achieving a well-mixed and stable nanoemulsion required high shear homogenization, so the formulation's homogeneity was further enhanced by sonication. Additionally, the high solubility of meloxicam in the various components further confirmed their suitability as part of the formulation's composition, with solubility values of $216 \pm 11$ mg/mL in eucalyptus oil, $130 \pm 10$ mg/mL in Tween® 80, $68 \pm 6$ mg/mL in Transcutol® P, and $146 \pm 5$ mg/mL in PEG 400. In the pursuit of optimal formulations, several nanoemulsions incorporating MX were developed, with different excipient proportions. Remarkably, all formulations exhibited excellent stability at room temperature, showing no signs of phase separation. The developed nanoemulsions displayed particle sizes within the nanometric range, ranging from $139 \pm 2.31$ to $257 \pm 3.61$ nm. Predictably, an increase in oil concentration led to larger particle sizes, likely due to a corresponding increase in the dispersed phase. Conversely, when the surfactant concentration was increased, while keeping the oil concentration constant, the formulation's particle size decreased, a finding that aligned with previous studies, which reported an inverse relationship between nanoemulsion particle size and surfactant concentration. The results of the in vitro drug release studies (Figure 12B) performed on the preliminary nanoemulsions showed that the percentage of MX released from the NE formulations varied between 55.0 ± 2.8% and 87.6 ± 3.9% after 6 h, hence exhibiting a controlled release profile. Notably, increasing the oil concentration in the preparation led to a decrease in MX release, likely due to the larger particle size, resulting in a smaller surface area available for drug release. The optimized nanoemulsion formulation was then transformed into a nanoemulgel by integrating HPMC as the gelling agent. A conventional HPMC gel formulation, containing MX, was also developed, for comparison purposes. Both the MX-loaded nanoemulgel and the MX-gel formulation appeared smooth, homogenous, and physically stable, demonstrating drug distributions above 99.3%, indicating a uniform dispersion of the drug. Additionally, both formulations, and especially the MX-nanoemulgel formulation, showed suitable viscosity for easy skin application (Figure 12C) and satisfactory spreadability (Figure 12D). Moreover, stability assessment, over 1 and 3 months, showed nonsignificant variations in the MX-nanoemulgel formulation's characterization parameters. In vitro drug release comparison revealed that the MX-nanoemulgel exhibited a lower percentage of drug release (39.4 ± 3.7%) compared to the MX-gel (52.1 ± 4.2%) and the optimized preliminary nanoemulsion (83.9 ± 2.41%), hence exhibiting a more controlled release pattern. This could be attributed to the incorporation of HPMC into the formulation's external phase, resulting in a slower drug release due to increased viscosity. Nevertheless, the results of the ex vivo skin permeation study (Figure 12E), across rat skin, supported the superiority of the developed nanoemulgel when compared to the other formulations, since it was observed that after 6 h a significantly greater amount of MX permeated from the nanoemulgel formulation, with a steady-state transdermal flux value of $141.28 \pm 9.17$ μg/cm$^2$, compared to the permeation from the MX-gel formulation, which showed a value of $84.28 \pm 10.83$ μg/cm$^2$. The permeation of the MX-nanoemulgel was found to be enhanced by 1.68-fold when compared to the conventional MX-gel. This increased permeation of the drug from the nanoemulgel formulation can be attributed to its small particle size, leading to a larger surface area than the conventional gel formulation, since by incorporating the drug into nanosized globules its permeation through the skin layer is facilitated. Furthermore, the presence of Tween® 80, PEG 400, and Transcutol® P in the formulation also played a vital role in enhancing MX permeation from the nanoemulgel, due to these surfactants/cosurfactants' capacity for increasing drug permeation through biological barriers. Moreover, according to the skin irritation tests performed on animals treated with the developed topical MX-nanoemulgel to ensure the safety of the formulation, no irritation, erythema, or edema were observed, indicating its safety. In addition, an in vivo anti-inflammatory study (Figure 12F), in which edema was induced in the rats' hind-paw, was also performed. The thickness of the edema directly correlated with the percentage of inflammation. After 12 h, inflammation percentages were 81.3 ± 6.4%, 59.8 ± 6.4%, 41.5 ± 4.6%, and 27.8 ± 5.7% for the control group, placebo (nanoemulgel vehicle, with no drug), MX-gel, and MX-nanoemulgel groups, respectively. The placebo group's reduced inflammation percentage, when compared to the control group, confirmed the anti-inflammatory effects of eucalyptus oil. Furthermore, the MX-nanoemulgel-treated group demonstrated a significantly higher anti-inflammatory effect when compared to the MX-gel group, suggesting the nanoemulgel's greater anti-inflammatory effect due to improved skin permeability. Hence, the study provided evidence of eucalyptus oil's anti-inflammatory effects, and its synergistic action with MX, making the developed nanoemulgel a promising option for treating inflammation-related skin conditions. In conclusion, an MX-loaded nanoemulgel was successfully developed and optimized, offering a promising nanocarrier topical formulation for enhanced drug delivery, good adhesion, and easy application onto the skin.

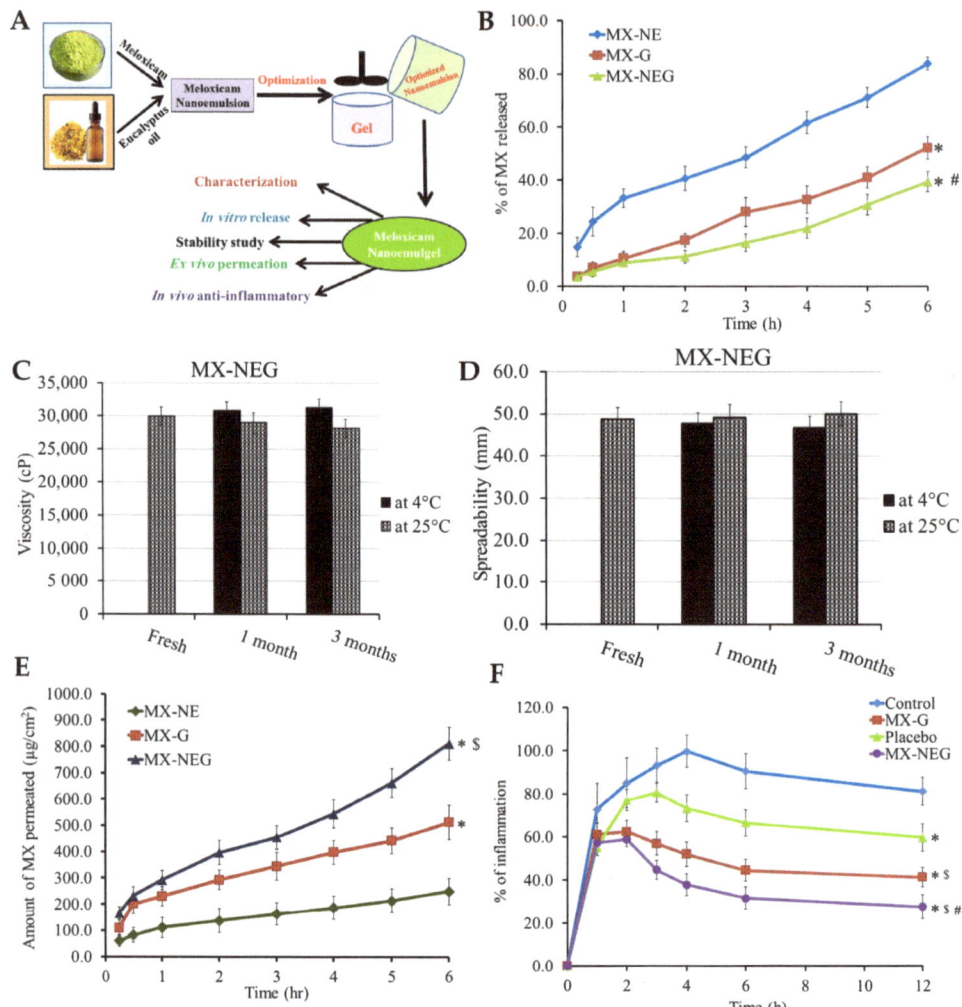

**Figure 12.** (**A**) Schematic representation of the developed MX and eucalyptus oil nanoemulgel, including performed physicochemical and efficacy characterization studies; (**B**) in vitro drug release profiles of the developed preliminary MX-nanoemulsion (MX-NE), conventional MX-gel (MX-G), and MX-nanoemulgel (MX-NEG), with * $p < 0.05$ compared to the preliminary MX-nanoemulsion, and # $p < 0.05$ compared to the conventional MX-gel; (**C**,**D**) stability profiles of the developed MX-nanoemulgel formulation, after 1 and 3 months, under storage at 4 °C and 25 °C, in what concerns viscosity (C) and spreadability (D); (**E**) ex vivo drug permeation profiles of the developed preliminary MX-nanoemulsion (MX-NE), conventional MX-gel (M-G), and MX-nanoemulgel (MG-NEG), with * $p < 0.05$ compared to the preliminary MX-nanoemulsion, and \$ $p < 0.05$ compared to the conventional MX-gel; (**F**) anti-inflammatory effects of various formulations on rat hind-paw edema, including the developed MX-nanoemulgel (MG-NEG), a conventional MX-gel (M-G), a placebo formulation (nanoemulgel vehicle with no drug), and a control group (no treatment), with * $p < 0.05$ compared to the control group, \$ $p < 0.05$ compared to the placebo group, and # $p < 0.05$ compared to the conventional MX-gel group; MX—meloxicam; adapted from Shehata et al. [160], reproduced with permission from MDPI (Creative Commons CC BY 4.0 470 license).

A different study, by Abdallah et al. [161], focused on the preparation, characterization, and evaluation of a topical nanoemulgel containing brucine as an active substance (Figure 13A). Brucine is a natural alkaloid produced from *Strychnos nux-vomica* L. seeds. This plant-derived molecule has been used in traditional medicine to relieve arthritis and traumatic pain [232,233]. Aside from having antitumor and antiangiogenic activities, it has relevant anti-inflammatory and antinociceptive effects. Nevertheless, its clinical use is limited due to high lipophilicity, and consequent low water solubility, and to high toxicity, especially in oral administration [234–236]. Hence, the study aimed to develop a nanoemulgel containing brucine for anti-inflammatory and antinociceptive topical effects, hence increasing its targeted delivery and decreasing systemic side effects. A preliminary nanoemulsion was formulated first by mixing brucine with a suitable oil (myrrh oil), surfactant (Tween® 80), cosurfactant (PEG 400), and cosolvent (ethyl alcohol), and then adding water. This formulation was then transformed into a nanoemulgel by incorporating it into a gel matrix, with sodium carboxymethyl cellulose as gelling agent, followed by a high-energy homogenization technique—high-pressure homogenization plus sonication. The droplet size of the developed nanoemulgel was evaluated and compared to that of an emulgel (same composition, but without high-energy homogenization), with results showing that the droplet size of the nanoemulgel (151 ± 12 nm) was 10-fold smaller than the droplet size of the emulgel (1621 ± 77 nm), hence showing significant improvement and displaying the relevance of the use of the high-pressure homogenization plus sonication in order to obtain the adequate nanometric size. The PDI values were also determined, which were 0.631 for the emulgel and 0.243 for the nanoemulgel, hence indicating monodispersity for the nanometric formulation (PDI value of less than 0.3), and further confirming the superiority of the nanoemulgel when compared to the emulgel formulation. Furthermore, a morphological evaluation of the nanoemulgel did not detect crystals of brucine, which indicated adequate drug solubility. Moreover, the brucine-loaded nanoemulgel showed adequate spreadability (48.3 ± 1.8 mm) and viscosity (62650 ± 700 cP) for topical application, comparable to those obtained for the emulgel and a conventional gel formulation (containing carboxymethyl cellulose). The in vitro drug release studies (Figure 13B) showed that brucine release from all developed formulations was lower and slower than that from a drug solution (99.5 ± 1.3%), hence exhibiting controlled release profiles. The amount of brucine released from the conventional gel formulation was greater (69.07 ± 2.54%) than that released from the emulgel (47.0 ± 4.32%) or nanoemulgel (58.6 ± 4.3%), probably due to a higher aqueous content in the gel formulation. Furthermore, stability studies showed that there were no changes in the color, homogeneity, spreadability, viscosity, or in vitro drug release of the nanoemulgel under the studied conditions (4 °C and 25 °C), hence making the developed formulation highly stable. Nevertheless, despite having a sustained drug release, the developed nanoemulgel formulation showed a superior drug permeation to all other formulations in ex vivo rat skin permeation studies using Franz diffusion cells (Figure 13C). Results showed that the best permeation through the skin was in fact attributed to the brucine-loaded nanoemulgel, and the worst to the brucine solution. Additionally, the cumulative amount of drug permeated from the nanoemulgel was substantially higher than from the other tested dosage forms. Moreover, the anti-inflammatory effects of the brucine-loaded nanoemulgel were evaluated using a carrageenan-induced paw edema animal model (Figure 13D). The animals were treated with the nanoemulgel formulation, and the degree of paw swelling was measured. Results showed a significant increase in the inflammation of the control and placebo groups after 4 h. The group treated with an oral drug solution reached the highest inflammation after 3 h, although it was lower than the control or placebo groups. Additionally, there were no significant differences between the brucine-loaded topical gel application and oral administration groups. Nevertheless, the brucine-loaded topical nanoemulgel group showed maximum inflammation after 1 h of treatment, with a remarkable decrease in swelling after 12 h. This result could be attributed to the nanoscale size of the nanoemulsion and the high penetration of the nanoemulgel through rat skin. Hence, a significant reduction in paw edema compared to the control groups indicated potent anti-inflammatory activity of the developed nanoemulgel. The antinociceptive effects of the developed formulations were also assessed using the acetic-acid-induced writhing test and hot plate test (Figure 13D). In the acetic-acid-induced writhing test, the formulations were administered to the animals and then the number of abdominal writhing movements recorded. A reduction in writhing movements indicated the antinociceptive potential of a given formulation, and results showed that the developed nanoemulgel led to a significant reduction in the number of abdominal writhing movements, leading to a greater inhibition than all other groups. In the hot plate test, the animals were placed on a hot plate, at a temperature of 55 °C. Then the time that elapsed from when the animal was put on the hot plate until any reaction was produced was measured. Antinociceptive effects, produced from formulation administration, will increase the time until the animals react. Results showed that the maximum antinociceptive effect by oral administration was reached after 1.5 h and after 3 h for topical application. The brucine-loaded nanoemulgel showed greater antinociceptive effects than gel and emulgel dosage forms, which, again, can be due to the larger surface area and nanosized particles, which improve the penetration of brucine through the *stratum corneum*, and hence enhance its absorption to reach better antinociceptive effect. Skin irritation studies were also performed, in order to assess the safety of the developed formulations. Sensitivity reactions, such as edema or erythema, were monitored in the animals at 1, 8, and 24 h post-treatment. Results showed that there were no signs of erythema or edema after 24 h of treatment with the different brucine formulations, hence meaning that they were in fact well-tolerated and not skin-sensitizing. Therefore, the developed brucine-loaded nanoemulgel may be a safe and well-tolerated drug delivery vehicle for topical application, also having high efficacy in permeating the skin, and producing anti-inflammatory and antinociceptive effects.

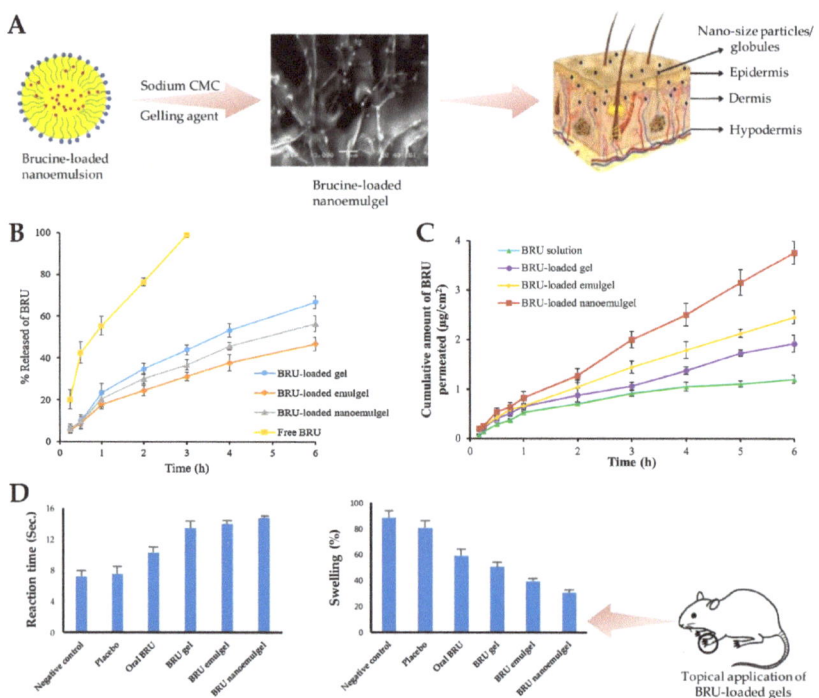

**Figure 13.** (**A**) Schematic representation of the developed brucine-loaded topical nanoemulgel, including a scanning electron microscopy image; (**B**) in vitro drug release profiles from different BRU-loaded

formulations, namely, a drug solution (Free BRU), a conventional gel, an emulgel, and a nanoemulgel; (**C**) ex vivo drug permeation profiles from different BRU-loaded formulations, namely, a drug solution, a conventional gel, an emulgel, and a nanoemulgel; (**D**) antinociceptive (reaction time) and anti-inflammatory (swelling) effects of different brucine-loaded formulations after administration to mice; BRU—brucine; adapted from Abdallah et al. [161], reproduced with permission from Elsevier (license number 5672000458127).

*2.5. Other Applications: Neuropathy and Antiaging Effects*

Diabetic neuropathy is a common complication of diabetes characterized by nerve damage that leads to pain, numbness, and tingling sensations [237–239]. Capsaicin, a natural compound found in chili peppers, has shown promise in alleviating neuropathic pain [240–242]. The analgesic effect of capsaicin is attributed to its potential for activation of transient potential vanilloid type 1 receptors, leading to the depletion of inflammatory neuropeptides and neuronal desensitization [240,243,244]. Additionally, capsaicin exhibits anti-inflammatory properties by inhibiting the activity of proinflammatory mediators [245–247]. Nevertheless, despite its efficacy, the clinical use of capsaicin through oral or intravenous administration has limitations due to its short half-life, pungency, and first-pass metabolism, and noncompliance [248–250]. Moreover, its effective delivery through the skin remains a challenge. In this context, Saab et al. [162] aimed to optimize a nanoemulgel formulation for the transdermal delivery of capsaicin, evaluating its skin permeation properties and assessing its in vivo potential for diabetic neuropathy treatment. Initially, preliminary nanoemulsions were developed containing eucalyptus oil as the oil, Tween® 80 as the surfactant, and propylene glycol, ethanol, and isopropyl alcohol as cosurfactants. Several different nanoemulsions were prepared, with different ratios of these excipients, and the formula exhibiting the best droplet size ($28.15 \pm 0.24$ nm) and PDI ($0.27 \pm 0.05$) values was selected for further studies, consisting of 8% oil phase, an oil to surfactant mixture ratio of 2.5-to-7.5 (Tween® 80/cosurfactant mixture), and a final capsaicin concentration of 0.05%. Then, by combining the chosen nanoemulsion formula with a gel base, using Carbopol® 940 as the gelling polymer, the nanoemulgel was successfully formed. The developed nanoemulgel was examined under transmission electron microscopy, alongside its corresponding nanoemulsion, with both showing spherical emulsion droplets within the nanosize range. Additionally, the emulsion droplets seemed to remain stable after gel incorporation, indicating resistance to destabilization processes, such as Oswald ripening, which was evident by it showing no significant changes in droplet size, PDI, transmittance percentage, or appearance after long-term storage (at 4 °C, 25 °C, and 32 °C), centrifugation, heating–cooling cycles, and freeze–thaw cycles. The rheological assessment of the nanoemulgel established its pseudoplastic behavior, with this formulation exhibiting a predictably higher viscosity (5580 mPa/s) than the correspondent nanoemulsion (11.5 mPa/s). The skin permeation study showed that the capsaicin-loaded nanoemulgel demonstrated enhanced (188.12 µg/cm$^2$ cumulative permeation) and accelerated (0.19 µg/cm$^2$.s$^{-1}$ permeation flux) permeation compared to a conventional gel (46.38 µg/cm$^2$ and 0.11 µg/cm$^2$.s$^{-1}$, respectively). The antinociceptive potential of the developed formulation was also evaluated, in an animal model, with the application of the capsaicin-loaded nanoemulgel having showed significant improvements in thermal latency, as demonstrated in the hot plate and tail withdrawal tests in alloxan-induced diabetic mice. Compared to the placebo-treated mice, the nanoemulgel treatment resulted in a remarkable increase in thermal latency in the 8th week. Conversely, treatment with conventional capsaicin gel showed less significant improvement. Furthermore, the capsaicin nanoemulgel exhibited superior efficacy in alleviating mechanical allodynia. The nanoemulgel treatment group showed a substantial improvement compared to the placebo-treated animals, approaching the level of improvement seen in the tramadol-treated group. These positive outcomes can be attributed to the successful transdermal delivery of capsaicin from the nanoemulgel, as previously confirmed in the skin permeation study. Hence, the optimized developed formulation exhibited favorable physicochemical properties and enhanced skin permeation, and the in vivo assays' results demonstrated its efficacy in reducing pain-related behaviors and improving sensory function. Given this, the developed capsaicin-loaded nanoemulgel holds promise as a treatment option for diabetic neuropathy, offering a noninvasive, convenient, and effective approach to alleviate neuropathic pain. Further studies, including clinical trials, are necessary to validate its safety, efficacy, and long-term effects in human subjects; nevertheless, these findings highlight the potential of a novel capsaicin-loaded nanoemulgel formulation for transdermal delivery in the management of the diabetic neuropathy.

Retinyl palmitate (RT), a derivate of vitamin A, has shown potential for skin rejuvenation and the treatment of various dermatological conditions, such as acne, psoriasis, ichthyosis, wrinkles, dark spots, and skin aging [251–253]. This molecule has the capability of exfoliating the surface layer of the skin and speeding up cell turnover, increasing skin moisture, and decreasing skin wrinkles, resulting in the skin looking fresher, smoother, and younger, and also acting as an antioxidant when applied topically, preventing tissue atrophy, and having anti-inflammatory effects [254–256]. Nevertheless, in addition to its good properties, RT also has side effects, such as skin irritation, redness, excessive peeling and dryness, being toxic in higher concentrations, while also being a lipophilic compound, with poor aqueous solubility, and limited skin permeation, which all hinder its therapeutic efficacy [257–259]. To overcome these limitations, Algahtani et al. [163] focused on the development and evaluation of a nanoemulgel formulation for enhanced topical delivery of RT (Figure 14A), to improve its solubility, stability, and permeation for antiaging effects. First, preliminary RT-loaded nanoemulsions were prepared, using a low-energy emulsification technique, by mixing an optimized oil and surfactant mixture phase with an aqueous phase, using a vortex mixer. RT showed good solubilization in different oils, surfactants, and cosurfactants, which were then selected based on their good miscibility with each other, but also on the resulting hydrophile–lipophile balance value. This was found to be optimum with an oil phase combining medium-chain monoglycerides (Capryol® 90) and medium-chain triglycerides (Captex® 355) in a 2:1 ratio, and a surfactant/cosurfactant mixture made of Kolliphor® EL and Transcutol® HP. The oil concentration affected the mean droplet size of the produced nanoemulsions, which increased with increasing oil concentration from 10% to 20%. Oil percentage also affected the PDI value, which also increased when increasing the oil phase percentage. On the other hand, variations in oil and surfactant/cosurfactant concentrations did not significantly affect zeta potential values. Viscosity values ranged from 77 to 89 cP, which showed that the viscosity remained constant and the nanoemulsions behaved as fluids. The selected preliminary formulations were further evaluated in in vitro drug diffusion studies (Figure 14B) using the dialysis bag method, with aliquots being collected at different time intervals for 24 h and analyzed by UV-spectroscopy. The drug release from the nanoemulsion systems was high and almost complete, ranging from 89% to 94%, while the release from the aqueous drug dispersion ended up being around 10 times smaller. Nanoemulsion storage stability was also tested, with droplet size and PDI values being monitored. Formulations remained stable for a period of 90 days, with the measured parameters remaining relatively constant (Figure 14D,E). The selected optimized nanoemulsion had a droplet size of 16.71 nm (Figure 14A), a PDI value of 0.015, and a zeta potential of $-20.6$ mV, with a measured drug percentage of approximately 99.0% and the highest cumulative drug release, and was selected for converting into a nanoemulgel system. The nanoemulgel formulation was prepared by incorporating the optimized nanoemulsion into a gel matrix, using Carbopol® 940 (0.5% $w/w$) as the gelling agent, and with the addition of glycerin as humectant and to provide a smooth and soothing sensation upon topical application. The pH of the nanoemulgel was $5.53 \pm 0.06$, which is similar to the skin's pH and is optimal for not causing skin irritation. The developed nanoemulgel exhibited non-Newtonian, pseudoplastic behavior, shear thinning, and thixotropic properties, which are convenient characteristics for topical application. Good spreadability is also important in achieving straightforward application of topical formulations, and results showed good spreadability for the developed nanoemulgel, with spreadability increasing with a higher applied force. Additionally, a stability evaluation of the developed nanoemulgel was performed, under different storage conditions, including temperature variations and exposure to light. The formulation exhibited excellent physical stability, with no significant changes observed in visual aspect, particle size, viscosity, or pH over the study period, indicating its robustness and long-term stability. Moreover, UV stability studies were done by measuring the percentage of RT remaining after an exposure to UV-A radiation (Figure 14C). A significant decrease in the amount of RT was evident after only 2 h, with the quantified amount reducing down to 19.46% in the control formulation, with no encapsulation of the drug. In contrast, 95.24% of the RT remained in the nanoemulgel formulation after 2 h of exposure, and 82.96% after 6 h, hence showing the potential of the developed formulation to encapsulate the drug and successfully protect it from photodegradation. Furthermore, the Franz diffusion cells were used again, this time to evaluate the ex vivo drug permeation of the developed nanoemulgel through excised dorsal rat skin. The nanoemulgel formulation demonstrated significantly enhanced skin permeation when compared to a conventional gel formation, indicating improved drug delivery and potentially enhanced therapeutic efficacy, with higher permeation coefficient.

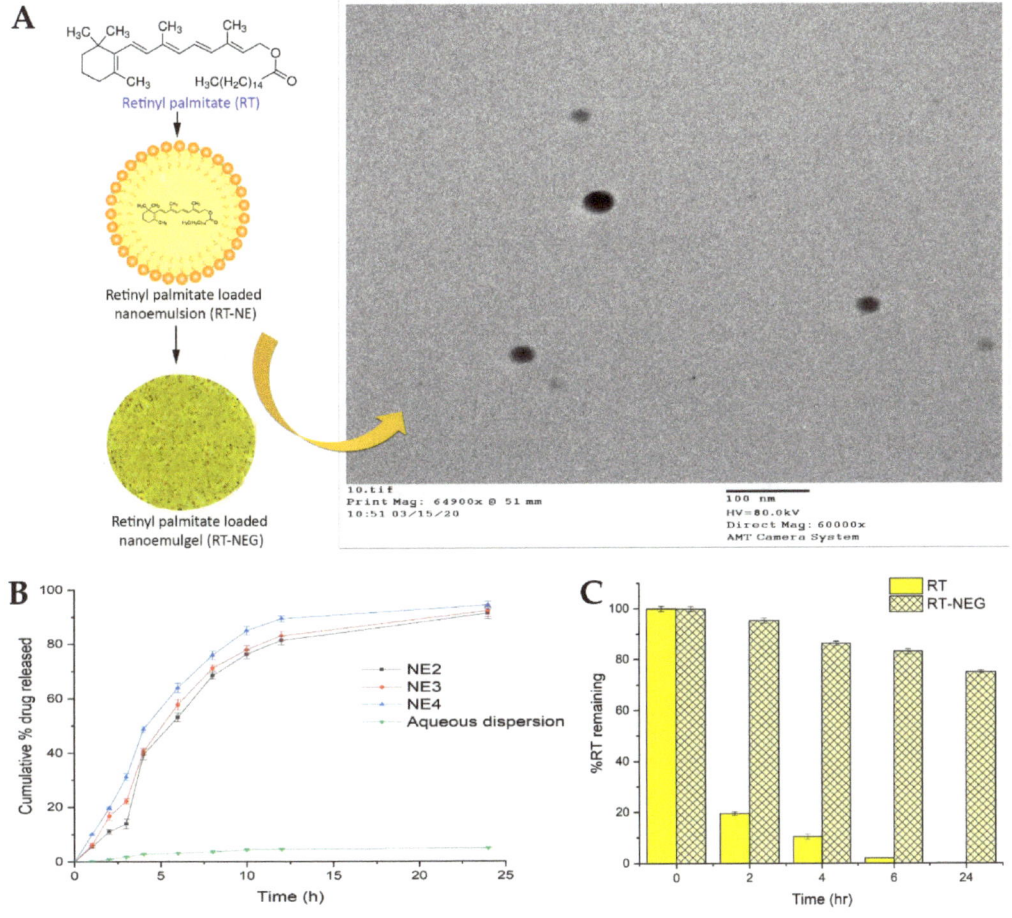

**Figure 14.** Cont.

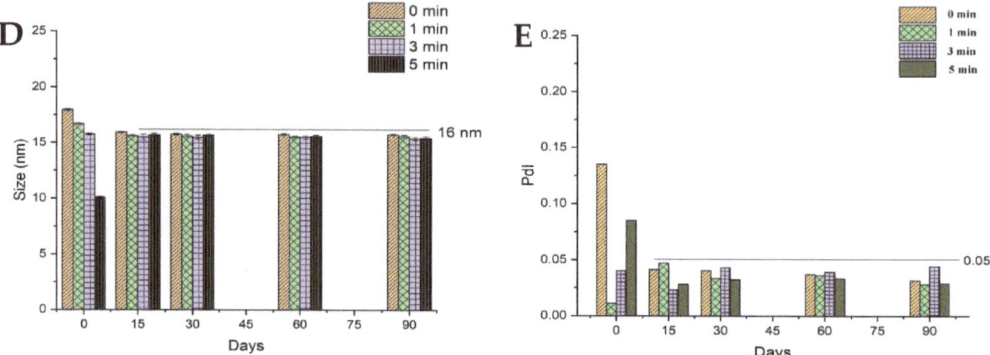

**Figure 14.** (**A**) Schematic representation of the developed retinyl palmitate topical nanoemulgel (left), with corresponding transmission electron microscopy image showing droplet morphology and size (right); (**B**) in vitro drug release profiles of different retinyl palmitate-loaded preliminary nanoemulsions, compared to an aqueous dispersion of the drug; (**C**) UV stability of the developed retinyl palmitate nanoemulgel, compared to the nonencapsulated drug; (**D**) mean droplet size variations of the preliminary optimized retinyl palmitate-loaded nanoemulsions as effect of storage time; (**E**) PDI variations of the preliminary optimized retinyl palmitate-loaded nanoemulsions as effect of storage time; NE—nanoemulsion; NEG—nanoemulgel; PDI—polydispersity index; RT—retinyl palmitate; adapted from Algahtani et al. [163], reproduced with permission from MDPI (Creative Commons CC BY 4.0 470 license).

## 3. Novel Nanoemulgels for Skin Application: The Future of Topical and Transdermal Drug Delivery?

Molecules that have been recently incorporated into nanoemulgels for skin application include repurposed marketed drugs, such as atorvastatin for wound healing [149], omeprazole for skin infections [155], and leflunomide for skin inflammatory diseases [146]. Natural-derived compounds have also had increased attention, with eucalyptol [150], naringenin [151], thymoquinone [152], and curcumin [153] being formulated for wound healing purposes, curcumin [153] and chrysin [142] for skin cancer, brucine [161] for skin inflammatory diseases, and capsaicin for neuropathy [162]. Other encapsulated compounds include marketed drugs such as miconazole [154], ketoconazole [157], and fusidic acid [158] for the treatment of skin and skin appendage infections, imiquimod [159] for skin cancer, and meloxicam [160] for skin inflammatory diseases, as well as the synthetic compound ebselen [156] as an anti-infection agent, vitamin E derivative tocotrienols for wound healing [151], and retinyl palmitate [163] for antiaging purposes. In what concerns formulation aspects, although spontaneous emulsification was sometimes used in the analyzed studies, most had to resort to high-pressure or high-speed homogenization, sometimes followed by ultrasonication, probably due to the formulations' components not having ideal miscibility and molecular compatibility, and hence not being able to form nanoemulsions without the help of high-energy methods. In what concerns formulation components, various oils were used, with eucalyptus oil, black seed oil, and myrrh oil being the most applied, followed by other natural-derived oils, such as almond oil or olive oil, as well as oleic acid, liquid paraffin, and Captex 300 EP/NF or Captex 355. The oils were either selected based on potential biocompatibility, highest drug/active compound solubilization capacity, or both. Hydrophobic surfactants were also at times used as oils, replacing the latter, with Capryol 90 being the most used, followed by Span 80 or Span 60, and Labrafac PG or Labrafac Lipophile WL1349. The advantages of using hydrophobic surfactants with the additional function of oils reside in the fact that generally a highest drug solubilization is achieved, hence increasing the obtainable final formulation drug strength, and also in the fact that the absence of a hydrophobic component with no surfactant capability increases the chances of obtaining a nanoemulsion/nanoemulgel with lower droplet size and PDI, also augmenting the chances for spontaneous emulsification to be possible (instead of having to resort to high-energy fabrication methods). In the hydrophilic surfactants category, Tween 80 was by far the most applied, followed by Kolliphor EL, with Tween 20 and Solutol HS15 also being used. All these surfactants have been proven to have the capacity to increase skin drug permeation and retention by potentially being able to increase the skin barrier lipids' fluidity and hence reduce the *stratum corneum*'s barrier function, hence making them adequate for skin drug delivery [260–263]. In what concerns cosolvents/cosurfactants, Transcutol was the most utilized. This potent cosolvent has also been reported to readily penetrate the *stratum corneum*, and once there, to modify its molecular mobility, namely, the protein and lipids that are part of its composition, hence decreasing the skin's barrier function [264,265]. Other cosolvents/cosurfactants included PEG 400, propylene glycol, ethyl alcohol, isopropyl alcohol, Soluplus, and dimethylacetamide. And, finally, in what concerns gelling agents, the most used were in fact Carbopol varieties, such as Carbopol 940, Carbopol 934, or Carbopol Ultrez 21, followed by cellulose derivatives, such as CMC or HPMC, and with Pluronic F127 and chitosan also being applied. These gelling agents increase the formulations' viscosity, hence resulting in an adequate consistency for topical application, with some also having bioadhesive properties, hence increasing the time of formulation retention at the application site, and with others even having drug permeation enhancement capacity as well, therefore contributing to a potentially better therapeutic outcome [266–268].

Regardless of formulation composition or preparation method, all developed nanoemulgels exhibited small droplet size and PDI values, resulting in good stability during storage or under stress conditions, adequate viscosity and spreadability, pH values compatible with the skin, controlled drug release profiles, and adequate ex vivo skin drug permeation and/or retention, at times even performing better than currently marketed formulations. Furthermore, in vitro and/or in vivo studies confirmed the safety and efficacy of the developed formulations for anti-infectious, anti-inflammatory, antitumor, wound-healing, antinociceptive, and/or antiaging effects, confirming their efficacy and therapeutic potential, and making them promising platforms for the replacement of current therapies, or as possible adjuvant treatments.

While clinical application is still further down the road, these systems have proven to have great potential for skin drug delivery. More work should be done to assess these formulations' true safety and long-term applicability, in the context of the current regulatory framework, in order to determine the ability of these nanoformulations for bioaccumulation and interference with the body's immune system (phagocytosis by immune cells), although this might be a bigger concern for transdermal drug delivery than for topical drug delivery [269,270]. Additionally, scale-up potential should also be determined, in order to assess the limitations facing any transition from a laboratory bench scale to an industrial manufacturing scale, so that these promising preparations can one day reach the pharmaceutical market [108,271–273].

## 4. Conclusions

Novel topical and transdermal nanoemulgels have been developed for skin application, encapsulating a wide variety of molecules, such as already marketed drugs (miconazole, ketoconazole, fusidic acid, imiquimod, meloxicam), repurposed marketed drugs (atorvastatin, omeprazole, leflunomide), natural-derived compounds (eucalyptol, naringenin, thymoquinone, curcumin, chrysin, brucine, capsaicin), and other synthetic molecules (ebselen, tocotrienols, retinyl palmitate), for wound healing, skin and skin appendage infections, skin inflammatory diseases, skin cancer, neuropathy, or antiaging purposes. All developed nanoemulgels had adequate droplet size, PDI, viscosity, spreadability, pH, stability, drug release, and drug permeation and/or retention capacity, having more advantageous characteristics than currently marketed formulations. In vitro and/or in vivo studies confirmed the safety and efficacy of the developed formulations, confirming their therapeutic potential and making them promising platforms for the replacement of current therapies, or as possible adjuvant treatments. Further studies will tell if these novel functional platforms for drug delivery might someday effectively reach the market to help fight highly incident skin or systemic diseases and conditions.

**Author Contributions:** Conceptualization, P.C.P., A.C.P.-S. and F.V.; methodology, P.C.P.; validation, A.C.P.-S. and F.V.; formal analysis, A.C.P.-S. and F.V.; investigation, P.C.P.; writing—original draft preparation, K.S., M.M., P.C.P. and A.C.P.-S.; writing—review and editing, F.V., P.C.P. and A.C.P.-S. All authors have read and agreed to the published version of the manuscript.

**Funding:** This research received no external funding.

**Institutional Review Board Statement:** Not applicable.

**Informed Consent Statement:** Not applicable.

**Data Availability Statement:** Not applicable.

**Conflicts of Interest:** The authors declare no conflicts of interest.

**Abbreviations**

ATR—atorvastatin; CMC—carboxymethyl cellulose; CNTs—carbon nanotubes; CUR—curcumin; DMA—dimethylacetamide; EB—ebselen; FA—fusidic acid; GRAS—generally regarded as safe; HPMC—hydroxypropyl methylcellulose; IMQ—imiquimod; MIC—minimum inhibitory concentration; MSNs—mesoporous silica nanoparticles; MX—meloxicam; NPs—nanoparticles; O/W—oil-in-water; PBS—phosphate buffer saline; PCL—polycaprolactone; PDI—polydispersity index; PEG—polyethylene glycol; PLGA—poly(lactic-co-glycolic acid; PMs—polymeric micelles; PNPs—polymeric nanoparticles; RT—retinyl palmitate; SEM—scanning electron microscopy; SLNs—solid lipid nanoparticles; TMQ—thymoquinone; W/O—water-in-oil; UV—ultraviolet.

## References

1. Arda, O.; Göksügür, N.; Tüzün, Y. Basic Histological Structure and Functions of Facial Skin. *Clin. Dermatol.* **2014**, *32*, 3–13. [CrossRef] [PubMed]
2. Boer, M.; Duchnik, E.; Maleszka, R.; Marchlewicz, M. Structural and Biophysical Characteristics of Human Skin in Maintaining Proper Epidermal Barrier Function. *Adv. Dermatol. Allergol.* **2016**, *1*, 1–5. [CrossRef] [PubMed]
3. Abdo, J.M.; Sopko, N.A.; Milner, S.M. The Applied Anatomy of Human Skin: A Model for Regeneration. *Wound Med.* **2020**, *28*, 100179. [CrossRef]
4. Baroni, A.; Buommino, E.; De Gregorio, V.; Ruocco, E.; Ruocco, V.; Wolf, R. Structure and Function of the Epidermis Related to Barrier Properties. *Clin. Dermatol.* **2012**, *30*, 257–262. [CrossRef] [PubMed]
5. Xie, J.; Hao, T.; Li, C.; Wang, X.; Yu, X.; Liu, L. Automatic Evaluation of Stratum Basale and Dermal Papillae Using Ultrahigh Resolution Optical Coherence Tomography. *Biomed. Signal Process Control* **2019**, *53*, 101527. [CrossRef]
6. Maynard, R.L.; Downes, N. The Skin or the Integument. In *Anatomy and Histology of the Laboratory Rat in Toxicology and Biomedical Research*; Elsevier: Amsterdam, The Netherlands, 2019; pp. 303–315.
7. Barbieri, J.S.; Wanat, K.; Seykora, J. Skin: Basic Structure and Function. In *Pathobiology of Human Disease*; Elsevier: Amsterdam, The Netherlands, 2014; pp. 1134–1144.
8. McBain, A.J.; O'Neill, C.A.; Oates, A. Skin Microbiology. In *Reference Module in Biomedical Sciences*; Elsevier: Amsterdam, The Netherlands, 2016.
9. Proksch, E.; Brandner, J.M.; Jensen, J. The Skin: An Indispensable Barrier. *Exp. Dermatol.* **2008**, *17*, 1063–1072. [CrossRef] [PubMed]
10. Nishifuji, K.; Yoon, J.S. The Stratum Corneum: The Rampart of the Mammalian Body. *Vet. Dermatol.* **2013**, *24*, 60. [CrossRef]
11. Tagami, H. Location-related Differences in Structure and Function of the Stratum Corneum with Special Emphasis on Those of the Facial Skin. *Int. J. Cosmet. Sci.* **2008**, *30*, 413–434. [CrossRef]
12. Rippa, A.L.; Kalabusheva, E.P.; Vorotelyak, E.A. Regeneration of Dermis: Scarring and Cells Involved. *Cells* **2019**, *8*, 607. [CrossRef]
13. Carroll, R.G. The Integument. In *Elsevier's Integrated Physiology*; Elsevier: Amsterdam, The Netherlands, 2007; pp. 11–17.
14. Carlson, B.M. Integumentary, Skeletal, and Muscular Systems. In *Human Embryology and Developmental Biology*; Elsevier: Amsterdam, The Netherlands, 2014; pp. 156–192.
15. Wong, R.; Geyer, S.; Weninger, W.; Guimberteau, J.; Wong, J.K. The Dynamic Anatomy and Patterning of Skin. *Exp. Dermatol.* **2016**, *25*, 92–98. [CrossRef]

16. Woo, W. Skin Structure and Biology. In *Imaging Technologies and Transdermal Delivery in Skin Disorders*; Wiley: Hoboken, NJ, USA, 2019; pp. 1–14.
17. Roberts, W. Air Pollution and Skin Disorders. *Int. J. Womens Dermatol.* **2021**, *7*, 91–97. [CrossRef] [PubMed]
18. Ju, Q.; Zouboulis, C.C. Endocrine-Disrupting Chemicals and Skin Manifestations. *Rev. Endocr. Metab. Disord.* **2016**, *17*, 449–457. [CrossRef] [PubMed]
19. Gromkowska-Kępka, K.J.; Puścion-Jakubik, A.; Markiewicz-Żukowska, R.; Socha, K. The Impact of Ultraviolet Radiation on Skin Photoaging—Review of in Vitro Studies. *J. Cosmet. Dermatol.* **2021**, *20*, 3427–3431. [CrossRef] [PubMed]
20. Cao, C.; Xiao, Z.; Wu, Y.; Ge, C. Diet and Skin Aging—From the Perspective of Food Nutrition. *Nutrients* **2020**, *12*, 870. [CrossRef] [PubMed]
21. Ortiz, A.; Grando, S.A. Smoking and the Skin. *Int. J. Dermatol.* **2012**, *51*, 250–262. [CrossRef]
22. Liu, S.W.; Lien, M.H.; Fenske, N.A. The Effects of Alcohol and Drug Abuse on the Skin. *Clin. Dermatol.* **2010**, *28*, 391–399. [CrossRef]
23. Lyu, F.; Wu, T.; Bian, Y.; Zhu, K.; Xu, J.; Li, F. Stress and Its Impairment of Skin Barrier Function. *Int. J. Dermatol.* **2023**, *62*, 621–630. [CrossRef]
24. Dąbrowska, A.K.; Spano, F.; Derler, S.; Adlhart, C.; Spencer, N.D.; Rossi, R.M. The Relationship between Skin Function, Barrier Properties, and Body-dependent Factors. *Ski. Res. Technol.* **2018**, *24*, 165–174. [CrossRef]
25. Ita, K.; Silva, M.; Bassey, R. Mechanical Properties of the Skin: What Do We Know? *Curr. Cosmet. Sci.* **2022**, *1*, e070122200109. [CrossRef]
26. Mortazavi, S.M.; Moghimi, H.R. Skin Permeability, a Dismissed Necessity for Anti-wrinkle Peptide Performance. *Int. J. Cosmet. Sci.* **2022**, *44*, 232–248. [CrossRef]
27. Parhi, R.; Mandru, A. Enhancement of Skin Permeability with Thermal Ablation Techniques: Concept to Commercial Products. *Drug Deliv. Transl. Res.* **2021**, *11*, 817–841. [CrossRef] [PubMed]
28. Lundborg, M.; Wennberg, C.L.; Narangifard, A.; Lindahl, E.; Norlén, L. Predicting Drug Permeability through Skin Using Molecular Dynamics Simulation. *J. Control. Release* **2018**, *283*, 269–279. [CrossRef] [PubMed]
29. Alkilani, A.; McCrudden, M.T.; Donnelly, R. Transdermal Drug Delivery: Innovative Pharmaceutical Developments Based on Disruption of the Barrier Properties of the Stratum Corneum. *Pharmaceutics* **2015**, *7*, 438–470. [CrossRef] [PubMed]
30. Yu, Y.-Q.; Yang, X.; Wu, X.-F.; Fan, Y.-B. Enhancing Permeation of Drug Molecules Across the Skin via Delivery in Nanocarriers: Novel Strategies for Effective Transdermal Applications. *Front. Bioeng. Biotechnol.* **2021**, *9*, 646554. [CrossRef] [PubMed]
31. Narasimha Murthy, S.; Shivakumar, H.N. Topical and Transdermal Drug Delivery. In *Handbook of Non-Invasive Drug Delivery Systems*; Elsevier: Amsterdam, The Netherlands, 2010; pp. 1–36.
32. Kathe, K.; Kathpalia, H. Film Forming Systems for Topical and Transdermal Drug Delivery. *Asian J. Pharm. Sci.* **2017**, *12*, 487–497. [CrossRef] [PubMed]
33. Leppert, W.; Malec–Milewska, M.; Zajaczkowska, R.; Wordliczek, J. Transdermal and Topical Drug Administration in the Treatment of Pain. *Molecules* **2018**, *23*, 681. [CrossRef] [PubMed]
34. Prausnitz, M.R.; Langer, R. Transdermal Drug Delivery. *Nat. Biotechnol.* **2008**, *26*, 1261–1268. [CrossRef]
35. Majdi, A.; Sadigh-Eteghad, S.; Gjedde, A. Effects of Transdermal Nicotine Delivery on Cognitive Outcomes: A Meta-analysis. *Acta Neurol. Scand.* **2021**, *144*, 179–191. [CrossRef]
36. Buster, J.E. Transdermal Menopausal Hormone Therapy: Delivery through Skin Changes the Rules. *Expert Opin. Pharmacother.* **2010**, *11*, 1489–1499. [CrossRef]
37. Sittl, R. Transdermal Buprenorphine in the Treatment of Chronic Pain. *Expert Rev. Neurother.* **2005**, *5*, 315–323. [CrossRef]
38. Rehman, K.; Zulfakar, M.H. Recent Advances in Gel Technologies for Topical and Transdermal Drug Delivery. *Drug Dev. Ind. Pharm.* **2014**, *40*, 433–440. [CrossRef] [PubMed]
39. Waghule, T.; Singhvi, G.; Dubey, S.K.; Pandey, M.M.; Gupta, G.; Singh, M.; Dua, K. Microneedles: A Smart Approach and Increasing Potential for Transdermal Drug Delivery System. *Biomed. Pharmacother.* **2019**, *109*, 1249–1258. [CrossRef] [PubMed]
40. Sawarkar, S.; Ashtekar, A. Transdermal Vitamin D Supplementation—A Potential Vitamin D Deficiency Treatment. *J. Cosmet. Dermatol.* **2020**, *19*, 28–32. [CrossRef] [PubMed]
41. Houck, C.S.; Sethna, N.F. Transdermal Analgesia with Local Anesthetics in Children: Review, Update and Future Directions. *Expert Rev. Neurother.* **2005**, *5*, 625–634. [CrossRef] [PubMed]
42. Touitou, E.; Natsheh, H. Topical Administration of Drugs Incorporated in Carriers Containing Phospholipid Soft Vesicles for the Treatment of Skin Medical Conditions. *Pharmaceutics* **2021**, *13*, 2129. [CrossRef] [PubMed]
43. Singh Malik, D.; Mital, N.; Kaur, G. Topical Drug Delivery Systems: A Patent Review. *Expert Opin. Ther. Pat.* **2016**, *26*, 213–228. [CrossRef]
44. Roberts, M.S.; Cheruvu, H.S.; Mangion, S.E.; Alinaghi, A.; Benson, H.A.E.; Mohammed, Y.; Holmes, A.; van der Hoek, J.; Pastore, M.; Grice, J.E. Topical Drug Delivery: History, Percutaneous Absorption, and Product Development. *Adv. Drug Deliv. Rev.* **2021**, *177*, 113929. [CrossRef]
45. Bonamonte, D.; De Marco, A.; Giuffrida, R.; Conforti, C.; Barlusconi, C.; Foti, C.; Romita, P. Topical Antibiotics in the Dermatological Clinical Practice: Indications, Efficacy, and Adverse Effects. *Dermatol. Ther.* **2020**, *33*, e13824. [CrossRef]
46. Lé, A.M.; Torres, T. New Topical Therapies for Psoriasis. *Am. J. Clin. Dermatol.* **2022**, *23*, 13–24. [CrossRef]

47. Lax, S.J.; Harvey, J.; Axon, E.; Howells, L.; Santer, M.; Ridd, M.J.; Lawton, S.; Langan, S.; Roberts, A.; Ahmed, A.; et al. Strategies for Using Topical Corticosteroids in Children and Adults with Eczema. *Cochrane Database Syst. Rev.* **2022**, *2022*, CD013356. [CrossRef]
48. Piraccini, B.M.; Blume-Peytavi, U.; Scarci, F.; Jansat, J.M.; Falqués, M.; Otero, R.; Tamarit, M.L.; Galván, J.; Tebbs, V.; Massana, E. Efficacy and Safety of Topical Finasteride Spray Solution for Male Androgenetic Alopecia: A Phase III, Randomized, Controlled Clinical Trial. *J. Eur. Acad. Dermatol. Venereol.* **2022**, *36*, 286–294. [CrossRef] [PubMed]
49. Ivens, U.I.; Steinkjer, B.; Serup, J.; Tetens, V. Ointment Is Evenly Spread on the Skin, in Contrast to Creams and Solutions. *Br. J. Dermatol.* **2001**, *145*, 264–267. [CrossRef] [PubMed]
50. Ridd, M.J.; Santer, M.; MacNeill, S.J.; Sanderson, E.; Wells, S.; Webb, D.; Banks, J.; Sutton, E.; Roberts, A.; Liddiard, L.; et al. Effectiveness and Safety of Lotion, Cream, Gel, and Ointment Emollients for Childhood Eczema: A Pragmatic, Randomised, Phase 4, Superiority Trial. *Lancet Child. Adolesc. Health* **2022**, *6*, 522–532. [CrossRef] [PubMed]
51. Zhang, Q.; Grice, J.; Wang, G.; Roberts, M. Cutaneous Metabolism in Transdermal Drug Delivery. *Curr. Drug Metab.* **2009**, *10*, 227–235. [CrossRef] [PubMed]
52. Svensson, C.K. Biotransformation of Drugs in Human Skin. *Drug Metab. Dispos.* **2009**, *37*, 247–253. [CrossRef] [PubMed]
53. Mugglestone, C.J.; Mariz, S.; Lane, M.E. The Development and Registration of Topical Pharmaceuticals. *Int. J. Pharm.* **2012**, *435*, 22–26. [CrossRef] [PubMed]
54. Patra, J.K.; Das, G.; Fraceto, L.F.; Campos, E.V.R.; Rodriguez-Torres, M.D.P.; Acosta-Torres, L.S.; Diaz-Torres, L.A.; Grillo, R.; Swamy, M.K.; Sharma, S.; et al. Nano Based Drug Delivery Systems: Recent Developments and Future Prospects. *J. Nanobiotechnol.* **2018**, *16*, 71. [CrossRef]
55. Sultana, A.; Zare, M.; Thomas, V.; Kumar, T.S.S.; Ramakrishna, S. Nano-Based Drug Delivery Systems: Conventional Drug Delivery Routes, Recent Developments and Future Prospects. *Med. Drug Discov.* **2022**, *15*, 100134. [CrossRef]
56. Demetzos, C.; Pippa, N. Advanced Drug Delivery Nanosystems (ADDnSs): A Mini-Review. *Drug Deliv.* **2014**, *21*, 250–257. [CrossRef]
57. Pires, P.C.; Santos, A.O. Nanosystems in Nose-to-Brain Drug Delivery: A Review of Non-Clinical Brain Targeting Studies. *J. Control. Release* **2018**, *270*, 89–100. [CrossRef]
58. Zhang, L.; Gu, F.; Chan, J.; Wang, A.; Langer, R.; Farokhzad, O. Nanoparticles in Medicine: Therapeutic Applications and Developments. *Clin. Pharmacol. Ther.* **2008**, *83*, 761–769. [CrossRef] [PubMed]
59. Rezić, I. Nanoparticles for Biomedical Application and Their Synthesis. *Polymers* **2022**, *14*, 4961. [CrossRef] [PubMed]
60. Mitchell, M.J.; Billingsley, M.M.; Haley, R.M.; Wechsler, M.E.; Peppas, N.A.; Langer, R. Engineering Precision Nanoparticles for Drug Delivery. *Nat. Rev. Drug Discov.* **2021**, *20*, 101–124. [CrossRef] [PubMed]
61. Afzal, O.; Altamimi, A.S.A.; Nadeem, M.S.; Alzarea, S.I.; Almalki, W.H.; Tariq, A.; Mubeen, B.; Murtaza, B.N.; Iftikhar, S.; Riaz, N.; et al. Nanoparticles in Drug Delivery: From History to Therapeutic Applications. *Nanomaterials* **2022**, *12*, 4494. [CrossRef] [PubMed]
62. Abdel-Mageed, H.M.; AbuelEzz, N.Z.; Radwan, R.A.; Mohamed, S.A. Nanoparticles in Nanomedicine: A Comprehensive Updated Review on Current Status, Challenges and Emerging Opportunities. *J. Microencapsul.* **2021**, *38*, 414–436. [CrossRef] [PubMed]
63. Ferreira, M.D.; Duarte, J.; Veiga, F.; Paiva-Santos, A.C.; Pires, P.C. Nanosystems for Brain Targeting of Antipsychotic Drugs: An Update on the Most Promising Nanocarriers for Increased Bioavailability and Therapeutic Efficacy. *Pharmaceutics* **2023**, *15*, 678. [CrossRef] [PubMed]
64. Yusuf, A.; Almotairy, A.R.Z.; Henidi, H.; Alshehri, O.Y.; Aldughaim, M.S. Nanoparticles as Drug Delivery Systems: A Review of the Implication of Nanoparticles' Physicochemical Properties on Responses in Biological Systems. *Polymers* **2023**, *15*, 1596. [CrossRef]
65. Sechi, M.; Sanna, V.; Pala, N. Targeted Therapy Using Nanotechnology: Focus on Cancer. *Int. J. Nanomed.* **2014**, *9*, 467–483. [CrossRef]
66. Khan, I.; Saeed, K.; Khan, I. Nanoparticles: Properties, Applications and Toxicities. *Arab. J. Chem.* **2019**, *12*, 908–931. [CrossRef]
67. Christian, P.; Von der Kammer, F.; Baalousha, M.; Hofmann, T. Nanoparticles: Structure, Properties, Preparation and Behaviour in Environmental Media. *Ecotoxicology* **2008**, *17*, 326–343. [CrossRef]
68. Hore, M.J.A. Polymers on Nanoparticles: Structure & Dynamics. *Soft Matter* **2019**, *15*, 1120–1134. [CrossRef] [PubMed]
69. Moradifar, N.; Kiani, A.A.; Veiskaramian, A.; Karami, K. Role of Organic and Inorganic Nanoparticles in the Drug Delivery System for Hypertension Treatment: A Systematic Review. *Curr. Cardiol. Rev.* **2022**, *18*, e110621194025. [CrossRef] [PubMed]
70. Khalid, K.; Tan, X.; Mohd Zaid, H.F.; Tao, Y.; Lye Chew, C.; Chu, D.-T.; Lam, M.K.; Ho, Y.-C.; Lim, J.W.; Chin Wei, L. Advanced in Developmental Organic and Inorganic Nanomaterial: A Review. *Bioengineered* **2020**, *11*, 328–355. [CrossRef] [PubMed]
71. Alshammari, B.H.; Lashin, M.M.A.; Mahmood, M.A.; Al-Mubaddel, F.S.; Ilyas, N.; Rahman, N.; Sohail, M.; Khan, A.; Abdullaev, S.S.; Khan, R. Organic and Inorganic Nanomaterials: Fabrication, Properties and Applications. *RSC Adv.* **2023**, *13*, 13735–13785. [CrossRef] [PubMed]
72. Jafari, S.; Derakhshankhah, H.; Alaei, L.; Fattahi, A.; Varnamkhasti, B.S.; Saboury, A.A. Mesoporous Silica Nanoparticles for Therapeutic/Diagnostic Applications. *Biomed. Pharmacother.* **2019**, *109*, 1100–1111. [CrossRef]
73. Li, Z.; Zhang, Y.; Feng, N. Mesoporous Silica Nanoparticles: Synthesis, Classification, Drug Loading, Pharmacokinetics, Biocompatibility, and Application in Drug Delivery. *Expert Opin. Drug Deliv.* **2019**, *16*, 219–237. [CrossRef]

74. Jain, P.; Hassan, N.; Iqbal, Z.; Dilnawaz, F. Mesoporous Silica Nanoparticles: A Versatile Platform for Biomedical Applications. *Recent. Pat. Drug Deliv. Formul.* **2019**, *12*, 228–237. [CrossRef]
75. Porrang, S.; Davaran, S.; Rahemi, N.; Allahyari, S.; Mostafavi, E. How Advancing Are Mesoporous Silica Nanoparticles? A Comprehensive Review of the Literature. *Int. J. Nanomed.* **2022**, *17*, 1803–1827. [CrossRef]
76. Huang, R.; Shen, Y.-W.; Guan, Y.-Y.; Jiang, Y.-X.; Wu, Y.; Rahman, K.; Zhang, L.-J.; Liu, H.-J.; Luan, X. Mesoporous Silica Nanoparticles: Facile Surface Functionalization and Versatile Biomedical Applications in Oncology. *Acta Biomater.* **2020**, *116*, 1–15. [CrossRef]
77. He, H.; Pham-Huy, L.A.; Dramou, P.; Xiao, D.; Zuo, P.; Pham-Huy, C. Carbon Nanotubes: Applications in Pharmacy and Medicine. *Biomed. Res. Int.* **2013**, *2013*, 578290. [CrossRef]
78. Rahamathulla, M.; Bhosale, R.R.; Osmani, R.A.M.; Mahima, K.C.; Johnson, A.P.; Hani, U.; Ghazwani, M.; Begum, M.Y.; Alshehri, S.; Ghoneim, M.M.; et al. Carbon Nanotubes: Current Perspectives on Diverse Applications in Targeted Drug Delivery and Therapies. *Materials* **2021**, *14*, 6707. [CrossRef] [PubMed]
79. Zhang, C.; Wu, L.; de Perrot, M.; Zhao, X. Carbon Nanotubes: A Summary of Beneficial and Dangerous Aspects of an Increasingly Popular Group of Nanomaterials. *Front. Oncol.* **2021**, *11*, 693814. [CrossRef] [PubMed]
80. Zare, H.; Ahmadi, S.; Ghasemi, A.; Ghanbari, M.; Rabiee, N.; Bagherzadeh, M.; Karimi, M.; Webster, T.J.; Hamblin, M.R.; Mostafavi, E. Carbon Nanotubes: Smart Drug/Gene Delivery Carriers. *Int. J. Nanomed.* **2021**, *16*, 1681–1706. [CrossRef] [PubMed]
81. Pu, Z.; Wei, Y.; Sun, Y.; Wang, Y.; Zhu, S. Carbon Nanotubes as Carriers in Drug Delivery for Non-Small Cell Lung Cancer, Mechanistic Analysis of Their Carcinogenic Potential, Safety Profiling and Identification of Biomarkers. *Int. J. Nanomed.* **2022**, *17*, 6157–6180. [CrossRef] [PubMed]
82. Din, F.U.; Aman, W.; Ullah, I.; Qureshi, O.S.; Mustapha, O.; Shafique, S.; Zeb, A. Effective Use of Nanocarriers as Drug Delivery Systems for the Treatment of Selected Tumors. *Int. J. Nanomed.* **2017**, *12*, 7291–7309. [CrossRef]
83. Francis, A.P.; Devasena, T. Toxicity of Carbon Nanotubes: A Review. *Toxicol. Ind. Health* **2018**, *34*, 200–210. [CrossRef]
84. Kobayashi, N.; Izumi, H.; Morimoto, Y. Review of Toxicity Studies of Carbon Nanotubes. *J. Occup. Health* **2017**, *59*, 394–407. [CrossRef]
85. López-Dávila, V.; Seifalian, A.M.; Loizidou, M. Organic Nanocarriers for Cancer Drug Delivery. *Curr. Opin. Pharmacol.* **2012**, *12*, 414–419. [CrossRef]
86. Palazzolo, S.; Bayda, S.; Hadla, M.; Caligiuri, I.; Corona, G.; Toffoli, G.; Rizzolio, F. The Clinical Translation of Organic Nanomaterials for Cancer Therapy: A Focus on Polymeric Nanoparticles, Micelles, Liposomes and Exosomes. *Curr. Med. Chem.* **2018**, *25*, 4224–4268. [CrossRef]
87. Calzoni, E.; Cesaretti, A.; Polchi, A.; Di Michele, A.; Tancini, B.; Emiliani, C. Biocompatible Polymer Nanoparticles for Drug Delivery Applications in Cancer and Neurodegenerative Disorder Therapies. *J. Funct. Biomater.* **2019**, *10*, 4. [CrossRef]
88. Hadinoto, K.; Sundaresan, A.; Cheow, W.S. Lipid–Polymer Hybrid Nanoparticles as a New Generation Therapeutic Delivery Platform: A Review. *Eur. J. Pharm. Biopharm.* **2013**, *85*, 427–443. [CrossRef]
89. Bulbake, U.; Doppalapudi, S.; Kommineni, N.; Khan, W. Liposomal Formulations in Clinical Use: An Updated Review. *Pharmaceutics* **2017**, *9*, 12. [CrossRef] [PubMed]
90. Nsairat, H.; Khater, D.; Sayed, U.; Odeh, F.; Al Bawab, A.; Alshaer, W. Liposomes: Structure, Composition, Types, and Clinical Applications. *Heliyon* **2022**, *8*, e09394. [CrossRef] [PubMed]
91. Liu, P.; Chen, G.; Zhang, J. A Review of Liposomes as a Drug Delivery System: Current Status of Approved Products, Regulatory Environments, and Future Perspectives. *Molecules* **2022**, *27*, 1372. [CrossRef] [PubMed]
92. Pires, P.C.; Paiva-Santos, A.C.; Veiga, F. Liposome-Derived Nanosystems for the Treatment of Behavioral and Neurodegenerative Diseases: The Promise of Niosomes, Transfersomes, and Ethosomes for Increased Brain Drug Bioavailability. *Pharmaceuticals* **2023**, *16*, 1424. [CrossRef] [PubMed]
93. Sercombe, L.; Veerati, T.; Moheimani, F.; Wu, S.Y.; Sood, A.K.; Hua, S. Advances and Challenges of Liposome Assisted Drug Delivery. *Front. Pharmacol.* **2015**, *6*, 286. [CrossRef] [PubMed]
94. Taher, M.; Susanti, D.; Haris, M.S.; Rushdan, A.A.; Widodo, R.T.; Syukri, Y.; Khotib, J. PEGylated Liposomes Enhance the Effect of Cytotoxic Drug: A Review. *Heliyon* **2023**, *9*, e13823. [CrossRef]
95. Luo, W.-C.; Lu, X. Solid Lipid Nanoparticles for Drug Delivery. *Methods Mol. Biol.* **2023**, *2622*, 139–146. [CrossRef]
96. Paliwal, R.; Paliwal, S.R.; Kenwat, R.; Das Kurmi, B.; Sahu, M.K. Solid Lipid Nanoparticles: A Review on Recent Perspectives and Patents. *Expert Opin. Ther. Pat.* **2020**, *30*, 179–194. [CrossRef]
97. zur Mühlen, A.; Schwarz, C.; Mehnert, W. Solid Lipid Nanoparticles (SLN) for Controlled Drug Delivery—Drug Release and Release Mechanism. *Eur. J. Pharm. Biopharm.* **1998**, *45*, 149–155. [CrossRef]
98. Corzo, C.; Meindl, C.; Lochmann, D.; Reyer, S.; Salar-Behzadi, S. Novel Approach for Overcoming the Stability Challenges of Lipid-Based Excipients. Part 3: Application of Polyglycerol Esters of Fatty Acids for the next Generation of Solid Lipid Nanoparticles. *Eur. J. Pharm. Biopharm.* **2020**, *152*, 44–55. [CrossRef] [PubMed]
99. Jain, A.; Bhardwaj, K.; Bansal, M. Polymeric Micelles as Drug Delivery System: Recent Advances, Approaches, Applications and Patents. *Curr. Drug Saf.* **2024**, *19*, 163–171. [CrossRef] [PubMed]
100. Hwang, D.; Ramsey, J.D.; Kabanov, A.V. Polymeric Micelles for the Delivery of Poorly Soluble Drugs: From Nanoformulation to Clinical Approval. *Adv. Drug Deliv. Rev.* **2020**, *156*, 80–118. [CrossRef] [PubMed]

101. Ghosh, B.; Biswas, S. Polymeric Micelles in Cancer Therapy: State of the Art. *J. Control. Release* **2021**, *332*, 127–147. [CrossRef] [PubMed]
102. Zielińska, A.; Carreiró, F.; Oliveira, A.M.; Neves, A.; Pires, B.; Venkatesh, D.N.; Durazzo, A.; Lucarini, M.; Eder, P.; Silva, A.M.; et al. Polymeric Nanoparticles: Production, Characterization, Toxicology and Ecotoxicology. *Molecules* **2020**, *25*, 3731. [CrossRef]
103. Begines, B.; Ortiz, T.; Pérez-Aranda, M.; Martínez, G.; Merinero, M.; Argüelles-Arias, F.; Alcudia, A. Polymeric Nanoparticles for Drug Delivery: Recent Developments and Future Prospects. *Nanomaterials* **2020**, *10*, 1403. [CrossRef]
104. Makadia, H.K.; Siegel, S.J. Poly Lactic-Co-Glycolic Acid (PLGA) as Biodegradable Controlled Drug Delivery Carrier. *Polymers* **2011**, *3*, 1377–1397. [CrossRef]
105. Sharma, S.; Parveen, R.; Chatterji, B.P. Toxicology of Nanoparticles in Drug Delivery. *Curr. Pathobiol. Rep.* **2021**, *9*, 133–144. [CrossRef]
106. Perumal, S.; Atchudan, R.; Lee, W. A Review of Polymeric Micelles and Their Applications. *Polymers* **2022**, *14*, 2510. [CrossRef]
107. Desai, N. Challenges in Development of Nanoparticle-Based Therapeutics. *AAPS J.* **2012**, *14*, 282–295. [CrossRef]
108. Herdiana, Y.; Wathoni, N.; Shamsuddin, S.; Muchtaridi, M. Scale-up Polymeric-Based Nanoparticles Drug Delivery Systems: Development and Challenges. *OpenNano* **2022**, *7*, 100048. [CrossRef]
109. Muthu, M.S.; Wilson, B. Challenges Posed by the Scale-up of Nanomedicines. *Nanomedicine* **2012**, *7*, 307–309. [CrossRef] [PubMed]
110. Yukuyama, M.N.; Kato, E.T.M.; Lobenberg, R.; Bou-Chacra, N.A. Challenges and Future Prospects of Nanoemulsion as a Drug Delivery System. *Curr. Pharm. Des.* **2017**, *23*, 495–508. [CrossRef] [PubMed]
111. Singh, Y.; Meher, J.G.; Raval, K.; Khan, F.A.; Chaurasia, M.; Jain, N.K.; Chourasia, M.K. Nanoemulsion: Concepts, Development and Applications in Drug Delivery. *J. Control. Release* **2017**, *252*, 28–49. [CrossRef] [PubMed]
112. Pires, P.C.; Paiva-Santos, A.C.; Veiga, F. Nano and Microemulsions for the Treatment of Depressive and Anxiety Disorders: An Efficient Approach to Improve Solubility, Brain Bioavailability and Therapeutic Efficacy. *Pharmaceutics* **2022**, *14*, 2825. [CrossRef] [PubMed]
113. Mariyate, J.; Bera, A. A Critical Review on Selection of Microemulsions or Nanoemulsions for Enhanced Oil Recovery. *J. Mol. Liq.* **2022**, *353*, 118791. [CrossRef]
114. Azeem, A.; Rizwan, M.; Ahmad, F.J.; Iqbal, Z.; Khar, R.K.; Aqil, M.; Talegaonkar, S. Nanoemulsion Components Screening and Selection: A Technical Note. *AAPS PharmSciTech* **2009**, *10*, 69–76. [CrossRef] [PubMed]
115. Almeida, F.; Corrêa, M.; Zaera, A.M.; Garrigues, T.; Isaac, V. Influence of Different Surfactants on Development of Nanoemulsion Containing Fixed Oil from an Amazon Palm Species. *Colloids Surf. A Physicochem. Eng. Asp.* **2022**, *643*, 128721. [CrossRef]
116. Koroleva, M.; Nagovitsina, T.; Yurtov, E. Nanoemulsions Stabilized by Non-Ionic Surfactants: Stability and Degradation Mechanisms. *Phys. Chem. Chem. Phys.* **2018**, *20*, 10369–10377. [CrossRef]
117. Mushtaq, A.; Mohd Wani, S.; Malik, A.R.; Gull, A.; Ramniwas, S.; Ahmad Nayik, G.; Ercisli, S.; Alina Marc, R.; Ullah, R.; Bari, A. Recent Insights into Nanoemulsions: Their Preparation, Properties and Applications. *Food Chem. X* **2023**, *18*, 100684. [CrossRef]
118. Ashaolu, T.J. Nanoemulsions for Health, Food, and Cosmetics: A Review. *Environ. Chem. Lett.* **2021**, *19*, 3381–3395. [CrossRef]
119. Kotta, S.; Khan, A.W.; Ansari, S.H.; Sharma, R.K.; Ali, J. Formulation of Nanoemulsion: A Comparison between Phase Inversion Composition Method and High-Pressure Homogenization Method. *Drug Deliv.* **2015**, *22*, 455–466. [CrossRef]
120. Espitia, P.J.P.; Fuenmayor, C.A.; Otoni, C.G. Nanoemulsions: Synthesis, Characterization, and Application in Bio-Based Active Food Packaging. *Compr. Rev. Food Sci. Food Saf.* **2019**, *18*, 264–285. [CrossRef]
121. Yukuyama, M.N.; Ghisleni, D.D.M.; Pinto, T.J.A.; Bou-Chacra, N.A. Nanoemulsion: Process Selection and Application in Cosmetics—A Review. *Int. J. Cosmet. Sci.* **2016**, *38*, 13–24. [CrossRef]
122. Anton, N.; Vandamme, T.F. The Universality of Low-Energy Nano-Emulsification. *Int. J. Pharm.* **2009**, *377*, 142–147. [CrossRef]
123. Sadurní, N.; Solans, C.; Azemar, N.; García-Celma, M.J. Studies on the Formation of O/W Nano-Emulsions, by Low-Energy Emulsification Methods, Suitable for Pharmaceutical Applications. *Eur. J. Pharm. Sci.* **2005**, *26*, 438–445. [CrossRef]
124. Rao, J.; McClements, D.J. Stabilization of Phase Inversion Temperature Nanoemulsions by Surfactant Displacement. *J. Agric. Food Chem.* **2010**, *58*, 7059–7066. [CrossRef]
125. Ren, G.; Sun, Z.; Wang, Z.; Zheng, X.; Xu, Z.; Sun, D. Nanoemulsion Formation by the Phase Inversion Temperature Method Using Polyoxypropylene Surfactants. *J. Colloid. Interface Sci.* **2019**, *540*, 177–184. [CrossRef]
126. Chuesiang, P.; Siripatrawan, U.; Sanguandeekul, R.; McLandsborough, L.; Julian McClements, D. Optimization of Cinnamon Oil Nanoemulsions Using Phase Inversion Temperature Method: Impact of Oil Phase Composition and Surfactant Concentration. *J. Colloid. Interface Sci.* **2018**, *514*, 208–216. [CrossRef]
127. Bouchemal, K.; Briançon, S.; Perrier, E.; Fessi, H. Nano-Emulsion Formulation Using Spontaneous Emulsification: Solvent, Oil and Surfactant Optimisation. *Int. J. Pharm.* **2004**, *280*, 241–251. [CrossRef]
128. Akram, S.; Anton, N.; Omran, Z.; Vandamme, T. Water-in-Oil Nano-Emulsions Prepared by Spontaneous Emulsification: New Insights on the Formulation Process. *Pharmaceutics* **2021**, *13*, 1030. [CrossRef]
129. Lefebvre, G.; Riou, J.; Bastiat, G.; Roger, E.; Frombach, K.; Gimel, J.-C.; Saulnier, P.; Calvignac, B. Spontaneous Nano-Emulsification: Process Optimization and Modeling for the Prediction of the Nanoemulsion's Size and Polydispersity. *Int. J. Pharm.* **2017**, *534*, 220–228. [CrossRef]
130. Gupta, A. Nanoemulsions. In *Nanoparticles for Biomedical Applications*; Elsevier: Amsterdam, The Netherlands, 2020; pp. 371–384.
131. Mehanna, M.M.; Mneimneh, A.T. Formulation and Applications of Lipid-Based Nanovehicles: Spotlight on Self-Emulsifying Systems. *Adv. Pharm. Bull.* **2020**, *11*, 56–67. [CrossRef]

132. Preeti; Sambhakar, S.; Malik, R.; Bhatia, S.; Al Harrasi, A.; Rani, C.; Saharan, R.; Kumar, S.; Geeta; Sehrawat, R. Nanoemulsion: An Emerging Novel Technology for Improving the Bioavailability of Drugs. *Scientifica* **2023**, *2023*, 6640103. [CrossRef]
133. Gupta, A.; Eral, H.B.; Hatton, T.A.; Doyle, P.S. Nanoemulsions: Formation, Properties and Applications. *Soft Matter* **2016**, *12*, 2826–2841. [CrossRef]
134. Jaiswal, M.; Dudhe, R.; Sharma, P.K. Nanoemulsion: An Advanced Mode of Drug Delivery System. *3 Biotech* **2015**, *5*, 123–127. [CrossRef]
135. Sabjan, K.B.; Munawar, S.M.; Rajendiran, D.; Vinoji, S.K.; Kasinathan, K. Nanoemulsion as Oral Drug Delivery—A Review. *Curr. Drug Res. Rev.* **2020**, *12*, 4–15. [CrossRef]
136. Aithal, G.C.; Narayan, R.; Nayak, U.Y. Nanoemulgel: A Promising Phase in Drug Delivery. *Curr. Pharm. Des.* **2020**, *26*, 279–291. [CrossRef]
137. Choudhury, H.; Gorain, B.; Pandey, M.; Chatterjee, L.A.; Sengupta, P.; Das, A.; Molugulu, N.; Kesharwani, P. Recent Update on Nanoemulgel as Topical Drug Delivery System. *J. Pharm. Sci.* **2017**, *106*, 1736–1751. [CrossRef]
138. Anand, K.; Ray, S.; Rahman, M.; Shaharyar, A.; Bhowmik, R.; Bera, R.; Karmakar, S. Nano-Emulgel: Emerging as a Smarter Topical Lipidic Emulsion-Based Nanocarrier for Skin Healthcare Applications. *Recent. Pat. Antiinfect. Drug Discov.* **2019**, *14*, 16–35. [CrossRef]
139. Donthi, M.R.; Munnangi, S.R.; Krishna, K.V.; Saha, R.N.; Singhvi, G.; Dubey, S.K. Nanoemulgel: A Novel Nano Carrier as a Tool for Topical Drug Delivery. *Pharmaceutics* **2023**, *15*, 164. [CrossRef]
140. Sengupta, P.; Chatterjee, B. Potential and Future Scope of Nanoemulgel Formulation for Topical Delivery of Lipophilic Drugs. *Int. J. Pharm.* **2017**, *526*, 353–365. [CrossRef]
141. Salem, H.F.; Kharshoum, R.M.; Abou-Taleb, H.A.; Naguib, D.M. Nanosized Nasal Emulgel of Resveratrol: Preparation, Optimization, in Vitro Evaluation and in Vivo Pharmacokinetic Study. *Drug Dev. Ind. Pharm.* **2019**, *45*, 1624–1634. [CrossRef]
142. Nagaraja, S.; Basavarajappa, G.M.; Attimarad, M.; Pund, S. Topical Nanoemulgel for the Treatment of Skin Cancer: Proof-of-Technology. *Pharmaceutics* **2021**, *13*, 902. [CrossRef]
143. Vichare, R.; Crelli, C.; Liu, L.; Das, A.C.; McCallin, R.; Zor, F.; Kulahci, Y.; Gorantla, V.S.; Janjic, J.M. A Reversibly Thermoresponsive, Theranostic Nanoemulgel for Tacrolimus Delivery to Activated Macrophages: Formulation and In Vitro Validation. *Pharmaceutics* **2023**, *15*, 2372. [CrossRef]
144. Ansari, M.N.; Soliman, G.A.; Rehman, N.U.; Anwer, M.K. Crisaborole Loaded Nanoemulsion Based Chitosan Gel: Formulation, Physicochemical Characterization and Wound Healing Studies. *Gels* **2022**, *8*, 318. [CrossRef]
145. Jeengar, M.K.; Rompicharla, S.V.K.; Shrivastava, S.; Chella, N.; Shastri, N.R.; Naidu, V.G.M.; Sistla, R. Emu Oil Based Nano-Emulgel for Topical Delivery of Curcumin. *Int. J. Pharm.* **2016**, *506*, 222–236. [CrossRef]
146. Pund, S.; Pawar, S.; Gangurde, S.; Divate, D. Transcutaneous Delivery of Leflunomide Nanoemulgel: Mechanistic Investigation into Physicomechanical Characteristics, in Vitro Anti-Psoriatic and Anti-Melanoma Activity. *Int. J. Pharm.* **2015**, *487*, 148–156. [CrossRef]
147. Aggarwal, G.; Dhawan, B.; Harikumar, S. Enhanced Transdermal Permeability of Piroxicam through Novel Nanoemulgel Formulation. *Int. J. Pharm. Investig.* **2014**, *4*, 65. [CrossRef]
148. Lee, J.Y.; Lee, S.H.; Hwangbo, S.A.; Lee, T.G. A Comparison of Gelling Agents for Stable, Surfactant-Free Oil-in-Water Emulsions. *Materials* **2022**, *15*, 6462. [CrossRef]
149. Morsy, M.A.; Abdel-Latif, R.G.; Nair, A.B.; Venugopala, K.N.; Ahmed, A.F.; Elsewedy, H.S.; Shehata, T.M. Preparation and Evaluation of Atorvastatin-Loaded Nanoemulgel on Wound-Healing Efficacy. *Pharmaceutics* **2019**, *11*, 609. [CrossRef]
150. Rehman, A.; Iqbal, M.; Khan, B.A.; Khan, M.K.; Huwaimel, B.; Alshehri, S.; Alamri, A.H.; Alzhrani, R.M.; Bukhary, D.M.; Safhi, A.Y.; et al. Fabrication, In Vitro, and In Vivo Assessment of Eucalyptol-Loaded Nanoemulgel as a Novel Paradigm for Wound Healing. *Pharmaceutics* **2022**, *14*, 1971. [CrossRef]
151. Yeo, E.; Yew Chieng, C.J.; Choudhury, H.; Pandey, M.; Gorain, B. Tocotrienols-Rich Naringenin Nanoemulgel for the Management of Diabetic Wound: Fabrication, Characterization and Comparative in Vitro Evaluations. *Curr. Res. Pharmacol. Drug Discov.* **2021**, *2*, 100019. [CrossRef]
152. Algahtani, M.S.; Ahmad, M.Z.; Shaikh, I.A.; Abdel-Wahab, B.A.; Nourein, I.H.; Ahmad, J. Thymoquinone Loaded Topical Nanoemulgel for Wound Healing: Formulation Design and In-Vivo Evaluation. *Molecules* **2021**, *26*, 3863. [CrossRef]
153. Algahtani, M.S.; Ahmad, M.Z.; Nourein, I.H.; Albarqi, H.A.; Alyami, H.S.; Alyami, M.H.; Alqahtani, A.A.; Alasiri, A.; Algahtani, T.S.; Mohammed, A.A.; et al. Preparation and Characterization of Curcumin Nanoemulgel Utilizing Ultrasonication Technique for Wound Healing: In Vitro, Ex Vivo, and In Vivo Evaluation. *Gels* **2021**, *7*, 213. [CrossRef]
154. Tayah, D.Y.; Eid, A.M. Development of Miconazole Nitrate Nanoparticles Loaded in Nanoemulgel to Improve Its Antifungal Activity. *Saudi Pharm. J.* **2023**, *31*, 526–534. [CrossRef]
155. Ullah, I.; Alhodaib, A.; Naz, I.; Ahmad, W.; Ullah, H.; Amin, A.; Nawaz, A. Fabrication of Novel Omeprazole-Based Chitosan Coated Nanoemulgel Formulation for Potential Anti-Microbia; In Vitro and Ex Vivo Characterizations. *Polymers* **2023**, *15*, 1298. [CrossRef]
156. Vartak, R.; Menon, S.; Patki, M.; Billack, B.; Patel, K. Ebselen Nanoemulgel for the Treatment of Topical Fungal Infection. *Eur. J. Pharm. Sci.* **2020**, *148*, 105323. [CrossRef]
157. Mahtab, A.; Anwar, M.; Mallick, N.; Naz, Z.; Jain, G.K.; Ahmad, F.J. Transungual Delivery of Ketoconazole Nanoemulgel for the Effective Management of Onychomycosis. *AAPS PharmSciTech* **2016**, *17*, 1477–1490. [CrossRef]

158. Almostafa, M.M.; Elsewedy, H.S.; Shehata, T.M.; Soliman, W.E. Novel Formulation of Fusidic Acid Incorporated into a Myrrh-Oil-Based Nanoemulgel for the Enhancement of Skin Bacterial Infection Treatment. *Gels* **2022**, *8*, 245. [CrossRef] [PubMed]
159. Algahtani, M.S.; Ahmad, M.Z.; Nourein, I.H.; Ahmad, J. Co-Delivery of Imiquimod and Curcumin by Nanoemugel for Improved Topical Delivery and Reduced Psoriasis-Like Skin Lesions. *Biomolecules* **2020**, *10*, 968. [CrossRef] [PubMed]
160. Shehata, T.M.; Elnahas, H.M.; Elsewedy, H.S. Development, Characterization and Optimization of the Anti-Inflammatory Influence of Meloxicam Loaded into a Eucalyptus Oil-Based Nanoemulgel. *Gels* **2022**, *8*, 262. [CrossRef] [PubMed]
161. Abdallah, M.H.; Abu Lila, A.S.; Unissa, R.; Elsewedy, H.S.; Elghamry, H.A.; Soliman, M.S. Preparation, Characterization and Evaluation of Anti-Inflammatory and Anti-Nociceptive Effects of Brucine-Loaded Nanoemulgel. *Colloids Surf. B Biointerfaces* **2021**, *205*, 111868. [CrossRef] [PubMed]
162. Saab, M.; Raafat, K.; El-Maradny, H. Transdermal Delivery of Capsaicin Nanoemulgel: Optimization, Skin Permeation and in Vivo Activity Against Diabetic Neuropathy. *Adv. Pharm. Bull.* **2021**, *12*, 780. [CrossRef] [PubMed]
163. Algahtani, M.S.; Ahmad, M.Z.; Ahmad, J. Nanoemulgel for Improved Topical Delivery of Retinyl Palmitate: Formulation Design and Stability Evaluation. *Nanomaterials* **2020**, *10*, 848. [CrossRef]
164. Childs, D.R.; Murthy, A.S. Overview of Wound Healing and Management. *Surg. Clin. N. Am.* **2017**, *97*, 189–207. [CrossRef]
165. Cañedo-Dorantes, L.; Cañedo-Ayala, M. Skin Acute Wound Healing: A Comprehensive Review. *Int. J. Inflam.* **2019**, *2019*, 3706315. [CrossRef]
166. Wang, P.-H.; Huang, B.-S.; Horng, H.-C.; Yeh, C.-C.; Chen, Y.-J. Wound Healing. *J. Chin. Med. Assoc.* **2018**, *81*, 94–101. [CrossRef]
167. Tottoli, E.M.; Dorati, R.; Genta, I.; Chiesa, E.; Pisani, S.; Conti, B. Skin Wound Healing Process and New Emerging Technologies for Skin Wound Care and Regeneration. *Pharmaceutics* **2020**, *12*, 735. [CrossRef]
168. Velnar, T.; Bailey, T.; Smrkolj, V. The Wound Healing Process: An Overview of the Cellular and Molecular Mechanisms. *J. Int. Med. Res.* **2009**, *37*, 1528–1542. [CrossRef]
169. Almadani, Y.H.; Vorstenbosch, J.; Davison, P.G.; Murphy, A.M. Wound Healing: A Comprehensive Review. *Semin. Plast. Surg.* **2021**, *35*, 141–144. [CrossRef] [PubMed]
170. Gordts, S.; Muthuramu, I.; Amin, R.; Jacobs, F.; De Geest, B. The Impact of Lipoproteins on Wound Healing: Topical HDL Therapy Corrects Delayed Wound Healing in Apolipoprotein E Deficient Mice. *Pharmaceuticals* **2014**, *7*, 419–432. [CrossRef] [PubMed]
171. Bogachkov, Y.Y.; Chen, L.; Le Master, E.; Fancher, I.S.; Zhao, Y.; Aguilar, V.; Oh, M.-J.; Wary, K.K.; DiPietro, L.A.; Levitan, I. LDL Induces Cholesterol Loading and Inhibits Endothelial Proliferation and Angiogenesis in Matrigels: Correlation with Impaired Angiogenesis during Wound Healing. *Am. J. Physiol. Cell Physiol.* **2020**, *318*, C762–C776. [CrossRef] [PubMed]
172. Hata, Y.; Iida, O.; Okamoto, S.; Ishihara, T.; Nanto, K.; Tsujimura, T.; Higashino, N.; Toyoshima, T.; Nakao, S.; Fukunaga, M.; et al. Clinical Outcomes of Patients with Cholesterol Crystal Embolism Accompanied by Lower Extremity Wound. *Angiology* **2023**, 00033197231195671. [CrossRef] [PubMed]
173. Farsaei, S.; Khalili, H.; Farboud, E.S. Potential Role of Statins on Wound Healing: Review of the Literature. *Int. Wound J.* **2012**, *9*, 238–247. [CrossRef] [PubMed]
174. Fitzmaurice, G.J.; McWilliams, B.; Nölke, L.; Redmond, J.M.; McGuinness, J.G.; O'Donnell, M.E. Do Statins Have a Role in the Promotion of Postoperative Wound Healing in Cardiac Surgical Patients? *Ann. Thorac. Surg.* **2014**, *98*, 756–764. [CrossRef] [PubMed]
175. Toker, S.; Gulcan, E.; Çaycı, M.K.; Olgun, E.G.; Erbilen, E.; Özay, Y. Topical Atorvastatin in the Treatment of Diabetic Wounds. *Am. J. Med. Sci.* **2009**, *338*, 201–204. [CrossRef]
176. Falagas, M.E.; Makris, G.C.; Matthaiou, D.K.; Rafailidis, P.I. Statins for Infection and Sepsis: A Systematic Review of the Clinical Evidence. *J. Antimicrob. Chemother.* **2008**, *61*, 774–785. [CrossRef]
177. Suzuki-Banhesse, V.F.; Azevedo, F.F.; Araujo, E.P.; do Amaral, M.E.C.; Caricilli, A.M.; Saad, M.J.A.; Lima, M.H.M. Effect of Atorvastatin on Wound Healing in Rats. *Biol. Res. Nurs.* **2015**, *17*, 159–168. [CrossRef]
178. Raziyeva, K.; Kim, Y.; Zharkinbekov, Z.; Kassymbek, K.; Jimi, S.; Saparov, A. Immunology of Acute and Chronic Wound Healing. *Biomolecules* **2021**, *11*, 700. [CrossRef]
179. Hurlow, J.; Bowler, P.G. Acute and Chronic Wound Infections: Microbiological, Immunological, Clinical and Therapeutic Distinctions. *J. Wound Care* **2022**, *31*, 436–445. [CrossRef] [PubMed]
180. Mulyaningsih, S.; Sporer, F.; Reichling, J.; Wink, M. Antibacterial Activity of Essential Oils from Eucalyptus and of Selected Components against Multidrug-Resistant Bacterial Pathogens. *Pharm. Biol.* **2011**, *49*, 893–899. [CrossRef] [PubMed]
181. Vijayakumar, K.; Manigandan, V.; Jeyapragash, D.; Bharathidasan, V.; Anandharaj, B.; Sathya, M. Eucalyptol Inhibits Biofilm Formation of Streptococcus Pyogenes and Its Mediated Virulence Factors. *J. Med. Microbiol.* **2020**, *69*, 1308–1318. [CrossRef] [PubMed]
182. Baltzis, D.; Eleftheriadou, I.; Veves, A. Pathogenesis and Treatment of Impaired Wound Healing in Diabetes Mellitus: New Insights. *Adv. Ther.* **2014**, *31*, 817–836. [CrossRef] [PubMed]
183. Okur, M.E.; Bülbül, E.Ö.; Mutlu, G.; Eleftheriadou, K.; Karantas, I.D.; Okur, N.Ü.; Siafaka, P.I. An Updated Review for the Diabetic Wound Healing Systems. *Curr. Drug Targets* **2022**, *23*, 393–419. [CrossRef] [PubMed]
184. Jais, S. Various Types of Wounds That Diabetic Patients Can Develop: A Narrative Review. *Clin. Pathol.* **2023**, *16*, 2632010X231205366. [CrossRef]
185. Ahsan, H.; Ahad, A.; Iqbal, J.; Siddiqui, W.A. Pharmacological Potential of Tocotrienols: A Review. *Nutr. Metab.* **2014**, *11*, 52. [CrossRef]

186. Zainal, Z.; Khaza'ai, H.; Kutty Radhakrishnan, A.; Chang, S.K. Therapeutic Potential of Palm Oil Vitamin E-Derived Tocotrienols in Inflammation and Chronic Diseases: Evidence from Preclinical and Clinical Studies. *Food Res. Int.* **2022**, *156*, 111175. [CrossRef]
187. Kandhare, A.D.; Alam, J.; Patil, M.V.K.; Sinha, A.; Bodhankar, S.L. Wound Healing Potential of Naringin Ointment Formulation via Regulating the Expression of Inflammatory, Apoptotic and Growth Mediators in Experimental Rats. *Pharm. Biol.* **2016**, *54*, 419–432. [CrossRef]
188. Kandhare, A.D.; Ghosh, P.; Bodhankar, S.L. Naringin, a Flavanone Glycoside, Promotes Angiogenesis and Inhibits Endothelial Apoptosis through Modulation of Inflammatory and Growth Factor Expression in Diabetic Foot Ulcer in Rats. *Chem. Biol. Interact.* **2014**, *219*, 101–112. [CrossRef]
189. Kmail, A.; Said, O.; Saad, B. How Thymoquinone from *Nigella sativa* Accelerates Wound Healing through Multiple Mechanisms and Targets. *Curr. Issues Mol. Biol.* **2023**, *45*, 9039–9059. [CrossRef] [PubMed]
190. Sallehuddin, N.; Nordin, A.; Bt Hj Idrus, R.; Fauzi, M.B. Nigella Sativa and Its Active Compound, Thymoquinone, Accelerate Wound Healing in an In Vivo Animal Model: A Comprehensive Review. *Int. J. Environ. Res. Public. Health* **2020**, *17*, 4160. [CrossRef] [PubMed]
191. Rajabian, A.; Hosseinzadeh, H. Dermatological Effects of Nigella Sativa and Its Constituent, Thymoquinone. In *Nuts and Seeds in Health and Disease Prevention*; Elsevier: Amsterdam, The Netherlands, 2020; pp. 329–355.
192. Kumari, A.; Raina, N.; Wahi, A.; Goh, K.W.; Sharma, P.; Nagpal, R.; Jain, A.; Ming, L.C.; Gupta, M. Wound-Healing Effects of Curcumin and Its Nanoformulations: A Comprehensive Review. *Pharmaceutics* **2022**, *14*, 2288. [CrossRef] [PubMed]
193. Tejada, S.; Manayi, A.; Daglia, M.; Nabavi, S.F.; Sureda, A.; Hajheydari, Z.; Gortzi, O.; Pazoki-Toroudi, H.; Nabavi, S.M. Wound Healing Effects of Curcumin: A Short Review. *Curr. Pharm. Biotechnol.* **2016**, *17*, 1002–1007. [CrossRef] [PubMed]
194. Khatun, M.; Nur, M.A.; Biswas, S.; Khan, M.; Amin, M.Z. Assessment of the Anti-Oxidant, Anti-Inflammatory and Anti-Bacterial Activities of Different Types of Turmeric (Curcuma Longa) Powder in Bangladesh. *J. Agric. Food Res.* **2021**, *6*, 100201. [CrossRef]
195. Vitiello, A.; Ferrara, F.; Boccellino, M.; Ponzo, A.; Cimmino, C.; Comberiati, E.; Zovi, A.; Clemente, S.; Sabbatucci, M. Antifungal Drug Resistance: An Emergent Health Threat. *Biomedicines* **2023**, *11*, 1063. [CrossRef] [PubMed]
196. Salam, M.A.; Al-Amin, M.Y.; Salam, M.T.; Pawar, J.S.; Akhter, N.; Rabaan, A.A.; Alqumber, M.A.A. Antimicrobial Resistance: A Growing Serious Threat for Global Public Health. *Healthcare* **2023**, *11*, 1946. [CrossRef]
197. Fothergill, A.W. Miconazole: A Historical Perspective. *Expert Rev. Anti Infect. Ther.* **2006**, *4*, 171–175. [CrossRef]
198. Quatresooz, P.; Vroome, V.; Borgers, M.; Cauwenbergh, G.; Piérard, G.E. Novelties in the Multifaceted Miconazole Effects on Skin Disorders. *Expert Opin. Pharmacother.* **2008**, *9*, 1927–1934. [CrossRef]
199. Gatta, L. Antimicrobial Activity of Esomeprazole versus Omeprazole against Helicobacter Pylori. *J. Antimicrob. Chemother.* **2003**, *51*, 439–442. [CrossRef]
200. Anagnostopoulos, G.K.; Tsiakos, S.; Margantinis, G.; Kostopoulos, P.; Arvanitidis, D. Esomeprazole versus Omeprazole for the Eradication of Helicobacter Pylori Infection. *J. Clin. Gastroenterol.* **2004**, *38*, 503–506. [CrossRef] [PubMed]
201. Kumar, L.; Verma, S.; Bhardwaj, A.; Vaidya, S.; Vaidya, B. Eradication of Superficial Fungal Infections by Conventional and Novel Approaches: A Comprehensive Review. *Artif. Cells Nanomed. Biotechnol.* **2014**, *42*, 32–46. [CrossRef] [PubMed]
202. Wu, Y.; Hu, S.; Wu, C.; Gu, F.; Yang, Y. Probiotics: Potential Novel Therapeutics Against Fungal Infections. *Front. Cell Infect. Microbiol.* **2022**, *11*, 793419. [CrossRef] [PubMed]
203. Rauseo, A.M.; Coler-Reilly, A.; Larson, L.; Spec, A. Hope on the Horizon: Novel Fungal Treatments in Development. *Open Forum Infect. Dis.* **2020**, *7*, ofaa016. [CrossRef] [PubMed]
204. Garland, M.; Hryckowian, A.J.; Tholen, M.; Oresic Bender, K.; Van Treuren, W.W.; Loscher, S.; Sonnenburg, J.L.; Bogyo, M. The Clinical Drug Ebselen Attenuates Inflammation and Promotes Microbiome Recovery in Mice after Antibiotic Treatment for CDI. *Cell Rep. Med.* **2020**, *1*, 100005. [CrossRef] [PubMed]
205. Sarma, B.K.; Mugesh, G. Antioxidant Activity of the Anti-Inflammatory Compound Ebselen: A Reversible Cyclization Pathway via Selenenic and Seleninic Acid Intermediates. *Chem. Eur. J.* **2008**, *14*, 10603–10614. [CrossRef]
206. Maślanka, M.; Mucha, A. Antibacterial Activity of Ebselen. *Int. J. Mol. Sci.* **2023**, *24*, 1610. [CrossRef]
207. Leung, A.K.C.; Lam, J.M.; Leong, K.F.; Hon, K.L.; Barankin, B.; Leung, A.A.M.; Wong, A.H.C. Onychomycosis: An Updated Review. *Recent. Pat. Inflamm. Allergy Drug Discov.* **2020**, *14*, 32–45. [CrossRef]
208. Gupta, A.K.; Stec, N.; Summerbell, R.C.; Shear, N.H.; Piguet, V.; Tosti, A.; Piraccini, B.M. Onychomycosis: A Review. *J. Eur. Acad. Dermatol. Venereol.* **2020**, *34*, 1972–1990. [CrossRef]
209. Ahmed, I.S.; Elnahas, O.S.; Assar, N.H.; Gad, A.M.; El Hosary, R. Nanocrystals of Fusidic Acid for Dual Enhancement of Dermal Delivery and Antibacterial Activity: In Vitro, Ex Vivo and In Vivo Evaluation. *Pharmaceutics* **2020**, *12*, 199. [CrossRef]
210. Pfaller, M.A.; Castanheira, M.; Sader, H.S.; Jones, R.N. Evaluation of the Activity of Fusidic Acid Tested against Contemporary Gram-Positive Clinical Isolates from the USA and Canada. *Int. J. Antimicrob. Agents* **2010**, *35*, 282–287. [CrossRef] [PubMed]
211. Curbete, M.M.; Salgado, H.R.N. A Critical Review of the Properties of Fusidic Acid and Analytical Methods for Its Determination. *Crit. Rev. Anal. Chem.* **2016**, *46*, 352–360. [CrossRef] [PubMed]
212. Algarin, Y.A.; Jambusaria-Pahlajani, A.; Ruiz, E.; Patel, V.A. Advances in Topical Treatments of Cutaneous Malignancies. *Am. J. Clin. Dermatol.* **2023**, *24*, 69–80. [CrossRef] [PubMed]
213. Hidalgo, L.; Saldías-Fuentes, C.; Carrasco, K.; Halpern, A.C.; Mao, J.J.; Navarrete-Dechent, C. Complementary and Alternative Therapies in Skin Cancer a Literature Review of Biologically Active Compounds. *Dermatol. Ther.* **2022**, *35*, e15842. [CrossRef] [PubMed]

214. Conforti, C.; Corneli, P.; Harwood, C.; Zalaudek, I. Evolving Role of Systemic Therapies in Non-Melanoma Skin Cancer. *Clin. Oncol.* **2019**, *31*, 759–768. [CrossRef] [PubMed]
215. Salari, N.; Faraji, F.; Jafarpour, S.; Faraji, F.; Rasoulpoor, S.; Dokaneheifard, S.; Mohammadi, M. Anti-Cancer Activity of Chrysin in Cancer Therapy: A Systematic Review. *Indian J. Surg. Oncol.* **2022**, *13*, 681–690. [CrossRef]
216. Kasala, E.R.; Bodduluru, L.N.; Madana, R.M.; Athira, K.V.; Gogoi, R.; Barua, C.C. Chemopreventive and Therapeutic Potential of Chrysin in Cancer: Mechanistic Perspectives. *Toxicol. Lett.* **2015**, *233*, 214–225. [CrossRef]
217. Talebi, M.; Talebi, M.; Farkhondeh, T.; Simal-Gandara, J.; Kopustinskiene, D.M.; Bernatoniene, J.; Samarghandian, S. Emerging Cellular and Molecular Mechanisms Underlying Anticancer Indications of Chrysin. *Cancer Cell Int.* **2021**, *21*, 214. [CrossRef]
218. Chang, S.-H.; Wu, C.-Y.; Chuang, K.-C.; Huang, S.-W.; Li, Z.-Y.; Wang, S.-T.; Lai, Z.-L.; Chang, C.-C.; Chen, Y.-J.; Wong, T.-W.; et al. Imiquimod Accelerated Antitumor Response by Targeting Lysosome Adaptation in Skin Cancer Cells. *J. Investig. Dermatol.* **2021**, *141*, 2219–2228. [CrossRef]
219. Bubna, A. Imiquimod—Its Role in the Treatment of Cutaneous Malignancies. *Indian J. Pharmacol.* **2015**, *47*, 354. [CrossRef]
220. Lelli, D.; Pedone, C.; Sahebkar, A. Curcumin and Treatment of Melanoma: The Potential Role of MicroRNAs. *Biomed. Pharmacother.* **2017**, *88*, 832–834. [CrossRef] [PubMed]
221. Phillips, J.M.; Clark, C.; Herman-Ferdinandez, L.; Moore-Medlin, T.; Rong, X.; Gill, J.R.; Clifford, J.L.; Abreo, F.; Nathan, C.O. Curcumin Inhibits Skin Squamous Cell Carcinoma Tumor Growth In Vivo. *Otolaryngol. Head Neck Surg.* **2011**, *145*, 58–63. [CrossRef] [PubMed]
222. van der Fits, L.; Mourits, S.; Voerman, J.S.A.; Kant, M.; Boon, L.; Laman, J.D.; Cornelissen, F.; Mus, A.-M.; Florencia, E.; Prens, E.P.; et al. Imiquimod-Induced Psoriasis-Like Skin Inflammation in Mice Is Mediated via the IL-23/IL-17 Axis. *J. Immunol.* **2009**, *182*, 5836–5845. [CrossRef] [PubMed]
223. Carlos, E.C.D.S.; Cristovão, G.A.; Silva, A.A.; de Santos Ribeiro, B.C.; Romana-Souza, B. Imiquimod-induced Ex Vivo Model of Psoriatic Human Skin via Interleukin-17A Signalling of T Cells and Langerhans Cells. *Exp. Dermatol.* **2022**, *31*, 1791–1799. [CrossRef] [PubMed]
224. Griffiths, C.E.M.; Armstrong, A.W.; Gudjonsson, J.E.; Barker, J.N.W.N. Psoriasis. *Lancet* **2021**, *397*, 1301–1315. [CrossRef] [PubMed]
225. Raharja, A.; Mahil, S.K.; Barker, J.N. Psoriasis: A Brief Overview. *Clin. Med.* **2021**, *21*, 170–173. [CrossRef] [PubMed]
226. Armstrong, A.W.; Read, C. Pathophysiology, Clinical Presentation, and Treatment of Psoriasis. *JAMA* **2020**, *323*, 1945. [CrossRef] [PubMed]
227. Herrmann, M.L.; Schleyerbach, R.; Kirschbaum, B.J. Leflunomide: An Immunomodulatory Drug for the Treatment of Rheumatoid Arthritis and Other Autoimmune Diseases. *Immunopharmacology* **2000**, *47*, 273–289. [CrossRef]
228. Alamri, R.D.; Elmeligy, M.A.; Albalawi, G.A.; Alquayr, S.M.; Alsubhi, S.S.; El-Ghaiesh, S.H. Leflunomide an Immunomodulator with Antineoplastic and Antiviral Potentials but Drug-Induced Liver Injury: A Comprehensive Review. *Int. Immunopharmacol.* **2021**, *93*, 107398. [CrossRef]
229. Boehncke, W.-H. Immunomodulatory Drugs for Psoriasis. *BMJ* **2003**, *327*, 634–635. [CrossRef]
230. Nagai, N.; Ogata, F.; Otake, H.; Kawasaki, N. Oral Administration System Based on Meloxicam Nanocrystals: Decreased Dose Due to High Bioavailability Attenuates Risk of Gastrointestinal Side Effects. *Pharmaceutics* **2020**, *12*, 313. [CrossRef] [PubMed]
231. Rostom, A.; Goldkind, L.; Laine, L. Nonsteroidal Anti-Inflammatory Drugs and Hepatic Toxicity: A Systematic Review of Randomized Controlled Trials in Arthritis Patients. *Clin. Gastroenterol. Hepatol.* **2005**, *3*, 489–498. [CrossRef] [PubMed]
232. Jain, B.; Jain, N.; Jain, S.; Teja, P.K.; Chauthe, S.K.; Jain, A. Exploring Brucine Alkaloid: A Comprehensive Review on Pharmacology, Therapeutic Applications, Toxicity, Extraction and Purification Techniques. *Phytomed. Plus* **2023**, *3*, 100490. [CrossRef]
233. Song, X.; Wang, Y.; Chen, H.; Jin, Y.; Wang, Z.; Lu, Y.; Wang, Y. Dosage-Efficacy Relationship and Pharmacodynamics Validation of Brucine Dissolving Microneedles against Rheumatoid Arthritis. *J. Drug Deliv. Sci. Technol.* **2021**, *63*, 102537. [CrossRef]
234. Lu, L.; Huang, R.; Wu, Y.; Jin, J.-M.; Chen, H.-Z.; Zhang, L.-J.; Luan, X. Brucine: A Review of Phytochemistry, Pharmacology, and Toxicology. *Front. Pharmacol.* **2020**, *11*, 377. [CrossRef] [PubMed]
235. Yin, W.; Wang, T.-S.; Yin, F.-Z.; Cai, B.-C. Analgesic and Anti-Inflammatory Properties of Brucine and Brucine N-Oxide Extracted from Seeds of Strychnos Nux-Vomica. *J. Ethnopharmacol.* **2003**, *88*, 205–214. [CrossRef] [PubMed]
236. Tang, M.; Zhu, W.; Yang, Z.; He, C. Brucine Inhibits TNF-α-induced HFLS-RA Cell Proliferation by Activating the JNK Signaling Pathway. *Exp. Ther. Med.* **2019**, *18*, 735–740. [CrossRef]
237. Elafros, M.A.; Andersen, H.; Bennett, D.L.; Savelieff, M.G.; Viswanathan, V.; Callaghan, B.C.; Feldman, E.L. Towards Prevention of Diabetic Peripheral Neuropathy: Clinical Presentation, Pathogenesis, and New Treatments. *Lancet Neurol.* **2022**, *21*, 922–936. [CrossRef]
238. Jensen, T.S.; Karlsson, P.; Gylfadottir, S.S.; Andersen, S.T.; Bennett, D.L.; Tankisi, H.; Finnerup, N.B.; Terkelsen, A.J.; Khan, K.; Themistocleous, A.C.; et al. Painful and Non-Painful Diabetic Neuropathy, Diagnostic Challenges and Implications for Future Management. *Brain* **2021**, *144*, 1632–1645. [CrossRef]
239. Cernea, S.; Raz, I. Management of Diabetic Neuropathy. *Metabolism* **2021**, *123*, 154867. [CrossRef]
240. Sharma, S.K.; Vij, A.S.; Sharma, M. Mechanisms and Clinical Uses of Capsaicin. *Eur. J. Pharmacol.* **2013**, *720*, 55–62. [CrossRef] [PubMed]
241. Lu, M.; Chen, C.; Lan, Y.; Xiao, J.; Li, R.; Huang, J.; Huang, Q.; Cao, Y.; Ho, C.-T. Capsaicin—The Major Bioactive Ingredient of Chili Peppers: Bio-Efficacy and Delivery Systems. *Food Funct.* **2020**, *11*, 2848–2860. [CrossRef] [PubMed]

242. Fernandes, E.S.; Cerqueira, A.R.A.; Soares, A.G.; Costa, S.K.P. Capsaicin and Its Role in Chronic Diseases. In *Drug Discovery from Mother Nature*; Springer: Berlin/Heidelberg, Germany, 2016; pp. 91–125.
243. Basith, S.; Cui, M.; Hong, S.; Choi, S. Harnessing the Therapeutic Potential of Capsaicin and Its Analogues in Pain and Other Diseases. *Molecules* 2016, *21*, 966. [CrossRef] [PubMed]
244. Abdel-Salam, O.M.E.; Mózsik, G. Capsaicin, The Vanilloid Receptor TRPV1 Agonist in Neuroprotection: Mechanisms Involved and Significance. *Neurochem. Res.* 2023, *48*, 3296–3315. [CrossRef] [PubMed]
245. Fattori, V.; Hohmann, M.; Rossaneis, A.; Pinho-Ribeiro, F.; Verri, W. Capsaicin: Current Understanding of Its Mechanisms and Therapy of Pain and Other Pre-Clinical and Clinical Uses. *Molecules* 2016, *21*, 844. [CrossRef] [PubMed]
246. Frydas, S.; Varvara, G.; Murmura, G.; Saggini, A.; Caraffa, A.; Antinolfi, P.; Tetè, S.; Tripodi, D.; Conti, F.; Cianchetti, E.; et al. Impact of Capsaicin on Mast Cell Inflammation. *Int. J. Immunopathol. Pharmacol.* 2013, *26*, 597–600. [CrossRef] [PubMed]
247. Kim, C.-S.; Kawada, T.; Kim, B.-S.; Han, I.-S.; Choe, S.-Y.; Kurata, T.; Yu, R. Capsaicin Exhibits Anti-Inflammatory Property by Inhibiting IkB-a Degradation in LPS-Stimulated Peritoneal Macrophages. *Cell Signal* 2003, *15*, 299–306. [CrossRef]
248. Srinivasan, K. Biological Activities of Red Pepper (Capsicum Annuum) and Its Pungent Principle Capsaicin: A Review. *Crit. Rev. Food Sci. Nutr.* 2016, *56*, 1488–1500. [CrossRef]
249. Rollyson, W.D.; Stover, C.A.; Brown, K.C.; Perry, H.E.; Stevenson, C.D.; McNees, C.A.; Ball, J.G.; Valentovic, M.A.; Dasgupta, P. Bioavailability of Capsaicin and Its Implications for Drug Delivery. *J. Control. Release* 2014, *196*, 96–105. [CrossRef]
250. Babbar, S.; Marier, J.-F.; Mouksassi, M.-S.; Beliveau, M.; Vanhove, G.F.; Chanda, S.; Bley, K. Pharmacokinetic Analysis of Capsaicin after Topical Administration of a High-Concentration Capsaicin Patch to Patients with Peripheral Neuropathic Pain. *Ther. Drug Monit.* 2009, *31*, 502–510. [CrossRef]
251. Oliveira, M.B.; do Prado, A.H.; Bernegossi, J.; Sato, C.S.; Lourenço Brunetti, I.; Scarpa, M.V.; Leonardi, G.R.; Friberg, S.E.; Chorilli, M. Topical Application of Retinyl Palmitate-Loaded Nanotechnology-Based Drug Delivery Systems for the Treatment of Skin Aging. *Biomed. Res. Int.* 2014, *2014*, 632570. [CrossRef] [PubMed]
252. Salem, H.F.; Kharshoum, R.M.; Awad, S.M.; Ahmed Mostafa, M.; Abou-Taleb, H.A. Tailoring of Retinyl Palmitate-Based Ethosomal Hydrogel as a Novel Nanoplatform for Acne Vulgaris Management: Fabrication, Optimization, and Clinical Evaluation Employing a Split-Face Comparative Study. *Int. J. Nanomed.* 2021, *16*, 4251–4276. [CrossRef] [PubMed]
253. Milosheska, D.; Roškar, R. Use of Retinoids in Topical Antiaging Treatments: A Focused Review of Clinical Evidence for Conventional and Nanoformulations. *Adv. Ther.* 2022, *39*, 5351–5375. [CrossRef] [PubMed]
254. Nandy, A.; Lee, E.; Mandal, A.; Saremi, R.; Sharma, S. Microencapsulation of Retinyl Palmitate by Melt Dispersion for Cosmetic Application. *J. Microencapsul.* 2020, *37*, 205–219. [CrossRef]
255. Gholizadeh, M.; Basafa Roodi, P.; Abaj, F.; Shab-Bidar, S.; Saedisomeolia, A.; Asbaghi, O.; Lak, M. Influence of Vitamin A Supplementation on Inflammatory Biomarkers in Adults: A Systematic Review and Meta-Analysis of Randomized Clinical Trials. *Sci. Rep.* 2022, *12*, 21384. [CrossRef] [PubMed]
256. Farooq, U.; Mahmood, T.; Shahzad, Y.; Yousaf, A.M.; Akhtar, N. Comparative Efficacy of Two Anti-aging Products Containing Retinyl Palmitate in Healthy Human Volunteers. *J. Cosmet. Dermatol.* 2018, *17*, 454–460. [CrossRef] [PubMed]
257. Fu, P.P.; Cheng, S.-H.; Coop, L.; Xia, Q.; Culp, S.J.; Tolleson, W.H.; Wamer, W.G.; Howard, P.C. Photoreaction, Phototoxicity, and Photocarcinogenicity of Retinoids. *J. Environ. Sci. Health Part. C* 2003, *21*, 165–197. [CrossRef]
258. Maugard, T.; Rejasse, B.; Legoy, M.D. Synthesis of Water-Soluble Retinol Derivatives by Enzymatic Method. *Biotechnol. Prog.* 2002, *18*, 424–428. [CrossRef]
259. Suh, D.-C.; Kim, Y.; Kim, H.; Ro, J.; Cho, S.-W.; Yun, G.; Choi, S.-U.; Lee, J. Enhanced In Vitro Skin Deposition Properties of Retinyl Palmitate through Its Stabilization by Pectin. *Biomol. Ther.* 2014, *22*, 73–77. [CrossRef]
260. Strati, F.; Neubert, R.H.H.; Opálka, L.; Kerth, A.; Brezesinski, G. Non-Ionic Surfactants as Innovative Skin Penetration Enhancers: Insight in the Mechanism of Interaction with Simple 2D Stratum Corneum Model System. *Eur. J. Pharm. Sci.* 2021, *157*, 105620. [CrossRef]
261. Ren, Q.; Deng, C.; Meng, U.; Chen, Y.; Chen, U.; Sha, X.; Fang, X. In Vitro, Ex Vivo, and In Vivo Evaluation of the Effect of Saturated Fat Acid Chain Length on the Transdermal Behavior of Ibuprofen-Loaded Microemulsions. *J. Pharm. Sci.* 2014, *103*, 1680–1691. [CrossRef] [PubMed]
262. Erdal, M.S.; Özhan, G.; Mat, C.; Özsoy, Y.; Güngör, S. Colloidal Nanocarriers for the Enhanced Cutaneous Delivery of Naftifine: Characterization Studies and in Vitro and in Vivo Evaluations. *Int. J. Nanomed.* 2016, *1027*, 1027–1037. [CrossRef] [PubMed]
263. Kim, B.S.; Won, M.; Yang; Lee, K.M.; Kim, C.S. In Vitro Permeation Studies of Nanoemulsions Containing Ketoprofen as a Model Drug. *Drug Deliv.* 2008, *15*, 465–469. [CrossRef] [PubMed]
264. Osborne, D.W.; Musakhanian, J. Skin Penetration and Permeation Properties of Transcutol®—Neat or Diluted Mixtures. *AAPS PharmSciTech* 2018, *19*, 3512–3533. [CrossRef] [PubMed]
265. Godwin, D.A.; Kim, N.-H.; Felton, L.A. Influence of Transcutol® CG on the Skin Accumulation and Transdermal Permeation of Ultraviolet Absorbers. *Eur. J. Pharm. Biopharm.* 2002, *53*, 23–27. [CrossRef]
266. Antunes, F.E.; Gentile, L.; Oliviero Rossi, C.; Tavano, L.; Ranieri, G.A. Gels of Pluronic F127 and Nonionic Surfactants from Rheological Characterization to Controlled Drug Permeation. *Colloids Surf. B Biointerfaces* 2011, *87*, 42–48. [CrossRef]
267. Zheng, Y.; Ouyang, W.-Q.; Wei, Y.-P.; Syed, S.; Hao, C.-S.; Wang, B.-Z.; Shang, Y.-H. Effects of Carbopol 934 Proportion on Nanoemulsion Gel for Topical and Transdermal Drug Delivery: A Skin Permeation Study. *Int. J. Nanomed.* 2016, *11*, 5971–5987. [CrossRef]

268. Babu, R.J.; Pandit, J.K. Effect of Penetration Enhancers on the Release and Skin Permeation of Bupranolol from Reservoir-Type Transdermal Delivery Systems. *Int. J. Pharm.* **2005**, *288*, 325–334. [CrossRef]
269. De Jong, W.H.; Geertsma, R.E.; Borchard, G. Regulatory Safety Evaluation of Nanomedical Products: Key Issues to Refine. *Drug Deliv. Transl. Res.* **2022**, *12*, 2042–2047. [CrossRef]
270. Foulkes, R.; Man, E.; Thind, J.; Yeung, S.; Joy, A.; Hoskins, C. The Regulation of Nanomaterials and Nanomedicines for Clinical Application: Current and Future Perspectives. *Biomater. Sci.* **2020**, *8*, 4653–4664. [CrossRef]
271. Liu, L.; Bagia, C.; Janjic, J.M. The First Scale-Up Production of Theranostic Nanoemulsions. *Biores Open Access* **2015**, *4*, 218–228. [CrossRef] [PubMed]
272. Adena, S.K.R.; Herneisey, M.; Pierce, E.; Hartmeier, P.R.; Adlakha, S.; Hosfeld, M.A.I.; Drennen, J.K.; Janjic, J.M. Quality by Design Methodology Applied to Process Optimization and Scale up of Curcumin Nanoemulsions Produced by Catastrophic Phase Inversion. *Pharmaceutics* **2021**, *13*, 880. [CrossRef] [PubMed]
273. Paliwal, R.; Babu, R.J.; Palakurthi, S. Nanomedicine Scale-up Technologies: Feasibilities and Challenges. *AAPS PharmSciTech* **2014**, *15*, 1527–1534. [CrossRef] [PubMed]

**Disclaimer/Publisher's Note:** The statements, opinions and data contained in all publications are solely those of the individual author(s) and contributor(s) and not of MDPI and/or the editor(s). MDPI and/or the editor(s) disclaim responsibility for any injury to people or property resulting from any ideas, methods, instructions or products referred to in the content.

*Review*

# Exploring the Potential of Artificial Intelligence for Hydrogel Development—A Short Review

Irina Negut [1] and Bogdan Bita [1,2,*]

[1] National Institute for Laser, Plasma and Radiation Physics, 409 Atomistilor Street, 077125 Magurele, Romania; negut.irina@inflpr.ro
[2] Faculty of Physics, University of Bucharest, 077125 Magurele, Romania
* Correspondence: bogdan.bita@inflpr.ro

**Abstract:** AI and ML have emerged as transformative tools in various scientific domains, including hydrogel design. This work explores the integration of AI and ML techniques in the realm of hydrogel development, highlighting their significance in enhancing the design, characterisation, and optimisation of hydrogels for diverse applications. We introduced the concept of AI train hydrogel design, underscoring its potential to decode intricate relationships between hydrogel compositions, structures, and properties from complex data sets. In this work, we outlined classical physical and chemical techniques in hydrogel design, setting the stage for AI/ML advancements. These methods provide a foundational understanding for the subsequent AI-driven innovations. Numerical and analytical methods empowered by AI/ML were also included. These computational tools enable predictive simulations of hydrogel behaviour under varying conditions, aiding in property customisation. We also emphasised AI's impact, elucidating its role in rapid material discovery, precise property predictions, and optimal design. ML techniques like neural networks and support vector machines that expedite pattern recognition and predictive modelling using vast datasets, advancing hydrogel formulation discovery are also presented. AI and ML's have a transformative influence on hydrogel design. AI and ML have revolutionised hydrogel design by expediting material discovery, optimising properties, reducing costs, and enabling precise customisation. These technologies have the potential to address pressing healthcare and biomedical challenges, offering innovative solutions for drug delivery, tissue engineering, wound healing, and more. By harmonising computational insights with classical techniques, researchers can unlock unprecedented hydrogel potentials, tailoring solutions for diverse applications.

**Keywords:** hydrogel design; artificial intelligence; machine learning

## 1. Introduction: The Concept of AI in Hydrogel Design

Hydrogels are three-dimensional, crosslinked polymer networks that can absorb and retain large amounts of water or biological fluids. They have gained considerable attention in the medical field due to their unique properties, such as high-water content, biocompatibility, and the ability to be tailored to specific applications. The term "hydrogel" is derived from the combination of "hydro", meaning water, and "gel", indicating a semi-solid, jelly-like state. This defining characteristic of hydrogels allows them to mimic the soft and hydrated environment of living tissues, making them highly compatible with biological systems. They can be fabricated from polymer chains linked by physical interactions or chemical bonds, allowing for precise control of the degradation rate, porosity, and release profile [1]. Moreover, hydrogels can undergo self-assembly using self-complementary amphiphilic peptides, enabling customisation to achieve optimal geometry for implantation or injection. These appealing features position hydrogels as attractive therapeutic delivery materials, with the potential to encapsulate agents within their water-swollen network. Hydrogels are extensively employed as drug delivery vehicles due to their capacity to encapsulate and release therapeutic agents in a controlled and sustained manner. This controlled drug release enhances treatment efficacy, reduces side effects, and improves patient compliance, particularly in chronic conditions [2]. Additionally, certain hydrogel types possess inherent antibacterial properties [3]. In tissue engineering, hydrogels serve as scaffolds for growing cells and regenerate damaged or lost tissues [4]. Their high-water content and biocompatibility mimic the natural extracellular environment, facilitating cell growth, proliferation, and differentiation [5]. This is particularly significant in the development of artificial organs and in repairing damaged tissues [4]. Moreover, hydrogels play a vital role in wound care and healing. They can maintain a moist environment, which accelerates the wound-healing process, reduces the risk of infection, and minimises scarring [6].

Hydrogels are the material of choice for contact lenses due to their water-retaining properties, ensuring comfort and optical clarity for the wearer [7].

Other applications of hydrogels include diagnostic assays and biosensors [8]. Their ability to undergo volume changes in response to specific analytes, such as glucose or pH, makes them valuable components in various diagnostic applications [8].

While the potential applications of hydrogels in the biomedical and pharmaceutical fields are vast, designing hydrogels with precise properties tailored to each application is a complex endeavour. Researchers must consider factors such as biocompatibility, mechanical strength, degradation rates, and drug release profiles. The interplay of these variables makes hydrogel design challenging and often reliant on time-consuming trial-and-error approaches. Big data generated from experiments, simulations, and computational calculations has provided potential for applying data-driven methodologies in material science, which shows promise for expediting the discovery and design of new materials. This approach harnesses the power of vast datasets generated through experiments, simulations, or observations to gain insights, make predictions, and guide decision-making in the field of materials science and in the context of hydrogel design.

In the 1990s and early 2000s, the integration of AI in hydrogel research began with the application of computational simulations and modelling techniques. Researchers started to use computational methods to simulate hydrogel behaviour under various conditions, aiding in predicting swelling properties, mechanical responses, and drug release kinetics. Hydrogel theoretical modelling is based on continuum mechanical concepts such as balancing laws, kinematics, and constitutive equations. The Flory–Rehner theory, which explains the swelling equilibrium of gels [9], represents a suitable example. As a result of the advancement of computing capabilities, the computational science paradigm gained enormous popularity. Simulations on both the macro- and micro-scales, such as those using the finite element and volume methods, are now possible [10].

These early efforts laid the groundwork for using AI to unravel the complex interactions within hydrogel networks.

As Machine Learning (ML) algorithms advanced, the hydrogel research community recognised the potential of AI to revolutionise material discovery. The utilisation of ML techniques gained traction. These approaches enabled researchers to analyse large datasets, correlate structure–property relationships, and accurately predict hydrogel behaviour. AI algorithms can analyse patient-specific data, such as genetic information, metabolism rates, and medical histories [11], to design hydrogels that deliver drugs with precision. For instance, in cancer treatment, AI-driven hydrogel formulations can adapt drug release rates based on the tumour's response to therapy, minimising side effects and maximising effectiveness [12]. AI models can simulate and predict the behaviour of hydrogels under various conditions, saving researchers significant time and resources. For example, they can predict how a hydrogel will swell, degrade, or release drugs in response to changes in pH, temperature, or biological factors [13]. AI-driven material design can identify the ideal combination of polymers, crosslinkers, and additives to create hydrogels with specific mechanical, thermal, and chemical properties. This accelerates the development of hydrogels tailored for applications such as wound healing, contact lenses, or tissue scaffolds. High-throughput screening, guided by AI algorithms, enables the evaluation of vast libraries of hydrogel formulations [14]. This accelerates the discovery of hydrogels with desirable properties, reducing the time needed to bring innovative materials to market.

Therefore, this era marked a significant shift from traditional empirical methods to data-driven approaches, offering faster and more informed decision-making in hydrogel design.

Artificial Intelligence (AI) represents a field that involves the development of algorithms and models that enable computers to mimic human intelligence. In hydrogel development, AI offers a novel approach to tackle the challenges associated with hydrogel properties and performance by leveraging data-driven insights, predictive modelling, and optimisation techniques. These challenges are multifaceted and include tailoring hydrogels for precise applications, optimising their mechanical and chemical characteristics, and navigating the complex interplay of material variables. First and foremost, AI expedites the process of discovering new hydrogel formulations. Traditional methods often involve extensive trial-and-error experimentation, consuming considerable time and resources. In contrast, AI-driven algorithms can rapidly analyse vast datasets, predict material properties, and recommend optimal compositions. This acceleration of the research process is particularly crucial in the field of hydrogels, where materials must meet precise criteria for applications in medicine and biology.

AI also offers a level of precision and customisation that was previously unattainable. Researchers can input specific characteristics they desire in a hydrogel—such as mechanical strength, porosity, or biodegradability—into ML models. AI then provides tailored recommendations for material compositions and processing techniques to achieve these desired properties. This level of precision is invaluable when designing hydrogels for diverse applications, from drug delivery systems to tissue engineering scaffolds.

Hydrogel research is inherently complex, involving intricate relationships among various factors. AI can help in this aspect by handling multidimensional data effectively. It identifies patterns and correlations that may elude traditional analysis, allowing researchers to make informed decisions about material design.

In addition, AI integration offers cost and resource efficiencies. It reduces the need for extensive laboratory experimentation, thus saving costs related to materials, equipment, and personnel. Moreover, it minimises material wastage, aligning research practices with environmental sustainability goals.

Interdisciplinary collaboration is another hallmark of AI-driven hydrogel research. This technology bridges the expertise of materials scientists, chemists, biologists, and computer scientists. Their collective knowledge and insights foster innovative solutions to complex challenges in hydrogel development.

AI also enables data-driven insights that can lead to breakthroughs and innovations. By analysing extensive datasets, it uncovers hidden patterns and relationships, guiding researchers toward novel solutions that might otherwise remain undiscovered.

In the pursuit of personalised medicine, AI plays a pivotal role by recommending hydrogel formulations tailored to individual medical needs. This approach promises more effective treatments with fewer side effects, marking a significant advancement in patient care.

The integration of AI in hydrogel development encompasses various stages, each contributing to a holistic framework for material design and optimisation. The AI begins by acquiring comprehensive datasets containing information about hydrogel compositions, fabrication methods, and resulting properties. These datasets serve as the foundation for training AI models. Preprocessing techniques ensure data quality, handling missing values and normalising variables for accurate analyses.

ML algorithms, a subset of AI, play a pivotal role in hydrogel development. Supervised learning techniques, such as regression and classification, enable the prediction of hydrogel properties based on input variables—unsupervised learning, like clustering and dimensionality reduction, aids in identifying patterns and relationships within complex datasets.

AI facilitates the creation of predictive models that map input parameters to specific hydrogel outcomes. These models can forecast properties like swelling behaviour, mechanical strength, and drug release kinetics, allowing researchers to make informed decisions during material design.

AI-driven optimisation methods, including genetic algorithms and Bayesian optimisation, guide the search for optimal hydrogel formulations within vast chemical and structural spaces. These techniques expedite the identification of compositions that meet predefined performance criteria.

While the potential of AI in hydrogel development is vast, several challenges warrant consideration. The availability of high-quality, well-curated datasets is crucial for training robust AI models. Data privacy, the integration of domain knowledge, and the interpretability of AI-generated models are also critical concerns that researchers must address. Additionally, the complexity of hydrogel behaviour, which arises from intricate molecular interactions and crosslinking mechanisms, poses a challenge for accurate predictive modelling.

We would like to point out that this manuscript was prepared with the help of AI-assisted technology. We think that the text's clarity and coherence were considerably improved via AI-driven language technologies. With the use of these tools, grammatical and structural errors were found and fixed, ensuring that our article meets the strictest requirements for scientific writing. This open disclosure demonstrates our commitment to rigorous and high-level scientific communication. We have given this manuscript the spirit of originality and correctness, which was made possible in part by the cooperation of human expertise and AI-driven language technologies.

This review aims to provide a comprehensive understanding of the potential benefits of AI in designing hydrogels. By shedding light on this innovative approach, we hope to inspire continued research and development, paving the way for more effective targeted therapies that can improve patient outcomes and transform the research of hydrogels in the future.

## 2. Physical and Chemical Methods for Designing Hydrogels

Hydrogels, three-dimensional polymeric networks capable of retaining large amounts of water, have emerged as remarkable materials with a diverse range of applications, particularly in the biomedical and pharmaceutical fields [15]. Their unique properties, such as high water content, biocompatibility, and tuneable mechanical characteristics, make them invaluable for various purposes.

In biomedicine, hydrogels have gained prominence as versatile materials for drug delivery, tissue engineering, wound healing, and diagnostics [15]. These hydrophilic networks can be engineered to mimic the extracellular matrix [16], providing an ideal environment for cell growth and tissue regeneration [17]. Moreover, their ability to encapsulate and release bioactive compounds in a controlled manner has revolutionised drug delivery systems.

In pharmaceutical sciences, hydrogels find applications in drug formulation, where they serve as carriers for poorly water-soluble drugs, enhancing their bioavailability. Additionally, their mucoadhesive properties make them suitable for mucosal drug delivery, opening avenues for novel drug administration routes.

Hydrogels can be prepared using both natural and synthetic materials as precursors. Raw materials, such as cellulose, gelatine, alginate, chitosan (CS), and silk fibroin, are directly sourced from nature and are known for their biocompatibility and bioactive properties. On the other hand, synthetic materials, including polymethylmethacrylate (PMMA), polyurethane (PU), poly(N-isopropylacrylamide) (PNIPAM), poly(lactic acid) (PLA), and poly(lactic-co-glycolic acid) (PLGA), are produced through chemical reactions, offering the advantage of tuneable mechanical properties.

While homopolymeric hydrogels serve a specific purpose, their functionality can be limited. Other biomaterials, such as bioceramics, are often incorporated to enhance the mechanical strength, biodegradability, and/or stimuli-responsiveness of hydrogel matrices. The combination of various biomaterials enables the creation of multifunctional hydrogels that cater to diverse biomedical applications.

This section refers to the physical and chemical methods employed in the design and development of hydrogels. Understanding these methodologies is crucial for harnessing the full potential of hydrogels in addressing contemporary challenges in healthcare and drug delivery. We explore the techniques utilised to tailor hydrogel properties, ensuring they meet the stringent requirements of various biomedical and pharmaceutical applications.

## 2.1. Physical Crosslinking

Physical crosslinking is versatile for creating hydrogels without chemical reactions or covalent bonds. Instead, physical crosslinking relies on non-covalent interactions to form a 3D network, resulting in hydrogels with reversible and dynamic properties. This method offers several advantages, such as ease of preparation, injectability, and the ability to respond to external stimuli, making it suitable for various biomedical and pharmaceutical applications [18].

Some polymers exhibit a temperature-dependent sol-gel transition, forming a gel at a specific temperature range. As the temperature is lowered, these polymers undergo self-assembly, forming a hydrogel network [19]. Common polymers that undergo temperature-induced gelation include thermoresponsive polymers like poly(N-isopropylacrylamide) (PNIPAAm). These hydrogels are particularly attractive for drug delivery applications, as they can respond to changes in body temperature and release drugs accordingly.

Ionic gelation involves using ionic interactions between charged polymer chains and counterions to form a hydrogel network [20]. Examples include alginate and chitosan hydrogels, which can form crosslinks through interactions with divalent cations like calcium ions [20]. The reversibility of these ionic interactions makes them suitable for cell encapsulation and tissue engineering applications.

Certain amphiphilic polymers can self-assemble into hydrogels through hydrophobic interactions or hydrogen bonding [21]. Lipid-based hydrogels, for instance, can spontaneously form through the self-assembly of amphiphilic molecules into nanostructures, resulting in a hydrogel network [22]. These hydrogels have applications in drug delivery, as they can encapsulate hydrophobic drugs and release them in a controlled manner.

Photocrosslinking involves using light to induce the crosslinking of photoreactive molecules or polymers [23]. Photocrosslinkable hydrogels are prepared with photoinitiators that initiate the crosslinking reaction upon exposure to specific wavelengths of light [23]. This method offers precise spatial and temporal control over hydrogel formation, making it valuable for tissue engineering and 3D bioprinting applications.

The freeze–thaw method involves the freezing and thawing of a mixture containing different components, typically polymers and other functional materials, to create a gel-like structure [24]. The process starts by preparing a solvent solution or suspension of the desired components. This mixture is then subjected to a freezing step, where it is cooled to a low temperature, usually below the solvent's freezing point. During freezing, ice crystals form and the solute molecules, including polymers and other functional materials, are excluded from the ice lattice, increasing their concentration in the unfrozen portion of the solution [25]. After freezing, the sample is thawed, allowing the ice crystals to melt and the components to redistribute in the liquid phase. This process promotes the formation of a gel network as the polymers and other functional materials interact and crosslink, creating a three-dimensional structure that retains a large amount of solvent within its matrix [24,25]. The freeze–thaw cycles can be repeated multiple times to improve the gel's stability and mechanical properties. By adjusting the composition and freezing–thawing conditions, it is possible to control the hydrophilic/hydrophobic character and other properties of the resulting hybrid hydrogel [25]. This process creates physical crosslinks between the polymer chains, producing a hydrogel.

## 2.2. Chemical Crosslinking

Chemical crosslinking is a versatile and widely used method for obtaining hydrogels with excellent mechanical stability and structural integrity [26]. This process involves the formation of covalent bonds between polymer chains, resulting in a stable 3D network that retains water and forms a hydrogel. Covalent bonds are a type of chemical bond that occurs when two atoms share electrons to achieve a stable electron configuration [27]. In the context of hydrogel formation, when hydrogel precursors, such as polymer chains or monomers, contain functional groups that are capable of forming covalent bonds, a chemical crosslinking process can be initiated [28]. During this process, these functional groups react with one another, establishing covalent bonds between the polymer chains or monomers [29].

Chemical crosslinking is especially suitable for creating hydrogels with controlled porosity, swelling behaviour, and degradation rates, making them ideal for various biomedical and pharmaceutical applications.

In the radical polymerisation method, monomers containing double bonds are polymerised by a crosslinking agent and a radical initiator. The polymerisation reaction generates free radicals, which initiate the chain-growth polymerisation, forming covalent bonds between monomer units [30]. Typical radical initiators include azo compounds and peroxides. The polymerisation process can be carried out in solution or in the presence of a template to create 3D structures [30].

The Michael addition reaction involves the reaction between a Michael donor, typically a thiol group (-SH), and a Michael acceptor, such as an $\alpha,\beta$-unsaturated carbonyl compound [31]. This reaction results in the formation of a stable covalent bond, creating crosslinks between polymer chains. Hydrogels formed through Michael's addition are highly biocompatible and find applications in drug delivery, tissue engineering, and wound healing [32].

Click chemistry refers to a set of high-yield and selective reactions that can efficiently form covalent bonds. Common click chemistry reactions include azide-alkyne cycloaddition, thiol-ene, and tetrazine-norbornene reactions [33]. These reactions are advantageous for hydrogel formation due to their rapid reaction kinetics, high yields, and bioorthogonality, meaning they can be performed in the presence of biological molecules without interfering with cellular processes [34].

Schiff base formation is a chemical reaction between an aldehyde and a primary amine or hydrazine group. This reaction results in the formation of a covalent imine bond, creating crosslinks between polymer chains [35]. Schiff-based hydrogels are often used for drug delivery, as they can release drugs in response to specific environmental cues or triggers [35].

Bioprinting for hybrid hydrogels represents a cutting-edge approach that combines the advantages of both bioprinting and hybrid hydrogel materials [36]. Hybrid hydrogels blend or incorporate multiple materials, such as natural and synthetic polymers [36] or inorganic nanoparticles [37], to create novel hydrogel formulations with enhanced properties. When combined with bioprinting, this approach allows for the precise and controlled deposition of complex 3D structures containing living cells and multifunctional materials [38]. In bioprinting for hybrid hydrogels, the careful selection of bioinks and hybrid materials is crucial. The choice of bioinks is essential to ensure cellular viability, biocompatibility, and mechanical stability [39]. Hybrid materials may include natural polymers like collagen and gelatine, synthetic polyethylene glycol (PEG) and polyvinyl alcohol (PVA), and inorganic nanoparticles like calcium phosphate or gold nanoparticles [40].

Hybrid hydrogels offer a wide range of tuneable mechanical properties. By incorporating different materials with varying stiffness or elasticity, it is possible to create hydrogels that mimic the mechanical properties of native tissues [41]. This is especially important in tissue engineering, where bioprinted constructs must match the target tissue's mechanical environment for proper cell function and growth [41].

Hybrid hydrogels can be functionalised with bioactive molecules, growth factors, or peptides to promote cell adhesion, proliferation, and differentiation [42]. Bioprinting allows for the precise spatial distribution of these bioactive components within the hydrogel, creating complex microenvironments that can support tissue regeneration and repair.

One of the critical challenges in bioprinting tissue constructs is the lack of vascularisation. Hybrid hydrogels offer a promising solution by incorporating bioactive factors that promote the formation of blood vessels (angiogenesis) [43]. Additionally, hybrid hydrogels can be engineered to contain channels or networks to facilitate the diffusion of nutrients and oxygen, enabling the survival of bioprinted cells within thick tissue constructs [44].

Bioprinting for hybrid hydrogels is a rapidly evolving field with immense potential for advancing tissue engineering, regenerative medicine, and drug development. As researchers continue to innovate in materials science, bioprinting technologies, and tissue engineering, the applications of bioprinted hybrid hydrogels are expected to expand, ultimately leading to groundbreaking advancements in healthcare and personalised medicine. For instance, in regenerative medicine, bioprinted hybrid hydrogels are anticipated to revolutionise the development of patient-specific organoids [45], facilitating drug testing and disease modelling with unparalleled accuracy. Moreover, these hydrogels are set to play a pivotal role in orthopaedics, enabling the creation of custom-designed scaffolds for bone and cartilage repair [46]. In the realm of dermatology, they hold the potential to transform wound healing, with hydrogel-based dressings tailored to individual patient needs [47]. In the pharmaceutical industry, bioprinted hydrogels are envisioned to streamline drug formulation testing, ensuring it becomes safer and more effective [48]. Beyond healthcare, these hydrogels are also expected to find applications in environmental science, such as in the removal of contaminants from water sources [49]. As these innovations gather momentum, they are poised to reshape multiple facets of our lives, promising a future where healthcare and various industries benefit from the limitless potential of bioprinted hybrid hydrogels.

The choice of method depends on the desired properties, functionality, and intended application of the hydrogel. Each technique offers unique advantages and can be tailored to suit specific research or medical needs.

## 3. Numerical and Analytical Methods in Hydrogel Design

Numerical and analytical methods play crucial roles in the design and characterisation of hydrogels. They provide insights into the complex behaviours, properties, and interactions within hydrogel systems.

### 3.1. Numerical Simulations

Numerical simulation methods have emerged as invaluable tools in hydrogel design, revolutionising how researchers approach the development and analysis of these versatile materials. Hydrogels, three-dimensional networks of hydrophilic polymers that can absorb and retain large amounts of water, have found applications in various fields, from biomedical engineering and drug delivery to tissue regeneration. As the demand for hydrogels with tailored properties continues to grow, the integration of computational techniques has become essential for expediting the design process, optimising performance, and predicting behaviour under varying conditions.

The intricate nature of hydrogels, influenced by factors such as polymer composition, crosslinking density, and environmental conditions, presents challenges in accurately characterising and predicting their behaviour solely through traditional experimental approaches. Numerical simulations bridge this gap by providing a virtual laboratory where researchers can explore the intricate interplay between molecular structures, mechanical forces, fluid dynamics, and other critical variables that dictate hydrogel performance. These simulations offer a deeper understanding of hydrogel behaviour, enabling informed design decisions and accelerating the development of hydrogel-based solutions.

Some standard numerical simulation methods used for hydrogel design include Finite Element Analysis (FEA), Computational Fluid Dynamics (CFD), Molecular Dynamics simulations (MD), Monte Carlo Simulations, etc.

FEA is widely used to simulate the mechanical behaviour of hydrogels, including their deformation, stress distribution, and responses to external forces [50,51]. It helps us understand how hydrogels will behave in different loading conditions and assists in designing hydrogels with specific mechanical properties for applications such as tissue engineering, drug delivery, and medical devices.

CFD is used to model the flow of fluids through hydrogel structures [52]. It is important for hydrogels used in drug delivery systems or tissue engineering scaffolds where transporting nutrients, oxygen, and waste products is critical [53]. CFD simulations can provide insights into mass transport phenomena and guide the design of hydrogel structures with optimised fluid flow patterns.

MD simulations are used to study the behaviour of individual molecules within hydrogel networks [54]. They provide insights into the interactions between polymer chains, solvent molecules, and solutes at the atomic level. MD simulations can predict swelling behaviour [55], diffusion rates [56], and biomolecule interactions.

Monte Carlo methods are often employed to model the statistical behaviour of hydrogel systems [57]. These simulations help predict the macroscopic properties of hydrogels based on the behaviour of individual molecules or particles within the system [57]. They can be applied to study phenomena such as the swelling equilibrium, polymer chain conformations, and gel network structure.

Hydrogels are often subjected to multiple physical phenomena simultaneously, such as mechanical deformation, fluid flow, and heat transfer. Multiphysics simulations combine different numerical approaches to model these coupled effects and provide a comprehensive understanding of hydrogel behaviour [58] in complex environments such as tumours [59].

Numerical simulations can be coupled with *optimisation algorithms* to search for the best combination of material properties or structural configurations that meet specific design criteria. This approach is valuable for tailoring hydrogel properties to achieve desired outcomes.

Some simulations combine multiple techniques, such as coupling MD with FEA, to simultaneously capture different aspects of hydrogel behaviour. These hybrid methods allow a comprehensive understanding of complex interactions within hydrogel systems [60].

These numerical simulation methods expedite the design process and enable researchers to explore a vast design space, optimise material properties, and predict hydrogel behaviour across various environments. In combination with experimental data, inverse modelling techniques refine simulations and enhance the accuracy of predictions. Furthermore, the fusion of numerical simulations with optimisation algorithms empowers researchers to identify optimal hydrogel compositions and structures that align with specific performance criteria.

*3.2. Analytical Methods*

The design of hydrogels involves a multifaceted approach that relies heavily on analytical methods to characterise their physical, chemical, and mechanical properties. These methods not only aid in understanding the fundamental behaviour of hydrogels but also drive the optimisation and tailoring of their properties for specific applications. This article delves into the pivotal role of analytical methods in hydrogel design, from characterisation techniques to advanced imaging modalities.

Analysing the chemical composition of hydrogel precursors and networks is crucial for understanding their structure–property relationships. Fourier-transform infrared spectroscopy (FTIR) [61] and nuclear magnetic resonance (NMR) spectroscopy [62] provide insights into functional groups and molecular structures within hydrogel matrices. These methods allow researchers to verify the successful incorporation of desired monomers and crosslinkers and monitor the progress of polymerisation reactions.

Mechanical properties heavily influence the performance of hydrogels in various applications. Compression testing, tensile testing, and rheological analysis quantify parameters like compressive strength, Young's modulus, and viscosity [63]. These data guide the selection of suitable hydrogel formulations for specific uses, ensuring that mechanical properties align with intended functions.

Hydrogels' ability to absorb water and swell is a fundamental characteristic that impacts applications such as drug delivery and wound dressings. Swelling behaviour is studied by immersing hydrogels in different solvents and measuring weight changes over time [64]. Analytical balances and swelling ratio calculations provide insights into hydrogel responsiveness to environmental changes, influencing design choices for optimal performance.

Microscopic analysis techniques like scanning electron microscopy (SEM) [65] and atomic force microscopy (AFM) [66] allow researchers to visualise hydrogel surfaces and internal structures. These images reveal information about pore size, distribution, and interconnectedness, which are crucial for applications involving cell adhesion, growth, and the diffusion of therapeutic agents [67].

Thermal properties impact hydrogel stability and behaviour at different temperatures. Differential scanning calorimetry (DSC) [68] and thermogravimetric analysis (TGA) [69] enable the assessment of glass transition temperatures, melting points, and thermal degradation profiles. Such data aid in determining suitable processing conditions and the temperature ranges within which hydrogels maintain their integrity.

Analytical methods are pivotal in studying drug release kinetics from hydrogel matrices. UV-Vis spectroscopy [70] and high-performance liquid chromatography (HPLC) monitor the concentration of released substances over time. These data are crucial for designing hydrogel-based drug delivery systems with controlled and sustained release profiles [71].

Recent advancements have introduced sophisticated imaging methods such as confocal microscopy and magnetic resonance imaging (MRI) to further probe hydrogel behaviour [72]. These techniques allow the in-depth visualisation of hydrogel interactions with cells, tissues, and drugs, providing insights into real-time responses and interactions.

Statistical Data Analysis

Statistical data analysis involves applying various statistical techniques to process, interpret, and draw meaningful conclusions from experimental data. In hydrogel design, statistical analysis plays a crucial role in understanding the relationships between different variables, optimising formulations, and ensuring the reproducibility of results. Statistical data analysis methods are applied at various stages of hydrogel design; see Figure 1.

Before conducting experiments, researchers use statistical tools to design experiments effectively. Techniques like Design of Experiments (DOE) help to determine which variables to control, which to manipulate, and how many experiments to perform to gather sufficient data.

Statistical analysis starts with the collection of data from experiments. These data can include information about hydrogel composition, structure, mechanical properties, swelling behaviour, drug release profiles, and more.

Descriptive statistics provide a summary of the collected data. Measures like mean, median, standard deviation, and range give an overview of the central tendency and variability of the data.

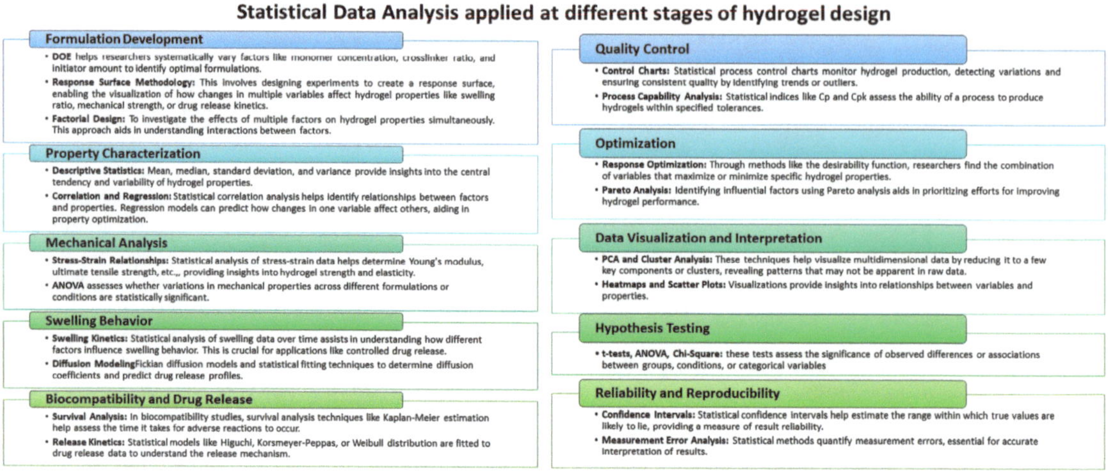

## Statistical Data Analysis applied at different stages of hydrogel design

**Formulation Development**
- **DOE** helps researchers systematically vary factors like monomer concentration, crosslinker ratio, and initiator amount to identify optimal formulations.
- **Response Surface Methodology:** This involves designing experiments to create a response surface, enabling the visualization of how changes in multiple variables affect hydrogel properties like swelling ratio, mechanical strength, or drug release kinetics.
- **Factorial Design:** To investigate the effects of multiple factors on hydrogel properties simultaneously. This approach aids in understanding interactions between factors.

**Property Characterization**
- **Descriptive Statistics:** Mean, median, standard deviation, and variance provide insights into the central tendency and variability of hydrogel properties.
- **Correlation and Regression:** Statistical correlation analysis helps identify relationships between factors and properties. Regression models can predict how changes in one variable affect others, aiding in property optimization.

**Mechanical Analysis**
- **Stress-Strain Relationships:** Statistical analysis of stress-strain data helps determine Young's modulus, ultimate tensile strength, etc., providing insights into hydrogel strength and elasticity.
- **ANOVA** assesses whether variations in mechanical properties across different formulations or conditions are statistically significant.

**Swelling Behavior**
- **Swelling Kinetics:** Statistical analysis of swelling data over time assists in understanding how different factors influence swelling behavior. This is crucial for applications like controlled drug release.
- **Diffusion Modeling:** Fickian diffusion models and statistical fitting techniques to determine diffusion coefficients and predict drug release profiles.

**Biocompatibility and Drug Release**
- **Survival Analysis:** In biocompatibility studies, survival analysis techniques like Kaplan-Meier estimation help assess the time it takes for adverse reactions to occur.
- **Release Kinetics:** Statistical models like Higuchi, Korsmeyer-Peppas, or Weibull distribution are fitted to drug release data to understand the release mechanism.

**Quality Control**
- **Control Charts:** Statistical process control charts monitor hydrogel production, detecting variations and ensuring consistent quality by identifying trends or outliers.
- **Process Capability Analysis:** Statistical indices like Cp and Cpk assess the ability of a process to produce hydrogels within specified tolerances.

**Optimization**
- **Response Optimization:** Through methods like the desirability function, researchers find the combination of variables that maximize or minimize specific hydrogel properties.
- **Pareto Analysis:** Identifying influential factors using Pareto analysis aids in prioritizing efforts for improving hydrogel performance.

**Data Visualization and Interpretation**
- **PCA and Cluster Analysis:** These techniques help visualize multidimensional data by reducing it to a few key components or clusters, revealing patterns that may not be apparent in raw data.
- **Heatmaps and Scatter Plots:** Visualizations provide insights into relationships between variables and properties.

**Hypothesis Testing**
- **t-tests, ANOVA, Chi-Square:** these tests assess the significance of observed differences or associations between groups, conditions, or categorical variables

**Reliability and Reproducibility**
- **Confidence Intervals:** Statistical confidence intervals help estimate the range within which true values are likely to lie, providing a measure of result reliability.
- **Measurement Error Analysis:** Statistical methods quantify measurement errors, essential for accurate interpretation of results.

**Figure 1.** Statistical data analysis at different stages of hydrogel design.

Researchers use correlation analysis to identify relationships between different variables. For example, it can reveal if there is a correlation between the composition of the hydrogel and its mechanical strength [73].

Regression models help establish mathematical relationships between variables. Researchers can use linear or nonlinear regression to predict one variable based on the values of others. This is useful for predicting hydrogel behaviour under different conditions [74].

Analysis of Variance (ANOVA) assesses the variance between different groups or conditions. It helps determine if the variations observed in the data are significant and whether they are due to manipulated variables or random chance [75].

Principal Component Analysis (PCA) is a dimensionality reduction technique that transforms complex data into a lower-dimensional space. It can help identify patterns and trends in multi-dimensional data sets, making it useful for analysing complex hydrogel datasets [76].

Multivariate analysis involves the analysis of multiple variables simultaneously to uncover hidden patterns and relationships that might not be apparent in individual studies [77].

Statistical methods are employed to ensure the quality and consistency of hydrogel production. Control charts, process capability analysis, and six sigma methodologies help maintain the desired quality standards.

Statistical analysis assesses the reliability and reproducibility of hydrogel properties allowing the calculation of confidence intervals, the assessment of experimental errors, and the determination of the precision of measurements.

Researchers use statistical optimisation techniques to find the optimal combination of hydrogel parameters that yield the desired properties. This is particularly useful in fine-tuning hydrogel formulations for specific applications.

Visualising data through plots, graphs, and charts helps us to understand trends and patterns intuitively. Visualisation tools (scatter plots, line charts, bar charts, heatmaps, 3D surface plots, radial charts, box plots, principal component analysis plots) enhance the communication of results and aid in decision-making.

Statistical tests, such as t-tests [78] or chi-square tests [79], are used to test hypotheses and determine whether observed differences between groups are statistically significant.

In hydrogel design, statistical data analysis aids in making informed decisions, optimising formulations, understanding the effects of variables, and ensuring the reliability of results. By applying appropriate statistical techniques, researchers can uncover insights that guide the development of hydrogels with tailored properties for a wide range of applications.

## 4. Leveraging Artificial Intelligence in Hydrogel Design

Leveraging AI in hydrogel design involves utilising advanced computational techniques to optimise and accelerate the development of hydrogel materials with specific properties and functionalities. AI-driven approaches revolutionise conventional trial-and-error methods, enabling rapid and informed material design. This leads to the development of hydrogels better tailored for diverse applications in medicine and biology [80]. Below, we present some expanded details and explore how AI is applied to the design and selection of hydrogels.

In the case of material design and selection, AI algorithms can predict and optimise hydrogel properties based on desired characteristics. ML models analyse large datasets of material properties to guide the selection of polymers, crosslinkers, and additives for specific applications. In addition, ML models can predict hydrogel behaviours, such as swelling ratios, mechanical strength, and degradation rates, based on formulation and environmental conditions.

AI can perform virtual screenings of vast chemical spaces to identify potential monomers, crosslinkers, and reaction conditions for synthesising hydrogels with desired properties, reducing the need for extensive experimental trial and error.

Also, AI-driven optimisation algorithms can enhance the hydrogel synthesis process by adjusting parameters such as temperature, pH, reaction time, and swelling to maximise yield and desired properties. For example, Islamkulov et al. used AI-supported optimisation applications (multilayer neural network sigmoid function model) for determining the swelling kinetics of hydrogel networks. In addition, the results of swelling behaviour under different experimental conditions, such as different crosslinker concentration temperatures and salt solutions, provided a deeper understanding of the physicochemical properties of the prepared hydrogels [81]. An important parameter for achieving a reproducible hydrogel is the gelation kinetics. AI models can predict hydrogel gelation kinetics by analysing the kinetics of polymerisation reactions, aiding in controlling gelation time and achieving reproducible results [82].

AI-enabled image analysis and spectroscopic techniques assist in the characterisation of hydrogel structures, porosity, and mechanical properties, ensuring quality and consistency in production. These analyses can be performed by using advanced imaging techniques such as scanning microscopy or NMR spectroscopy.

AI algorithms can assess the biocompatibility and functionality of hydrogels for specific biological applications, guiding the design of hydrogels for drug delivery, tissue engineering, and wound healing. Boztepe and colleagues [12] introduced an innovative hydrogel for the controlled release of doxorubicin. This research addressed a notable gap in the exploration of AI-driven hydrogel systems. Their study revealed the remarkable performance of the AI-based model in accurately predicting the drug release behaviour of the hydrogels they developed. These findings underscore the significance of such investigations in advancing novel materials while building upon empirical knowledge.

AI-driven data analysis can reveal previously unnoticed patterns (correlations between components, optimal manufacturing conditions, material interactions, performance over time, cost-effective formulations, biological response) and relationships within hydrogel datasets, leading to the discovery of novel hydrogel formulations and applications. These unnoticed patterns are often buried within vast datasets and can be challenging for humans to discern. AI's strength lies in its ability to sift through immense amounts of data, and to make predictions or recommendations based on these findings.

Another essential aspect of the research and development of hydrogels is the number of experimental iterations required. AI can accelerate hydrogel development, leading to the faster translation of hydrogel-based technologies, from research to practical applications.

AI-driven materials informatics platforms organise and analyse hydrogel-related data, facilitating collaboration and knowledge-sharing among researchers. Also, AI can help design hydrogels with tailored properties for specific patient needs, such as wound dressings or drug delivery systems.

AI-driven predictive models assist in predicting hydrogel performance under different regulatory conditions, aiding in compliance with safety and efficacy standards.

Integrating AIs into hydrogel research and development streamlines processes; it accelerates innovation and enhances the capabilities of hydrogels for diverse biomedical applications.

## 5. Machine Learning Techniques in Hydrogel Development

ML techniques have sparked a paradigm shift in the realm of hydrogel development. These sophisticated computational tools are reshaping how researchers approach materials design and expediting the entire innovation lifecycle of hydrogels. By harnessing the power of data-driven insights and predictive modelling, ML techniques have established themselves as indispensable assets at various crucial stages of hydrogel development, paving the way for accelerated discovery, enhanced precision, and the creation of novel and tailored hydrogel materials.

### 5.1. Machine Learning Subsets

ML subsets can be applied to various hydrogel research and development aspects, offering innovative solutions and accelerating progress in this field. ML consists of several subsets or branches [83], each with its own focus and techniques. Some of these subsets include:

Supervised Learning: In hydrogel research, supervised learning involves training a model on labelled data, where inputs (e.g., polymer type, crosslinking density) are associated with desired outputs (e.g., swelling ratio, degradation rate). This approach enables the prediction of hydrogel properties based on known relationships, aiding in efficiently screening potential formulations and optimising synthesis conditions [84].

Unsupervised Learning: Unsupervised learning techniques like clustering can uncover hidden patterns within complex hydrogel datasets. By grouping similar hydrogels based on structural and functional attributes, researchers can identify novel categories or classes of hydrogels with distinct behaviours, facilitating targeted investigations and customised designs [85].

Semi-Supervised Learning: When hydrogel datasets have limited labelled samples, semi-supervised learning combines labelled and unlabelled data. This approach can enhance predictions by leveraging the broader dataset, providing valuable insights into hydrogel behaviour even with a scarcity of labelled samples.

Reinforcement Learning: Hydrogel design can benefit from reinforcement learning by treating it as a sequential decision-making process. Algorithms can optimise synthesis parameters over multiple iterations to achieve the desired properties, such as mechanical strength or drug release profiles, while considering the feedback received from previous experiments [86].

Deep Learning: Deep neural networks, a subset of AI [87], can capture intricate relationships between input variables and hydrogel properties. By training on a diverse range of hydrogel compositions and experimental outcomes, deep learning models can predict complex behaviours, guiding the design of new hydrogel formulations.

Transfer Learning: Transfer learning allows models pre-trained on one hydrogel dataset to be fine-tuned for a different application. For instance, a neural network initially trained to predict swelling behaviour in one type of hydrogel can be adapted to predict degradation in another, saving time and computational resources [88].

Generative Adversarial Networks (GANs): GANs can aid in the design of new hydrogel structures by generating molecular configurations that meet specific performance criteria. This approach is promising for creating unique hydrogel formulations optimised for biomedical applications [89].

Applying these ML subsets to hydrogel research offers a multidimensional approach to understanding, designing, and optimising hydrogel materials for diverse biomedical, pharmaceutical, and industrial purposes. A schematic of AI-ML context and some of the application areas in the field of hydrogels are shown in Figure 2.

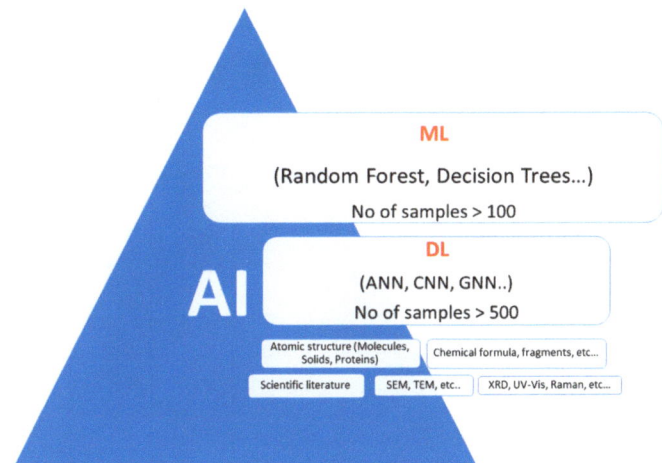

**Figure 2.** Schematic showing an overview of AI, ML, and DL methods.

*5.2. Machine Learning Algorithms*

5.2.1. Random Forest

RF stands out as a powerful and versatile ML algorithm that has revolutionised the field of hydrogel development [10]. Built upon the principles of ensemble learning, RF offers a sophisticated solution for tackling complex challenges in materials design, property prediction, and optimisation. Its unique characteristics make it an invaluable asset in pursuing innovative and tailored hydrogel materials.

At its core, RF is a collection of decision trees that operate collectively as a cohesive unit. Each decision tree is constructed using a random subset of the available data and features, making them diverse and distinct [90]. This diversity is a crucial strength of RF, enabling the model to capture a wide range of relationships within the data, from simple to complex. When a prediction is required, each decision tree contributes its output, and the final result is determined by aggregating these outputs—usually through voting for classification tasks or averaging for regression tasks (Figure 3). This ensemble approach enhances RF's accuracy, stability, and robustness, enabling it to handle the noisy or incomplete datasets that often characterise hydrogel research.

RF's ability to handle high-dimensional data and complex interactions is advantageous in hydrogel development. As researchers work with intricate combinations of variables—from monomer types and crosslinking ratios to environmental conditions and desired material properties—RF excels at identifying non-linear relationships and interactions that could be challenging to discern through traditional methods. By analysing these relationships comprehensively, RF aids researchers in predicting how changes in one or more variables may impact the final hydrogel properties, thus guiding more informed and efficient decision-making.

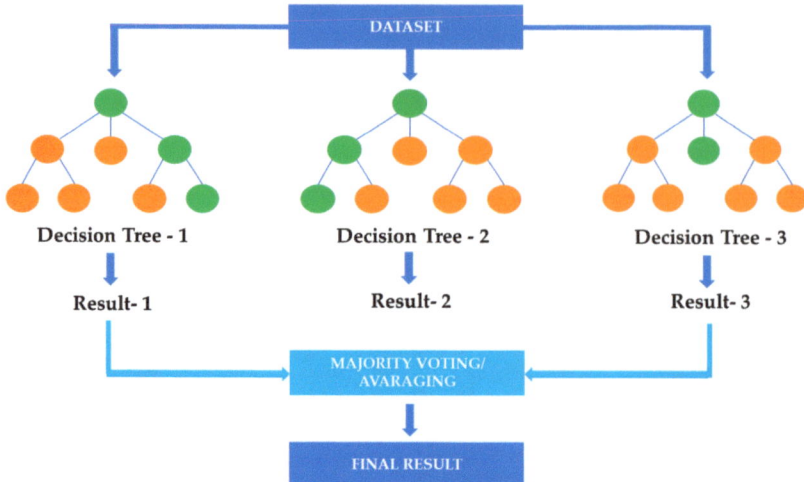

**Figure 3.** Simplified random forest.

Moreover, RF offers a degree of interpretability that differentiates it from other ML techniques. It can provide insights into feature importance, revealing which variables contribute most significantly to the model's predictions. This feature importance analysis in hydrogel research is invaluable for uncovering the key factors influencing specific material properties or behaviours. By identifying the most influential parameters, researchers can focus on fine-tuning these aspects of hydrogel design to achieve the desired outcomes effectively.

The RF method was recently applied for the hypothesis that polysaccharide hydrogels may feature fundamental separation criteria relevant to the permeability of compounds across the Gram-negative bacterial cell envelope, and that such permeability data could be used for predicting antibiotic accumulation in such bacteria [91]. Applying contemporary ML tools to the in vitro data, the same authors reported the first data on in bacterio accumulation of aminoglycosides and sulphonamides—essential classes of antibiotics used to treat Gram-negative infections. Expanding the investigations to antibiotic activity against highly relevant Gram-negative species gave evidence that in vitro permeability data may allow the exclusion of inactive substances at an early stage of antibiotic development [91].

5.2.2. Artificial Neural Network

Artificial Neural Networks (ANNs), inspired by the human brain's neural structure, excel at capturing intricate patterns in large datasets [83]. In hydrogel research, ANNs can model complex hydrogel–property relationships, allowing for accurate predictions of material behaviours. ANNs have been employed to optimise hydrogel formulations, predict drug release kinetics, and even simulate hydrogel–cell interactions, accelerating the understanding of hydrogel functionality.

For example, Brahima et al. used an ANN to model the nonlinear, multivariable, and complex drug delivery behaviour of poly(NIPAAm–co–AAc) IPN hydrogel systems. The developed ANN model was used to efficiently predict the drug release behaviours of hydrogels [92].

5.2.3. Support Vector Machines

Support Vector Machines (SVM) are a supervised learning method used for classification and regression tasks [93]. In hydrogel development, SVM can classify hydrogel formulations based on their properties or predict properties based on known compositions.

SVM aid in identifying relevant features that influence hydrogel performance, guiding researchers to prioritise specific components for achieving desired outcomes.

### 5.2.4. Deep Neural Networks

Deep Neural Networks (DNNs) can predict hydrogel properties based on their molecular structures [83,94]. By training on a dataset of known hydrogel compositions and their corresponding properties, DNNs can learn complex relationships between molecular features and hydrogel behaviour. This can aid in predicting properties like mechanical strength, swelling behaviour, and drug release profiles for new hydrogel formulations. DNNs can also be used for optimising hydrogel synthesis parameters. By setting up the DNN as an optimisation algorithm, it can iteratively suggest modifications to the formulation based on desired property outcomes. This can accelerate the process of finding optimal synthesis conditions.

### 5.2.5. Convolutional Neural Networks

Convolutional Neural Networks (CNNs) excel at analysing visual data, making them helpful in analysing hydrogel images, such as microstructure images captured through microscopy. CNNs can identify features, patterns, and structures in these images, providing insights into the internal structure of hydrogels [94].

In the context of 3D-printed hydrogels, CNNs can analyse the printing process and optimise printing parameters. By learning from 3D printing data, CNNs help the user adjust the printing speed, material deposition rate, and other variables to achieve the desired printing outcomes [95]. For example, a deep learning model using CNN was used to generate a model that differentiated between excellent and poor hydrogel prints. The CNN model was found to classify the bad and good images with an accuracy of 93.51%. The model achieved a validation accuracy of 90.244% [96]. Jin et al. [68] developed an anomaly detection system to classify imperfections for hydrogel-based bioink based on convolutional neural networks. Images were processed as small image patches for grid, gyroid, rectilinear, and honeycomb shapes. This research envisions high-quality tissue composition through real-time autonomous correction in the 3D bioprinting process [97].

CNNs can assist in designing microfluidic channels for hydrogel synthesis. By considering fluid flow dynamics, mixing efficiency, and gelation behaviour, CNNs can suggest optimised channel geometries for efficient and controlled synthesis.

CNNs can be employed to analyse and characterise the surface features of hydrogels. This includes identifying surface roughness, pore size distribution, and other topographical aspects influencing hydrogel performance.

Also, this algorithm can aid in quality control by identifying defects or inconsistencies in hydrogel products. This can ensure that the synthesised hydrogels meet the desired specifications and perform as expected.

The versatility of the above-mentioned algorithms can be extended. Combining some of the above-mentioned can aid researchers in identifying the most relevant variables among a large pool of options. This streamlines hydrogel development by focusing on the most impactful factors, reducing experimentation time and resources. For example, Pluronic F127, Pluronic F68, and Methocel K4M created and characterised enemas that deliver rectal protein. The concentrations of various polymers were utilised as input values to correlate with the final properties of the hydrogel using FormRules version 4.03, a commercial hybrid artificial intelligence tool platform that combines ANNs and fuzzy logic technologies. It is possible to assess the effects of each polymeric component in the hydrogel composition. For instance, it was discovered that F127 affected mucoadhesion and syringeability [98].

In a recent study conducted by Boztepe et al., ANNs were used to predict doxorubicin delivery from pH- and temperature-responsive poly(N–Isopropyl acrylamide-co-Acrylic acid)/Poly(ethylene glycol) (poly(NIPAAm-co-AAc)/PEG) interpenetrating polymer network (IPN) hydrogel [12]. In the same study conducted by Boztepe et al., derivates of

SVM were used to predict doxorubicin delivery from pH- and temperature-responsive IPN hydrogel [12].

Another study employed random forests, Gaussian process, and support vector machines as ML models to predict the cardiomyocyte (CM) content following the differentiation of human-induced pluripotent stem cells (hiPSCs) encapsulated in hydrogel microspheroids and to identify the main experimental variables affecting the CM yield. The models were built to predict two classes, sufficient and insufficient, for CM content on differentiation day 10. The best model predicted the sufficient class with an accuracy of 75% and a precision of 71%. This study showed that we can extract information from the experiments and build predictive models that could enhance the cell production by using ML techniques [99].

In a novel study, Li. et al. showed a combinational approach to generate a structurally diverse hydrogel library with more than 2000 peptides and evaluated their corresponding properties. The authors combined algorithms with the best precisions (54%, 57%, and 62% for RF, logistic regression, and gradient boosting, respectively, at the 50% recall). The authors correlated chemical variables and quantitative structure–property interactions with the self-assembly behaviour, and they were effective in identifying key structural elements influencing hydrogel formation [100].

Moreover, many of these algorithms can be combined for hydrogel synthesis. In a recent work, a CNN regression method was used to predict the Young's modulus and Poisson's ratio of BG-COL composites. First, 2000 images of BG-COL microstructures were generated. Then, the mechanical properties of the BG-COL composite were calculated using the finite element simulation. This numerical simulation software obtained data that were used to train a CNN regression model for predicting the mechanical properties of BG-COL based on its microstructural image. The authors demonstrated that the accepted CNN regression model could predict the mechanical properties of BG-COL. Hence, it can aid in overcoming the challenges of predicting these properties with traditional homogenisation methods. This work could guide the design of BG-COL and other composite hydrogels [50].

Zhu et al. used DNN and 3D CNN to reveal the implicit relationship between the network structure and mechanical properties of hydrogels to predict mechanical properties from different network structures. A modelling method for a single-network hydrogel network, that is, a self-avoiding walk network model which approximates the real polyacrylamide (PAAm) hydrogel structure at a mesoscopic scale, was proposed. After, the authors developed a DNN based on MLP and a 3D CNN containing the physical information of the network and utilised them to predict the nominal stress–stretch curves of hydrogels under uniaxial tension. By having a dataset of 2200 randomly generated network structures of PAAm hydrogel and their corresponding stress–stretch curves, the authors trained and evaluated the performance of the two models [94].

The selection of machine learning algorithms in hydrogel development is a dynamic and data-driven process. Researchers must consider the specific objectives, data types, and desired outcomes when choosing the most appropriate and particular machine learning algorithm or technique. This adaptability and versatility make AI a powerful tool in advancing the development of hydrogels for a wide range of biomedical applications.

## 6. Conclusions and Future Perspectives

Hydrogel-based strategies hold great promise and offer customisable solutions for various clinical scenarios. Hydrogels have immense promise for transforming hard and soft tissue treatments, but their successful implementation requires overcoming various obstacles through continued research, collaboration, and regulatory compliance.

Numerical simulations offer a virtual playground where hydrogel properties can be tailored, refined, and fine-tuned, creating a synergy between computational and experimental methodologies. As the field of hydrogel design continues to evolve and advance, the integration of numerical simulations can expand the frontiers of what is achievable with these remarkable materials.

Looking ahead, the fusion of AI with hydrogel development holds immense promise. Advanced AI techniques, including deep learning and reinforcement learning, will likely push the boundaries of predictive accuracy and material optimisation. Collaborations between materials scientists, chemists, and AI experts will foster interdisciplinary innovation, leading to the discovery of novel hydrogel formulations with tailored functionalities.

Across the hydrogel landscape, ML's transformative influence is undeniable. As researchers strive to design hydrogels with specific properties and functionalities, traditional methods often entail time-consuming trial-and-error approaches. In contrast, ML techniques are potent engines for pattern recognition, data analysis, and predictive modelling. They enable researchers to navigate the complex interplay of numerous variables—such as hydrogel composition, processing parameters, and end-use requirements—by generating comprehensive insights from large, intricate datasets. This capability propels hydrogel development beyond the boundaries of conventional experimentation, allowing researchers to extract valuable knowledge from raw data and make informed decisions with unprecedented efficiency. The continued collaboration between materials scientists, chemists, and AI/ML experts is instrumental in advancing hydrogel design. Together, they can harness the power of data-driven approaches, tackle complex problems, and create innovative hydrogel-based solutions that have a profound impact on healthcare and biomedical applications.

ML's impact is particularly profound in optimising hydrogel formulations. By rapidly evaluating an extensive range of chemical compositions and structural arrangements, ML algorithms guide researchers towards promising hydrogel candidates for specific applications. This ability to traverse multidimensional parameter spaces accelerates the identification of optimal formulations, expediting the path from conceptualisation to tangible hydrogel prototypes. Furthermore, ML techniques empower researchers to predict hydrogel properties with remarkable accuracy, sparing them the need for resource-intensive trial iterations. This predictive prowess shortens development timelines and empowers researchers to fine-tune hydrogel properties to precise specifications—a critical advantage in tailoring materials for diverse applications, from drug delivery systems to tissue engineering scaffolds.

Integrating ML techniques into hydrogel development represents a fundamental shift from conventional approaches to a dynamic, data-centric methodology. ML's capacity to handle complexity, recognise intricate patterns, and optimise outcomes positions it as an indispensable tool in the arsenal of modern hydrogel development. As the boundaries of ML continue to expand, its synergy with hydrogel research is poised to reshape the trajectory of biomaterials innovation.

As ML continues to evolve, advanced techniques like deep learning and reinforcement learning are promising for further pushing push the boundaries of hydrogel research. Integrating domain knowledge into ML models and addressing challenges such as data scarcity and interpretability will be crucial for realising the full potential of ML in hydrogel development.

In conclusion, integrating ML techniques such as RF, ANN, SVM, and LR has revolutionised the field of hydrogel development. These tools expedite the discovery process, optimise material properties, and pave the way for innovative applications in drug delivery, tissue engineering, and diagnostics. As ML technology advances, it is poised to reshape the landscape of hydrogel research, unlocking new possibilities and accelerating advancements in biomaterials science.

Even though incorporating AI into hydrogel design has opened a new era of precision and efficiency, significant challenges persist, necessitating innovative solutions and interdisciplinary collaboration. As we mentioned, AI models rely heavily on data. In hydrogel design, comprehensive and accurate datasets can be elusive. Materials scientists, chemists, and AI experts must collaborate to curate high-quality data [101,102].

Successful hydrogel design demands input from chemistry, biology, materials science, and AI. Effective cross-disciplinary collaboration is essential but can be challenging. Hydrogel properties are influenced by numerous factors. Modelling these interactions accurately remains a challenge [101,102].

As AI-driven hydrogel design advances, ethical concerns surrounding data privacy, bias in algorithms, and intellectual property rights become more prominent. Addressing these issues transparently and ethically is critical to maintaining trust and integrity in research. Transitioning from small-scale AI-optimised designs to large-scale production is challenging, particularly for medical and industrial applications [101,102].

Hydrogels developed using AI may face regulatory hurdles, particularly in the medical field. However, demonstrating their safety and efficacy to regulatory bodies is a complex and resource-intensive process [101,102].

As AI continues to evolve, its integration with hydrogel research holds the promise of unlocking new capabilities and applications, revolutionising the field of biomaterials and shaping the future of medical science and technology.

**Author Contributions:** Conceptualisation, I.N. and B.B.; methodology, I.N.; software, B.B.; validation, I.N.; resources, I.N.; data curation, I.N.; writing—original draft preparation, I.N. and B.B.; writing—review and editing, I.N. and B.B.; visualisation, I.N. and B.B.; supervision, I.N. and B.B.; project administration, I.N.; funding acquisition, I.N. All authors have read and agreed to the published version of the manuscript.

**Funding:** This research was funded by a grant from the Ministry of Research, Innovation and Digitization, CNCS—UEFISCDI, project number PN-III-P2-2.1-PED-2021-3178 within PNCDI III.

**Institutional Review Board Statement:** Not applicable.

**Informed Consent Statement:** Not applicable.

**Acknowledgments:** I.N. and B.B. acknowledge the support of a grant from the Ministry of Research, Innovation and Digitization, CNCS—UEFISCDI, project number PN-III-P2-2.1-PED-2021-3178. I.N. and B.B. also acknowledge support from the Romanian Ministry of Research, Innovation and Digitalization under the Romanian National Nucleu Program LAPLAS VII—contract no. 30N/2023. We acknowledge the use of https://quillbot.com/ to edit and improve our writing at the final stage of preparing the manuscript.

**Conflicts of Interest:** The authors declare no conflict of interest.

## References

1. Vermonden, T.; Klumperman, B. The past, present and future of hydrogels. *Eur. Polym. J.* **2015**, *72*, 341–343. [CrossRef]
2. Buwalda, S.J.; Vermonden, T.; Hennink, W.E. Hydrogels for Therapeutic Delivery: Current Developments and Future Directions. *Biomacromolecules* **2017**, *18*, 316–330. [CrossRef] [PubMed]
3. Liu, M.; Guo, R.; Ma, Y. Construction of a specific and efficient antibacterial agent against Pseudomonas aeruginosa based on polyethyleneimine cross-linked fucose. *J. Mater. Sci.* **2021**, *56*, 6083–6094. [CrossRef]
4. Revete, A.; Aparicio, A.; Cisterna, B.A.; Revete, J.; Luis, L.; Ibarra, E.; Segura González, E.A.; Molino, J.; Reginensi, D. Advancements in the Use of Hydrogels for Regenerative Medicine: Properties and Biomedical Applications. *Int. J. Biomater.* **2022**, *2022*, e3606765. [CrossRef] [PubMed]
5. Alka, A.; Verma, A.; Mishra, N.; Singh, N.; Singh, P.; Nisha, R.; Pal, R.R.; Saraf, S.A. Polymeric Gel Scaffolds and Biomimetic Environments for Wound Healing. *Curr. Pharm. Des.* **2023**, *29*, 1–19. [CrossRef] [PubMed]
6. Ko, A.; Liao, C. Hydrogel wound dressings for diabetic foot ulcer treatment: Status-quo, challenges, and future perspectives. *BMEMat* **2023**, *1*, e12037. [CrossRef]
7. Nie, L.; Li, Y.; Liu, Y.; Shi, L.; Chen, H. Recent Applications of Contact Lenses for Bacterial Corneal Keratitis Therapeutics: A Review. *Pharmaceutics* **2022**, *14*, 2635. [CrossRef]
8. Barhoum, A.; Sadak, O.; Ramirez, I.A.; Iverson, N. Stimuli-bioresponsive hydrogels as new generation materials for implantable, wearable, and disposable biosensors for medical diagnostics: Principles, opportunities, and challenges. *Adv. Colloid Interface Sci.* **2023**, *317*, 102920. [CrossRef]
9. Gerlach, G.; Arndt, K.-F. *Hydrogel Sensors and Actuators: Engineering and Technology*; Springer Science & Business Media: Berlin/Heidelberg, Germany, 2009.
10. Wang, Y.; Wallmersperger, T.; Ehrenhofer, A. Application of back propagation neural networks and random forest algorithms in material research of hydrogels. *PAMM* **2023**, *23*, e202200278. [CrossRef]
11. Vora, L.K.; Gholap, A.D.; Jetha, K.; Thakur, R.R.S.; Solanki, H.K.; Chavda, V.P. Artificial Intelligence in Pharmaceutical Technology and Drug Delivery Design. *Pharmaceutics* **2023**, *15*, 1916. [CrossRef]

12. Boztepe, C.; Künkül, A.; Yüceer, M. Application of artificial intelligence in modeling of the doxorubicin release behavior of pH and temperature responsive poly(NIPAAm-co-AAc)-PEG IPN hydrogel. *J. Drug Deliv. Sci. Technol.* **2020**, *57*, 101603. [CrossRef]
13. Soleimani, S.; Heydari, A.; Fattahi, M. Swelling prediction of calcium alginate/cellulose nanocrystal hydrogels using response surface methodology and artificial neural network. *Ind. Crops Prod.* **2023**, *192*, 116094. [CrossRef]
14. Mukherjee, S.; Kim, B.; Cheng, L.Y.; Doerfert, M.D.; Li, J.; Hernandez, A.; Liang, L.; Jarvis, M.I.; Rios, P.D.; Ghani, S.; et al. Screening hydrogels for antifibrotic properties by implanting cellularly barcoded alginates in mice and a non-human primate. *Nat. Biomed. Eng.* **2023**, *7*, 867–886. [CrossRef] [PubMed]
15. de Lima, C.S.A.; Balogh, T.S.; Varca, J.P.R.O.; Varca, G.H.C.; Lugão, A.B.; Camacho-Cruz, L.A.; Bucio, E.; Kadlubowski, S.S. An Updated Review of Macro, Micro, and Nanostructured Hydrogels for Biomedical and Pharmaceutical Applications. *Pharmaceutics* **2020**, *12*, 970. [CrossRef]
16. Lou, J.; Mooney, D.J. Chemical strategies to engineer hydrogels for cell culture. *Nat. Rev. Chem.* **2022**, *6*, 726–744. [CrossRef]
17. El-Husseiny, H.M.; Mady, E.A.; Hamabe, L.; Abugomaa, A.; Shimada, K.; Yoshida, T.; Tanaka, T.; Yokoi, A.; Elbadawy, M.; Tanaka, R. Smart/stimuli-responsive hydrogels: Cutting-edge platforms for tissue engineering and other biomedical applications. *Mater. Today Bio* **2022**, *13*, 100186. [CrossRef]
18. Yang, J.; Chen, Y.; Zhao, L.; Zhang, J.; Luo, H. Constructions and Properties of Physically Cross-Linked Hydrogels Based on Natural Polymers. *Polym. Rev.* **2023**, *63*, 574–612. [CrossRef]
19. Cui, K.; Yu, C.; Ye, Y.N.; Li, X.; Gong, J.P. Mechanism of temperature-induced asymmetric swelling and shrinking kinetics in self-healing hydrogels. *Proc. Natl. Acad. Sci. USA* **2022**, *119*, e2207422119. [CrossRef]
20. Gadziński, P.; Froelich, A.; Jadach, B.; Wojtyłko, M.; Tatarek, A.; Białek, A.; Krysztofiak, J.; Gackowski, M.; Otto, F.; Osmałek, T. Ionotropic Gelation and Chemical Crosslinking as Methods for Fabrication of Modified-Release Gellan Gum-Based Drug Delivery Systems. *Pharmaceutics* **2023**, *15*, 108. [CrossRef]
21. Omar, J.; Ponsford, D.; Dreiss, C.A.; Lee, T.-C.; Loh, X.J. Supramolecular Hydrogels: Design Strategies and Contemporary Biomedical Applications. *Chem.–Asian J.* **2022**, *17*, e202200081. [CrossRef]
22. Tripathi, M.; Sharma, R.; Hussain, A.; Kumar, I.; Sharma, A.K.; Sarkar, A. Chapter 17—Hydrogels and their combination with lipids and nucleotides. In *Sustainable Hydrogels*; Thomas, S., Sharma, B., Jain, P., Shekhar, S., Eds.; Elsevier: Amsterdam, The Netherlands, 2023; pp. 471–487. [CrossRef]
23. Liu, J.; Su, C.; Chen, Y.; Tian, S.; Lu, C.; Huang, W.; Lv, Q. Current Understanding of the Applications of Photocrosslinked Hydrogels in Biomedical Engineering. *Gels* **2022**, *8*, 216. [CrossRef] [PubMed]
24. Hao, M.; Wang, Y.; Li, L.; Liu, Y.; Bai, Y.; Zhou, W.; Lu, Q.; Sun, F.; Li, L.; Feng, S.; et al. Tough Engineering Hydrogels Based on Swelling-Freeze-Thaw Method for Artificial Cartilage. *ACS Appl. Mater. Interfaces* **2022**, *14*, 25093–25103. [CrossRef] [PubMed]
25. Varshney, N.; Sahi, A.K.; Poddar, S.; Vishwakarma, N.K.; Kavimandan, G.; Prakash, A.; Mahto, S.K. Freeze-Thaw-Induced Physically Cross-linked Superabsorbent Polyvinyl Alcohol/Soy Protein Isolate Hydrogels for Skin Wound Dressing: In Vitro and In Vivo Characterization. *ACS Appl. Mater. Interfaces* **2022**, *14*, 14033–14048. [CrossRef]
26. Xue, X.; Hu, Y.; Wang, S.; Chen, X.; Jiang, Y.; Su, J. Fabrication of physical and chemical crosslinked hydrogels for bone tissue engineering. *Bioact. Mater.* **2022**, *12*, 327–339. [CrossRef]
27. Pei, X.; Wang, J.; Cong, Y.; Fu, J. Recent progress in polymer hydrogel bioadhesives. *J. Polym. Sci.* **2021**, *59*, 1312–1337. [CrossRef]
28. García, J.M.; García, F.C.; Ruiz, J.A.R.; Vallejos, S.; Trigo-López, M. *Smart Polymers: Principles and Applications*; Walter de Gruyter GmbH & Co KG: Berlin, Germany, 2022.
29. Yang, J.; Bai, R.; Chen, B.; Suo, Z. Hydrogel Adhesion: A Supramolecular Synergy of Chemistry, Topology, and Mechanics. *Adv. Funct. Mater.* **2020**, *30*, 1901693. [CrossRef]
30. Seidi, F.; Zhao, W.; Xiao, H.; Jin, Y.; Reza Saeb, M.; Zhao, C. Radical polymerization as a versatile tool for surface grafting of thin hydrogel films. *Polym. Chem.* **2020**, *11*, 4355–4381. [CrossRef]
31. Guaresti, O.; Basasoro, S.; González, K.; Eceiza, A.; Gabilondo, N. In situ cross-linked chitosan hydrogels via Michael addition reaction based on water-soluble thiol-maleimide precursors. *Eur. Polym. J.* **2019**, *119*, 376–384. [CrossRef]
32. Khan, A.H.; Cook, J.K.; Wortmann III, W.J.; Kersker, N.D.; Rao, A.; Pojman, J.A.; Melvin, A.T. Synthesis and characterization of thiol-acrylate hydrogels using a base-catalyzed Michael addition for 3D cell culture applications. *J. Biomed. Mater. Res. B Appl. Biomater.* **2020**, *108*, 2294–2307. [CrossRef]
33. Li, X.; Xiong, Y. Application of "Click" Chemistry in Biomedical Hydrogels. *ACS Omega* **2022**, *7*, 36918–36928. [CrossRef]
34. Li, Y.; Wang, X.; Han, Y.; Sun, H.-Y.; Hilborn, J.; Shi, L. Click chemistry-based biopolymeric hydrogels for regenerative medicine. *Biomed. Mater.* **2021**, *16*, 022003. [CrossRef] [PubMed]
35. Xu, J.; Liu, Y.; Hsu, S. Hydrogels Based on Schiff Base Linkages for Biomedical Applications. *Molecules* **2019**, *24*, 3005. [CrossRef] [PubMed]
36. He, Y.; Wang, F.; Wang, X.; Zhang, J.; Wang, D.; Huang, X. A photocurable hybrid chitosan/acrylamide bioink for DLP based 3D bioprinting. *Mater. Des.* **2021**, *202*, 109588. [CrossRef]
37. Zhu, H.; Monavari, M.; Zheng, K.; Distler, T.; Ouyang, L.; Heid, S.; Jin, Z.; He, J.; Li, D.; Boccaccini, A.R. 3D Bioprinting of Multifunctional Dynamic Nanocomposite Bioinks Incorporating Cu-Doped Mesoporous Bioactive Glass Nanoparticles for Bone Tissue Engineering. *Small* **2022**, *18*, 2104996. [CrossRef]
38. Yang, Z.; Yi, P.; Liu, Z.; Zhang, W.; Mei, L.; Feng, C.; Tu, C.; Li, Z. Stem Cell-Laden Hydrogel-Based 3D Bioprinting for Bone and Cartilage Tissue Engineering. *Front. Bioeng. Biotechnol.* **2022**, *10*, 865770. [CrossRef]

39. Zhou, K.; Sun, Y.; Yang, J.; Mao, H.; Gu, Z. Hydrogels for 3D embedded bioprinting: A focused review on bioinks and support baths. *J. Mater. Chem. B* **2022**, *10*, 1897–1907. [CrossRef]
40. Xie, M.; Su, J.; Zhou, S.; Li, J.; Zhang, K. Application of Hydrogels as Three-Dimensional Bioprinting Ink for Tissue Engineering. *Gels* **2023**, *9*, 88. [CrossRef]
41. Ghandforoushan, P.; Alehosseini, M.; Golafshan, N.; Castilho, M.; Dolatshahi-Pirouz, A.; Hanaee, J.; Davaran, S.; Orive, G. Injectable hydrogels for cartilage and bone tissue regeneration: A review. *Int. J. Biol. Macromol.* **2023**, *246*, 125674. [CrossRef]
42. Deptuła, M.; Zawrzykraj, M.; Sawicka, J.; Banach-Kopeć, A.; Tylingo, R.; Pikuła, M. Application of 3D- printed hydrogels in wound healing and regenerative medicine. *Biomed. Pharmacother.* **2023**, *167*, 115416. [CrossRef]
43. Shahriari-Khalaji, M.; Sattar, M.; Cao, R.; Zhu, M. Angiogenesis, hemocompatibility and bactericidal effect of bioactive natural polymer-based bilayer adhesive skin substitute for infected burned wound healing. *Bioact. Mater.* **2023**, *29*, 177–195. [CrossRef]
44. Qi, J.; Zheng, S.; Zhao, N.; Li, Y.; Zhang, G.; Yin, W. A 3D bioprinted hydrogel multilevel arc vascular channel combined with an isomaltol core sacrificial process. *Mater. Today Commun.* **2023**, *36*, 106492. [CrossRef]
45. Zheng, F.; Xiao, Y.; Liu, H.; Fan, Y.; Dao, M. Patient-Specific Organoid and Organ-on-a-Chip: 3D Cell-Culture Meets 3D Printing and Numerical Simulation. *Adv. Biol.* **2021**, *5*, e2000024. [CrossRef] [PubMed]
46. Mei, Q.; Rao, J.; Bei, H.P.; Liu, Y.; Zhao, X. 3D Bioprinting Photo-Crosslinkable Hydrogels for Bone and Cartilage Repair. *Int. J. Bioprint.* **2021**, *7*, 367. [CrossRef] [PubMed]
47. Solanki, D.; Vinchhi, P.; Patel, M.M. Design Considerations, Formulation Approaches, and Strategic Advances of Hydrogel Dressings for Chronic Wound Management. *ACS Omega* **2023**, *8*, 8172–8189. [CrossRef] [PubMed]
48. Dutta, S.D.; Ganguly, K.; Hexiu, J.; Randhawa, A.; Moniruzzaman, M.; Lim, K.-T. A 3D Bioprinted Nanoengineered Hydrogel with Photoactivated Drug Delivery for Tumor Apoptosis and Simultaneous Bone Regeneration via Macrophage Immunomodulation. *Macromol. Biosci.* **2023**, *23*, 2300096. [CrossRef]
49. Jiang, H.; Hao, Z.; Zhang, J.; Tang, J.; Li, H. Bioinspired swelling enhanced hydrogels for underwater sensing. *Colloids Surf. Physicochem. Eng. Asp.* **2023**, *664*, 131197. [CrossRef]
50. Shokrollahi, Y.; Dong, P.; Gamage, P.T.; Patrawalla, N.; Kishore, V.; Mozafari, H.; Gu, L. Finite Element-Based Machine Learning Model for Predicting the Mechanical Properties of Composite Hydrogels. *Appl. Sci.* **2022**, *12*, 10835. [CrossRef]
51. Kakarla, A.B.; Kong, I.; Nukala, S.G.; Kong, W. Mechanical Behaviour Evaluation of Porous Scaffold for Tissue-Engineering Applications Using Finite Element Analysis. *J. Compos. Sci.* **2022**, *6*, 46. [CrossRef]
52. Teoh, J.H.; Abdul Shakoor, F.T.; Wang, C.-H. 3D Printing Methyl Cellulose Hydrogel Wound Dressings with Parameter Exploration Via Computational Fluid Dynamics Simulation. *Pharm. Res.* **2022**, *39*, 281–294. [CrossRef]
53. Li, S.; Liu, Y.; Li, Y.; Zhang, Y.; Hu, Q. Computational and experimental investigations of the mechanisms used by coaxial fluids to fabricate hollow hydrogel fibers. *Chem. Eng. Process. Process Intensif.* **2015**, *95*, 98–104. [CrossRef]
54. Wei, Q.; Yang, R.; Sun, D.; Zhou, J.; Li, M.; Zhang, Y.; Wang, Y. Design and evaluation of sodium alginate/polyvinyl alcohol blend hydrogel for 3D bioprinting cartilage scaffold: Molecular dynamics simulation and experimental method. *J. Mater. Res. Technol.* **2022**, *17*, 66–78. [CrossRef]
55. Shahshahani, S.; Shahgholi, M.; Karimipour, A. The thermal performance and mechanical stability of methacrylic acid porous hydrogels in an aqueous medium at different initial temperatures and hydrogel volume fraction using the molecular dynamics simulation. *J. Mol. Liq.* **2023**, *382*, 122001. [CrossRef]
56. Salahshoori, I.; Ramezani, Z.; Cacciotti, I.; Yazdanbakhsh, A.; Hossain, M.K.; Hassanzadeganroudsari, M. Cisplatin uptake and release assessment from hydrogel synthesized in acidic and neutral medium: An experimental and molecular dynamics simulation study. *J. Mol. Liq.* **2021**, *344*, 117890. [CrossRef]
57. Casalini, T.; Perale, G. From Microscale to Macroscale: Nine Orders of Magnitude for a Comprehensive Modeling of Hydrogels for Controlled Drug Delivery. *Gels* **2019**, *5*, 28. [CrossRef] [PubMed]
58. Gharehnazifam, Z.; Dolatabadi, R.; Baniassadi, M.; Shahsavari, H.; Kajbafzadeh, A.-M.; Abrinia, K.; Baghani, M. Computational analysis of vincristine loaded silk fibroin hydrogel for sustained drug delivery applications: Multiphysics modeling and experiments. *Int. J. Pharm.* **2021**, *609*, 121184. [CrossRef]
59. Gharehnazifam, Z.; Dolatabadi, R.; Baniassadi, M.; Shahsavari, H.; Kajbafzadeh, A.-M.; Abrinia, K.; Gharehnazifam, K.; Baghani, M. Multiphysics modeling and experiments on ultrasound-triggered drug delivery from silk fibroin hydrogel for Wilms tumor. *Int. J. Pharm.* **2022**, *621*, 121787. [CrossRef]
60. Liu, D.; Ma, S.; Yuan, H.; Markert, B. Modelling and simulation of coupled fluid transport and time-dependent fracture in fibre-reinforced hydrogel composites. *Comput. Methods Appl. Mech. Eng.* **2022**, *390*, 114470. [CrossRef]
61. Karvinen, J.; Kellomäki, M. Characterization of self-healing hydrogels for biomedical applications. *Eur. Polym. J.* **2022**, *181*, 111641. [CrossRef]
62. Li, P.; Malveau, C.; Zhu, X.X.; Wuest, J.D. Using Nuclear Magnetic Resonance Spectroscopy to Probe Hydrogels Formed by Sodium Deoxycholate. *Langmuir* **2022**, *38*, 5111–5118. [CrossRef]
63. Xing, W.; Tang, Y. On mechanical properties of nanocomposite hydrogels: Searching for superior properties. *Nano Mater. Sci.* **2022**, *4*, 83–96. [CrossRef]
64. Asy-Syifa, N.; Kusjuriansah; Waresindo, W.X.; Edikresnha, D.; Suciati, T.; Khairurrijal, K. The Study of the Swelling Degree of the PVA Hydrogel with varying concentrations of PVA. *J. Phys. Conf. Ser.* **2022**, *2243*, 012053. [CrossRef]

65. Martinez-Garcia, F.D.; van Dongen, J.A.; Burgess, J.K.; Harmsen, M.C. Matrix Metalloproteases from Adipose Tissue-Derived Stromal Cells Are Spatiotemporally Regulated by Hydrogel Mechanics in a 3D Microenvironment. *Bioengineering* **2022**, *9*, 340. [CrossRef] [PubMed]
66. Joshi, J.; Homburg, S.V.; Ehrmann, A. Atomic Force Microscopy (AFM) on Biopolymers and Hydrogels for Biotechnological Applications—Possibilities and Limits. *Polymers* **2022**, *14*, 1267. [CrossRef] [PubMed]
67. Jayawardena, I.; Turunen, P.; Cambraia Garms, B.; Rowan, A.; Corrie, S.; Grøndahl, L. Evaluation of techniques used for visualisation of hydrogel morphology and determination of pore size distributions. *Mater. Adv.* **2023**, *4*, 669–682. [CrossRef]
68. Sulaeman, A.S.; Putro, P.A.; Nikmatin, S. Thermal studies of hydrogels based on poly(acrylic acid) and its copolymers by differential scanning calorimetry: A systematic literature review. *Polym. Polym. Compos.* **2022**, *30*, 09673911221094022. [CrossRef]
69. Reguieg, F.; Ricci, L.; Bouyacoub, N.; Belbachir, M.; Bertoldo, M. Thermal characterization by DSC and TGA analyses of PVA hydrogels with organic and sodium MMT. *Polym. Bull.* **2020**, *77*, 929–948. [CrossRef]
70. Zhao, Y.; Li, H.; Wang, Y.; Zhang, Z.; Wang, Q. Preparation, characterization and release kinetics of a multilayer encapsulated Perilla frutescens L. essential oil hydrogel bead. *Int. J. Biol. Macromol.* **2023**, *249*, 124776. [CrossRef]
71. Vildanova, R.R.; Petrova, S.F.; Kolesov, S.V.; Khutoryanskiy, V.V. Biodegradable Hydrogels Based on Chitosan and Pectin for Cisplatin Delivery. *Gels* **2023**, *9*, 342. [CrossRef]
72. Wang, R.; Xin, J.; Ji, Z.; Zhu, M.; Yu, Y.; Xu, M. Spin-Space-Encoding Magnetic Resonance Imaging: A New Application for Rapid and Sensitive Monitoring of Dynamic Swelling of Confined Hydrogels. *Molecules* **2023**, *28*, 3116. [CrossRef]
73. Takayama, G.; Kondo, T. Quantitative evaluation of fiber network structure–property relationships in bacterial cellulose hydrogels. *Carbohydr. Polym.* **2023**, *321*, 121311. [CrossRef]
74. Xu, Y.; Wang, J.; Zhao, S.; Chen, D.; Zhang, H.; Fang, Y.; Kong, N.; Zhou, Z.; Li, W.; Wang, H. Accelerating the prediction and discovery of peptide hydrogels with human-in-the-loop. *Nat. Commun.* **2023**, *14*, 3880. [CrossRef] [PubMed]
75. Shu, J.; Wang, C.; Tao, Y.; Wang, S.; Cheng, F.; Zhang, Y.; Shi, K.; Xia, K.; Wang, R.; Wang, J.; et al. Thermosensitive hydrogel-based GPR124 delivery strategy for rebuilding blood-spinal cord barrier. *Bioeng. Transl. Med.* **2023**, *8*, e10561. [CrossRef] [PubMed]
76. Pannala, R.K.P.K.; Juyal, U.; Kodavaty, J. Optimization of hydrogel composition for effective release of drug. *Chem. Prod. Process Model.* **2023**. [CrossRef]
77. Taaca, K.L.M.; Nakajima, H.; Thumanu, K.; Prieto, E.I.; Vasquez, M.R. Network formation and differentiation of chitosan–acrylic acid hydrogels using X-ray absorption spectroscopy and multivariate analysis of Fourier transform infrared spectra. *J. Electron Spectrosc. Relat. Phenom.* **2023**, *267*, 147372. [CrossRef]
78. Nudell, V.; Wang, Y.; Pang, Z.; Lal, N.K.; Huang, M.; Shaabani, N.; Kanim, W.; Teijaro, J.; Maximov, A.; Ye, L. HYBRiD: Hydrogel-reinforced DISCO for clearing mammalian bodies. *Nat. Methods* **2022**, *19*, 479–485. [CrossRef] [PubMed]
79. Reindel, W.; Steffen, R.; Mosehauer, G.; Schafer, J.; Rah, M. Assessment of a novel silicone hydrogel daily disposable lens among physically active subjects. *Contact Lens Anterior Eye* **2022**, *45*, 101611. [CrossRef]
80. Choudhary, K.; DeCost, B.; Chen, C.; Jain, A.; Tavazza, F.; Cohn, R.; Park, C.W.; Choudhary, A.; Agrawal, A.; Billinge, S.J.L.; et al. Recent advances and applications of deep learning methods in materials science. *Npj Comput. Mater.* **2022**, *8*, 59. [CrossRef]
81. Islamkulov, M.; Karakuş, S.; Özeroğlu, C. Design artificial intelligence-based optimization and swelling behavior of novel crosslinked polymeric network hydrogels based on acrylamide-2-hydroxyethyl methacrylate and acrylamide-N-isopropylacrylamide. *Colloid Polym. Sci.* **2023**, *301*, 259–272. [CrossRef]
82. Martineau, R.; Bayles, A.V.; Hung, C.-S.; Reyes, K.G.; Helgeson, M.E.; Gupta, M.K. Engineering Gelation Kinetics in Living Silk Hydrogels by Differential Dynamic Microscopy Microrheology and Machine Learning. *Adv. Biol.* **2022**, *6*, 2101070. [CrossRef]
83. Wei, J.; Chu, X.; Sun, X.-Y.; Xu, K.; Deng, H.-X.; Chen, J.; Wei, Z.; Lei, M. Machine learning in materials science. *InfoMat* **2019**, *1*, 338–358. [CrossRef]
84. Zhang, J.; Liu, Y.; Durga, C.S.P.; Singh, M.; Tong, Y.; Kucukdeger, E.; Yoon, H.Y.; Haring, A.P.; Roman, M.; Kong, Z.; et al. Rapid, autonomous high-throughput characterization of hydrogel rheological properties via automated sensing and physics-guided machine learning. *Appl. Mater. Today* **2023**, *30*, 101720. [CrossRef]
85. Younes, K.; Kharboutly, Y.; Antar, M.; Chaouk, H.; Obeid, E.; Mouhtady, O.; Abu-samha, M.; Halwani, J.; Murshid, N. Application of Unsupervised Learning for the Evaluation of Aerogels' Efficiency towards Dye Removal—A Principal Component Analysis (PCA) Approach. *Gels* **2023**, *9*, 327. [CrossRef] [PubMed]
86. Tseng, B.-Y.; Cai, Y.-C.; Conan Guo, C.-W.; Zhao, E.; Yu, C.-H. Reinforcement learning design framework for nacre-like structures optimized for pre-existing crack resistance. *J. Mater. Res. Technol.* **2023**, *24*, 3502–3512. [CrossRef]
87. Lin, H.W.; Tegmark, M.; Rolnick, D. Why Does Deep and Cheap Learning Work So Well? *J. Stat. Phys.* **2017**, *168*, 1223–1247. [CrossRef]
88. Owh, C.; Ho, D.; Loh, X.J.; Xue, K. Towards machine learning for hydrogel drug delivery systems. *Trends Biotechnol.* **2023**, *41*, 476–479. [CrossRef]
89. Menon, D.; Ranganathan, R. A Generative Approach to Materials Discovery, Design, and Optimization. *ACS Omega* **2022**, *7*, 25958–25973. [CrossRef]
90. Huljanah, M.; Rustam, Z.; Utama, S.; Siswantining, T. Feature Selection using Random Forest Classifier for Predicting Prostate Cancer. *IOP Conf. Ser. Mater. Sci. Eng.* **2019**, *546*, 052031. [CrossRef]

91. Richter, R.; Kamal, M.A.M.; García-Rivera, M.A.; Kaspar, J.; Junk, M.; Elgaher, W.A.M.; Srikakulam, S.K.; Gress, A.; Beckmann, A.; Grißmer, A.; et al. A hydrogel-based in vitro assay for the fast prediction of antibiotic accumulation in Gram-negative bacteria. *Mater. Today Bio* **2020**, *8*, 100084. [CrossRef]
92. Brahima, S.; Boztepe, C.; Kunkul, A.; Yuceer, M. Modeling of drug release behavior of pH and temperature sensitive poly(NIPAAm-co-AAc) IPN hydrogels using response surface methodology and artificial neural networks. *Mater. Sci. Eng. C* **2017**, *75*, 425–432. [CrossRef]
93. Suykens, J.A.K. Support Vector Machines: A Nonlinear Modelling and Control Perspective. *Eur. J. Control* **2001**, *7*, 311–327. [CrossRef]
94. Zhu, J.-A.; Jia, Y.; Lei, J.; Liu, Z. Deep Learning Approach to Mechanical Property Prediction of Single-Network Hydrogel. *Mathematics* **2021**, *9*, 2804. [CrossRef]
95. Ning, H.; Zhou, T.; Joo, S.W. Machine learning boosts three-dimensional bioprinting. *Int. J. Bioprint.* **2023**, *9*, 739. [CrossRef] [PubMed]
96. Allencherry, J.; Pradeep, N.; Shrivastava, R.; Joy, L.; Imbriacco, F.; Özel, T. Investigation of Hydrogel and Gelatin Bath Formulations for Extrusion-Based 3D Bioprinting using Deep Learning. *Procedia CIRP* **2022**, *110*, 360–365. [CrossRef]
97. Ng, W.L.; Chan, A.; Ong, Y.S.; Chua, C.K. Deep learning for fabrication and maturation of 3D bioprinted tissues and organs. *Virtual Phys. Prototyp.* **2020**, *15*, 340–358. [CrossRef]
98. Garcia-del Rio, L.; Diaz-Rodriguez, P.; Landin, M. New tools to design smart thermosensitive hydrogels for protein rectal delivery in IBD. *Mater. Sci. Eng. C* **2020**, *106*, 110252. [CrossRef]
99. Mohammadi, S.; Hashemi, M.; Finklea, F.; Williams, B.; Lipke, E.; Cremaschi, S. Classification of cardiac differentiation outcome, percentage of cardiomyocytes on day 10 of differentiation, for hydrogel-encapsulated hiPSCs. *J. Adv. Manuf. Process.* **2023**, *5*, e10148. [CrossRef]
100. Li, F.; Han, J.; Cao, T.; Li, L. Design of self-assembly dipeptide hydrogels and machine learning via their chemical features. *Proc. Natl. Acad. Sci. USA* **2019**, *116*, 11259–11264. [CrossRef]
101. Nosrati, H.; Nosrati, M. Artificial Intelligence in Regenerative Medicine: Applications and Implications. *Biomimetics* **2023**, *8*, 442. [CrossRef]
102. Zicari, R.V.; Ahmed, S.; Amann, J.; Braun, S.A.; Brodersen, J.; Bruneault, F.; Brusseau, J.; Campano, E.; Coffee, M.; Dengel, A.; et al. Co-Design of a Trustworthy AI System in Healthcare: Deep Learning Based Skin Lesion Classifier. *Front. Hum. Dyn.* **2021**, *3*, 688152. [CrossRef]

**Disclaimer/Publisher's Note:** The statements, opinions and data contained in all publications are solely those of the individual author(s) and contributor(s) and not of MDPI and/or the editor(s). MDPI and/or the editor(s) disclaim responsibility for any injury to people or property resulting from any ideas, methods, instructions or products referred to in the content.

*Review*

# Dissolving and Swelling Hydrogel-Based Microneedles: An Overview of Their Materials, Fabrication, Characterization Methods, and Challenges

Bana Shriky *, Maksims Babenko and Ben R. Whiteside *

Faculty of Engineering and Digital Technologies, University of Bradford, Bradford BD7 1DP, UK; m.babenko1@bradford.ac.uk
* Correspondence: b.shriky1@bradford.ac.uk (B.S.); b.r.whiteside@bradford.ac.uk (B.R.W.)

**Abstract:** Polymeric hydrogels are a complex class of materials with one common feature—the ability to form three-dimensional networks capable of imbibing large amounts of water or biological fluids without being dissolved, acting as self-sustained containers for various purposes, including pharmaceutical and biomedical applications. Transdermal pharmaceutical microneedles are a pain-free drug delivery system that continues on the path to widespread adoption—regulatory guidelines are on the horizon, and investments in the field continue to grow annually. Recently, hydrogels have generated interest in the field of transdermal microneedles due to their tunable properties, allowing them to be exploited as delivery systems and extraction tools. As hydrogel microneedles are a new emerging technology, their fabrication faces various challenges that must be resolved for them to redeem themselves as a viable pharmaceutical option. This article discusses hydrogel microneedles from a material perspective, regardless of their mechanism of action. It cites the recent advances in their formulation, presents relevant fabrication and characterization methods, and discusses manufacturing and regulatory challenges facing these emerging technologies before their approval.

**Keywords:** hydrogels; microneedles; drug delivery; dissolving microneedles; swelling microneedles; hydrogel-forming microneedles; microneedle manufacturing

## 1. Introduction

Hydrogels are a collection of soft matter classes that include disparate varieties of polymeric materials from different origins that can form water-swollen three-dimensional networks via chemical or physical crosslinks in which the nature of the bonds formed determines the hydrogel reversibility. While there are a number of historical references to these materials—records documenting the use of the gel substances gelatin and aloe vera go back to Mesopotamia and ancient Egypt [1]—the term hydrogel was first used by Van Bemmelen to describe a colloid of inorganic salts in 1894 [2]. Since then, the definition has moved away from the original description with a huge body of work dedicated to quantifying the gelling threshold and the systems' mechanical properties.

Since the early development days, the popularity of hydrogels as carriers of physically, chemically, or biologically active ingredients has been continuously on the rise [2]. Their versatility and tunability make them ideal systems for a wide range of applications, ranging from energy [3] and agriculture to the food industry and bioengineering [4,5]. The first biomedical application can be traced to the 1950s, when the cross-linked PVA implant 'sponges' developed by Grindly and Waugh made their way to hospitals under the names of Ivalon (Clay Adams, US) and Prosthex (Ramer Chemical, UK) [6,7]. This was followed by the widespread manufacturing of poly hydroxyethyl methacrylate (PHEMA) gel contact lenses based on the work of Wichterle and Lim published in 1960 [8].

Pharmaceutically, hydrogels can be utilized as standalone dosage forms or as parts of drug delivery systems (DDSs) and medical devices. They can be incorporated into drug-eluting DDSs either as the carrier matrix or by integrating them as stabilizers or release controllers [9–11]. Such developments have paved the way for the development of customizable dosage forms aimed at reducing dosing frequency and improving patient compliance. Hydrogel systems have offered enhanced biocompatibility and responsiveness to many types of stimuli, such as temperature [12–15], light [16], pH [17–20], and electric and magnetic fields [21,22]. These 'smart' hydrogels have been delivered via different routes, including oral, ocular, topical, vaginal, and injection [23–28].

The transdermal delivery route utilizes access through the skin, mainly using patches, gels, and emulsions, allowing patients to self-administer their medications without the systemic loss encountered via the oral route or the pain associated with injection—see Figure 1 [29]. The transdermal route, however, is currently limited to low-molecular-weight (<500 Daltons) and moderately lipophilic active pharmaceutical ingredients (API). Skin permeation is restricted by penetration and diffusion through the epidermis layers to reach the blood vessels in the dermis (2000 μm). The outer layer of the epidermis, the hydrophobic stratum corneum, provides the principal barrier to absorption through the transdermal route; fewer than two dozen APIs are currently approved by the FDA for transdermal delivery [30].

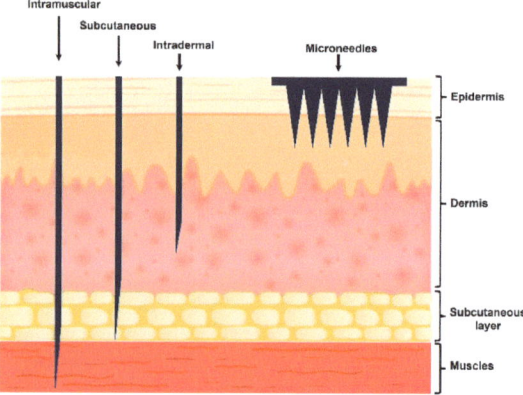

**Figure 1.** Schematic showing human skin layers and microneedle insertion compared to conventional (intramuscular, subcutaneous, and intradermal) injection methods.

Several physical and chemical solutions have been developed to overcome the stratum corneum barrier. Among the most promising options are the use of microneedle (MN) products [31].

In 1967, Gerstel and Place filed the first percutaneous MN drug delivery device patent [32]. After its expiration in the late 1990s, (and due to complementary new developments in microscale manufacturing technologies for MEMS devices), interest in MN research surged, and microneedles were to be celebrated as the future of transdermal drug delivery [31]. Around the same period, microneedling (also referred to as percutaneous collagen induction and needle dermabrasion) has been utilized as an experimental dermatological technique for treating scars [33]. The cosmetic treatment then evolved to use 1–3 mm needles attached to a roller paired with topical vitamin (A, C, and E) formulations to promote rapid skin healing and collagen production [34,35].

Microneedle arrays are micron-scaled (height $\leq$ 1000 μm with a tip diameter of 1–25 μm) projections assembled on a supporting base [30]. MNs are designed to disrupt the stratum corneum by creating micro-channels in the skin, enhancing skin permeability and drug delivery for local or systemic effects without causing pain, bleeding, or infection [36]. Additionally, they are easy to transport and store, and they offer reduced waste and manufacturing costs [37]. For certain applications, when compared to intramuscular injection for vaccine delivery, MNs were found to be more effective due to the large population of immune cells in the dermis and epidermis [29].

Microneedles are manufactured from a wide variety of materials, including glass, silicone, biodegradable polymers, and metals [38]. In the literature, MNs are usually classified into five categories: solid, coated, hollow, dissolving, and the most recent addition, hydrogel-forming.

From a material point of view, this classification system is restricting when describing hydrogel MNs since hydrogels could fit under both the dissolving and hydrogel-forming categories according to their components and properties.

The application of hydrogels on skin paired with pre-treatment using MN arrays or commercial derma rollers was found to enhance their permeation of larger doses of actives. The application of thermogelling systems to pre-treated skin was found to fill the MNs' created cavities, providing a sustained 72 h lasting release of methotrexate and fluorescein sodium [39,40].

The fabrication of hydrogel-based MNs combines physical microporation with the tunability of the residence time of hydrogels in one device.

Hydrogels can be manufactured from a broad range of polymeric materials, which are commonly classified according to the origin of their material components (either natural or synthetic) [41].

This review aims to highlight recent advances in the emerging field of hydrogel-based MNs, summarizing the latest formulation trends, manufacturing methods, and performance evaluation techniques.

## 2. Types of Hydrogels Used in MNs

The ideal drug delivery device should be able to (i) encapsulate active(s) in a biocompatible and mechanically robust system, (ii) allow the control of the release profile, and (iii) be easy to administer with minimal medical supervision [42]. Hydrogels (HG) provide compelling solutions for prerequisites (i) and (ii), and forming them into microneedle geometries is a very appealing way to provide prerequisite (iii). Therefore, it is not surprising that there is a significant research interest in hydrogen MN systems.

Hydrogel MNs are fabricated using a variety of methods and dried to obtain hard, solid MNs with sufficient mechanical properties for application. When embedded in the skin, the interstitial fluid hydrates the MNs via diffusion, and the dried matrix forms a hydrogel, which swells in place, imbibing large amounts of fluid withdrawn from its surroundings. The fluid movement within the polymeric network facilitates the API's diffusion and release through the MNs' created channels. For some systems, this could be followed by dissolution, depending on the matrix's internal structure. The great swelling ability of MNs could be additionally utilized to absorb fluids to be used for diagnostic purposes.

This article will discuss the most common hydrogel substrates used for MN systems from a material perspective, covering both dissolvable and swelling MN matrices. As highlighted in the recent publication by Moore et al. [43], hydrogel-based materials account for 80% of drug delivery and 65% of vaccine MN systems matrices—as displayed in Figure 2. This review looks into how these materials have been recently used, covering neat and mixed hydrogel MN systems to demonstrate the effect of additives on performance. The authors acknowledge the presence of other promising substrates in the literature; however, this article is dedicated to the types frequently used for MN production displayed in Figure 2. The discussed matrices are divided into three main polymeric classes: natural, covering both plant and animal sources, synthetic, and semi-synthetic, a hybrid of the previous two types.

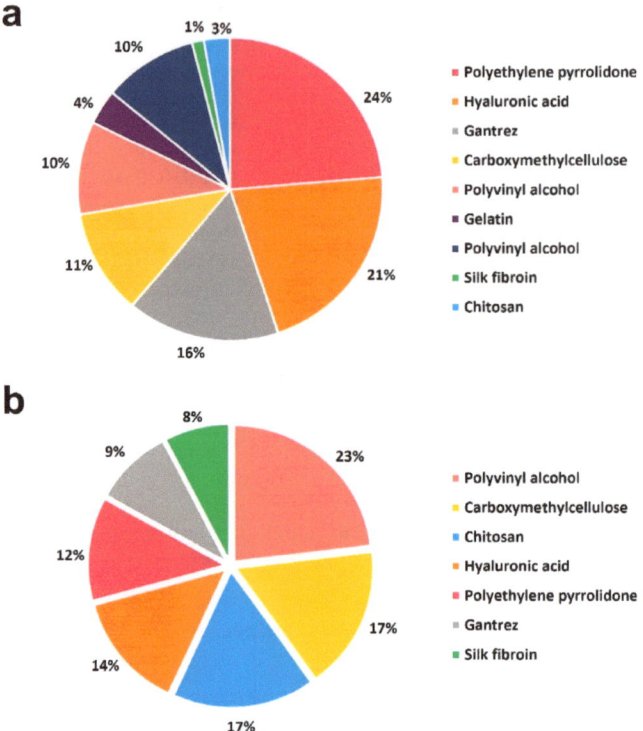

**Figure 2.** The collection of gelling agents used for the fabrication of dissolving MNs for (**a**) drug delivery and (**b**) vaccine delivery; the figure is based on data reported by Moore et al. [43].

## 2.1. Natural

### 2.1.1. Cellulose and Its Derivatives

Cellulose is a hydrophobic biodegradable polysaccharide linked by β-(1,4) glycosidic bonds. It forms the main component of plant cells' walls and can be produced by some bacteria. Its chemical structure can be modified (e.g., the esterification of hydroxyl groups) to produce a wide array of derivatives with diverse physicochemical properties that correlate with the degree of substitution (DoS). Popular pharmaceutical cellulose-based gelling agents include methylcellulose (MC), hydroxypropyl methylcellulose (HPMC), ethyl cellulose (EC), hydroxyethyl cellulose (HEC), and carboxymethylcellulose (CMC), some of which display environmental sensitivities that could be exploited to control the MN release profile [44].

Among the most widely used ones is CMC, a water-soluble anionic linear polysaccharide obtained by substituting some of the hydroxyl ends with carboxymethyl groups (DoS 1.2-2). The low cost and broad range of functionalities have seen the adoption of CMC for a diverse array of biomedical applications [45]. CMCs have been used in many MN systems, where they were found to be effective in stabilizing various protein molecules due to their reduced mobility in solid CMC, and the release kinetics of the molecules could be controlled via the payload and the MN patch design. The authors reported that adding a loaded backing layer may increase the MN encapsulation capacity and introduce longer-term sustained release functionality [42]. Valproic acid CMC MNs outperformed the conventional alopecia topical treatment; the arrays successfully delivered the designed dose at a higher accuracy, and their micro-incisions simulated the hair follicle stem cells [46]. The majority of neat CMC MN systems reported in the literature were fabricated using high Mw grades and casting methods at concentrations below 10% due to observed casting difficulties, as reported for diclofenac sodium, lactobacillus, and HIV-1 vaccine MNs [47–49]. Similar concentrations of low viscosity grades were found unsuitable for the casting of MNs due to the resulting low mechanical strength and high flexibility of the needles, which made them unsuitable for skin penetration and prompted their use as backing layers instead for insulin MNs [50] (Figure 3).

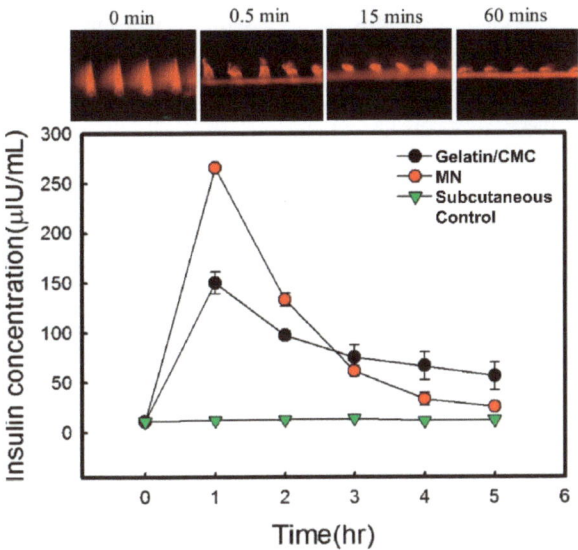

**Figure 3.** The swelling of loaded gelatin/CMC: (top) in vitro porcine skin after the insertion of gelatin/CMC MNs loaded with R6G and (bottom) in vivo drug release profiles of insulin-loaded MNs versus time modified from [50].

Higher concentrations of the low viscosity grades were found to give excellent results for the fabrication of systems prepared via droplet-borne air blowing (DAB) and centrifugal lithography (CL) processes that were employed in the manufacture of the promising lidocaine MN and typhus vaccine ScaA MNs [51,52]. Similar to the other gel-MN-forming materials discussed in this article, the introduction of excipients or other polymers to CMC MN formulations reduced their recrystallization, which led to an enhanced release from the MNs as observed dextrin blends [53].

HPC and HPMC are nonionic water-soluble cellulose derivatives. Ito et al. compared the performance of three types of HPC lidocaine-loaded MNs in a topical cream. All the MNs showed a faster onset than the cream. The whole HPC-lidocaine-loaded MNs were more effective than only using loaded tips or the loaded MNs paired with a reservoir to deliver the drug, for which the presence of the reservoir suppressed the faster release observed for the other MN types [54]. MNs containing only 5% HPC displayed sufficient strength for insertion while containing high loads of donepezil hydrochloride (78%) [55]. Lyophilized acyclovir MNs fabricated from neat 8% HPMC or mixed with 30% PVP showed 4.5–16 times higher skin–drug concentrations than topical applications. The shortest lag time (the time required for the API to reach circulation, determined by gel swelling, API diffusion, and permeability through the skin layers) was recorded for the HPMC–PVA blend, which was attributed to the improved penetration length displayed by the systems [56].

The selection of a suitable cellulose-based agent depends on two factors. The first relates to the targeted gel (and MNs after drying) properties, which is affected by the degree of substitution and Mw. The second is the formulation recipe and whether the matrix charge could have detrimental effects leading to instabilities.

### 2.1.2. Chitosan (CS)

Chitosan (CS) is a linear cationic amino-polysaccharide composed of β-(1, 4)-linked D-glucosamine and N-acetyl-D-glucosamine obtained via the alkaline deacetylation of chitin, the second-most-abundant natural biopolymer after cellulose. It is found in fungi, crustaceans, and insect shells. The biopolymer is soluble in dilute acidic solutions and can form stimuli-responsive gels at low concentrations when neutralized [57]. CS is known for its biocompatibility, biodegradability, cytocompatibility, and natural antimicrobial activity against both *Gram*-positive and *Gram*-negative bacteria. This versatility allowed its use in various formats, ranging from injectables and topical productions to transdermal delivery [58]. Its properties depend on the molecular weight (Mw), degree of deacetylation (DDA), and purity [59]. Studies on Caco-2 cells showed that high DDA enhanced absorption independently of Mw, while medium values (65–51%) only enhanced it at higher Mws [60]. CS gel preparation is often a two-step process; acid hydrolysis is followed by a dialysis method to refine the Mw, and adjusting the process conditions allows the optimization of both the MN mechanical strength and the matrix dissolution rate. Acetic acid is widely used for the hydrolysis step. It has been used to fabricate MNs to deliver bovine serum albumin (BSA) [61] and cetirizine hydrochloride [62]. Other acids used include hydrochloric acid during the fabrication of levonorgestrel [63], trifluoroacetic acid (TFA), and lactic acid for rhodamine-B- and diclofenac-sodium-loaded MNs, respectively [64,65]. Chemically modified CS blends have been used to fabricate MNs to enhance the stability of the encapsulated materials and alter the system's properties. Examples include the improved physical strength exhibited by CS/PVA blends [65,66], as well as the pH and electrical responsiveness achieved for MNs fabricated from CS/porous carbon nanocomposite [67].

### 2.1.3. Hyaluronic Acid (HA)

Hyaluronic acid (HA), also known as hyaluronan, is a liner anionic copolymer of non-sulfated glycosaminoglycan. It was extracted for the first time in 1934 from bovine vitreous fluid [44]. HA is a naturally occurring component of the extracellular matrix in animals and microorganisms, including human tissues, where 50% of the bodily content is distributed in the skin. HA's hydroxyl-rich chains allow bonding with large quantities of water (holding up to 1000 times its original weight) and bridges with surrounding HA chains resulting in secondary and tertiary structures. HA has been used for numerous applications, ranging from injectables to formulating gels and creams. It is proven to stimulate the proliferation of fibroblasts and adhesion to the wound site, promoting healing. When these benefits are paired with its favorable mechanical properties, HA makes an ideal matrix for transdermal and MN delivery systems [68]. Most of the reported HA-based MNs are uncross-linked and prepared via the two-step casting method of polymeric solutions [69]. In addition to its volumizing effect, HA-neat MNs were found to mildly inhibit human hypertrophic scar fibroblasts (hHSFs), and when loaded with bleomycin, the synergistic inhibitory effect was further increased by up to 40% and maintained over 6 weeks, providing a promising new method to treat hypertrophic scars [70]. The anti-diabetic peptide exenatide MNs have delivered similar outcomes to subcutaneously administered solutions in 2 min with minimal patient discomfort [71]. The morphology of 5-aminolevulinic acid (5-ALA) MNs was tailored according to the required clinical targets to be used effectively in photodynamic therapy. MNs with a length of 907 μm formed from HA with an MW of 10 kDa and a 30% concentration with API-loaded tips were reported to deliver 122 μg of 5-ALA for subcutaneous tumors [72], while shorter (650 μm) MNs fabricated using 50% HA of the same grade were used to deliver 610 μg of the active to superficial tumors [73]. For topical applications, shorter HA chains with an Mw of 5–50 kDa exhibited higher permeability than longer ones [74]. This, however, does not determine the transdermal behavior of HA MNs. Chi et al. studied the Mw effect on the performance of 5% HA MNs using 10, 74, and 290 kDa grades. Over two days, the results indicated that, under the same conditions, the 74 kDa MNs demonstrated the best mechanical–transdermal release balance, showing an intermediate mechanical strength, the best rhodamine B skin retention, and the release profiles shown in Figure 4 [75]. Adenosine-loaded high-Mw HA MNs were found to be more effective in reducing the density of wrinkles than the low-Mw counterparts [76].

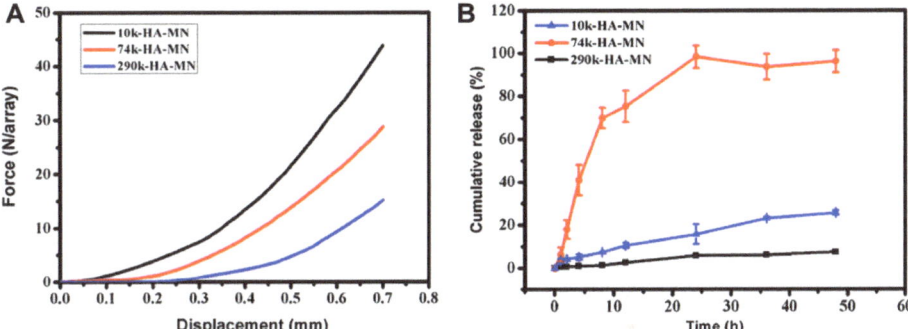

**Figure 4.** The (**A**) mechanical and (**B**) transdermal release profiles for 10, 74, and 290 kDa HA MNs modified from [75].

The hydrophilic nature of HA could pose challenges for fabricating systems to encapsulate hydrophobic loads or for those requiring slow release rates. Introducing stabilizers such as lipid-based nanoparticles (LNP), liposomes, and other polymers to HA matrices was recorded to overcome this limitation and enhance HA MNs' performance [43,77,78].

2.1.4. Silk Fibroin (SF)

Silk is a fibrous natural polymer obtained from silkworm cocoons, spiders, mites, and pseudoscorpions' silk fiber [79–81]. The most commonly used silk is a product of *Bombyx mori* cocoons, which consists of two proteins: sericin and fibroin. Silk fibroin is generally regarded as safe (GRAS) by the FDA, and it is a biocompatible and biodegradable polymer [82]. Pure SF extraction is a relatively low-cost and straightforward process that usually involves two steps: (i) stripping the immunogenic sericins in alkaline solutions, leaving behind SF fibers that are dried, and (ii) dissolution in concentrated electrolyte solutions and dialysis to remove the salt residue. SF solutions of 6–8% are commonly used for MN fabrication. The shelf life of SF can also be extended by lyophilizing the solutions [83]. SF is built from alternating hydrophilic and hydrophobic blocks that form crystalline β-sheet structures via hydrogen bonding [84]. Pure SF MNs are highly soluble, which makes them unsuitable for prolonged release; however, SF's gelation and properties could be easily modified to control solubility, release, and mechanical strength [85], which can be done during the formulation stage or post MN fabrication [86]. Post-fabrication methanol treatment was found to increase SF's surface crystallinity, which improved its strength by approx. 113% and delayed the payload release from neat SF MNs [87].

Zhu et al. compared three types of insulin-loaded SF systems: neat MN arrays, methanol-treated neat MNs, and MNs with a backing layer from proline amino acid. The authors reported that the neat MNs almost completely dissolved without undergoing any swelling, while the mixed-base systems experienced higher swelling and dissolution than the treated MNs—which barely released 40% of the payload after 9 h. In vitro tests showed that the mixed base system exhibited a faster release within the first two hours, and then the rate slowed down afterwards to achieve a complete release at around 9 h, providing a slower and smoother release profile when compared to insulin injections [88]. Similar to its effect on SF fibers [89], water vapor annealing effectively increased the SF crystallinity degree in MNs. This effect was enhanced when paired with a crosslinking agent (glutaraldehyde), increasing the fracture force by six times when compared to the neat untreated systems [90]. FS-loaded tips on a polymeric blend base were reported to provide a sustained in vivo release of influenza vaccine that outperformed intramuscular injections [91]. Yin et al. investigated the properties and the in vitro performance of four SF-modified systems with various additive concentrations and Mws of loaded FITC-dextran. The added 2-ethoxyethanol was found to enhance the swelling capacity (up to 800%), minimize dissolution, and increase the matrix pore dimension, resulting in a consistent release profile [92].

*2.2. Synthetic*

2.2.1. Polyvinyl Alcohol (PVA)

Polyvinyl alcohol (PVA) is a water-soluble synthetic polymer prepared via the hydrolysis (saponification) of polyvinyl acetate. The length of the polyvinyl acetate units and the hydrolysis conditions can be controlled to produce a large variety of molecular PVA weights, in which the hydrolysis percentage is inversely proportional to the solubility [93]. This is reflected in the huge disparity in the physicochemical properties that the polymer exhibits, including mechanical strength and water affinity [94]. PVA is a GRAS ingredient, according to the FDA, with excellent biocompatibility, biodegradability, and drug loading capability [95,96]. PVA has been used to fabricate both MNs themselves and their supporting base, in which higher concentrations were used for MNs to provide greater mechanical strength [97,98]. PVA concentration and Mw could be additionally used to tailor the payload release from the matrix through their effect on the solubility and swelling behavior [99–102]. PVA is often used in combination with other polymers to enhance the matrix properties and stability; examples include the mechanically improved PVA/PVP MNs' body or base [103,104], the high delivery rate of HA/PVA layered systems [105], the modified water retention of CS and CMC, and the tunability of SF blends [99,106–108]. Many blends containing gelatin, HA, polyalkylmethacrylates (PAMAs), and metal–organic frameworks (MOFs) have been formulated to optimize either MN performance or the molding process [109–112].

2.2.2. Polyethylene Pyrrolidone (PVP)

PVP is a synthetic nonionic water-soluble biocompatible polymer. The FDA-approved polymer is widely used as a plasma expander and in various medical applications due to its stability and good film-forming properties [113]. PVP's amphiphilic nature makes it an ideal drug stabilizer and controlled release carrier. PVP is cleared renally when the Mw is up to 40 kDa; however, when large quantities of high Mws are used, it is necessary to administer a degrading enzyme ($\gamma$ lactamase) [114,115]. PVP MNs' properties depend on the polymeric chain length and concentration, which could be altered according to the nature of the payload and desired release profiles. Thakur et al. reported that ocular macromolecules loaded into MNs fabricated from a low Mw (10 kDa) were found to be more brittle and dissolved faster than ones made from higher Mws (40 and 58 kDa) [116]. Systems containing only 10% of 40 kDa PVP successfully delivered $\alpha$-choriogonadotropin through the skin tissue without any mechanical failure even when forces as high as 13 N were used [117]. Although most PVP-based MNs are produced using casting PVP solutions [118], some studies have successfully synthesized them using an in-mold photopolymerization process of liquid vinyl pyrrolidone monomers. The method allows the tailoring of the PVP matrix to obtain custom functionalities and the ability to synthesize copolymers such as PEGDA–PVP and PVP–MAA in a one-step process [119–121].

2.2.3. Poly Methyl Vinyl Ether-Co-Maleic Acid (PMVE/MA)

Poly (methyl vinyl ether-co-maleic anhydride) is a synthetic-class biodegradable copolymer commercially available under the name Gantrez™. The relatively low-cost copolymer is widely used in cosmetics and personal care products [122]. Four grades are commonly used for MN production with Mws ranging from 130 to 2000 kDa. For MN fabrication, Gantrez® is either used as a stand-alone dissolving polymer matrix [123] or crossed-linked with polyethylene oxide (Mw of 10 kDa) at 80 °C to form matrices with enhanced mechanical strength [124]. For the uncross-linked systems, the API is dissolved with the polymer and molded, as reported for Gantrez® AN119—amphotericin B MNs [123], Gantrez™ AN139 for formulations containing aminolevulinic acid and nadroparin calcium [125,126], and Gantrez™ S97 for the delivery of acyclovir and vitamin K [127,128]. Cross-linked systems have been used for either extraction [129,130] or delivery purposes. For the latter, the actives elute from the bulk of the MNs or from an external drug reservoir, as recorded for metformin, atorvastatin, esketamine, and donepezil [131–134]. The formulations of both cross-linked and uncross-linked Gantrez™ S97 were tested as carriers of the model protein antigen ovalbumin. The results showed that the uncross-linked systems demonstrated an enhanced immunological response and a peak release after 5 h, while the cross-linked MN matrix barely released 27% of the available load even after 24 h [135].

## 2.3. Semi-Synthetics

### 2.3.1. Methacrylated HA (MeHA)

As previously mentioned, HA has exceptional biocompatibility; however, it is not suitable for prolonged release purposes due to high water solubility [111]. HA's hydroxyl groups can be easily modified to obtain copolymers with improved rigidity and a tailored dissolution profile. Methacrylated HA (MeHA) is one of the copolymers that has recently gained popularity. MeHA can be obtained via esterification under UV illumination after MNs' drying or demolding. The resultant gel properties will depend on the degree of substitution and UV curing time [136]. Crosslinking increases the gel's life in situ and allows MNs to be used for both delivery and extraction purposes. MeHA MN patches were reported to have excellent adhesion strength (~$0.20\,\text{N}\,\text{cm}^{-1}$), double the value recorded for the medical-grade Tegaderm™ films. Therefore, this could negate the need for additional supporting adhesive strips for microneedle patch retention [137]. The use of MeHA systems is recommended for the delivery of poorly soluble actives. When compared to neat dissolving HA MNs, the extended doxorubicin release profile exhibited by MeHA was attributed to the longer presence of the MN-generated channels, which lasted for a longer time due to the copolymer's slower dissolution. Images of the dissolution process and release profiles are presented in Figure 5 [138]. Chang et al. reported that their MNs' weight increased up to 9 times within minutes after skin insertion.

The volume of the interstitial fluid (ISF) extracted was found to be inversely proportional to the UV exposure time without any correlation with the observed mechanical strength [139]. Zheng et al. managed to increase the MeHA MNs' swelling capabilities by introducing an osmolyte (maltose) to the matrix. The authors investigated the influence of maltose concentration and UV exposure on the gel properties. They reported that the maltose containing MNs expanded to ×2.15 their size compared to the ×1.67 seen for MeHA MNs and that the extent of swelling depended on the maltose %. The authors confirmed the swelling–cure time (crosslinking correlation); however, they reported that that a minimum cure time is required to ensure the MNs' structural integrity [140]. For delivery applications, the swelling–cure time relationship determines the API loading capacity, in which longer times result in reduced loading [137]. Recently, MeHA systems were investigated as alternatives for subcutaneous injections (SC) in rheumatoid arthritis therapy. It was reported that MeHA MNs' performance was the equivalent of the SC delivery of etanercept [141] and that they maintained a sustained release profile of melittin [142]. Similar results were seen for MeHA-loaded tips for the co-delivery of tocilizumab (TCZ)—an inhibitor of interleukin-6 receptors (IL-6R) and an aptamer, Apt1-67. The inhibition rates of IL-6 and tumor necrosis factor-$\alpha$ (TNF-$\alpha$) were slightly higher for the SC administration; however, the loaded MNs were superior in relieving arthritis symptoms, including synovial hyperplasia, inflammatory cell infiltration, and joint cavity roughness [143].

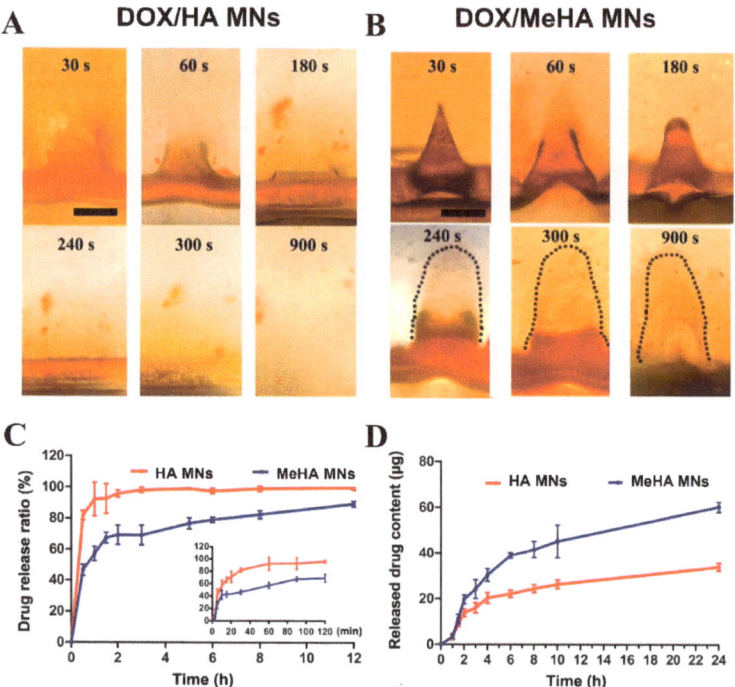

**Figure 5.** Representative images of doxorubicin (DOX) MNs: (**A**) DOX/DMNs dissolution or (**B**) DOX/SMNs swelling and drug release at different times after insertion into skin-mimicking gel (scale bar = 200 µm). (**C**) In vitro DOX release ratio of DOX/DMNs or DOX/SMNs patch (20 mm × 20 mm) with a dose of 0.5 mg in PBS (pH 7.4) at 37 °C ($n$ = 3). (**D**) Transdermally released content of DOX from DOX/DMNs or DOX/SMNs arrays (20 × 20 needles) with a dose of 0.125 mg using skin-mimicking gel mounted on Franz cells ($n$ = 3). The figure is modified from [138].

### 2.3.2. Gelatin-Methacryloyl (GelMA)

Gelatin is a hydrophilic biopolymer obtained via the hydrolysis of collagen, a natural protein occurring in extracellular matrices [144]. Gelatin is inexpensive, biodegradable, biocompatible, and non-immunogenic, but its applications are restricted by its poor mechanical properties [44]. The polymer has been modified by various chemical reactions to enhance its properties, including the introduction of methacryloyl groups to produce gelatine-methacryloyl (GelMA) [145], which could be cross-linked via photopolymerization, a rapid reaction completed within minutes under mild environmental conditions and visible or UV light, depending on the used photoinitiator type [146]. The methacrylation is limited to <5% of the amino acid residues, allowing GelMA to retain gelatin's cell adhesion and in vitro enzymatic degradation sensitivity [147]. Like the cross-linked MeHA, the polymer has excellent swellability that makes it ideal for both extraction and delivery applications. Cross-linked GelMA offers a tunable performance via curing time optimization [148]. When compared to the uncured GelMA and oral dosage form, the irradiated GelMA (for even as quick as 10 s) displayed a superior in vivo therapeutic effect for donepezil MNs [149]. Similar comparisons were performed for lidocaine-loaded systems, and it was reported that 15 s was sufficient to enhance swelling and, therefore, the release profile from the system [150]; for this reason, most of the studied GelMA systems are cross-linked. Different approaches were investigated to improve the system's adhesion and release profile. Fu et al. designed an MN system to encapsulate gemcitabine, an anticancer active, and adjusted the release profiles via tuning diffusion through polymer concentration in the cross-linked matrix [151].

Qiao et al. reported that their graphene-oxide-laden GelMA MN systems enabled the collection of large ISF volumes (21.34 µL in 30 min) and allowed the detection of multiple microRNA biomarkers to be used in psoriasis detection [152]. The unique adhesive property of the GelMA MN was exploited for the delivery of silicate nanoplatelets (SNs) and prompt rapid hemorrhage treatment. The researchers tested the system for both external and internal applications. They reported that SN MNs reduced liver bleeding by approximately 92%, and they were found to degrade in vivo after weeks without any signs of major inflammation [153].

## 3. Characterization of Hydrogel Microneedles

### 3.1. Chemical Composition

The composition of the matrices in MN systems is usually investigated at various fabrication stages to ensure the uniformity of content and validate the effectiveness of crosslinking or polymerization reactions [139,154,155]. The method selection will depend on the nature of the sample and the expected transitions.

The most utilized method is Fourier-transform infrared spectroscopy (FTIR) due to its robustness, flexibility, and broad range (4000~400 $cm^{-1}$) covering the absorption radiation of most organic compounds. In their CS cetirizine-loaded MNs, Arshad et al. noticed that the positions of CS's peaks (carbonyl, hydroxyl, and N-H of amine) shifted to higher wavelengths in the dried MNs, suggesting a looser molecular packing and lower transition energies. The MNs' spectra combined both CS and cetirizine peaks without any additional ones, indicating the absence of chemical interactions during casting [62]. However, the peak shifts appear to depend on the polymer itself; for CMC-casted MNs, their spectrum was found to be identical to the neat polymer without any drying effects [47]. The method was additionally used to assess the copolymerization and crosslinking in swelling hydrogel MNs [156], as shown in Figure 4B, for cured GelMA via the increased definition of the peaks in the region of 2800–3100 $cm^{-1}$, attributed to the stretching of $CH_2$ and formation of tertiary CH when compared to uncross-linked polymer [157]. Raman spectroscopy is another complementary vibrational method that is used to identify structural changes in MN; however, it is rarely utilized for the characterization of composition. Recently, surface-enhanced Raman scattering (SERS) has been embedded in systems with real-time sensors to analyze the extracted biomarkers [158,159].

Nuclear magnetic resonance (NMR) is heavily used to verify the degree of substitution in cross-linked GelMA and MeHA systems shown in Figure 6c,d. This is indicated by the appearance of the vinyl peaks of methacrylate group around 5.63 and 6.06 ppm, which were not previously observed for HA or gelatin backbones [148,160].

Other spectral methods that have been reported in the literature include X-ray photoelectron spectroscopy (XPS), which is used to assess the MNs surface composition and detect any contamination attributed to molding, as reported for silicon residues on Gantrez AN-119 BF and amphotericin B MNs from polydimethylsiloxane (PDMS) casting [123].

### 3.2. Thermal Stability

Thermal methods offer quick and easy gel quality control measurements. They allow the monitoring of stability and the concentration of the various systems' components and can identify any thermally driven transitions that could affect fabrication or release. Popular methods include the following.

#### 3.2.1. Differential Scanning Calorimetry

Differential Scanning Calorimetry (DSC) measures the difference in the heat flow rate of a sample compared to a reference as a function of time during heating at a constant rate (with a linear temperature rise). The resulting thermal spectra can be used to characterize phase transitions, which provide information about the gel structure [161]. For hydrogel-based MNs, the method is used to monitor crystallization or crosslinking, in which any variation in a manufacturing process could negatively affect the matrix drying kinetics, swelling/dissolution, and, therefore, performance. DSC thermograms were used to monitor the effects of methanol treatment on silk crystallites' formation [87] and relate the hydrolysis level of PVA to the matrix degradability [102].

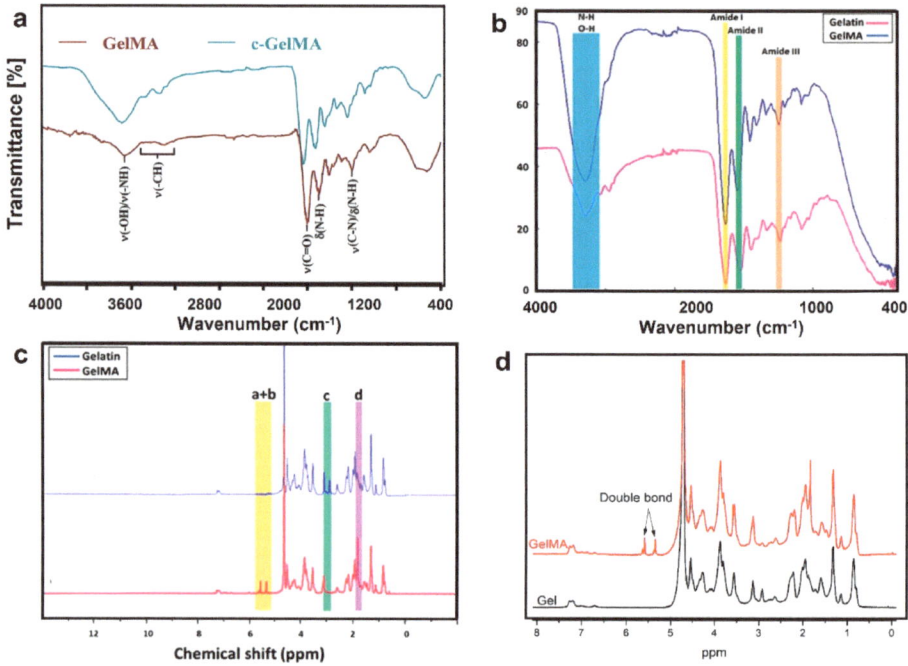

**Figure 6.** Spectral methods of MN characterization: (**a**) the FTIR-ATR spectra of GelMA and c-GelMA modified from [157], (**b**) the FTIR spectra of gelatin and GelMA, (**c**) the 1H NMR spectra of gelatin and GelMA modified from [156], and (**d**) the H NMR spectroscopy of gel and GelMA modified from [148].

### 3.2.2. Thermogravimetric Analysis (TGA)

TGA is a thermal quantitative technique in which the substance mass is monitored continuously as a function of temperature or time in a controlled atmosphere to elucidate thermal stability. TGA uses a high-resolution mass balance to enable the detection of weight losses attributed to free and bound water [162], dehydration [66], and polymer degradation that could take place during drying [47].

## 3.3. MN Morphology

Imaging techniques are regularly used to assess the geometric quality of MNs and indentation site inspection. The selection of the appropriate technique will depend on the purpose of imaging, whether it is quantitative or qualitative, the nature of the sample, and the required resolution. Common microscopic measurements are classified according to the illumination path and sensor type and include bright-field [163] and confocal laser scanning microscopy (CLSM) [164,165]. These methods are often paired with dyes or fluorescent probes to determine the loaded material distribution, its concentration, and its successful penetration. Scanning electron microscopes (SEM) use a focused electron beam, rather than visible light, which has a lower equivalent wavelength and diffraction limit, allowing them to achieve higher-resolution images. However, the charged electrons can collect on the surface of non-conductive materials, causing image artifacts. This problem is commonly mitigated by using gold-sputtering to apply a very thin (few nm) conductive layer on samples to conduct the charge to earth or by allowing a small amount of air in the chamber that is ionized by the electron beam and acts as a conduction path for the surface charging on the sample (commonly referred to as environmental SEM). Examples of imaging techniques to inspect polymeric MNs are displayed in Figure 7.

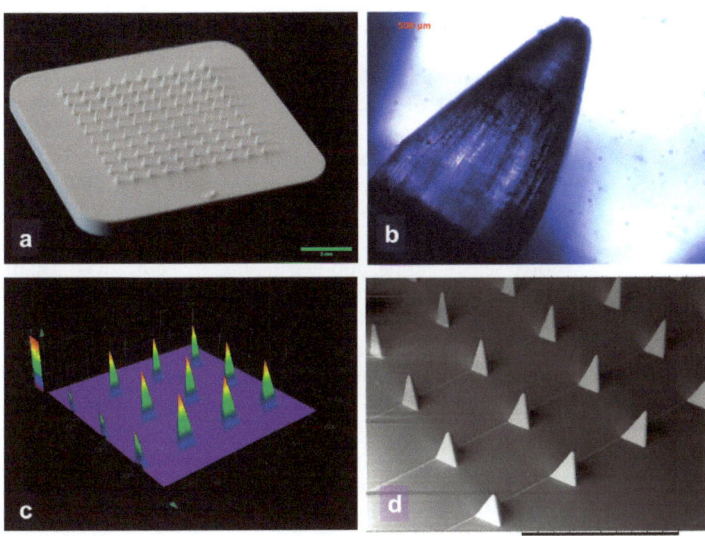

**Figure 7.** Examples of imaging techniques to inspect polymeric MNs: (**a**) optical image, (**b**) optical microscope paired with polarizers, (**c**) a 3D construction of a CLSM image, and (**d**) SEM image. The images were generated by the authors for this publication.

*3.4. Mechanical Properties*

MNs endure a spectrum of stresses, particularly during insertion, removal, general handling, and transportation [164,166], so mechanical characterization is an essential component of HG MN device development. Various mechanical tests can be explored to evaluate the mechanical properties of MNs, with the compression test being a commonly employed method that effectively simulates MN insertion into the skin. The stresses encountered during skin penetration can lead to multiple failure modes, such as the bending, buckling, and baseplate fracturing of the MNs [166,167]. Notably, the tips of the MNs must possess sufficient mechanical strength to penetrate the stratum corneum due to the skin's elasticity and heterogeneity [98]. The mechanical strength of HG MN arrays is influenced by several factors, including the type and concentration of the polymer used, the moisture content, drug type, and concentration encapsulated within the MNs, and the specific preparation methods employed [42,116,120]. Due to the diverse geometrical dimensions of MNs, the array of tests employed, and the various measuring equipment utilized, direct comparisons between different MN designs are challenging. To address this significant limitation, the MN technology field would greatly benefit from the establishment and standardization of mechanical tests to ensure consistent and reliable assessment across various MN designs [166].

3.4.1. Hydrogel MNs' Mechanical Strength

In many of the investigations, the mechanical strength of HG MNs has been assessed by compressing the entire patch using a flat, hard surface, such as a stainless steel plate or probe, as shown in Figure 8. The mechanical strength of individual microneedles is assessed by dividing the patch's compression force by the quantity of the microneedles. More often, a mechanical testing setup such as the Instron 5943 or a texture analyzer such as the TA-XT would be applied, allowing the control of velocity and force. Typical compression velocities vary between 0.01 mm/s and 10 mm/s, with 0.5 mm/s being the most prevalent choice. Researchers would ideally set a desired upper compression force, typically ranging from 0.5 to 50 N. If establishing a maximum compression force is unfeasible, the testing apparatus would operate under displacement control mode. In this scenario, the force/displacement curve would be monitored for a sudden decline in force. This drop in the curve indicates the significant deformation or fracture of the microneedle. The corresponding force is then documented as the failure force. Alternatively, the percentage of MNs' height reduction was calculated and plotted against the applied compression force [82,90,96,97,99,100,104,109,110,116,117,146,168–174].

With less frequency, researchers characterize the mechanical properties of individual microneedles. It is more challenging since it requires diminutive probes while ensuring the avoidance of contact with adjacent microneedles. In this regard, Du et al. (2021b) employed a glass rod measuring 100 µm in diameter, while Kim et al. (2013) employed a force sensor probe, and Lee et al. (2015) employed a stainless steel pillar; however, the detailed dimensional specifications of these probing instruments were not reported. An alternative approach involves sectioning samples to a size equivalent to that of a single microneedle, as demonstrated by Oh et al. (2022).

Researchers often use the minimum force of skin penetration as a parameter to validate the strength of MNs. A single value of 0.058 N/MN is used frequently in the literature [120,137,175]. The value originates from the work of Davis et al. (2004), in which it was demonstrated that the force required for insertion mostly depends on the area where the needle meets the skin at its tip, and other aspects of needle shape matter less. Park et al. [176] looked at three different needle shapes that had the same tip diameter of 25 µm, resulting in an effective contact area of 490 µm$^2$ where they touched the skin; they should all have demanded the same amount of force for insertion. This force was predicted to be around 0.058 N per needle.

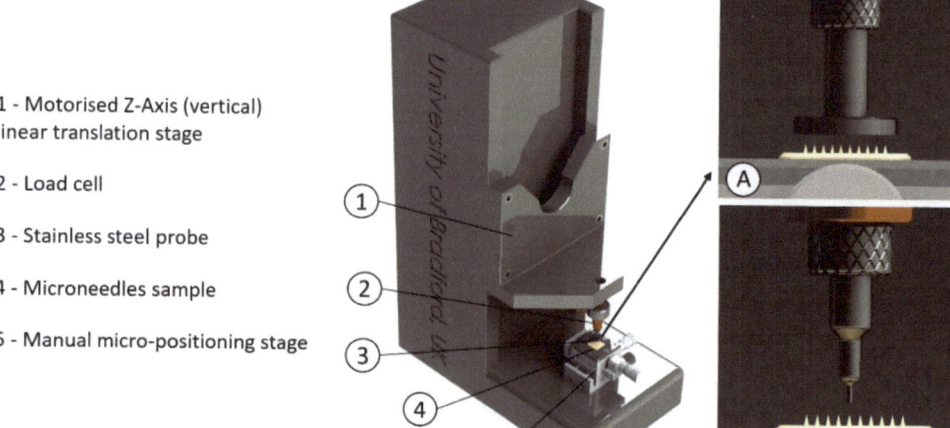

1 - Motorised Z-Axis (vertical) linear translation stage

2 - Load cell

3 - Stainless steel probe

4 - Microneedles sample

5 - Manual micro-positioning stage

**Figure 8.** Schematic of mechanical microneedle testing. (**A**) Compression of the entire patch with a 10-mm-diameter stainless steel probe. (**B**) Compression of individual microneedles with a 0.5-mm-diameter stainless steel probe.

A value of 0.1 N also often comes up in the literature. Yu et al. (2021) referred to it as the minimum force required to penetrate human skin. Wang et al. (2016) characterized it as the force required to puncture human skin. Zhu et al. (2019a) stated that the force needed to penetrate the skin is more than 0.1 N. An et al. (2022) referred to it as the force sufficient to insert into the skin. Qiao et al. (2022) used a similar value of 0.15 N/needle as a threshold value for the skin insertion of MNs. All these authors have made reference to the study conducted by Davis et al. in 2004, wherein a single MN was introduced into the skin of a human subject's hand, specifically within a 1 cm$^2$ area at the base of a knuckle. The microneedles employed in this experiment had a height of 720 µm, a radius ranging between 30 and 80 µm, and a wall thickness spanning from 5 to 58 µm. The insertion process was carried out at a rate of 1.1 mm/s, utilizing a maximum force of 500 g. The recorded forces necessary for insertion ranged from 0.08 to 3.04 N. Consequently, it would be incorrect to assert that a minimum force of 0.1 N per needle is sufficient to breach human skin.

Another study identified the force of insertion to be approximately 0.029 N/MN at an insertion speed of 0.5 mm/s and 0.021 N/MN at an insertion spend of 1 mm/s. MN arrays with varying numbers of MNs per array and different interspacing were inserted into an excised (700-µm-thick) and trimmed stillborn piglet's skin [177].

### 3.4.2. Hydrogel MNs' Penetration Efficiency

The typical method for evaluating the skin insertion ability of HG MNs involves applying pressure to MN samples onto the skin, followed by the removal of the sample. Staining agents such as trypan blue or methylene blue are then utilized to improve the visibility of the puncture sites. These staining solutions are applied to the skin for a period of 2 to 30 min, typically around 5 min. After this, any excess dye is wiped away, and the area where the MNs were applied is examined, as demonstrated in Figure 9 [138,141,142,153,174,178–180]. The application site can be inspected using a digital camera [177], optical microscope [141,174,181], digital microscope [109,169], bright-field microscope [110,142,172], and stereo microscope [121,163,174,178,182] or visually inspected with the eyes. The results can be expressed as an insertion ratio by dividing the number of punctures in the skin after insertion by the number of array needles [110,138,163] or as a percentage of the number of punctures observed/number of punctures expected ×100 [102,109,117,169,172,178]. More elaborate methods, such as electrical resistance measurements [101,183] or electrical impedance measurements, can be employed to study skin penetration. Due to its effective insulation against electricity, the stratum corneum enables the identification of penetration through this barrier [184]. Another alternative for examining skin penetration is transepidermal water loss measurements [101,126,185]. The techniques outlined earlier lack the capability to offer precise numerical data regarding the depth to which microneedles are inserted into the skin. In other words, these methods do not provide detailed measurements that quantify how deeply the microneedles penetrate the skin. Histological cryo-sectioning, coupled with additional staining, can be employed to acquire information about the depth at which the skin was penetrated, as demonstrated by Chew et al. (2020), Ling and Chen (2013), Lee et al. (2015), and Chen et al. (2015). Utilizing confocal microscopy along with fluorescent dyes also enables the assessment of penetration depth. Nonetheless, this approach requires the creation of three-dimensional reconstructions from fluorescent area images [98,101,163,170]. The application of optical coherence tomography (OCT) has also been employed to investigate the extent of the penetration depth of HG MNs [100,149,186,187]. The use of OCT offers the ability to deliver high-resolution, volumetric, non-intrusive, real-time images of the skin. It does not necessitate specific skin pre-treatment or cutting and has potential for studying microneedle insertions in vivo.

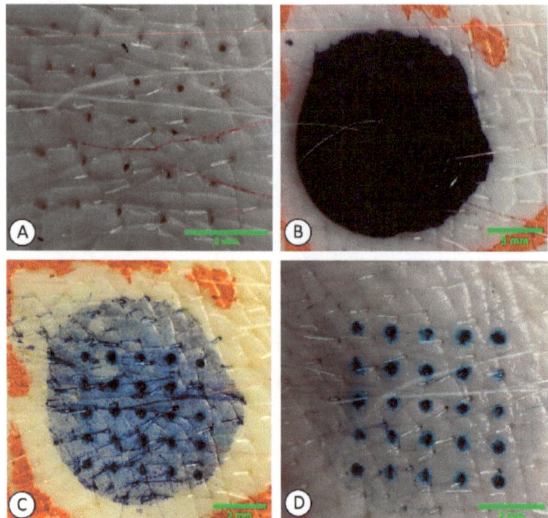

**Figure 9.** Microneedles' skin penetration efficiency in porcine excised skin (skin staining method). (**A**) Application site after microneedle sample was pressed against the skin for 30 s and removed, where insertion holes can be observed. (**B**) Methylene blue staining agent applied onto the site for 5 min. (**C**) After 5 min, the excess stain was removed. (**D**) Application site cleaned with 70% isopropyl alcohol, and penetration efficiency can be clearly observed. The penetration inspection images were generated by the authors for this publication.

Of course, human or animal skin tissue is essential for assessing the skin puncture capability of HG MNs. While human skin is the most suitable for conducting experiments involving HG MNs or any other MNs, there are associated challenges. The utilization of human skin in scientific investigations necessitates endorsement from ethical and regulatory organizations, and there exist safety apprehensions concerning its manipulation. The procurement, preparation, and upkeep of human skin for testing can incur substantial costs and demand significant resources [164,188]. Nevertheless, several investigations have been conducted involving human skin. The efficiency of penetration was investigated using human cadaver skin specimens obtained from a 92-year-old Caucasian woman, revealing penetration exceeding 33.9% [178]. Microneedles were inserted into ex vivo human skin from three different donors using a custom impact-insertion device, ensuring consistent piercing at a constant speed of 3 m/s [165]. Nguyen et al. (2018) demonstrated a 100% penetration efficiency of PVA microneedles into dermatomed human cadaver skin. In 2010, Gomaa et al. utilized dermatomed human cadaver skin to study the impacts of MN density, MN length, application frequency, and insertion duration. They employed a transepidermal water loss (TEWL) device to monitor alterations in human skin's barrier function. Sun et al. employed discarded neonatal foreskins from elective circumcisions to assess the penetrative potential of PVP microneedles in human skin [119].

Although human skin is widely considered the standard for in vitro investigations into drug penetration, the challenges discussed in the previous paragraph mean that animal models can provide attractive alternatives. Porcine skin is recognized as the preferred animal model because it shares many anatomical, histological, and physiological similarities with human skin, and its penetration behavior is comparable [188].

There are many reports of the ex vivo or in vitro introduction of MNs into porcine skin—as shown in Figure 9. However, the specific location on the pig's body or the thickness of the skin used was not indicated, nor was the age of the animals [82,87,90,96,103–105,120,121,149,153,154,163,172–174,182,189]. Olatunji and colleagues in 2013 performed the removal and adjustment of neonatal pig skin to achieve a thickness of 700 μm to perform their insertion experiments. In 2014, Larrañeta et al. utilized neonatal pig skin obtained from piglets that were stillborn with a complete skin thickness of around 0.5 mm. Aung et al. in 2020 used neonatal pig skin with an average thickness of $0.9 \pm 0.12$ mm, while Vora et al. (2020) in the same year worked with neonatal pig skin of approximately 350 µm thickness. In the study conducted by Cole and colleagues in 2017, neonatal pig cadaver skin was also employed for MN penetration research; however, no information regarding skin thickness was provided. The skin penetration ability of MNs was also confirmed in a porcine ear skin model [137,140].

Alternative animal skin models, such as those of mice, rats, or rabbits, have also been employed for HG MN penetration studies. When contrasted with pig skin models, utilizing rodent skin offers several benefits. The advantages of rodents are their compact size, straightforward handling, and relatively low cost [188]. Notably, rodent skin models have found applications as in vivo representations for studying HG microneedle penetration. In vivo MN penetration in mice skin models was mainly performed on depilated dorsal skin [139,141,143,152,181], whereas Lau et al. (2017) inserted dissolvable MNs in the abdominal skin of a mouse. Similarly, rat dorsal skin was utilized for in vivo skin insertions [62,86,118,163]. In vitro skin insertion capacity was evaluated in both cadaver mice [86,138,154,171,179] and rat skin models [142,190] using the dorsal or abdomen skin regions. The in vivo insertion of HG MNs into the back of depilated rabbit skin was performed to detect glucose levels by [99], whereas [88] used depilated rabbit skin to assess the ability of penetration of three types of HG MNs.

Several concepts of artificial skin models to study the penetration ability of MNs have been developed. A model introduced by Larrañeta et al. (2014) in which a sheet of Parafilm$^{TM}$ (a hydrophobic, flexible, semi-transparent sheet made of polyolefins and paraffin waxes) folded eight times to create an approximately 1-mm-thick membrane demonstrated equivalent insertion profiles to those compared with neonatal pig skin. Since the thickness of each layer of Parafilm$^{TM}$ was approximately 100 µm, the total depth of penetration could be estimated by inspecting each layer for puncture marks [186]. The advantage of this model is that it is inexpensive, widely available in many laboratories, and requires minimum manipulation. This method was adapted by Zhao et al. (2022a), Arshad et al. (2020), Kathuria et al. (2020), Larrañeta et al. (2015), Abdelghany et al. (2019), Nguyen et al. (2018), and Vora et al. (2020). Notably, Kathuria and colleagues (2020) documented a measurement of $154 \pm 6.8$ µm for the thickness of the Parafilm$^{TM}$ layer. Hence, it is recommended to assess the thickness of each layer before and after application.

As an alternative to Parafilm$^{TM}$, agarose gel or PDMS can be used. Agarose typically originates from specific varieties of red seaweed and is a type of carbohydrate polymer. Agarose gel is considered a suitable skin model because it can be fine-tuned to resemble part of the stress–strain relation of human skin by varying the agarose concentration. MNs could be introduced into the agarose hydrogel directly or into a layered system consisting of an agarose hydrogel base with a Parafilm$^{®}$ layer placed on top, imitating the stratum corneum. This approach has been utilized to examine both drug release and needle penetration simultaneously [38,139,140,152,154,157,190–193]. PDMS, a type of silicone rubber, can be used as an artificial material to emulate human skin mechanics due to its availability, affordability, simple fabrication, hydrophobic nature, transparency, and capacity to adjust mechanical characteristics across a broad range of Young's moduli by manipulating the material composition [194–196].

*3.5. Swelling and Dissolution*
3.5.1. Swelling Ability

Swellable microneedles, made from cross-linked hydrogels, expand in the skin without dissolving, facilitating ISF withdrawal and the controlled release of preloaded drugs [137,197]. The swelling process can be examined both qualitatively and quantitatively in vitro. Typically, the swelling ability is measured by immersing HG MNs in phosphate-buffered saline [62,88,99,106,137–139,149,152,166,179]; alternatively, they can be inserted into simulated substances such as agarose gel [138], biological tissue such as porcine skin [109,140], or even human skin [106].

Quantitively swelling is investigated by studying the mass change of MNs before and after incubation in PBS or insertion in tissue. The swelling ratio is calculated via the following equation:

$$SR\% = \frac{(W_t - W_0)}{W_0} \times 100\%$$

where SR is the swelling ratio, $W_0$ is the dry mass, and $W_t$ is the mass of swollen MNs, respectively [88,106,137,138].

The swelling rate, calculated using the same formula, involves periodically removing and measuring the mass of the microneedles [109,138,149,152,166]. Qualitatively, MNs can be inspected using a microscope or digital camera before and after swelling, and the real-time swelling behavior of MNs can be visualized directly in porcine skin using OCT imaging [139,140].

3.5.2. Dissolution

Following the insertion of a dissolvable MN array into the skin, MNs dissolve upon contact with interstitial fluid, leading to the release of the drug cargo. Depending on the composition and dissolution method, HG MNs would be inserted into the skin or immersed in a PBS solution and then examined at designated time intervals to observe the presence or absence of the remaining needles on the microneedle patches using optical, digital, or stereo microscopy. The kinetics of dissolution can be investigated in vitro by submerging MNs in distilled water [96,178], agarose gel [152,192], gelatin gel [98], porcine skin [103,121], or, most commonly, a PBS buffer solution [88,101,103,109,110,173,175]. Ex vivo MNs' dissolution was investigated in porcine and neonatal porcine skin [42,97,109,116,120,173,175,187], rodent skin [138,168], and human skin [119,165]. Considering the practical use of microneedles for drug delivery in humans, it is important to also take into account the in vivo kinetic dissolution of these microneedles within the skin. In vivo dissolution studies were conducted in dorsal or abdominal mice skin [98,109,141,172].

3.5.3. Drug Delivery

In addition to ensuring that the dissolution process can be reliably replicated in vivo, it is essential to accurately determine the quantity of a drug administered. The amount of a drug actually delivered can often fall significantly below the maximum theoretical dosage because MNs may not completely dissolve. The majority of drug permeability studies utilize the Franz diffusion cell method. This method comprises two chambers: a donor chamber, where the tested formulation is applied with an animal model membrane positioned between it, and the receptor chamber, ensuring that the stratum corneum faces the donor compartment while the dermis contacts the receptor compartment. In vitro dorsal or abdominal rat skin [117,138,149], mice skin [171], or porcine skin are used [106,109,169,172,173,175,186,187]. The receptor chamber is usually filled with PBS that is preheated and maintained at 37 °C and stirred with a magnetic bar. Sample solutions would be taken for analysis and the receptor would be filled with an equal amount of PBS.

After the microneedles have fully dissolved, it is necessary to select an appropriate method to determine the drug content based on the drug's characteristics. Typically, for chemical substances, various methods are used, such as HPLC, fluorescence spectroscopy techniques, ultraviolet spectroscopy, and other methods relying on their physical and chemical properties for measurement [106,109,166,169,171–173,187]. In the case of biological drugs, such as proteins, specific biological methods, like ELISA Kits and nucleic acid analysis using tools such as Nanodrop 2000, are necessary for quantification [88,163,175]. A less complex alternative to the Franz-diffusion cell is to submerge MNs in water [168], gelatin hydrogel [90], or PBS [49,101,143,150] and employ the same analytical techniques mentioned earlier for evaluation.

## 4. Manufacturing Methods for HG MNs

### 4.1. Molding-Based Technologies

Molding refers to any method that involves the replication of a master structure to produce MNs. The master mold is usually prepared from a metal or silicone substrate via any conventional tooling method capable of achieving the intricate MN geometry. Candidate techniques include micro-electro-discharge machining (μEDM) [71,163,198], laser-machining [169,199,200], 2-photon polymerization (2PP), and lithography [166,170,201]. Recently, additive manufacturing or '3D printing' technology has been utilized to fabricate master molds using either bottom-up stereolithography (SLA) and digital light processing (DLP) [202–206], or the higher-resolution continuous liquid interface production (CLIP) method [207,208].

For solution mold casting, a reusable PDMS mold is created from a master mold. The PDMS mold is filled with the gel aliquots of the matrix (termed 'drop casting'); the molds are usually vacuumed to remove air bubbles [209] and/or centrifuged to ensure filling and further remove trapped air [171]. The steps can be repeated as necessary to ensure that the mold cavities are void-free and filled completely (alternatively, a vacuum centrifuge would achieve similar effects in a single step [68]) and exploited to build multi-layered MN structures [172] or generate quick-release profile systems [165]. The centrifugation step has been substituted with the application of positive pressure [210], a vacuum [120,173,178], and sonication [179], as reported for PVA/PVP, HA, and GelMA matrices, respectively.

Atomized spraying and piezoelectric (inkjet) dispensing offer alternative filling formats that lower the interfacial tension inside a mold and negate the need for the additional packing steps used in solution casting. They promise to enhance packing and enable precise dosing into cavities. McGrath et al. fabricated seven types of MNs, including ones using CMC, HPMC, and PVA, using an atomizer to dispense 10–50 μm droplets with 0.25-bar compressed air into a PDMS mold. Skin penetration and ketoprofen release were found to vary according to the formulation [189].

Unlike spraying, piezoelectric dispensing has a higher targeting accuracy, and it allows the control of the volumes down to the picolitres level. Allen et al. demonstrated that the optimal influenza vaccine MNs' biological activity depends on both the formulation and actuation settings used during dispensing [211].

The MN cure conditions vary according to the materials used in a formulation. Chitosan- and Gantrez™ -based systems have been air-dried at room temperature, which is a considerably long process to obtain the optimal moisture levels [62,63,127]. The majority of reported arrays are usually cured using one or more of the following methods to speed up the process; heating [129], vacuum [70,192], microwave [212], and visible or UV irradiation for the photopolymerized GelMA and MeHA systems discussed above.

### 4.2. Surface Drawing Technologies

Surface drawing methods offer fast, mold-free fabrication that relies on the matrix's adhesion, viscosity, surface tension and the movement of two flat surfaces to shape the MNs. Four types have been reported to be used for MN manufacturing: droplet-borne air-blowing (DAB), centrifugal lithography (CL), drawing lithography (DL), and electro-drawing (ED) [111].

Droplet-borne air-blowing is performed at room temperature where two plates sandwich the hydrogel droplets, and the displacement of the two stacked plates draws the hydrogel droplets into biconcave-shaped pillars, which are solidified using blown air and then separated to produce MNs at each plate [213]. Kim et al. used DAB in under 10 min to produce insulin-loaded two-layer MN arrays using multiple formulations of 10% CMC (90 kDa), 25% HA (90 kDa), and 35% PVP (130 kDa) as matrices. The authors reported a reduction in the mechanical strength as the height of MNs increased; these values, however, always remained above the threshold required for skin penetration. CMC MNs were found to completely dissolve within an hour and achieved bioavailability of 96.6 ± 2.4%, making the systems an ideal replacement for SC injections [175]. A small-scale ($n$ = 20) clinical study reported on testing HA DAB-fabricated MNs. The system contained multiple actives with varied mechanisms of action (niacinamide, ascorbic acid 2-glucoside (AA2G), tranexamic acid, resveratrol, 4-n-butyl-resorcinol, and Halidrys siliquosa extract), and the multi-targeted approach was proven effective in treating melanogenesis [214].

For centrifugal lithography, the plates are loaded vertically into a centrifuge that elongates the drops and induces the hydrogel drops into shape via centrifugal evaporation. Once the structure is thinned enough, the MNs are solidified via the application of low temperatures that could be paired with vacuum [215]. Two groups manufactured adenosine-loaded MNs with two different Mws of HA by CL at 4 °C. Their results demonstrated that the potency of HA MNs was equivalent to a topical cream containing 140 times its dose [216], and the application of arrays resulted in wrinkle reduction without any adverse effects [76]. The two methods, DAB and CL were used to fabricate identical HA or CMC MNs to deliver growth factor and vitamin C to test the methods' ability to stabilize sensitive APIs. The prepared MNs displayed similar characteristics except for bioactivity; the CL-fabricated growth factor formulations displayed higher bioactivity levels. The authors attributed the high immunoreactivity to the shorter fabrication time and use of a lower temperature [217]. These results are supported by the performance of CMC MNs fabricated via CL to encapsulate the scrub typhus vaccine antigen. In addition to the enhanced effectiveness, the CL MNs maintained their stable immunogenicity for up to 4 weeks of storage at room temperature [51].

Drawing lithography uses a similar drawing technique to the methods previously discussed, with one difference: the matrix material is only drawn at its glass transition temperature ($T_g$), and curing is achieved by reducing the temperature below its $T_g$ [218]. DL fabrication technology was found to be effective in inhibiting caffeine crystallization in HA MNs, and the MNs demonstrated higher release and in vivo efficacy in obese mice when compared to the topical route [219].

Electro-drawing is a contact-free fabrication method performed at mild temperatures (20–40 °C) in which the gel droplet is drawn to be shaped from the plate through the application of an electro-hydrodynamic force; then, heat is applied to evaporate the solvent and solidify the MNs. ED is not widely used and has only been used to fabricate MNs from Poly(lactic-co-glycolic acid) [174,220]. Although there are no records on the use of ED for hydrogel-based MN fabrication, the method remains promising, especially for higher-viscosity matrices (large solid contents or high Mws) if the material properties are tailored for it. Figure 10 presents a schematic showing a summary of casting and surface drawing manufacturing methods.

### 4.3. Additive Manufacturing

Additive manufacturing (AM) has been touted as a promising manufacturing method for a myriad of MN types due to its high resolution, flexibility, and reasonable production costs [221,222]. However, for gel-based systems, its use currently remains restricted to making the master molds for casting methods. One of the early AM method's adaptations for gel-based systems was reported in 2014 by Boehm et al. The researchers used an Inkjet 3D printer to coat the surface of dehydrated and molded Gantrez® AN 169 MNs with miconazole [223]. DLP was recently utilized to fabricate MNs in the UV and visible light ranges. Amoxicillin-loaded GelMA (6% $w/v$) MNs were simultaneously UV-cured during the printing process and air-dried at room temperature to obtain the final product. The authors reported that the recorded tip sharpness and displacement forces were suitable for penetration. Moreover, the printed MNs showed promising in vitro release and antimicrobial activity levels [224]. Visible light DLP was used to cure and fabricate silk fibroin MNs from polymeric 6% solutions containing riboflavin as a photoinitiator. Shrinkage (horizontal contraction) was reported to be the main reason for deformation post-printing, which was attributed to the difference in surface area across the MN bodies, causing uneven dehydration [225].

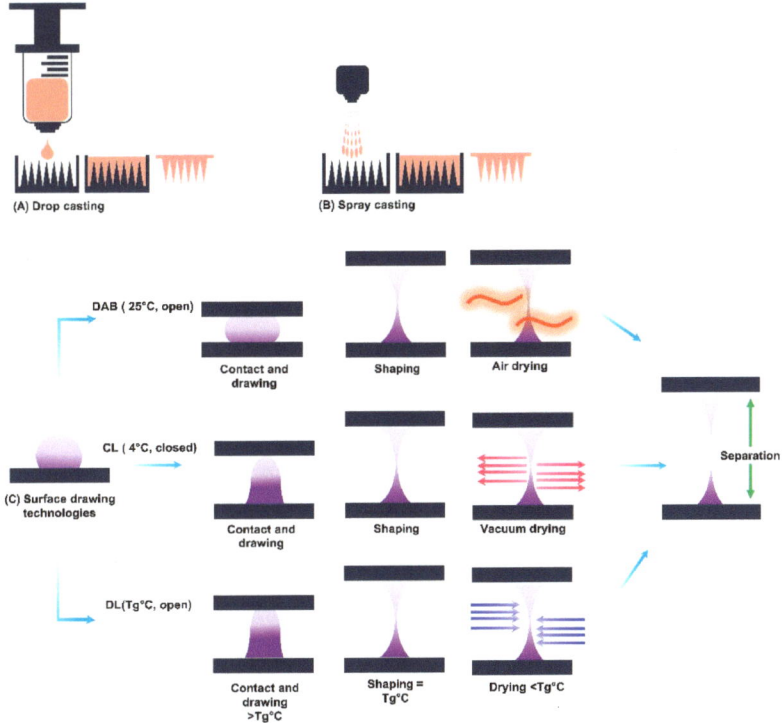

**Figure 10.** A schematic showing the manufacturing process of microneedles via casting, spraying, DAB, CL, and DL.

## 5. Challenges

The development of gel-based MN technologies still suffers from a multitude of challenges that need to be addressed for society to benefit from these technologies' full potential. These challenges can be addressed using a three-pronged approach to ensure optimal performance and wide adoptability.

### 5.1. Formulation

Sections 1 and 2 focused on the general properties and the materials used by numerous researchers to fabricate an effective gel-based MN system. The selection of candidate bulk materials depends on (a) the desired performance and method of delivery (dissolving or swelling), (b) the suitable fabrication methods, and (c) the affordability/manufacturability of the matrix polymers. The exploitation of adaptable molecules offers the customization of MN properties via altering the polymeric matrices to match application and manufacturing needs, such as solubility, Mw, concentration %, hydrolysis, or substitution degrees. HA remains a favorite due to its remarkable biocompatibility; however, its purification process remains expensive. Chitosan and cellulose derivatives are abundant and low-cost, but they can require further treatment to enhance their uniformity and properties. Molecular weight, distribution, and concentration all influence the key mechanical, flow, and dissolution properties of the material and can be selected to manipulate the properties to obtain the ideal matrix. To create a formulation suitable for an MN device, it is important to characterize the viscoelastic properties of both soft gel and hard solid material states.

Regardless of the final MNs' applications, the materials used need to be biocompatible and suitable for biomedical applications. Further studies will need to be performed on a formulation-by-formulation basis to elucidate the effects of the local prolonged use of the polymers used with a particular focus on their metabolism and excretion routes [226]. The MN systems and their ingredients' biocompatibility can be evaluated via the ISO 10993 toxicity assays recommended for medical devices [227]. The recently published document covers both the general biocompatibility and test-specific consideration. For tissues, such as skin in MNs' case, the evaluation techniques will depend on the MNs' contact duration, as shown in Table 1.

**Table 1.** Biocompatibility evaluation endpoint adopted from the FDA's (2023) ISO 10993-1:2018 recommended endpoints for considerations.

| Biological Effect | Contact Duration | | |
|---|---|---|---|
| | Limited (<24 h) | Prolonged (>24 h to 30 d) | Long-Term (> 30 d) |
| Cytotoxicity | X | X | X |
| Sensitization | X | X | X |
| Irritation or intracutaneous reactivity | X | X | X |
| Acute systemic toxicity | X | X | X |
| Material-mediated pyrogenicity | X | X | X |
| Subacute/subchronic toxicity | | X | X |
| Genotoxicity | X | X | X |
| Implantation | X | X | X |
| Hemocompatibility | X | X | X |
| Chronic toxicity | | | X |
| Carcinogenicity | | | X |

For approval, the degradation information should be provided for the device, its components, or the materials remaining in contact with tissue.

A 160-day trial was performed to evaluate the effects of administering daily PVA MNs (88% hydrolyzed, Mw of 10 kDa). The dorsal-skin bright-field images showed complete recovery within 60 min, and the hematologic and histological results indicated the absence of any toxicity in healthy female mice [228]. Gel-based MNs have a higher capacity than other MN types since actives can be included in the polymeric matrix or diffused through the external reservoir. Although drug loading could be difficult to increase without affecting the mechanical strength of the system, formulations can be tailored to encapsulate high loading and maintain the necessary mechanical strength for insertion. McCrudden et al. successfully formulated Gantrez$^{TM}$ dissolving MNs that contained 30% $w/w$ of ibuprofen, a low-potency active. The highly filled MNs still achieved 90–100% insertion rates, and the ibuprofen plasma levels in rats after 24 h were 20 times greater than the human therapeutic level [229]. However, a highly filled formulation can be challenging to mold; therefore, it will be critical to investigate the rheological behavior and ensure that the formulation flow profile is appropriate for the proposed manufacturing method.

### 5.2. Manufacturing

Fabrication challenges significantly depend on the gel properties and intended use of MN systems; however, common issues for hydrogels are scale-up issues such as ensuring material consistency, process repeatability, and the resulting microneedle quality for high-volume manufacturing runs whilst simultaneously ensuring dose uniformity and sterility, which will be critical for regulatory approval. Any of the methods discussed in Section 4 could be optimized for larger-scale production if an investment is available; nonetheless, fewer production steps, closed manufacturing loops, and wide processing windows would be ideal for reducing costs and contamination while enhancing the stability of sensitive actives. Although MNs are still considered new pharmaceutical systems, they have been marketed and sold in cosmetics and cosmeceuticals for some years. Even in the cosmetics industry, there is no universally accepted production method. Some of these manufacturers prefer to use mold casting (Microneedles Inc.-US, Mineed Technology- Thailand, CosMED Pharmaceutical Co. Ltd.-Japan), while others prefer DAB (Raphas-Korea) for the production of their gel-based systems [230–233]. These examples could be classified as luxury goods with high price points, so it is difficult to evaluate whether such approaches could deliver cost-effective therapeutic treatments for other markets, particularly in developing nations.

According to the ICH Q1A(R2) stability guidelines for drug products, thermal stability and moisture sensitivity should be tested for periods long enough to cover storage, shipment, and application [234]. For dehydrated and hygroscopic gel-based systems, relative humidity (RH) and temperature control are detrimental to the MNs' stability, and sensitivity to these environmental factors will depend on the formulation components; however, moisture-impermeable packaging and immediate use is always recommended. The extent of temperature and RH effects was found to differ from one report to another, which could be attributed to Mws and experimental differences.

Hiraishi et al. reported that HA MNs' mechanical failure force was inversely proportional to the RH and that its values should be maintained within 11–75% [232]. However, a more recent study reported that HA MNs (Mw of 150 kDa) were stable at all the tested RH (0, 60, and 82%) and temperature values (4, RT, 37, and 60 °C) [235]. Wang et al. investigated the effects of various RH conditions (20, 40, 60, and 80%) on the insertion ability into mice skin using PVA, HA, chitosan, and gelatin MNs. The lowest RH values had no effect on the MNs, and they performed similarly. An RH of 40% only had a significant effect on PVA MNs' performance, and the recommended application time was <10 min, while at 80%, all the polymeric MNs were suggested to be used immediately after unpacking (PVA <1 min and the rest <5 min) [236]. Therefore, packaging will be critical to maintain fabricated MNs' stability until use. Researchers have been investigating guideline-compliant solutions to overcome these environmental instabilities. Sealed Protect™ 470 foil packaging was found effective in maintaining the stability and mechanical capabilities of amoxicillin-loaded PVA/PVP MNs for 168 days [237]. Another moisture control solution that has been explored is PLA 3D-printed containers, which successfully prolonged the shelf life of unpacked PVA/PVP MNs during the study period for a duration of up to one month [238].

For the industrial-scale production of MNs, a good manufacturing practice (GMP) framework and standardized testing procedures need to be developed to ensure quality and performance, but cost optimization should also be a focus to drive down production costs to a point at which MN devices will be readily accepted in the marketplace. These procedures remain difficult to develop in the absence of regulatory requirements. All medical MN systems aim to penetrate the stratum corneum, therefore exceeding the legal restriction for the FDA's Class I medical devices. According to the FDA's latest regulatory considerations published in 2017 and revised in 2020, MN devices fall under Class II medical devices, with the possibility of granting De Novo requests and taking into consideration the intended use. The World Trade Organization and European Regulation followed the FDA's lead in classifying the devices as Class II (special controls) [239,240]. As with the regulations for similar classifications, effective sterilization will be essential for approval. The prefabrication filtration of the matrix solutions could be utilized, depending on the formulation viscosity; where it is only suitable for low-viscosity and moderately viscous solutions or gels. Conventional sterilization methods such as high temperature, dry heat, and steam are inappropriate for gel-based systems, and they can have a catastrophic effect on the integrity of the polymeric matrix or load [241,242]. Although aseptic production is always an option, it can be costly and complicate manufacturing environment requirements. Sterilization can be employed at different points of production, depending on the properties of MN systems used and their packaging. Non-destructive methods such as e-beam, ethylene oxide, and gamma irradiation were proven to be effective in sterilizing various systems [243–245], although gamma rays should be used with caution, as they have been observed to cause the degradation of sensitive actives ascorbic acid [246].

## 5.3. Bodily Application

Skin irritation is a natural, unavoidable immunogenic side effect of the application of MN systems. It appears visually as redness paired with minor swelling in the area, indicating the formation of local erythema. These temporary effects could facilitate the activity of some drugs, and MNs sensitivity is routinely assessed via pharmacodynamic/pharmacokinetic activity during clinical studies. Irritation can be minimized by adjusting the formulation and optimizing the design parameters [247]. Patient information leaflets and healthcare practitioners need to provide clear application instructions, explaining the anticipated transient irritation to manage patients' expectations [248]. Another safety concern is the introduction of pathogens into the sterile skin layers during the administration of MNs. Many hydrogel-MN-forming materials, such as Gantrez$^{TM}$ and chitosan, have intrinsic antimicrobial qualities, with the latter being angiogenic as well; this, however, does not guarantee sterility during application [249,250]. A study by Donnelly et al. has shown that solid silicon MNs' microbial penetration was less than that recorded for hypodermic needles, and no microorganisms were found to have crossed the viable epidermis [251]. The repeated application of MN patches containing Gantrez$^{TM}$ S-97, PEO (Mw of 10 kDa), and PVA (Mw of 58 kDa) on mice led to no changes in the skin's barrier functions. Additionally, the authors reported that infection and inflammation biomarkers remained statistically indifferent from the control over the span of the five-week study, regardless of the formulation, MN density, or number of applications [252]. More recent research using similar formulations on human volunteers confirmed the absence of adverse effects in a study that lasted five days [253].

## 6. Current Developments and Final Remarks

Only a few transdermal MNs have been registered in phase 2 and 3 clinical trials; none, however, were completed. Radius Health announced in 2021 that its abaloparatide MN system (wearABLe) trial did not meet its primary or secondary endpoints and ceased all work and development on the system in June 2022 [254]. Around the same time, after the FDA's concerns regarding the patches' bioavailability, Zosano Pharma decided to suspend its Qtrypta project (zolmitriptan) after reaching phase 3 trials [255]. PharmaTher's phase 2 clinical trial of GelMA MNs containing ketamine was suspended in late 2022 [256]. Currently, there are three gel-based MN systems registered for clinical trials on clinicaltrials.gov due to start during 2024/2025.

The standardization of the testing and clinical assessment process could significantly enhance the chances of MN systems' approval. As previously discussed and evident to any reader soon after delving into the MN literature, there is no universal evaluation process of MN performance, and only a few authors have pointed to their methods' limitations in the process, which contributes to many of the challenges encountered. A fundamental understanding of the materials' behaviors and how to quantify them is essential to develop these standard guidelines in order to optimize both the fabrication methods and end products.

For example, dehydrated MNs' mechanical assessments depend on many factors beyond the arbitrary force value. These factors include the experimental setup, the nature of the polymeric material, the tissue type, its state—its hydration or tension levels—and the MN application method. Although there is a direct correlation between the dried material property and its starting gel/solution rheological behavior, these measurements are rarely seen in the hydrogel-based MN literature. If conducting similar studies becomes customary, it could help avoid unsuitable fabrication techniques or formulations to enhance the quality of produced MNs.

## 7. Conclusions

The emerging field of MN delivery systems is steadily growing, and new technologies are continuously being developed to enhance formulation stability, mechanical performance, and biological effect.

The versatility and unique properties of hydrogels make them ideal systems for biomedical applications. When used in transdermal microneedle systems, they can provide mechanical strength for insertion, a carrier for preserving the API, and a controlled release to match the intended dose profile.

Regulatory uncertainties have impeded the routes to market. However, advances in formulation and manufacturing technologies are providing improved devices that are now demonstrating regulatory compliance. New guidelines for designers and manufacturers of hydrogel MN systems will be soon implemented, which will encourage further industrial adoption and provide new routes for the transdermal delivery of a broad range of actives.

**Author Contributions:** B.S. (conceptualization, original draft, review, and editing), M.B. (original draft, review, and editing), and B.R.W. (supervision, review, and editing). All authors have read and agreed to the published version of the manuscript.

**Funding:** This research received no external funding.

**Institutional Review Board Statement:** Not applicable.

**Informed Consent Statement:** Not applicable.

**Data Availability Statement:** Not applicable.

**Conflicts of Interest:** The authors declare that they have no known competing interest.

# References

1. Bacani, R.; Trindade, F.; Politi, M.J.; Triboni, E.R. *Nano Design for Smart Gels*; Elsevier Science: Amsterdam, The Netherlands, 2019.
2. Van Bemmelen, J. Das hydrogel und das krystallinische hydrat des kupferoxyds. *Z. Anorg. Chem.* **1894**, *5*, 466–483. [CrossRef]
3. Yang, P.; Yang, J.L.; Liu, K.; Fan, H.J. Hydrogels Enable Future Smart Batteries. *ACS Nano* **2022**, *16*, 15528–15536. [CrossRef]
4. Thakur, V.K.; Thakur, M.K. *Hydrogels: Recent Advances*; Springer: Singapore, 2018.
5. Qu, B.; Luo, Y. Chitosan-based hydrogel beads: Preparations, modifications and applications in food and agriculture sectors—A review. *Int. J. Biol. Macromol.* **2020**, *152*, 437–448. [CrossRef]
6. Grindlay, J.H.; Waugh, J.M. Plastic sponge which acts as a framework for living tissue: Experimental studies and preliminary report of use to reinforce abdominal aneurysms. *AMA Arch. Surg.* **1951**, *63*, 288–297. [CrossRef]
7. Owen, K. Polyvinyl alcohol sponge as an arterial substitute. *Proc. R. Soc. Med.* **1956**, *6*, 340–342. [CrossRef]
8. Kopeček, J.; Yang, J. Hydrogels as smart biomaterials. *Polym. Int.* **2007**, *56*, 1078–1098. [CrossRef]
9. Colombo, P.; Bettini, R.; Santi, P.; Peppas, N.A. Swellable matrices for controlled drug delivery: Gel-layer behaviour, mechanisms and optimal performance. *Pharm. Sci. Technol. Today* **2000**, *3*, 198–204. [CrossRef]
10. Alvarez-Rivera, F.; Concheiro, A.; Alvarez-Lorenzo, C. Epalrestat-loaded silicone hydrogels as contact lenses to address diabetic-eye complications. *Eur. J. Pharm. Biopharm.* **2018**, *122*, 126–136. [CrossRef]
11. Sheng, L.; Zhang, Z.; Zhang, Y.; Wang, E.; Ma, B.; Xu, Q.; Ma, L.; Zhang, M.; Pei, G.; Chang, J. A novel "hot spring"-mimetic hydrogel with excellent angiogenic properties for chronic wound healing. *Biomaterials* **2021**, *264*, 120414. [CrossRef]
12. Shriky, B.; Mahmoudi, N.; Kelly, A.; Isreb, M.; Gough, T. The effect of PEO homopolymers on the behaviours and structural evolution of Pluronic F127 smart hydrogels for controlled drug delivery systems. *Colloids Surf. A Physicochem. Eng. Asp.* **2022**, *645*, 128842. [CrossRef]
13. Shriky, B.; Kelly, A.; Isreb, M.; Babenko, M.; Mahmoudi, N.; Rogers, S.; Shebanova, O.; Snow, T.; Gough, T. Pluronic F127 thermosensitive injectable smart hydrogels for controlled drug delivery system development. *J. Colloid Interface Sci.* **2020**, *565*, 119–130. [CrossRef] [PubMed]
14. Nourbakhsh, M.; Zarrintaj, P.; Jafari, S.H.; Hosseini, S.M.; Aliakbari, S.; Pourbadie, H.G.; Naderi, N.; Zibaii, M.I.; Gholizadeh, S.S.; Ramsey, J.D.; et al. Fabricating an electroactive injectable hydrogel based on pluronic-chitosan/aniline-pentamer containing angiogenic factor for functional repair of the hippocampus ischemia rat model. *Mater. Sci. Eng. C* **2020**, *117*, 111328. [CrossRef]
15. Zhang, W.; Jin, X.; Li, H.; Zhang, R.R.; Wu, C.W. Injectable and body temperature sensitive hydrogels based on chitosan and hyaluronic acid for pH sensitive drug release. *Carbohydr. Polym.* **2018**, *186*, 82–90. [CrossRef]
16. Fan, L.; Zhang, X.; Liu, X.; Sun, B.; Li, L.; Zhao, Y. Responsive Hydrogel Microcarrier-Integrated Microneedles for Versatile and Controllable Drug Delivery. *Adv. Healthc. Mater.* **2021**, *10*, e2002249. [CrossRef]
17. Yoshida, T.; Lai, T.C.; Kwon, G.S.; Sako, K. pH- and ion-sensitive polymers for drug delivery. *Expert Opin. Drug Deliv.* **2013**, *10*, 1497–1513. [CrossRef]
18. Li, X.; Fan, D.; Ma, X.; Zhu, C.; Luo, Y.; Liu, B.; Chen, L. A Novel Injectable pH/Temperature Sensitive CS-HLC/β-GP Hydrogel: The Gelation Mechanism and Its Properties. *Soft Mater.* **2013**, *12*, 1–11. [CrossRef]
19. Farid-Ul-Haq, M.; Hussain, M.A.; Haseeb, M.T.; Ashraf, M.U.; Hussain, S.Z.; Tabassum, T.; Hussain, I.; Sher, M.; Bukhari, S.N.A.; Naeem-Ul-Hassan, M. A stimuli-responsive, superporous and non-toxic smart hydrogel from seeds of mugwort (Artemisia

20. Ahmad, N.; Mohd Amin, M.C.; Ismail, I.; Buang, F. Enhancement of oral insulin bioavailability: In vitro and in vivo assessment of nanoporous stimuli-responsive hydrogel microparticles. *Expert Opin. Drug Deliv.* **2016**, *13*, 621–632. [CrossRef]
21. Liu, T.Y.; Hu, S.H.; Liu, T.Y.; Liu, D.M.; Chen, S.Y. Magnetic-sensitive behavior of intelligent ferrogels for controlled release of drug. *Langmuir* **2006**, *22*, 5974–5978. [CrossRef]
22. Ge, J.; Neofytou, E.; Cahill, T.J., 3rd; Beygui, R.E.; Zare, R.N. Drug release from electric-field-responsive nanoparticles. *ACS Nano* **2012**, *6*, 227–233. [CrossRef]
23. Ashraf, M.U.; Hussain, M.A.; Bashir, S.; Haseeb, M.T.; Hussain, Z. Quince seed hydrogel (glucuronoxylan): Evaluation of stimuli responsive sustained release oral drug delivery system and biomedical properties. *J. Drug Deliv. Sci. Technol.* **2018**, *45*, 455–465. [CrossRef]
24. Torres-Luna, C.; Fan, X.; Domszy, R.; Hu, N.; Wang, N.S.; Yang, A. Hydrogel-based ocular drug delivery systems for hydrophobic drugs. *Eur. J. Pharm. Sci.* **2020**, *154*, 105503. [CrossRef] [PubMed]
25. Samchenko, Y.; Ulberg, Z.; Korotych, O. Multipurpose smart hydrogel systems. *Adv. Colloid Interface Sci.* **2011**, *168*, 247–262. [CrossRef] [PubMed]
26. Sa-Lima, H.; Caridade, S.G.; Mano, J.F.; Reis, R.L. Stimuli-responsive chitosan-starch injectable hydrogels combined with encapsulated adipose-derived stromal cells for articular cartilage regeneration. *Soft Matter* **2010**, *6*, 5184–5195. [CrossRef]
27. Dos Santos, A.M.; Carvalho, S.G.; Araujo, V.H.S.; Carvalho, G.C.; Gremião, M.P.D.; Chorilli, M. Recent advances in hydrogels as strategy for drug delivery intended to vaginal infections. *Int. J. Pharm.* **2020**, *590*, 119867. [CrossRef]
28. Ahsan, A.; Tian, W.-X.; Farooq, M.A.; Khan, D.H. An overview of hydrogels and their role in transdermal drug delivery. *Int. J. Polym. Mater. Polym. Biomater.* **2021**, *70*, 574–584. [CrossRef]
29. Alkilani, A.Z.; McCrudden, M.T.; Donnelly, R.F. Transdermal Drug Delivery: Innovative Pharmaceutical Developments Based on Disruption of the Barrier Properties of the stratum corneum. *Pharmaceutics* **2015**, *7*, 438–470. [CrossRef]
30. Kulkarni, D.; Damiri, F.; Rojekar, S.; Zehravi, M.; Ramproshad, S.; Dhoke, D.; Musale, S.; Mulani, A.A.; Modak, P.; Paradhi, R.; et al. Recent Advancements in Microneedle Technology for Multifaceted Biomedical Applications. *Pharmaceutics* **2022**, *14*, 1097. [CrossRef]
31. Henry, S.; McAllister, D.V.; Allen, M.G.; Prausnitz, M.R. Microfabricated microneedles: A novel approach to transdermal drug delivery. *J. Pharm. Sci.* **1998**, *87*, 922–925. [CrossRef]
32. Gerstel, M.; Place, V. Drug Delivery Device. U.S. Patent 3964482A, 22 June 1976.
33. Camirand, A.; Doucet, J. Needle dermabrasion. *Aesthetic Plast. Surg.* **1997**, *21*, 48–51. [CrossRef] [PubMed]
34. Fernandes, D. Minimally invasive percutaneous collagen induction. *Oral Maxillofac. Surg. Clin.* **2005**, *17*, 51–63. [CrossRef] [PubMed]
35. Fernandes, D.; Signorini, M. Combating photoaging with percutaneous collagen induction. *Clin. Dermatol.* **2008**, *26*, 192–199. [CrossRef] [PubMed]
36. Avcil, M.; Celik, A. Microneedles in Drug Delivery: Progress and Challenges. *Micromachines* **2021**, *12*, 1321. [CrossRef] [PubMed]
37. Kwon, K.M.; Lim, S.M.; Choi, S.; Kim, D.H.; Jin, H.E.; Jee, G.; Hong, K.J.; Kim, J.Y. Microneedles: Quick and easy delivery methods of vaccines. *Clin. Exp. Vaccine Res.* **2017**, *6*, 156–159. [CrossRef]
38. Sharma, D. Microneedles: An approach in transdermal drug delivery: A Review. *PharmaTutor* **2018**, *6*, 7–15. [CrossRef]
39. Sivaraman, A.; Banga, A.K. Novel in situ forming hydrogel microneedles for transdermal drug delivery. *Drug Deliv. Transl. Res.* **2017**, *7*, 16–26. [CrossRef]
40. Khan, S.; Minhas, M.U.; Tekko, I.A.; Donnelly, R.F.; Thakur, R.R.S. Evaluation of microneedles-assisted in situ depot forming poloxamer gels for sustained transdermal drug delivery. *Drug Deliv. Transl. Res.* **2019**, *9*, 764–782. [CrossRef]
41. Buwalda, S.J.; Boere, K.W.; Dijkstra, P.J.; Feijen, J.; Vermonden, T.; Hennink, W.E. Hydrogels in a historical perspective: From simple networks to smart materials. *J. Control. Release* **2014**, *190*, 254–273. [CrossRef]
42. Lee, J.W.; Park, J.H.; Prausnitz, M.R. Dissolving microneedles for transdermal drug delivery. *Biomaterials* **2008**, *29*, 2113–2124. [CrossRef]
43. Moore, L.E.; Vucen, S.; Moore, A.C. Trends in drug- and vaccine-based dissolvable microneedle materials and methods of fabrication. *Eur. J. Pharm. Biopharm.* **2022**, *173*, 54–72. [CrossRef]
44. Gyles, D.A.; Castro, L.D.; Silva, J.O.C., Jr.; Ribeiro-Costa, R.M. A review of the designs and prominent biomedical advances of natural and synthetic hydrogel formulations. *Eur. Polym. J.* **2017**, *88*, 373–392. [CrossRef]
45. Rahman, M.S.; Hasan, M.S.; Nitai, A.S.; Nam, S.; Karmakar, A.K.; Ahsan, M.S.; Shiddiky, M.J.; Ahmed, M.B. Recent developments of carboxymethyl cellulose. *Polymers* **2021**, *13*, 1345. [CrossRef] [PubMed]
46. Lahiji, S.F.; Seo, S.H.; Kim, S.; Dangol, M.; Shim, J.; Li, C.G.; Ma, Y.; Lee, C.; Kang, G.; Yang, H. Transcutaneous implantation of valproic acid-encapsulated dissolving microneedles induces hair regrowth. *Biomaterials* **2018**, *167*, 69–79. [CrossRef]
47. Silva, A.C.; Pereira, B.; Lameirinhas, N.S.; Costa, P.C.; Almeida, I.F.; Dias-Pereira, P.; Correia-Sá, I.; Oliveira, H.; Silvestre, A.J.; Vilela, C. Dissolvable Carboxymethylcellulose Microneedles for Noninvasive and Rapid Administration of Diclofenac Sodium. *Macromol. Biosci.* **2023**, *23*, 2200323. [CrossRef]

48. Zaric, M.; Becker, P.D.; Hervouet, C.; Kalcheva, P.; Yus, B.I.; Cocita, C.; O'Neill, L.A.; Kwon, S.-Y.; Klavinskis, L.S. Long-lived tissue resident HIV-1 specific memory CD8+ T cells are generated by skin immunization with live virus vectored microneedle arrays. *J. Control. Release* **2017**, *268*, 166–175. [CrossRef]
49. Chen, H.-J.; Lin, D.-A.; Liu, F.; Zhou, L.; Liu, D.; Lin, Z.; Yang, C.; Jin, Q.; Hang, T.; He, G. Transdermal delivery of living and biofunctional probiotics through dissolvable microneedle patches. *ACS Appl. Bio Mater.* **2018**, *1*, 374–381. [CrossRef]
50. Lee, I.C.; Lin, W.M.; Shu, J.C.; Tsai, S.W.; Chen, C.H.; Tsai, M.T. Formulation of two-layer dissolving polymeric microneedle patches for insulin transdermal delivery in diabetic mice. *J. Biomed. Mater. Res. Part A* **2017**, *105*, 84–93. [CrossRef]
51. Lee, C.; Kim, H.; Kim, S.; Lahiji, S.F.; Ha, N.Y.; Yang, H.; Kang, G.; Nguyen, H.Y.T.; Kim, Y.; Choi, M.S. Comparative Study of Two Droplet-Based Dissolving Microneedle Fabrication Methods for Skin Vaccination. *Adv. Healthc. Mater.* **2018**, *7*, 1701381. [CrossRef]
52. Lee, B.-M.; Lee, C.; Lahiji, S.F.; Jung, U.-W.; Chung, G.; Jung, H. Dissolving microneedles for rapid and painless local anesthesia. *Pharmaceutics* **2020**, *12*, 366. [CrossRef]
53. Hwa, K.-Y.; Chang, V.H.; Cheng, Y.-Y.; Wang, Y.-D.; Jan, P.-S.; Subramani, B.; Wu, M.-J.; Wang, B.-K. Analyzing polymeric matrix for fabrication of a biodegradable microneedle array to enhance transdermal delivery. *Biomed. Microdevices* **2017**, *19*, 84. [CrossRef] [PubMed]
54. Ito, Y.; Ohta, J.; Imada, K.; Akamatsu, S.; Tsuchida, N.; Inoue, G.; Inoue, N.; Takada, K. Dissolving microneedles to obtain rapid local anesthetic effect of lidocaine at skin tissue. *J. Drug Target.* **2013**, *21*, 770–775. [CrossRef] [PubMed]
55. Kim, J.-Y.; Han, M.-R.; Kim, Y.-H.; Shin, S.-W.; Nam, S.-Y.; Park, J.-H. Tip-loaded dissolving microneedles for transdermal delivery of donepezil hydrochloride for treatment of Alzheimer's disease. *Eur. J. Pharm. Biopharm.* **2016**, *105*, 148–155. [CrossRef]
56. Nagra, U.; Barkat, K.; Ashraf, M.U.; Shabbir, M. Feasibility of enhancing skin permeability of acyclovir through sterile topical lyophilized wafer on self-dissolving microneedle-treated skin. *Dose-Response* **2022**, *20*, 15593258221097594. [CrossRef]
57. Chen, X.G.; Zheng, L.; Wang, Z.; Lee, C.Y.; Park, H.J. Molecular affinity and permeability of different molecular weight chitosan membranes. *J. Agric. Food Chem.* **2002**, *50*, 5915–5918. [CrossRef]
58. Gorantla, S.; Dabholkar, N.; Sharma, S.; Rapalli, V.K.; Alexander, A.; Singhvi, G. Chitosan-based microneedles as a potential platform for drug delivery through the skin: Trends and regulatory aspects. *Int. J. Biol. Macromol.* **2021**, *184*, 438–453. [CrossRef] [PubMed]
59. Learoyd, T.P.; Burrows, J.L.; French, E.; Seville, P.C. Chitosan-based spray-dried respirable powders for sustained delivery of terbutaline sulfate. *Eur. J. Pharm. Biopharm.* **2008**, *68*, 224–234. [CrossRef]
60. Zhao, J.; Li, J.; Jiang, Z.; Tong, R.; Duan, X.; Bai, L.; Shi, J. Chitosan, N,N,N-trimethyl chitosan (TMC) and 2-hydroxypropyltrimethyl ammonium chloride chitosan (HTCC): The potential immune adjuvants and nano carriers. *Int. J. Biol. Macromol.* **2020**, *154*, 339–348. [CrossRef]
61. Chen, M.C.; Ling, M.H.; Lai, K.Y.; Pramudityo, E. Chitosan microneedle patches for sustained transdermal delivery of macromolecules. *Biomacromolecules* **2012**, *13*, 4022–4031. [CrossRef]
62. Arshad, M.S.; Hassan, S.; Hussain, A.; Abbas, N.; Kucuk, I.; Nazari, K.; Ali, R.; Ramzan, S.; Alqahtani, A.; Andriotis, E.G.; et al. Improved transdermal delivery of cetirizine hydrochloride using polymeric microneedles. *Daru* **2019**, *27*, 673–681. [CrossRef]
63. Yao, G.; Quan, G.; Lin, S.; Peng, T.; Wang, Q.; Ran, H.; Chen, H.; Zhang, Q.; Wang, L.; Pan, X.; et al. Novel dissolving microneedles for enhanced transdermal delivery of levonorgestrel: In vitro and in vivo characterization. *Int. J. Pharm.* **2017**, *534*, 378–386. [CrossRef] [PubMed]
64. Chandrasekharan, A.; Hwang, Y.J.; Seong, K.Y.; Park, S.; Kim, S.; Yang, S.Y. Acid-Treated Water-Soluble Chitosan Suitable for Microneedle-Assisted Intracutaneous Drug Delivery. *Pharmaceutics* **2019**, *11*, 209. [CrossRef] [PubMed]
65. Dathathri, E.; Lal, S.; Mittal, M.; Thakur, G.; De, S. Fabrication of low-cost composite polymer-based micro needle patch for transdermal drug delivery. *Appl. Nanosci.* **2020**, *10*, 371–377. [CrossRef]
66. Ryall, C.; Chen, S.; Duarah, S.; Wen, J. Chitosan-based microneedle arrays for dermal delivery of Centella asiatica. *Int. J. Pharm.* **2022**, *627*, 122221. [CrossRef]
67. Gaware, S.A.; Rokade, K.A.; Bala, P.; Kale, S.N. Microneedles of chitosan-porous carbon nanocomposites: Stimuli (pH and electric field)-initiated drug delivery and toxicological studies. *J. Biomed. Mater. Res. A* **2019**, *107*, 1582–1596. [CrossRef]
68. Saha, I.; Rai, V.K. Hyaluronic acid based microneedle array: Recent applications in drug delivery and cosmetology. *Carbohydr. Polym.* **2021**, *267*, 118168. [CrossRef]
69. Yang, H.; Wu, X.; Zhou, Z.; Chen, X.; Kong, M. Enhanced transdermal lymphatic delivery of doxorubicin via hyaluronic acid based transfersomes/microneedle complex for tumor metastasis therapy. *Int. J. Biol. Macromol.* **2019**, *125*, 9–16. [CrossRef]
70. Xie, Y.; Wang, H.; Mao, J.; Li, Y.; Hussain, M.; Zhu, J.; Li, Y.; Zhang, L.; Tao, J.; Zhu, J. Enhanced in vitro efficacy for inhibiting hypertrophic scar by bleomycin-loaded dissolving hyaluronic acid microneedles. *J. Mater. Chem. B* **2019**, *7*, 6604–6611. [CrossRef]
71. Zhu, Z.; Luo, H.; Lu, W.; Luan, H.; Wu, Y.; Luo, J.; Wang, Y.; Pi, J.; Lim, C.Y.; Wang, H. Rapidly dissolvable microneedle patches for transdermal delivery of exenatide. *Pharm. Res.* **2014**, *31*, 3348–3360. [CrossRef]
72. Zhu, J.; Dong, L.; Du, H.; Mao, J.; Xie, Y.; Wang, H.; Lan, J.; Lou, Y.; Fu, Y.; Wen, J.; et al. 5-Aminolevulinic Acid-Loaded Hyaluronic Acid Dissolving Microneedles for Effective Photodynamic Therapy of Superficial Tumors with Enhanced Long-Term Stability. *Adv. Healthc. Mater.* **2019**, *8*, e1900896. [CrossRef]
73. Zhao, X.; Li, X.; Zhang, P.; Du, J.; Wang, Y. Tip-loaded fast-dissolving microneedle patches for photodynamic therapy of subcutaneous tumor. *J. Control. Release* **2018**, *286*, 201–209. [CrossRef]

74. Jegasothy, S.M.; Zabolotniaia, V.; Bielfeldt, S. Efficacy of a New Topical Nano-hyaluronic Acid in Humans. *J. Clin. Aesthet. Dermatol.* **2014**, *7*, 27–29.
75. Chi, Y.; Huang, Y.; Kang, Y.; Dai, G.; Liu, Z.; Xu, K.; Zhong, W. The effects of molecular weight of hyaluronic acid on transdermal delivery efficiencies of dissolving microneedles. *Eur. J. Pharm. Sci.* **2022**, *168*, 106075. [CrossRef]
76. Jang, M.; Baek, S.; Kang, G.; Yang, H.; Kim, S.; Jung, H. Dissolving microneedle with high molecular weight hyaluronic acid to improve skin wrinkles, dermal density and elasticity. *Int. J. Cosmet. Sci.* **2020**, *42*, 302–309. [CrossRef] [PubMed]
77. Zhu, J.; Tang, X.; Jia, Y.; Ho, C.T.; Huang, Q. Applications and delivery mechanisms of hyaluronic acid used for topical/transdermal delivery—A review. *Int. J. Pharm.* **2020**, *578*, 119127. [CrossRef]
78. Chiu, Y.H.; Chen, M.C.; Wan, S.W. Sodium Hyaluronate/Chitosan Composite Microneedles as a Single-Dose Intradermal Immunization System. *Biomacromolecules* **2018**, *19*, 2278–2285. [CrossRef] [PubMed]
79. Im, D.S.; Kim, M.H.; Yoon, Y.I.; Park, W.H. Gelation Behaviors and Mechanism of Silk Fibroin According to the Addition of Nitrate Salts. *Int. J. Mol. Sci.* **2016**, *17*, 1697. [CrossRef]
80. Johari, N.; Khodaei, A.; Samadikuchaksaraei, A.; Reis, R.L.; Kundu, S.C.; Moroni, L. Ancient fibrous biomaterials from silkworm protein fibroin and spider silk blends: Biomechanical patterns. *Acta Biomater.* **2022**, *153*, 38–67. [CrossRef] [PubMed]
81. Gosline, J.M.; Guerette, P.A.; Ortlepp, C.S.; Savage, K.N. The mechanical design of spider silks: From fibroin sequence to mechanical function. *J. Exp. Biol.* **1999**, *202*, 3295–3303. [CrossRef] [PubMed]
82. You, X.Q.; Chang, J.H.; Ju, B.K.; Pak, J.J. Rapidly dissolving fibroin microneedles for transdermal drug delivery. *Mater. Sci. Eng. C* **2011**, *31*, 1632–1636. [CrossRef]
83. Phillips, D.M.; Drummy, L.F.; Conrady, D.G.; Fox, D.M.; Naik, R.R.; Stone, M.O.; Trulove, P.C.; De Long, H.C.; Mantz, R.A. Dissolution and regeneration of Bombyx mori silk fibroin using ionic liquids. *J. Am. Chem. Soc.* **2004**, *126*, 14350–14351. [CrossRef] [PubMed]
84. Cao, Y.; Wang, B. Biodegradation of silk biomaterials. *Int. J. Mol. Sci.* **2009**, *10*, 1514–1524. [CrossRef] [PubMed]
85. Wenk, E.; Merkle, H.P.; Meinel, L. Silk fibroin as a vehicle for drug delivery applications. *J. Control. Release* **2011**, *150*, 128–141. [CrossRef] [PubMed]
86. Tsioris, K.; Raja, W.K.; Pritchard, E.M.; Panilaitis, B.; Kaplan, D.L.; Omenetto, F.G. Fabrication of Silk Microneedles for Controlled-Release Drug Delivery. *Adv. Funct. Mater.* **2012**, *22*, 330–335. [CrossRef]
87. Lee, J.; Park, S.H.; Seo, I.H.; Lee, K.J.; Ryu, W. Rapid and repeatable fabrication of high A/R silk fibroin microneedles using thermally-drawn micromolds. *Eur. J. Pharm. Biopharm.* **2015**, *94*, 11–19. [CrossRef]
88. Zhu, M.; Liu, Y.; Jiang, F.; Cao, J.; Kundu, S.C.; Lu, S. Combined Silk Fibroin Microneedles for Insulin Delivery. *ACS Biomater. Sci. Eng.* **2020**, *6*, 3422–3429. [CrossRef]
89. Qiu, W.; Patil, A.; Hu, F.; Liu, X.Y. Hierarchical Structure of Silk Materials Versus Mechanical Performance and Mesoscopic Engineering Principles. *Small* **2019**, *15*, e1903948. [CrossRef]
90. Lin, Z.; Li, Y.; Meng, G.; Hu, X.; Zeng, Z.; Zhao, B.; Lin, N.; Liu, X.Y. Reinforcement of Silk Microneedle Patches for Accurate Transdermal Delivery. *Biomacromolecules* **2021**, *22*, 5319–5326. [CrossRef]
91. Stinson, J.A.; Boopathy, A.V.; Cieslewicz, B.M.; Zhang, Y.C.; Hartman, N.W.; Miller, D.P.; Dirckx, M.; Hurst, B.L.; Tarbet, E.B.; Kluge, J.A.; et al. Enhancing influenza vaccine immunogenicity and efficacy through infection mimicry using silk microneedles. *Vaccine* **2021**, *39*, 5410–5421. [CrossRef]
92. Yin, Z.; Kuang, D.; Wang, S.; Zheng, Z.; Yadavalli, V.K.; Lu, S. Swellable silk fibroin microneedles for transdermal drug delivery. *Int. J. Biol. Macromol.* **2018**, *106*, 48–56. [CrossRef]
93. Hassan, C.M.; Trakampan, P.; Peppas, N.A. Water solubility characteristics of poly (vinyl alcohol) and gels prepared by freezing/thawing processes. In *Water Soluble Polymers: Solutions Properties and Applications*; Springer: Boston, MA, USA, 2020; pp. 31–40, ISBN 0-306-46915-4/0-306-45931-0.
94. Halima, N.B. Poly (vinyl alcohol): Review of its promising applications and insights into biodegradation. *RSC Adv.* **2016**, *6*, 39823–39832. [CrossRef]
95. Chong, S.F.; Smith, A.A.A.; Zelikin, A.N. Microstructured, Functional PVA Hydrogels through Bioconjugation with Oligopeptides under Physiological Conditions. *Small* **2013**, *9*, 942–950. [CrossRef] [PubMed]
96. Zhu, D.D.; Zhang, X.P.; Shen, C.B.; Cui, Y.; Guo, X.D. The maximum possible amount of drug in rapidly separating microneedles. *Drug Deliv. Transl. Res.* **2019**, *9*, 1133–1142. [CrossRef] [PubMed]
97. Abdelghany, S.; Tekko, I.A.; Vora, L.; Larraneta, E.; Permana, A.D.; Donnelly, R.F. Nanosuspension-Based Dissolving Microneedle Arrays for Intradermal Delivery of Curcumin. *Pharmaceutics* **2019**, *11*, 308. [CrossRef]
98. Lau, S.; Fei, J.; Liu, H.; Chen, W.; Liu, R. Multilayered pyramidal dissolving microneedle patches with flexible pedestals for improving effective drug delivery. *J. Control. Release* **2017**, *265*, 113–119. [CrossRef]
99. He, R.; Niu, Y.; Li, Z.; Li, A.; Yang, H.; Xu, F.; Li, F. A Hydrogel Microneedle Patch for Point-of-Care Testing Based on Skin Interstitial Fluid. *Adv. Healthc. Mater.* **2020**, *9*, e1901201. [CrossRef]
100. Cole, G.; McCaffrey, J.; Ali, A.A.; McBride, J.W.; McCrudden, C.M.; Vincente-Perez, E.M.; Donnelly, R.F.; McCarthy, H.O. Dissolving microneedles for DNA vaccination: Improving functionality via polymer characterization and RALA complexation. *Hum. Vaccin. Immunother.* **2017**, *13*, 50–62. [CrossRef]

101. Nguyen, H.X.; Bozorg, B.D.; Kim, Y.; Wieber, A.; Birk, G.; Lubda, D.; Banga, A.K. Poly (vinyl alcohol) microneedles: Fabrication, characterization, and application for transdermal drug delivery of doxorubicin. *Eur. J. Pharm. Biopharm.* **2018**, *129*, 88–103. [CrossRef]
102. Oh, N.G.; Hwang, S.Y.; Na, Y.H. Fabrication of a PVA-Based Hydrogel Microneedle Patch. *ACS Omega* **2022**, *7*, 25179–25185. [CrossRef]
103. Chen, M.-C.; Ling, M.-H.; Kusuma, S.J. Poly-γ-glutamic acid microneedles with a supporting structure design as a potential tool for transdermal delivery of insulin. *Acta Biomater.* **2015**, *24*, 106–116. [CrossRef]
104. Song, G.; Jiang, G.; Liu, T.; Zhang, X.; Zeng, Z.; Wang, R.; Li, P.; Yang, Y. Separable Microneedles for Synergistic Chemo-Photothermal Therapy against Superficial Skin Tumors. *ACS Biomater. Sci. Eng.* **2020**, *6*, 4116–4125. [CrossRef] [PubMed]
105. Wang, Q.L.; Zhang, X.P.; Chen, B.Z.; Guo, X.D. Dissolvable layered microneedles with core-shell structures for transdermal drug delivery. *Mater. Sci. Eng. C* **2018**, *83*, 143–147. [CrossRef] [PubMed]
106. Yang, S.; Feng, Y.; Zhang, L.; Chen, N.; Yuan, W.; Jin, T. A scalable fabrication process of polymer microneedles. *Int. J. Nanomed.* **2012**, *7*, 1415–1422. [CrossRef]
107. Raja, W.K.; Maccorkle, S.; Diwan, I.M.; Abdurrob, A.; Lu, J.; Omenetto, F.G.; Kaplan, D.L. Transdermal delivery devices: Fabrication, mechanics and drug release from silk. *Small* **2013**, *9*, 3704–3713. [CrossRef]
108. Dandekar, A.A.; Garimella, H.T.; German, C.L.; Banga, A.K. Microneedle mediated iontophoretic delivery of tofacitinib citrate. *Pharm. Res.* **2023**, *40*, 735–747. [CrossRef]
109. Aung, N.N.; Ngawhirunpat, T.; Rojanarata, T.; Patrojanasophon, P.; Pamornpathomkul, B.; Opanasopit, P. Fabrication, characterization and comparison of alpha-arbutin loaded dissolving and hydrogel forming microneedles. *Int. J. Pharm.* **2020**, *586*, 119508. [CrossRef]
110. Chen, B.Z.; Ashfaq, M.; Zhang, X.P.; Zhang, J.N.; Guo, X.D. In vitro and in vivo assessment of polymer microneedles for controlled transdermal drug delivery. *J. Drug Target.* **2018**, *26*, 720–729. [CrossRef]
111. Turner, J.G.; White, L.R.; Estrela, P.; Leese, H.S. Hydrogel-Forming Microneedles: Current Advancements and Future Trends. *Macromol. Biosci.* **2021**, *21*, e2000307. [CrossRef]
112. Sun, Y.; Liu, J.L.; Wang, H.Y.; Li, S.S.; Pan, X.T.; Xu, B.L.; Yang, H.L.; Wu, Q.Y.; Li, W.X.; Su, X.; et al. NIR Laser-Triggered Microneedle-Based Liquid Band-Aid for Wound Care. *Adv. Funct. Mater.* **2021**, *31*, 2100218. [CrossRef]
113. Teodorescu, M.; Bercea, M. Poly(vinylpyrrolidone)—A Versatile Polymer for Biomedical and Beyond Medical Applications. *Polym.-Plast. Technol. Eng.* **2015**, *54*, 923–943. [CrossRef]
114. Zhang, L.; Guo, R.; Wang, S.; Yang, X.; Ling, G.; Zhang, P. Fabrication, evaluation and applications of dissolving microneedles. *Int. J. Pharm.* **2021**, *604*, 120749. [CrossRef] [PubMed]
115. Mangang, K.N.; Thakran, P.; Halder, J.; Yadav, K.S.; Ghosh, G.; Pradhan, D.; Rath, G.; Rai, V.K. PVP-microneedle array for drug delivery: Mechanical insight, biodegradation, and recent advances. *J. Biomater. Sci. Polym. Ed.* **2023**, *34*, 986–1017. [CrossRef]
116. Thakur, R.R.; Tekko, I.A.; Al-Shammari, F.; Ali, A.A.; McCarthy, H.; Donnelly, R.F. Rapidly dissolving polymeric microneedles for minimally invasive intraocular drug delivery. *Drug Deliv. Transl. Res.* **2016**, *6*, 800–815. [CrossRef] [PubMed]
117. Shah, V.; Choudhury, B.K. Fabrication, Physicochemical Characterization, and Performance Evaluation of Biodegradable Polymeric Microneedle Patch System for Enhanced Transcutaneous Flux of High Molecular Weight Therapeutics. *AAPS PharmSciTech* **2017**, *18*, 2936–2948. [CrossRef] [PubMed]
118. Di Natale, C.; De Rosa, D.; Profeta, M.; Jamaledin, R.; Attanasio, A.; Lagreca, E.; Scognamiglio, P.L.; Netti, P.A.; Vecchione, R. Design of biodegradable bi-compartmental microneedles for the stabilization and the controlled release of the labile molecule collagenase for skin healthcare. *J. Mater. Chem. B* **2021**, *9*, 392–403. [CrossRef]
119. Sun, W.; Araci, Z.; Inayathullah, M.; Manickam, S.; Zhang, X.; Bruce, M.A.; Marinkovich, M.P.; Lane, A.T.; Milla, C.; Rajadas, J.; et al. Polyvinylpyrrolidone microneedles enable delivery of intact proteins for diagnostic and therapeutic applications. *Acta Biomater.* **2013**, *9*, 7767–7774. [CrossRef]
120. Sullivan, S.P.; Koutsonanos, D.G.; Del Pilar Martin, M.; Lee, J.W.; Zarnitsyn, V.; Choi, S.O.; Murthy, N.; Compans, R.W.; Skountzou, I.; Prausnitz, M.R. Dissolving polymer microneedle patches for influenza vaccination. *Nat. Med.* **2010**, *16*, 915–920. [CrossRef]
121. Sullivan, S.P.; Murthy, N.; Prausnitz, M.R. Minimally invasive protein delivery with rapidly dissolving polymer microneedles. *Adv. Mater.* **2008**, *20*, 933–938. [CrossRef]
122. Burnett, C.L.; Bergfeld, W.F.; Belsito, D.V.; Hill, R.A.; Klaassen, C.D.; Liebler, D.C.; Marks Jr, J.G.; Shank, R.C.; Slaga, T.J.; Snyder, P.W. Final report of the Amended Safety Assessment of PVM/MA copolymer and its related salts and esters as used in cosmetics. *Int. J. Toxicol.* **2011**, *30*, 128S–144S. [CrossRef]
123. Azizi Machekposhti, S.; Nguyen, A.K.; Vanderwal, L.; Stafslien, S.; Narayan, R.J. Micromolding of Amphotericin-B-Loaded Methoxyethylene-Maleic Anhydride Copolymer Microneedles. *Pharmaceutics* **2022**, *14*, 1551. [CrossRef]
124. Reyna, D.; Bejster, I.; Chadderdon, A.; Harteg, C.; Anjani, Q.K.; Sabri, A.H.B.; Brown, A.N.; Drusano, G.L.; Westover, J.; Tarbet, E.B. A five-day treatment course of zanamivir for the flu with a single, self-administered, painless microneedle array patch: Revolutionizing delivery of poorly membrane-permeable therapeutics. *Int. J. Pharm.* **2023**, *641*, 123081. [CrossRef] [PubMed]
125. Requena, M.B.; Permana, A.D.; Vollet-Filho, J.D.; González-Vázquez, P.; Garcia, M.R.; De Faria, C.M.G.; Pratavieira, S.; Donnelly, R.F.; Bagnato, V.S. Dissolving microneedles containing aminolevulinic acid improves protoporphyrin IX distribution. *J. Biophotonics* **2021**, *14*, e202000128. [CrossRef] [PubMed]

126. Gomaa, Y.A.; Garland, M.J.; McInnes, F.; El-Khordagui, L.K.; Wilson, C.; Donnelly, R.F. Laser-engineered dissolving microneedles for active transdermal delivery of nadroparin calcium. *Eur. J. Pharm. Biopharm.* **2012**, *82*, 299–307. [CrossRef]
127. Pamornpathomkul, B.; Ngawhirunpat, T.; Tekko, I.A.; Vora, L.; McCarthy, H.O.; Donnelly, R.F. Dissolving polymeric microneedle arrays for enhanced site-specific acyclovir delivery. *Eur. J. Pharm. Sci.* **2018**, *121*, 200–209. [CrossRef]
128. Hutton, A.R.; Quinn, H.L.; McCague, P.J.; Jarrahian, C.; Rein-Weston, A.; Coffey, P.S.; Gerth-Guyette, E.; Zehrung, D.; Larrañeta, E.; Donnelly, R.F. Transdermal delivery of vitamin K using dissolving microneedles for the prevention of vitamin K deficiency bleeding. *Int. J. Pharm.* **2018**, *541*, 56–63. [CrossRef] [PubMed]
129. Sabri, A.H.B.; Anjani, Q.K.; Donnelly, R.F. Synthesis and characterization of sorbitol laced hydrogel-forming microneedles for therapeutic drug monitoring. *Int. J. Pharm.* **2021**, *607*, 121049. [CrossRef]
130. Caffarel-Salvador, E.; Brady, A.J.; Eltayib, E.; Meng, T.; Alonso-Vicente, A.; Gonzalez-Vazquez, P.; Torrisi, B.M.; Vicente-Perez, E.M.; Mooney, K.; Jones, D.S. Hydrogel-forming microneedle arrays allow detection of drugs and glucose in vivo: Potential for use in diagnosis and therapeutic drug monitoring. *PLoS ONE* **2015**, *10*, e0145644.
131. Migdadi, E.M.; Courtenay, A.J.; Tekko, I.A.; McCrudden, M.T.; Kearney, M.-C.; McAlister, E.; McCarthy, H.O.; Donnelly, R.F. Hydrogel-forming microneedles enhance transdermal delivery of metformin hydrochloride. *J. Control. Release* **2018**, *285*, 142–151. [CrossRef]
132. Naser, Y.A.; Tekko, I.A.; Vora, L.K.; Peng, K.; Anjani, Q.K.; Greer, B.; Elliott, C.; McCarthy, H.O.; Donnelly, R.F. Hydrogel-forming microarray patches with solid dispersion reservoirs for transdermal long-acting microdepot delivery of a hydrophobic drug. *J. Control. Release* **2023**, *356*, 416–433. [CrossRef]
133. Kearney, M.-C.; Caffarel-Salvador, E.; Fallows, S.J.; McCarthy, H.O.; Donnelly, R.F. Microneedle-mediated delivery of donepezil: Potential for improved treatment options in Alzheimer's disease. *Eur. J. Pharm. Biopharm.* **2016**, *103*, 43–50. [CrossRef]
134. Courtenay, A.J.; McAlister, E.; McCrudden, M.T.; Vora, L.; Steiner, L.; Levin, G.; Levy-Nissenbaum, E.; Shterman, N.; Kearney, M.-C.; McCarthy, H.O. Hydrogel-forming microneedle arrays as a therapeutic option for transdermal esketamine delivery. *J. Control. Release* **2020**, *322*, 177–186. [CrossRef]
135. Courtenay, A.J.; McCrudden, M.T.; McAvoy, K.J.; McCarthy, H.O.; Donnelly, R.F. Microneedle-mediated transdermal delivery of bevacizumab. *Mol. Pharm.* **2018**, *15*, 3545–3556. [CrossRef] [PubMed]
136. Poldervaart, M.T.; Goversen, B.; De Ruijter, M.; Abbadessa, A.; Melchels, F.P.; Öner, F.C.; Dhert, W.J.; Vermonden, T.; Alblas, J. 3D bioprinting of methacrylated hyaluronic acid (MeHA) hydrogel with intrinsic osteogenicity. *PLoS ONE* **2017**, *12*, e0177628.
137. Chew, S.W.; Shah, A.H.; Zheng, M.; Chang, H.; Wiraja, C.; Steele, T.W.; Xu, C. A self-adhesive microneedle patch with drug loading capability through swelling effect. *Bioeng. Transl. Med.* **2020**, *5*, e10157. [CrossRef] [PubMed]
138. Yu, M.; Lu, Z.; Shi, Y.; Du, Y.; Chen, X.; Kong, M. Systematic comparisons of dissolving and swelling hyaluronic acid microneedles in transdermal drug delivery. *Int. J. Biol. Macromol.* **2021**, *191*, 783–791. [CrossRef]
139. Chang, H.; Zheng, M.; Yu, X.; Than, A.; Seeni, R.Z.; Kang, R.; Tian, J.; Khanh, D.P.; Liu, L.; Chen, P. A swellable microneedle patch to rapidly extract skin interstitial fluid for timely metabolic analysis. *Adv. Mater.* **2017**, *29*, 1702243. [CrossRef]
140. Zheng, M.; Wang, Z.; Chang, H.; Wang, L.; Chew, S.W.; Lio, D.C.S.; Cui, M.; Liu, L.; Tee, B.C.; Xu, C. Osmosis-powered hydrogel microneedles for microliters of skin interstitial fluid extraction within minutes. *Adv. Healthc. Mater.* **2020**, *9*, 1901683. [CrossRef]
141. Cao, J.; Zhang, N.; Wang, Z.; Su, J.; Yang, J.; Han, J.; Zhao, Y. Microneedle-assisted transdermal delivery of etanercept for rheumatoid arthritis treatment. *Pharmaceutics* **2019**, *11*, 235. [CrossRef]
142. Du, G.; He, P.; Zhao, J.; He, C.; Jiang, M.; Zhang, Z.; Zhang, Z.; Sun, X. Polymeric microneedle-mediated transdermal delivery of melittin for rheumatoid arthritis treatment. *J. Control. Release* **2021**, *336*, 537–548. [CrossRef]
143. An, M.; Shi, M.; Su, J.; Wei, Y.; Luo, R.; Sun, P.; Zhao, Y. Dual-drug loaded separable microneedles for efficient rheumatoid arthritis therapy. *Pharmaceutics* **2022**, *14*, 1518. [CrossRef]
144. Dabholkar, N.; Gorantla, S.; Waghule, T.; Rapalli, V.K.; Kothuru, A.; Goel, S.; Singhvi, G. Biodegradable microneedles fabricated with carbohydrates and proteins: Revolutionary approach for transdermal drug delivery. *Int. J. Biol. Macromol.* **2021**, *170*, 602–621. [CrossRef] [PubMed]
145. Van Den Bulcke, A.I.; Bogdanov, B.; De Rooze, N.; Schacht, E.H.; Cornelissen, M.; Berghmans, H. Structural and rheological properties of methacrylamide modified gelatin hydrogels. *Biomacromolecules* **2000**, *1*, 31–38. [CrossRef] [PubMed]
146. Noshadi, I.; Hong, S.; Sullivan, K.E.; Sani, E.S.; Portillo-Lara, R.; Tamayol, A.; Shin, S.R.; Gao, A.E.; Stoppel, W.L.; Black III, L.D. In vitro and in vivo analysis of visible light crosslinkable gelatin methacryloyl (GelMA) hydrogels. *Biomater. Sci.* **2017**, *5*, 2093–2105. [CrossRef] [PubMed]
147. Yue, K.; Trujillo-de Santiago, G.; Alvarez, M.M.; Tamayol, A.; Annabi, N.; Khademhosseini, A. Synthesis, properties, and biomedical applications of gelatin methacryloyl (GelMA) hydrogels. *Biomaterials* **2015**, *73*, 254–271. [CrossRef]
148. Liu, Y.; Long, L.; Zhang, F.; Hu, X.; Zhang, J.; Hu, C.; Wang, Y.; Xu, J. Microneedle-mediated vascular endothelial growth factor delivery promotes angiogenesis and functional recovery after stroke. *J. Control. Release* **2021**, *338*, 610–622. [CrossRef]
149. Zhao, Z.Q.; Liang, L.; Hu, L.F.; He, Y.T.; Jing, L.Y.; Liu, Y.; Chen, B.Z.; Guo, X.D. Subcutaneous Implantable Microneedle System for the Treatment of Alzheimer's Disease by Delivering Donepezil. *Biomacromolecules* **2022**, *23*, 5330–5339. [CrossRef]
150. Zhao, Z.Q.; Zhang, B.L.; Chu, H.Q.; Liang, L.; Chen, B.Z.; Zheng, H.; Guo, X.D. A high-dosage microneedle for programmable lidocaine delivery and enhanced local long-lasting analgesia. *Biomater. Adv.* **2022**, *133*, 112620. [CrossRef]
151. Fu, X.; Zhang, X.; Huang, D.; Mao, L.; Qiu, Y.; Zhao, Y. Bioinspired adhesive microneedle patch with gemcitabine encapsulation for pancreatic cancer treatment. *Chem. Eng. J.* **2022**, *431*, 133362. [CrossRef]

152. Qiao, Y.; Du, J.; Ge, R.; Lu, H.; Wu, C.; Li, J.; Yang, S.; Zada, S.; Dong, H.; Zhang, X. A sample and detection microneedle patch for psoriasis MicroRNA biomarker analysis in interstitial fluid. *Anal. Chem.* **2022**, *94*, 5538–5545. [CrossRef]
153. Haghniaz, R.; Kim, H.-J.; Montazerian, H.; Baidya, A.; Tavafoghi, M.; Chen, Y.; Zhu, Y.; Karamikamkar, S.; Sheikhi, A.; Khademhosseini, A. Tissue adhesive hemostatic microneedle arrays for rapid hemorrhage treatment. *Bioact. Mater.* **2023**, *23*, 314–327. [CrossRef]
154. Yang, Q.; Wang, Y.; Liu, T.; Wu, C.; Li, J.; Cheng, J.; Wei, W.; Yang, F.; Zhou, L.; Zhang, Y. Microneedle array encapsulated with programmed DNA hydrogels for rapidly sampling and sensitively sensing of specific microRNA in dermal interstitial fluid. *ACS Nano* **2022**, *16*, 18366–18375. [CrossRef] [PubMed]
155. Tekko, I.A.; Chen, G.; Domínguez-Robles, J.; Thakur, R.R.S.; Hamdan, I.M.; Vora, L.; Larrañeta, E.; McElnay, J.C.; McCarthy, H.O.; Rooney, M. Development and characterisation of novel poly (vinyl alcohol)/poly (vinyl pyrrolidone)-based hydrogel-forming microneedle arrays for enhanced and sustained transdermal delivery of methotrexate. *Int. J. Pharm.* **2020**, *586*, 119580. [CrossRef] [PubMed]
156. Baykara, D.; Bedir, T.; Ilhan, E.; Mutlu, M.E.; Gunduz, O.; Narayan, R.; Ustundag, C.B. Fabrication and optimization of 3D printed gelatin methacryloyl microneedle arrays based on vat photopolymerization. *Front. Bioeng. Biotechnol.* **2023**, *11*, 1157541. [CrossRef] [PubMed]
157. Fonseca, D.F.; Costa, P.C.; Almeida, I.F.; Dias-Pereira, P.; Correia-Sá, I.; Bastos, V.; Oliveira, H.; Vilela, C.; Silvestre, A.J.; Freire, C.S. Swellable gelatin methacryloyl microneedles for extraction of interstitial skin fluid toward minimally invasive monitoring of urea. *Macromol. Biosci.* **2020**, *20*, 2000195. [CrossRef]
158. Yang, J.; Yang, J.; Gong, X.; Zheng, Y.; Yi, S.; Cheng, Y.; Li, Y.; Liu, B.; Xie, X.; Yi, C. Recent progress in microneedles-mediated diagnosis, therapy, and theranostic systems. *Adv. Healthc. Mater.* **2022**, *11*, 2102547. [CrossRef]
159. Ju, J.; Hsieh, C.-M.; Tian, Y.; Kang, J.; Chia, R.; Chang, H.; Bai, Y.; Xu, C.; Wang, X.; Liu, Q. Surface enhanced Raman spectroscopy based biosensor with a microneedle array for minimally invasive in vivo glucose measurements. *ACS Sens.* **2020**, *5*, 1777–1785. [CrossRef]
160. Xu, W.; Zhang, M.; Du, W.; Ling, G.; Yuan, Y.; Zhang, P. Engineering a naturally-derived wound dressing based on bio-ionic liquid conjugation. *Eur. Polym. J.* **2023**, *191*, 112055. [CrossRef]
161. Qiang, N.; Liu, Z.; Lu, M.; Yang, Y.; Liao, F.; Feng, Y.; Liu, G.; Qiu, S. Preparation and Properties of Polyvinylpyrrolidone/Sodium Carboxymethyl Cellulose Soluble Microneedles. *Materials* **2023**, *16*, 3417. [CrossRef]
162. Hawkins, N. Thermal Solutions Separation of Free and Bound Water in Pharmaceuticals. TS-17A. Available online: https://www.tainstruments.com/pdf/literature/TS17.pdf (accessed on 24 July 2023).
163. Ling, M.-H.; Chen, M.-C. Dissolving polymer microneedle patches for rapid and efficient transdermal delivery of insulin to diabetic rats. *Acta Biomater.* **2013**, *9*, 8952–8961. [CrossRef]
164. Makvandi, P.; Kirkby, M.; Hutton, A.R.; Shabani, M.; Yiu, C.K.; Baghbantaraghdari, Z.; Jamaledin, R.; Carlotti, M.; Mazzolai, B.; Mattoli, V. Engineering microneedle patches for improved penetration: Analysis, skin models and factors affecting needle insertion. *Nano-Micro Lett.* **2021**, *13*, 1–41.
165. Mönkäre, J.; Nejadnik, M.R.; Baccouche, K.; Romeijn, S.; Jiskoot, W.; Bouwstra, J.A. IgG-loaded hyaluronan-based dissolving microneedles for intradermal protein delivery. *J. Control. Release* **2015**, *218*, 53–62. [CrossRef] [PubMed]
166. Larrañeta, E.; Lutton, R.E.; Woolfson, A.D.; Donnelly, R.F. Microneedle arrays as transdermal and intradermal drug delivery systems: Materials science, manufacture and commercial development. *Mater. Sci. Eng. R Rep.* **2016**, *104*, 1–32.
167. Zahn, J.D.; Talbot, N.H.; Liepmann, D.; Pisano, A.P. Microfabricated polysilicon microneedles for minimally invasive biomedical devices. *Biomed. Microdevices* **2000**, *2*, 295–303. [CrossRef]
168. Champeau, M.; Jary, D.; Mortier, L.; Mordon, S.; Vignoud, S. A facile fabrication of dissolving microneedles containing 5-aminolevulinic acid. *Int. J. Pharm.* **2020**, *586*, 119554. [CrossRef] [PubMed]
169. Donnelly, R.F.; Majithiya, R.; Singh, T.R.R.; Morrow, D.I.; Garland, M.J.; Demir, Y.K.; Migalska, K.; Ryan, E.; Gillen, D.; Scott, C.J. Design, optimization and characterisation of polymeric microneedle arrays prepared by a novel laser-based micromoulding technique. *Pharm. Res.* **2011**, *28*, 41–57. [CrossRef] [PubMed]
170. Faraji Rad, Z.; Nordon, R.E.; Anthony, C.J.; Bilston, L.; Prewett, P.D.; Arns, J.-Y.; Arns, C.H.; Zhang, L.; Davies, G.J. High-fidelity replication of thermoplastic microneedles with open microfluidic channels. *Microsyst. Nanoeng.* **2017**, *3*, 17034. [CrossRef]
171. Chen, Y.; Xian, Y.; Carrier, A.J.; Youden, B.; Servos, M.; Cui, S.; Luan, T.; Lin, S.; Zhang, X. A simple and cost-effective approach to fabricate tunable length polymeric microneedle patches for controllable transdermal drug delivery. *RSC Adv.* **2020**, *10*, 15541–15546. [CrossRef]
172. Wang, Q.L.; Zhu, D.D.; Liu, X.B.; Chen, B.Z.; Guo, X.D. Microneedles with controlled bubble sizes and drug distributions for efficient transdermal drug delivery. *Sci. Rep.* **2016**, *6*, 38755. [CrossRef]
173. Park, Y.; Kim, B. Skin permeability of compounds loaded within dissolving microneedles dependent on composition of sodium hyaluronate and carboxymethyl cellulose. *Korean J. Chem. Eng.* **2017**, *34*, 133–138. [CrossRef]
174. Onesto, V.; Di Natale, C.; Profeta, M.; Netti, P.A.; Vecchione, R. Engineered PLGA-PVP/VA based formulations to produce electro-drawn fast biodegradable microneedles for labile biomolecule delivery. *Prog. Biomater.* **2020**, *9*, 203–217. [CrossRef]
175. Kim, J.D.; Kim, M.; Yang, H.; Lee, K.; Jung, H. Droplet-born air blowing: Novel dissolving microneedle fabrication. *J. Control. Release* **2013**, *170*, 430–436. [CrossRef] [PubMed]

176. Park, J.-H.; Allen, M.G.; Prausnitz, M.R. Biodegradable polymer microneedles: Fabrication, mechanics and transdermal drug delivery. *J. Control. Release* **2005**, *104*, 51–66. [CrossRef]
177. Olatunji, O.; Das, D.B.; Garland, M.J.; Belaid, L.; Donnelly, R.F. Influence of array interspacing on the force required for successful microneedle skin penetration: Theoretical and practical approaches. *J. Pharm. Sci.* **2013**, *102*, 1209–1221. [CrossRef] [PubMed]
178. Kathuria, H.; Kang, K.; Cai, J.; Kang, L. Rapid microneedle fabrication by heating and photolithography. *Int. J. Pharm.* **2020**, *575*, 118992. [CrossRef]
179. Luo, Z.; Sun, W.; Fang, J.; Lee, K.; Li, S.; Gu, Z.; Dokmeci, M.R.; Khademhosseini, A. Biodegradable gelatin methacryloyl microneedles for transdermal drug delivery. *Adv. Healthc. Mater.* **2019**, *8*, 1801054. [CrossRef]
180. Leone, M.; Monkare, J.; Bouwstra, J.A.; Kersten, G. Dissolving Microneedle Patches for Dermal Vaccination. *Pharm. Res.* **2017**, *34*, 2223–2240. [CrossRef] [PubMed]
181. Chen, S.-X.; Ma, M.; Xue, F.; Shen, S.; Chen, Q.; Kuang, Y.; Liang, K.; Wang, X.; Chen, H. Construction of microneedle-assisted co-delivery platform and its combining photodynamic/immunotherapy. *J. Control. Release* **2020**, *324*, 218–227. [CrossRef]
182. Zhu, D.D.; Chen, B.Z.; He, M.C.; Guo, X.D. Structural optimization of rapidly separating microneedles for efficient drug delivery. *J. Ind. Eng. Chem.* **2017**, *51*, 178–184. [CrossRef]
183. Davis, S.P.; Landis, B.J.; Adams, Z.H.; Allen, M.G.; Prausnitz, M.R. Insertion of microneedles into skin: Measurement and prediction of insertion force and needle fracture force. *J. Biomech.* **2004**, *37*, 1155–1163. [CrossRef]
184. Roxhed, N.; Gasser, T.C.; Griss, P.; Holzapfel, G.A.; Stemme, G. Penetration-enhanced ultrasharp microneedles and prediction on skin interaction for efficient transdermal drug delivery. *J. Microelectromech. Syst.* **2007**, *16*, 1429–1440. [CrossRef]
185. Gomaa, Y.A.; Morrow, D.I.; Garland, M.J.; Donnelly, R.F.; El-Khordagui, L.K.; Meidan, V.M. Effects of microneedle length, density, insertion time and multiple applications on human skin barrier function: Assessments by transepidermal water loss. *Toxicol. Vitr.* **2010**, *24*, 1971–1978. [CrossRef] [PubMed]
186. Larrañeta, E.; Moore, J.; Vicente-Pérez, E.M.; González-Vázquez, P.; Lutton, R.; Woolfson, A.D.; Donnelly, R.F. A proposed model membrane and test method for microneedle insertion studies. *Int. J. Pharm.* **2014**, *472*, 65–73. [CrossRef] [PubMed]
187. Vora, L.K.; Courtenay, A.J.; Tekko, I.A.; Larrañeta, E.; Donnelly, R.F. Pullulan-based dissolving microneedle arrays for enhanced transdermal delivery of small and large biomolecules. *Int. J. Biol. Macromol.* **2020**, *146*, 290–298. [CrossRef] [PubMed]
188. Flaten, G.E.; Palac, Z.; Engesland, A.; Filipović-Grčić, J.; Vanić, Ž.; Škalko-Basnet, N. In vitro skin models as a tool in optimization of drug formulation. *Eur. J. Pharm. Sci.* **2015**, *75*, 10–24.
189. McGrath, M.G.; Vucen, S.; Vrdoljak, A.; Kelly, A.; O'Mahony, C.; Crean, A.M.; Moore, A. Production of dissolvable microneedles using an atomised spray process: Effect of microneedle composition on skin penetration. *Eur. J. Pharm. Biopharm.* **2014**, *86*, 200–211. [CrossRef]
190. Zhu, J.; Zhou, X.; Kim, H.J.; Qu, M.; Jiang, X.; Lee, K.; Ren, L.; Wu, Q.; Wang, C.; Zhu, X. Gelatin methacryloyl microneedle patches for minimally invasive extraction of skin interstitial fluid. *Small* **2020**, *16*, 1905910. [CrossRef]
191. Koelmans, W.; Krishnamoorthy, G.; Heskamp, A.; Wissink, J.; Misra, S.; Tas, N. Microneedle characterization using a double-layer skin simulant. *Mech. Eng. Res.* **2013**, *3*, 51. [CrossRef]
192. Bonfante, G.; Lee, H.; Bao, L.; Park, J.; Takama, N.; Kim, B. Comparison of polymers to enhance mechanical properties of microneedles for bio-medical applications. *Micro Nano Syst. Lett.* **2020**, *8*, 13.
193. Zheng, L.; Zhu, D.; Xiao, Y.; Zheng, X.; Chen, P. Microneedle coupled epidermal sensor for multiplexed electrochemical detection of kidney disease biomarkers. *Biosens. Bioelectron.* **2023**, *237*, 115506. [CrossRef]
194. Ranamukhaarachchi, S.A.; Schneider, T.; Lehnert, S.; Sprenger, L.; Campbell, J.R.; Mansoor, I.; Lai, J.C.; Rai, K.; Dutz, J.; Häfeli, U.O. Development and validation of an artificial mechanical skin model for the study of interactions between skin and microneedles. *Macromol. Mater. Eng.* **2016**, *301*, 306–314. [CrossRef]
195. Sakamoto, M.; Hasegawa, Y.; Shikida, M. Development of spear-shaped microneedle and applicator for tip insertion into artificial skin. *Microsyst. Technol.* **2021**, *27*, 3907–3916. [CrossRef]
196. Nishino, R.; Aoyagi, S.; Suzuki, M.; Ueda, A.; Okumura, Y.; Takahashi, T.; Hosomi, R.; Fukunaga, K.; Uta, D.; Takazawa, T. Development of artificial skin using keratin film for evaluation of puncture performance of microneedle. *J. Robot. Mechatron.* **2020**, *32*, 351–361. [CrossRef]
197. Wang, M.; Hu, L.; Xu, C. Recent advances in the design of polymeric microneedles for transdermal drug delivery and biosensing. *Lab. Chip.* **2017**, *17*, 1373–1387. [CrossRef] [PubMed]
198. Sharma, S.; Huang, Z.; Rogers, M.; Boutelle, M.; Cass, A.E. Evaluation of a minimally invasive glucose biosensor for continuous tissue monitoring. *Anal. Bioanal. Chem.* **2016**, *408*, 8427–8435. [CrossRef] [PubMed]
199. Evens, T.; Malek, O.; Castagne, S.; Seveno, D.; Van Bael, A. A novel method for producing solid polymer microneedles using laser ablated moulds in an injection moulding process. *Manuf. Lett.* **2020**, *24*, 29–32. [CrossRef]
200. Gülçür, M.; Romano, J.-M.; Penchev, P.; Gough, T.; Brown, E.; Dimov, S.; Whiteside, B. A cost-effective process chain for thermoplastic microneedle manufacture combining laser micro-machining and micro-injection moulding. *CIRP J. Manuf. Sci. Technol.* **2021**, *32*, 311–321. [CrossRef]
201. Faraji Rad, Z.; Prewett, P.D.; Davies, G.J. High-resolution two-photon polymerization: The most versatile technique for the fabrication of microneedle arrays. *Microsyst. Nanoeng.* **2021**, *7*, 71. [CrossRef]
202. Ghanbariamin, D.; Samandari, M.; Ghelich, P.; Shahbazmohamadi, S.; Schmidt, T.A.; Chen, Y.; Tamayol, A. Cleanroom-Free Fabrication of Microneedles for Multimodal Drug Delivery. *Small* **2023**, *19*, 2207131. [CrossRef]

203. Krieger, K.J.; Bertollo, N.; Dangol, M.; Sheridan, J.T.; Lowery, M.M.; O'Cearbhaill, E.D. Simple and customizable method for fabrication of high-aspect ratio microneedle molds using low-cost 3D printing. *Microsyst. Nanoeng.* **2019**, *5*, 42. [CrossRef]
204. Balmert, S.C.; Carey, C.D.; Falo, G.D.; Sethi, S.K.; Erdos, G.; Korkmaz, E.; Falo, L.D., Jr. Dissolving undercut microneedle arrays for multicomponent cutaneous vaccination. *J. Control. Release* **2020**, *317*, 336–346. [CrossRef]
205. He, C.; Chen, X.; Sun, Y.; Xie, M.; Yu, K.; He, J.; Lu, J.; Gao, Q.; Nie, J.; Wang, Y. Rapid and mass manufacturing of soft hydrogel microstructures for cell patterns assisted by 3D printing. *Bio-Des. Manuf.* **2022**, *5*, 641–659. [CrossRef]
206. Cordeiro, A.S.; Tekko, I.A.; Jomaa, M.H.; Vora, L.; McAlister, E.; Volpe-Zanutto, F.; Nethery, M.; Baine, P.T.; Mitchell, N.; McNeill, D.W. Two-photon polymerisation 3D printing of microneedle array templates with versatile designs: Application in the development of polymeric drug delivery systems. *Pharm. Res.* **2020**, *37*, 1–15.
207. Johnson, A.R.; Procopio, A.T. Low cost additive manufacturing of microneedle masters. *3D Print. Med.* **2019**, *5*, 1–10. [CrossRef] [PubMed]
208. Johnson, A.R.; Caudill, C.L.; Tumbleston, J.R.; Bloomquist, C.J.; Moga, K.A.; Ermoshkin, A.; Shirvanyants, D.; Mecham, S.J.; Luft, J.C.; DeSimone, J.M. Single-step fabrication of computationally designed microneedles by continuous liquid interface production. *PLoS ONE* **2016**, *11*, e0162518.
209. Arshad, M.S.; Fatima, S.; Nazari, K.; Ali, R.; Farhan, M.; Muhammad, S.A.; Abbas, N.; Hussain, A.; Kucuk, I.; Chang, M.-W. Engineering and characterisation of BCG-loaded polymeric microneedles. *J. Drug Target.* **2020**, *28*, 525–532. [CrossRef]
210. Permana, A.D.; Paredes, A.J.; Volpe-Zanutto, F.; Anjani, Q.K.; Utomo, E.; Donnelly, R.F. Dissolving microneedle-mediated dermal delivery of itraconazole nanocrystals for improved treatment of cutaneous candidiasis. *Eur. J. Pharm. Biopharm.* **2020**, *154*, 50–61. [CrossRef]
211. Allen, E.A.; O'Mahony, C.; Cronin, M.; O'Mahony, T.; Moore, A.C.; Crean, A.M. Dissolvable microneedle fabrication using piezoelectric dispensing technology. *Int. J. Pharm.* **2016**, *500*, 1–10. [CrossRef]
212. Larraneta, E.; Lutton, R.E.; Brady, A.J.; Vicente-Pérez, E.M.; Woolfson, A.D.; Thakur, R.R.S.; Donnelly, R.F. Microwave-assisted preparation of hydrogel-forming microneedle arrays for transdermal drug delivery applications. *Macromol. Mater. Eng.* **2015**, *300*, 586–595. [CrossRef]
213. Tarbox, T.N.; Watts, A.B.; Cui, Z.; Williams, R.O. An update on coating/manufacturing techniques of microneedles. *Drug Deliv. Transl. Res.* **2018**, *8*, 1828–1843.
214. Avcil, M.; Akman, G.; Klokkers, J.; Jeong, D.; Çelik, A. Clinical efficacy of dissolvable microneedles armed with anti-melanogenic compounds to counter hyperpigmentation. *J. Cosmet. Dermatol.* **2021**, *20*, 605–614. [CrossRef]
215. Yang, H.; Kim, S.; Kang, G.; Lahiji, S.F.; Jang, M.; Kim, Y.M.; Kim, J.M.; Cho, S.N.; Jung, H. Centrifugal lithography: Self-shaping of polymer microstructures encapsulating biopharmaceutics by centrifuging polymer drops. *Adv. Healthc. Mater.* **2017**, *6*, 1700326. [CrossRef] [PubMed]
216. Kang, G.; Tu, T.; Kim, S.; Yang, H.; Jang, M.; Jo, D.; Ryu, J.; Baek, J.; Jung, H. Adenosine-loaded dissolving microneedle patches to improve skin wrinkles, dermal density, elasticity and hydration. *Int. J. Cosmet. Sci.* **2018**, *40*, 199–206. [CrossRef] [PubMed]
217. Huh, I.; Kim, S.; Yang, H.; Jang, M.; Kang, G.; Jung, H. Effects of two droplet-based dissolving microneedle manufacturing methods on the activity of encapsulated epidermal growth factor and ascorbic acid. *Eur. J. Pharm. Sci.* **2018**, *114*, 285–292. [CrossRef] [PubMed]
218. Lee, K.; Jung, H. Drawing lithography for microneedles: A review of fundamentals and biomedical applications. *Biomaterials* **2012**, *33*, 7309–7326. [CrossRef]
219. Dangol, M.; Kim, S.; Li, C.G.; Lahiji, S.F.; Jang, M.; Ma, Y.; Huh, I.; Jung, H. Anti-obesity effect of a novel caffeine-loaded dissolving microneedle patch in high-fat diet-induced obese C57BL/6J mice. *J. Control. Release* **2017**, *265*, 41–47. [CrossRef] [PubMed]
220. Vecchione, R.; Coppola, S.; Esposito, E.; Casale, C.; Vespini, V.; Grilli, S.; Ferraro, P.; Netti, P.A. Electro-drawn drug-loaded biodegradable polymer microneedles as a viable route to hypodermic injection. *Adv. Funct. Mater.* **2014**, *24*, 3515–3523. [CrossRef]
221. Detamornrat, U.; McAlister, E.; Hutton, A.R.; Larrañeta, E.; Donnelly, R.F. The role of 3D printing technology in microengineering of microneedles. *Small* **2022**, *18*, 2106392. [CrossRef] [PubMed]
222. Rajesh, N.U.; Coates, I.; Driskill, M.M.; Dulay, M.T.; Hsiao, K.; Ilyin, D.; Jacobson, G.B.; Kwak, J.W.; Lawrence, M.; Perry, J. 3D-printed microarray patches for transdermal applications. *JACS Au* **2022**, *2*, 2426–2445. [CrossRef] [PubMed]
223. Boehm, R.D.; Miller, P.R.; Daniels, J.; Stafslien, S.; Narayan, R.J. Inkjet printing for pharmaceutical applications. *Mater. Today* **2014**, *17*, 247–252. [CrossRef]
224. Erkus, H.; Bedir, T.; Kaya, E.; Tinaz, G.B.; Gunduz, O.; Chifiriuc, M.-C.; Ustundag, C.B. Innovative transdermal drug delivery system based on amoxicillin-loaded gelatin methacryloyl microneedles obtained by 3D printing. *Materialia* **2023**, *27*, 101700. [CrossRef]
225. Shin, D.; Hyun, J. Silk fibroin microneedles fabricated by digital light processing 3D printing. *J. Ind. Eng. Chem.* **2021**, *95*, 126–133. [CrossRef]
226. Quinn, H.L.; Bonham, L.; Hughes, C.M.; Donnelly, R.F. Design of a dissolving microneedle platform for transdermal delivery of a fixed-dose combination of cardiovascular drugs. *J. Pharm. Sci.* **2015**, *104*, 3490–3500. [CrossRef] [PubMed]
227. FDA. *Use of International Standard ISO 10993-1, "Biological Evaluation of Medical Devices—Part 1: Evaluation and Testing within a Risk Management Process". Guidance for Industry and Food and Drug Administration Staff*; Center for Devices and Radiological Health, Food and Drug Administration: Silver Spring, MD, USA, 2023.

228. Zhang, X.P.; Wang, B.B.; Li, W.X.; Fei, W.M.; Cui, Y.; Guo, X.D. In vivo safety assessment, biodistribution and toxicology of polyvinyl alcohol microneedles with 160-day uninterruptedly applications in mice. *Eur. J. Pharm. Biopharm.* **2021**, *160*, 1–8. [CrossRef] [PubMed]
229. McCrudden, M.T.; Alkilani, A.Z.; McCrudden, C.M.; McAlister, E.; McCarthy, H.O.; Woolfson, A.D.; Donnelly, R.F. Design and physicochemical characterisation of novel dissolving polymeric microneedle arrays for transdermal delivery of high dose, low molecular weight drugs. *J. Control. Release* **2014**, *180*, 71–80. [CrossRef]
230. Kim, J.; Jeong, D. Dissolvable microneedles: Applications and opportunities. *ONdrugDelivery Mag.* **2018**, *84*, 24–29.
231. Mineedtech. Mineed Tech Publications. Available online: https://mineed.tech/publications/ (accessed on 3 August 2023).
232. Hiraishi, Y.; Nakagawa, T.; Quan, Y.-S.; Kamiyama, F.; Hirobe, S.; Okada, N.; Nakagawa, S. Performance and characteristics evaluation of a sodium hyaluronate-based microneedle patch for a transcutaneous drug delivery system. *Int. J. Pharm.* **2013**, *441*, 570–579. [CrossRef]
233. Microneedles, I. Production Technology. Available online: https://microneedle.tech/ (accessed on 3 August 2023).
234. ICH. Q1A (R2) Stability Testing of New Drug Substances and Products. 2003, pp. 1–25. Docket Number: FDA-2002-D-0222. Available online: https://www.fda.gov/regulatory-information/search-fda-guidance-documents/q1ar2-stability-testing-new-drug-substances-and-products (accessed on the 4 August 2023).
235. Leone, M.; Priester, M.I.; Romeijn, S.; Nejadnik, M.R.; Mönkäre, J.; O'Mahony, C.; Jiskoot, W.; Kersten, G.; Bouwstra, J.A. Hyaluronan-based dissolving microneedles with high antigen content for intradermal vaccination: Formulation, physicochemical characterization and immunogenicity assessment. *Eur. J. Pharm. Biopharm.* **2019**, *134*, 49–59. [CrossRef]
236. Wang, Q.L.; Ren, J.W.; Chen, B.Z.; Jin, X.; Zhang, C.Y.; Guo, X.D. Effect of humidity on mechanical properties of dissolving microneedles for transdermal drug delivery. *J. Ind. Eng. Chem.* **2018**, *59*, 251–258. [CrossRef]
237. McAlister, E.; Kearney, M.-C.; Martin, E.L.; Donnelly, R.F. From the laboratory to the end-user: A primary packaging study for microneedle patches containing amoxicillin sodium. *Drug Deliv. Transl. Res.* **2021**, *11*, 2169–2185. [CrossRef]
238. Anjani, Q.K.; Cárcamo-Martínez, Á.; Wardoyo, L.A.H.; Moreno-Castellanos, N.; Sabri, A.H.B.; Larrañeta, E.; Donnelly, R.F. MAP-box: A novel, low-cost and easy-to-fabricate 3D-printed box for the storage and transportation of dissolving microneedle array patches. *Drug Deliv. Transl. Res.* **2023**, 1–15. [CrossRef]
239. Dalvi, M.; Kharat, P.; Thakor, P.; Bhavana, V.; Singh, S.B.; Mehra, N.K. Panorama of dissolving microneedles for transdermal drug delivery. *Life Sci.* **2021**, *284*, 119877. [PubMed]
240. Kulkarni, D.; Gadade, D.; Chapaitkar, N.; Shelke, S.; Pekamwar, S.; Aher, R.; Ahire, A.; Avhale, M.; Badgule, R.; Bansode, R. Polymeric Microneedles: An Emerging Paradigm for Advanced Biomedical Applications. *Sci. Pharm.* **2023**, *91*, 27.
241. Barbucci, R.; Lamponi, S.; Borzacchiello, A.; Ambrosio, L.; Fini, M.; Torricelli, P.; Giardino, R. Hyaluronic acid hydrogel in the treatment of osteoarthritis. *Biomaterials* **2002**, *23*, 4503–4513. [CrossRef] [PubMed]
242. Huerta-Angeles, G.; Nesporova, K.; Ambrozova, G.; Kubala, L.; Velebny, V. An Effective Translation: The Development of Hyaluronan-Based Medical Products From the Physicochemical, and Preclinical Aspects. *Front. Bioeng. Biotechnol.* **2018**, *6*, 62. [CrossRef]
243. McCrudden, M.T.; Alkilani, A.Z.; Courtenay, A.J.; McCrudden, C.M.; McCloskey, B.; Walker, C.; Alshraiedeh, N.; Lutton, R.E.; Gilmore, B.F.; Woolfson, A.D. Considerations in the sterile manufacture of polymeric microneedle arrays. *Drug Deliv. Transl. Res.* **2015**, *5*, 3–14. [CrossRef]
244. Castilla-Casadiego, D.A.; Miranda-Muñoz, K.A.; Roberts, J.L.; Crowell, A.D.; Gonzalez-Nino, D.; Choudhury, D.; Aparicio-Solis, F.O.; Servoss, S.L.; Rosales, A.M.; Prinz, G. Biodegradable microneedle patch for delivery of meloxicam for managing pain in cattle. *PLoS ONE* **2022**, *17*, e0272169.
245. Swathi, H.P.; Anusha Matadh, V.; Paul Guin, J.; Narasimha Murthy, S.; Kanni, P.; Varshney, L.; Suresh, S.; Shivakumar, H.N. Effect of gamma sterilization on the properties of microneedle array transdermal patch system. *Drug Dev. Ind. Pharm.* **2020**, *46*, 606–620. [CrossRef]
246. Kim, S.; Lee, J.; Shayan, F.L.; Kim, S.; Huh, I.; Ma, Y.; Yang, H.; Kang, G.; Jung, H. Physicochemical study of ascorbic acid 2-glucoside loaded hyaluronic acid dissolving microneedles irradiated by electron beam and gamma ray. *Carbohydr. Polym.* **2018**, *180*, 297–303. [CrossRef]
247. Waghule, T.; Singhvi, G.; Dubey, S.K.; Pandey, M.M.; Gupta, G.; Singh, M.; Dua, K. Microneedles: A smart approach and increasing potential for transdermal drug delivery system. *Biomed. Pharmacother.* **2019**, *109*, 1249–1258.
248. Kirkby, M.; Hutton, A.R.; Donnelly, R.F. Microneedle mediated transdermal delivery of protein, peptide and antibody based therapeutics: Current status and future considerations. *Pharm. Res.* **2020**, *37*, 117.
249. Donnelly, R.F.; Singh, T.R.R.; Alkilani, A.Z.; McCrudden, M.T.; O'Neill, S.; O'Mahony, C.; Armstrong, K.; McLoone, N.; Kole, P.; Woolfson, A.D. Hydrogel-forming microneedle arrays exhibit antimicrobial properties: Potential for enhanced patient safety. *Int. J. Pharm.* **2013**, *451*, 76–91. [CrossRef] [PubMed]
250. Chi, J.; Zhang, X.; Chen, C.; Shao, C.; Zhao, Y.; Wang, Y. Antibacterial and angiogenic chitosan microneedle array patch for promoting wound healing. *Bioact. Mater.* **2020**, *5*, 253–259. [CrossRef] [PubMed]
251. Donnelly, R.F.; Singh, T.R.R.; Tunney, M.M.; Morrow, D.I.; McCarron, P.A.; O'Mahony, C.; Woolfson, A.D. Microneedle arrays allow lower microbial penetration than hypodermic needles in vitro. *Pharm. Res.* **2009**, *26*, 2513–2522. [CrossRef] [PubMed]

252. Vicente-Perez, E.M.; Larrañeta, E.; McCrudden, M.T.; Kissenpfennig, A.; Hegarty, S.; McCarthy, H.O.; Donnelly, R.F. Repeat application of microneedles does not alter skin appearance or barrier function and causes no measurable disturbance of serum biomarkers of infection, inflammation or immunity in mice in vivo. *Eur. J. Pharm. Biopharm.* **2017**, *117*, 400–407. [CrossRef]
253. Al-Kasasbeh, R.; Brady, A.J.; Courtenay, A.J.; Larrañeta, E.; McCrudden, M.T.; O'Kane, D.; Liggett, S.; Donnelly, R.F. Evaluation of the clinical impact of repeat application of hydrogel-forming microneedle array patches. *Drug Deliv. Transl. Res.* **2020**, *10*, 690–705. [CrossRef]
254. Radius Health, Inc. FORM 10-Q Quarterly Report Pursuant to Sections 13 or 15(d). 2022, Volume 153. Available online: http://edgar.secdatabase.com/2126/162828022021751/filing-main.htm (accessed on 24 July 2023).
255. Taylor, N.P. No-Go for Zosano as FDA Knockback Triggers Suspension of Transdermal Migraine Patch Program. Available online: https://www.fiercepharma.com/pharma/no-go-zosano-fda-knockback-triggers-suspension-transdermal-migraine-patch-program (accessed on 9 August 2023).
256. NIH. *Randomized, Double-Blind, Active Placebo-Controlled Study of Ketamine to Treat Levodopa-Induced Dyskinesia*; ClinicalTrials.gov Identifier: NCT04912115; NIH: Bethesda, MD, USA, 2022. Available online: https://classic.clinicaltrials.gov/ct2/show/NCT04912115 (accessed on 9 August 2023).

**Disclaimer/Publisher's Note:** The statements, opinions and data contained in all publications are solely those of the individual author(s) and contributor(s) and not of MDPI and/or the editor(s). MDPI and/or the editor(s) disclaim responsibility for any injury to people or property resulting from any ideas, methods, instructions or products referred to in the content.

Review

# Advancements and Applications of Injectable Hydrogel Composites in Biomedical Research and Therapy

Hossein Omidian * and Sumana Dey Chowdhury

Barry and Judy Silverman College of Pharmacy, Nova Southeastern University, Fort Lauderdale, FL 33328, USA; sd2236@mynsu.nova.edu
* Correspondence: omidian@nova.edu

**Abstract:** Injectable hydrogels have gained popularity for their controlled release, targeted delivery, and enhanced mechanical properties. They hold promise in cardiac regeneration, joint diseases, postoperative analgesia, and ocular disorder treatment. Hydrogels enriched with nano-hydroxyapatite show potential in bone regeneration, addressing challenges of bone defects, osteoporosis, and tumor-associated regeneration. In wound management and cancer therapy, they enable controlled release, accelerated wound closure, and targeted drug delivery. Injectable hydrogels also find applications in ischemic brain injury, tissue regeneration, cardiovascular diseases, and personalized cancer immunotherapy. This manuscript highlights the versatility and potential of injectable hydrogel nanocomposites in biomedical research. Moreover, it includes a perspective section that explores future prospects, emphasizes interdisciplinary collaboration, and underscores the promising future potential of injectable hydrogel nanocomposites in biomedical research and applications.

**Keywords:** injectable hydrogel nanocomposites; biomedical applications; controlled release; tissue engineering; therapeutic outcomes

Citation: Omidian, H.; Chowdhury, S.D. Advancements and Applications of Injectable Hydrogel Composites in Biomedical Research and Therapy. *Gels* 2023, 9, 533. https://doi.org/10.3390/gels9070533

Academic Editors: Diana Silva, Ana Paula Serro and María Vivero-Lopez

Received: 30 May 2023
Revised: 23 June 2023
Accepted: 28 June 2023
Published: 30 June 2023

**Copyright:** © 2023 by the authors. Licensee MDPI, Basel, Switzerland. This article is an open access article distributed under the terms and conditions of the Creative Commons Attribution (CC BY) license (https://creativecommons.org/licenses/by/4.0/).

## 1. Introduction

Injectable hydrogel systems have witnessed noteworthy progress in recent years within the realm of biomedical applications. These biomaterials present a multitude of advantages, including controlled release, targeted delivery, and enhanced mechanical properties [1]. Their potential has been demonstrated in various therapeutic areas, such as cardiac regeneration [1], joint diseases [2], postoperative analgesia [3], and the treatment of ocular disorders [4]. Particularly in the field of tissue engineering, hydrogels play a vital role by promoting crucial aspects such as cell viability, adhesion, differentiation, and host integration [5]. By mimicking the native extracellular matrix in bone and cartilage tissue engineering, hydrogels provide a biocompatible and regenerative environment [5]. Notably, injectable hydrogels enriched with nano-hydroxyapatite have shown promise in bone regeneration [6]. Additionally, collagen-based hydrogels, carboxymethyl-chitosan gels, gelatin, and nano-hydroxyapatite exhibit potential for bone tissue engineering [7,8].

Addressing the clinical challenges associated with bone defects, osteoporosis, and tumor-related bone regeneration represents a significant aspect of injectable hydrogel systems. These biomaterials enhance bone formation and repair by promoting osteogenic differentiation, angiogenesis, and controlled release of bioactive molecules [9–27]. Various composite biomaterials and injectable hydrogels, such as the RADA16 peptide hydrogel with calcium sulfate/nano-hydroxyapatite cement and the GelMA-HAMA/nHAP composite hydrogel, have demonstrated efficacy in bone healing and regeneration [9,11]. Furthermore, injectable hydrogels incorporating bio-responsive drug-loaded nanoparticles exhibit dual functions and hold promise for targeted delivery, tumor suppression, and bone regeneration [17,18].

Injectable hydrogel systems also find application in wound management and therapy, as they facilitate controlled release of bioactive substances, accelerated wound closure, and tissue regeneration. The incorporation of nanoparticles further enhances their antibacterial activity and bioactivity, advancing the field of wound care [28,29].

In cancer therapy, injectable hydrogel systems offer promising solutions for targeted drug delivery, chemo-photothermal therapy, and combination therapies. These hydrogels' versatility, controlled-release capabilities, and biocompatibility render them ideal platforms for enhancing therapeutic outcomes [30–39]. By delivering therapeutic agents and enabling targeted treatments, injectable hydrogels have the potential to significantly advance cancer therapy and improve patient outcomes [40,41].

In addition to their aforementioned applications, injectable hydrogel systems hold great potential in several other areas of biomedical research. These include ischemic brain injury, tissue regeneration, cardiovascular diseases, and personalized cancer immunotherapy [42–45]. Notably, in the treatment of ischemic brain injury, injectable hydrogels have demonstrated the ability to facilitate neuronal proliferation, angiogenesis, and tissue regeneration, offering promising therapeutic prospects [42]. Similarly, in the realm of cardiovascular diseases, hydrogels combined with stem cells, nanoparticles, or genetic material have shown promise in improving cardiac function and promoting tissue regeneration [44]. Moreover, the integration of nanotechnology and biomaterials in injectable hydrogels has enabled the development of personalized cancer immunotherapy approaches, allowing for targeted and effective treatment modalities [46–50].

Recent studies have also focused on the development of injectable hydrogels for cartilage regeneration [51–53]. These hydrogels have shown the capacity to promote cell migration, chondrogenesis, and the expression of cartilage-specific genes, presenting potential solutions for cartilage tissue engineering. Furthermore, functionalization strategies such as kartogenin (KGN)-conjugated nanoparticles and chondroitin sulfate nanoparticles have been employed to enhance the regenerative capabilities of these hydrogels [51,52]. Additionally, the incorporation of nanocrystalline hydroxyapatite has been found to support cell viability and differentiation, further advancing the field of cartilage regeneration [53]. These recent advancements hold promise for improved therapies and treatments in various cartilage-related conditions.

This manuscript provides an overview of the recent advancements and applications of injectable hydrogel nanocomposites in various biomedical fields. It explores their crucial role in drug delivery, tissue engineering, bone regeneration, wound management, and cancer therapy, highlighting the significant contributions these biomaterials make to the advancement of biomedical research and clinical practice.

## 2. Drug Delivery

Over the past few years, hydrogel-based drug delivery systems have made significant strides in providing targeted and controlled release of drugs, thereby holding great potential for enhancing therapeutic outcomes in diverse medical conditions [1]. Hydrogels, with their wide range of properties and synthesis routes, play a crucial role in the realm of controlled drug delivery, rendering them suitable carriers in the field of medicine [54]. For instance, injectable nano-enabled thermogels have demonstrated promise in achieving controlled release of anti-angiogenic peptides for the treatment of ocular disorders, offering extended-release capabilities and biocompatibility [4].

Cardiac regeneration following myocardial infarction (MI) poses challenges due to limited regenerative capacity and scar formation [1]. Tissue engineering approaches, encompassing nano-carriers, controlled release matrices, injectable hydrogels, and cardiac patches, have shown promise in addressing this issue [1]. Similarly, improved drug delivery strategies are required for joint diseases such as osteoarthritis and rheumatoid arthritis. Advanced drug delivery systems, including nano- and microcarriers, have been developed to enhance efficacy and minimize side effects in these conditions [2]. Specific nano-drug delivery systems have been devised for postoperative analgesia, resulting in improved pain relief and better postoperative outcomes [3]. Innovative approaches, such as maleimide-functionalized polyethylene glycol hydrogels loaded with nanoparticles, exhibit potential in preventing cartilage degradation and inhibiting osteophyte formation in the treatment of post-traumatic osteoarthritis [55].

Researchers have explored various drug delivery systems, yielding promising results. For instance, nano hybrid silk hydrogel systems have been developed for localized and targeted delivery of anticancer drugs, demonstrating slow and sustained release as well as active targeting of cancer cells [40]. pH- and temperature-responsive hydrogels incorporating chondroitin sulfate nanogels have exhibited selective binding to lung carcinoma cells and effective inhibition of cancer cell growth [41]. Injectable hydrogel systems equipped with tumor-targeting nano-micelles and glutathione-responsive drug release have demonstrated significant tumor growth inhibition and enhanced antitumor efficacy [56]. These examples underscore the potential of hydrogel-based drug delivery systems in the realm of cancer treatment.

Moreover, hydrogel systems have shown promise in the domain of intravaginal drug delivery for gynecological drugs, contributing to improved drug solubility and distribution [57]. Intravesical liposome-in-gel systems have been developed to enhance drug retention in the bladder while reducing systemic levels, presenting a promising approach for intravesical applications [58]. Similarly, liposome-in-gel-paclitaxel systems have been investigated for regional delivery in chemoradiotherapy, resulting in increased cytotoxicity and reduction in tumor volume [59].

In recent studies, researchers have delved into the exploration of injectable thermosensitive photothermal-network hydrogels capable of near-infrared (NIR)-triggered drug delivery and synergistic photothermal-chemotherapy, showcasing remarkable efficacy in tumor eradication [60]. Additionally, smart and biomimetic 3D- and 4D-printed composite hydrogels have been investigated for their potential in tissue engineering and controlled drug release, offering versatile applications in the field [61]. Ultrasoft polymeric DNA networks (pDNets) with variable crystallinities have been developed as a means of controlled release of anticancer drugs, enabling efficient localized drug delivery and demonstrating significant antitumor efficacy [62]. Moreover, injectable thermoresponsive hydrogels based on graft copolymers have been synthesized to facilitate controlled drug delivery and promote bone cell growth, thus holding promise for applications in regenerative medicine [63]. Furthermore, injectable and degradable polysaccharide-based hydrogels embedded with nanoparticles have exhibited potential in drug delivery and bone tissue engineering endeavors [64].

Smart stimuli-responsive injectable gels and hydrogels have emerged as highly promising tools in the realm of drug delivery and tissue engineering [65]. These systems have shown considerable potential in delivering protein therapeutics for chronic and autoimmune diseases, exemplified by their successful administration of insulin for patients with conditions such as type 1 diabetes mellitus [66]. Moreover, pH-sensitive drug release systems utilizing folic-acid-conjugated graphene oxide have demonstrated targeted delivery of doxorubicin (DOX) in breast cancer therapy. These systems exhibit pH-responsive drug release, enhanced cytotoxicity in vitro, and significant reduction in tumor volume in animal studies, thereby showcasing their potential as effective treatment modalities [67].

Injectable hydrogel-based drug delivery systems have also yielded promising results in various medical applications. For instance, hydrogel nanomaterials employed in continuous subcutaneous insulin infusion (CSII) have shown improvements in blood glucose control and reduced therapeutic time in pediatric patients with type 1 diabetes mellitus [68]. Nanocomposite hydrogels have been investigated for their potential in promoting the healing of diabetic ulcers, capitalizing on their high drug loading capacity and stability, thus offering valuable therapeutic avenues [69]. Disease-responsive drug delivery systems have garnered attention for their improved targeting capabilities and controlled release profiles, contributing to advancements in the field of personalized medicine [70].

Researchers have made notable advancements in the development of injectable hydrogels with diverse properties and functionalities. For instance, an injectable, self-healing, and pH-responsive nanocomposite hydrogel has been devised, displaying potential applications in cancer therapy, wound healing, and infection treatment, thereby presenting a versatile platform for various medical interventions [71]. Similarly, an injectable liquid metal nanoflake hydrogel has been designed to enhance postsurgical tumor recurrence suppression, displaying long-term antitumor effects while mitigating systemic toxicity [72]. Furthermore, a bio-inspired fluorescent nano-injectable hydrogel with sequential drug release capabilities has been proposed, offering potential applications in visualization and dual drug delivery, thus opening new avenues for therapeutic interventions [73]. Lastly, an injectable micromotor@hydrogel drug delivery system has been developed for antibacterial therapy, exhibiting antibacterial effects and enhanced activity near bacteria without the need for exogenous hydrogen peroxide, showcasing its potential in combating bacterial infections [74].

Other research studies have concentrated on specific therapeutic domains, aiming to address specific medical needs. To facilitate angiogenic drug delivery, a nano polydopamine (PDA) crosslinked thiol-functionalized hyaluronic acid (HA) hydrogel has been developed, showcasing an injectable nature and sustained drug release capabilities [75]. For the repair of spinal cord injuries (SCI), drug delivery systems have been optimized utilizing sustained-release microspheres loaded with melatonin (Mel) and Laponite hydrogels, enabling stable and prolonged release for neural function restoration [76].

In the localized treatment of non-small cell lung cancer (NSCLC), an injectable thermosensitive hydrogel composed of the amphiphilic triblock copolymer poly(d, l-lactide)-poly(ethylene glycol)-poly(d, l-lactide) loaded with erlotinib-loaded hollow mesoporous silica nanoparticles (ERT@HMSNs/gel) has exhibited promising outcomes (Figure 1) [77]. This investigation involved determining the sol-to-gel transition temperature, which corresponds to the body temperature, as well as evaluating in vitro drug release from HMSNs with or without the gel. Moreover, an in vivo comparative study of different ERT formulations and the commercially available drug Tarceva was conducted using NSCLC xenograft models (Figure 2). The findings revealed that the ERT-loaded HMSNs/gel system displays substantial potential as an in-situ treatment approach for NSCLC, offering prolonged drug retention along with efficacy and safety. Similarly, an injectable silk fibroin nanofiber hydrogel for vancomycin delivery has demonstrated exceptional antibacterial properties and biocompatibility [78].

Polysaccharide-based nanoscale drug delivery systems have garnered significant attention within the field of tissue engineering due to their potential applications. These systems leverage the unique properties of polysaccharides to mimic the extracellular matrix and modulate cellular functions [79]. A recent perspective article discussed the clinical relevance and future prospects of an injectable hydrogel and nanoparticle system designed for microRNA (miR) delivery to the heart, emphasizing its potential in cardiac therapy [80]. In another study, researchers explored a composite hydrogel incorporating P24 peptide-loaded microspheres and nano-hydroxyapatite, demonstrating promising outcomes for bone regeneration in tissue engineering [81]. Similarly, a thermosensitive injectable hydrogel based on PLGA-PEG-PLGA copolymer, integrated with hydroxyapatite particles, showcased controlled release of calcium cations, thus exhibiting potential for calcium delivery in bone regeneration [82].

Silica microparticles encapsulating triptorelin acetate formed an injectable depot with sustained release characteristics, exhibiting pharmacodynamic effects comparable to those of commercially available products [83]. The controlled release of bone morphogenetic protein-2 (BMP-2) from a three-dimensional tissue-engineered nano-scaffold resulted in significant ectopic bone formation, further underscoring its potential for tissue regeneration [84]. A multifunctional sustainable delivery system for calcitriol, employing a thermosensitive hydrogel, nano-hydroxyapatite, and calcitriol-loaded micelles, stimulated osteogenesis and bone regeneration both in vitro and in vivo, displaying low cytotoxicity and an appropriate degradation rate [85]. Additionally, a thermosensitive micellar hydrogel composed of PELT triblock copolymer exhibited potential as an injectable nanomedicine reservoir and platform for co-delivery of therapeutic agents [86]. Lastly, a macroscale thermosensitive micellar-hydrogel depot demonstrated sustained drug release, stable immobilization of radioisotopes, and enhanced antitumor effects, thus offering a promising approach for combined chemoradiotherapy [87].

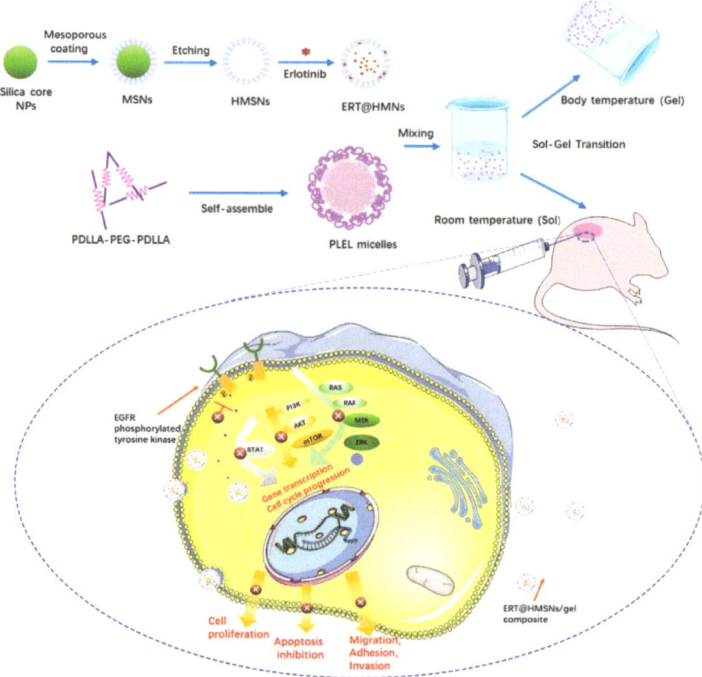

**Figure 1.** Schematic representation of ERT@HMSNs/gel composite to treat NSCLC. Hollow mesoporous silica nanoparticles (HMSNs) act as a carrier for encapsulating erlotinib, aiming to enhance its therapeutic efficacy and mitigate drug-related toxicity. To endure homogeneity and stability of the injectable matrix, PLEL added to the ERT@HMSNs solution. The evaluation included assessing transition from sol to gel phase and the sustained release of the drug. Adopted with permission [77].

In the realm of drug delivery systems, innovative approaches have been explored, such as a cell-penetrable nano-polyplex hydrogel system that enables localized siRNA delivery, thereby facilitating long-term and site-specific gene silencing through a single injection [88]. Furthermore, a moldable and biodegradable colloidal nano-network was developed for the protection and localized delivery of antimicrobial peptides, highlighting its potential in preserving bioactivity and effectively eliminating bacteria [89].

Efforts to promote neuroregeneration following traumatic spinal cord injury have been a major focus in therapeutic delivery strategies. Injectable hydrogel-based systems and scaffold architectures have demonstrated promising results in facilitating neuroregeneration [90]. Graphene oxide (GO) has been extensively investigated as a nanofiller in self-assembling peptide hydrogels for intervertebral disc repair, exhibiting mechanical properties akin to those of the nucleus pulposus while supporting cell viability [91]. Alginate nanohydrogels loaded with bone morphogenetic protein-2 (BMP-2) have been proposed as a controlled release system to enhance osteoblastic growth and differentiation [92].

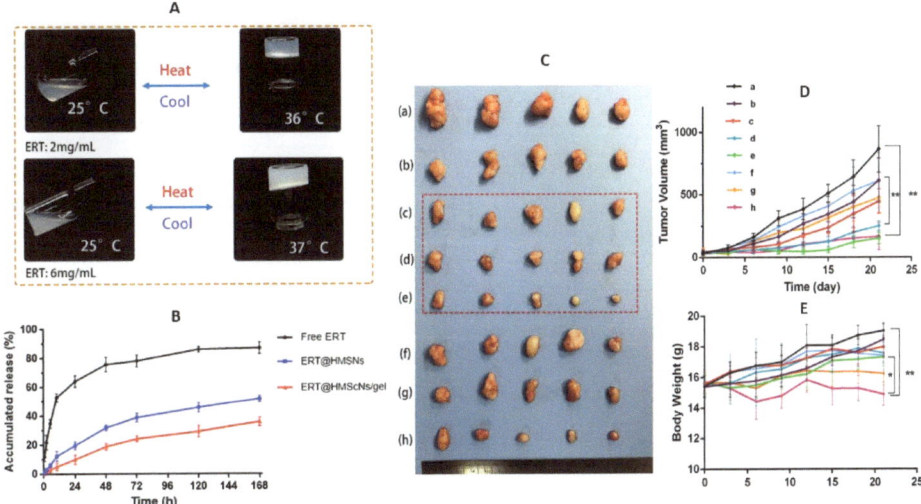

**Figure 2.** (**A**) Reversible sol–gel phase transition of ERT@HMSNs/gel composite; (**B**) In vitro drug release profile; (**C**) In vivo antitumor efficiency of different ERT formulations and Terceva on NSCLC xenograft models. (a) NS; (b) ERT@HMSNs; (c–e) different concentrations of ERT@HMSNs/gel (25, 50, and 100 mg/kg); (f–h) different concentrations of marketed drug Tarceva (25, 50, and 100 mg/kg). (**D**) The tumor growth curves of each group. (**E**) Body weight changes of mice as a function of time in each group. All quantitative data are given as mean ± SD (n = 5). "*" mean $p < 0.05$ and "**" mean $p < 0.01$. Adopted with permission [77].

In the realm of antioxidant therapy, a hydrogel delivery system incorporating curcumin into oligo-conjugated linoleic acid vesicles (OCLAVs) and a chitosan (CS) hydrogel demonstrated sustained release of curcumin over a span of 96 h, accompanied by enhanced antioxidant activity [93]. Similarly, a study focused on an injectable hydrogel composed of chitosan, collagen, hydroxypropyl-gamma-cyclodextrin, and polyethylene glycol, which showcased a two-step forming process and controlled release of bioactive substances [94]. These findings underscore the potential of hydrogel systems in antioxidant therapy, offering prolonged release and improved therapeutic effects.

Advances in the field of bone regeneration have led to the synthesis of poly(phosphazene) hydrogels with specific release rates, demonstrating their efficacy in promoting bone regeneration [95]. Additionally, a noteworthy development is the injectable hydrogel depot system that employs sustained release of exendin 4 (Ex-4) for the treatment of type 2 diabetes mellitus [96]. These studies exemplify the versatility of hydrogel-based systems in controlled release and their potential in bone regeneration and diabetes treatment.

Enhancing the mechanical properties of injectable hydrogels has been a significant focus in the context of bone regeneration. The incorporation of nano-hydroxyapatite or strontium hydroxyapatite, along with dopamine modification, has been shown to augment the mechanical properties of alginate-based hydrogels [97]. Furthermore, the injectable bone regeneration composite (IBRC), which facilitates controlled release of recombinant human bone morphogenetic protein-2 (rhBMP-2), has demonstrated potential for clinical applications in bone defect repair [98]. These advancements underscore the promise of injectable hydrogel systems in the realms of bone regeneration and tissue engineering.

In the pursuit of improved drug delivery, nanoparticle-based systems have been explored for the enhanced delivery of nitric oxide (NO) in cancer treatment [99]. Injectable nano-apatite scaffolds, capable of delivering osteogenic cells and growth factors, have exhibited promise in promoting bone regeneration [100]. Moreover, injectable quadruple-functional hydrogels have demonstrated enhanced tumor targeting and significant reduction in tumor volume through sustained targeting and combined therapy approaches [101]. Additionally, the development of a multiple magnetic hyperthermia (MHT)-mediated release system utilizing an injectable, thermosensitive polymeric hydrogel has proven effective in combination cancer therapy [102]. These studies shed light on the potential of injectable hydrogel systems in cancer therapy, bone regeneration, and controlled release of therapeutic agents.

Overall, the ongoing advancements in hydrogel-based systems continue to drive innovation in the field of drug delivery, providing versatile platforms for controlled release, targeted delivery, and improved therapeutic outcomes in various biomedical applications. These advancements pave the way for enhanced patient care and treatment outcomes. Table 1 provides an overview of the injectable hydrogel materials employed primarily in drug delivery systems.

**Table 1.** Different hydrogel materials used for drug delivery.

| Hydrogel Composition | Outcomes | Ref. |
|---|---|---|
| Polyurethane hydrogel and copper-substituted bioactive mesoporous glasses (Cu-MBGs) | Injectable hybrid formulations based on polyurethane hydrogel and Cu-MBGs enable simultaneous localized co-delivery of functional ions and drugs with sustained release profiles and tunable residence time at the pathological site. | [103] |
| Dexmedetomidine-loaded nano-hydrogel | Injectable nano-drug delivery system combined with Dexmedetomidine for thoracic paravertebral block significantly relieved pain, improved sleep quality, and reduced the need for remedial analgesia and side effects after thoracic surgery. | [3] |
| Nano-thermogel system of polyethylene glycol-polycaprolactone-polyethylene glycol (PEG-PCL-PEG) triblock with poly(lactic-co-glycolic acid) (PLGA) nanoparticles loaded p11 peptide | Controlled release of p11 peptide achieved with nano-thermogel system, showing potential for effective treatment of ocular disorders characterized by angiogenesis. | [4] |
| Hybrid silk hydrogel with carbon nanotubes | Hybrid silk hydrogel with carbon nanotubes enables localized, targeted, and on-demand delivery of anticancer drugs, reducing systemic side effects. | [40] |
| pH- and temperature-responsive hydrogels poly(ethylene glycol)-poly(beta-aminoester urethane) | Chondroitin sulfate nanogels incorporated into pH- and temperature-responsive hydrogels deliver cisplatin selectively to cancer cells, improving targeted therapy. | [41] |
| Urothelium-adherent, ion-triggered liposome-in-gel system | Liposome-in-gel system enhances drug penetration and adhesion in the bladder, showing prolonged drug retention and potential use in intravesical applications. | [58] |
| Composite liposome-in-gel system (gellan hydrogel) | Liposome-in-gel system delivers radiosensitizer paclitaxel to tumor site, enhancing the effect of concurrent radiotherapy and improving tumor volume reduction and animal survival. | [59] |
| Four-arm maleimide-functionalized polyethylene glycol (PEG-4MAL) hydrogel system | PEG-4MAL hydrogel acted as a mechanical pillow to protect the knee joint, inhibit cartilage degradation, and prevent osteophyte formation in an in vivo load-induced osteoarthritis mouse model. | [55] |
| Injectable hydrogel (amphiphilic polymers) system with tumor-targeting nano-micelles | The injectable hydrogel system sustainedly released tumor-targeting nano-micelles, which exhibited GSH-responsive drug release behavior, leading to enhanced antitumor efficiency and improved bioavailability of the drug. | [56] |
| Injectable thermosensitive photothermal-network hydrogel | The thermosensitive photothermal-network hydrogel demonstrated high photothermal conversion efficiency, reversible gel–sol transition, and on-demand drug release, enabling effective near-infrared-triggered photothermal-chemotherapy for breast cancer treatment. | [60] |

Table 1. Cont.

| Hydrogel Composition | Outcomes | Ref. |
| --- | --- | --- |
| Ultrasoft polymeric DNA networks of variable crystallinities | Ultrasoft self-supporting polymerized DNA networks with variable crystallinities showed tunable mechanical properties, pH-responsive drug release, and crystallinity-dependent antitumor efficacy, providing a favorable microenvironment for demand-localized drug delivery. | [62] |
| Sugar-based injectable thermoresponsive hydrogel | Injectable thermoresponsive hydroxypropyl guar-graft-poly(N-vinylcaprolactam) (HPG-g-PNVCL) hydrogel and its composite with nano-hydroxyapatite (n-HA) showed biocompatibility, thermoreversibility, slow drug release, and supported osteoblastic cell growth, making them potential scaffolds for bone tissue engineering. | [63] |
| Injectable polysaccharide hydrogel with hydroxyapatite and calcium carbonate | Injectable and degradable polysaccharide-based hydrogels integrated with hydroxyapatite and calcium carbonate show controlled gelation, enhanced mechanical properties, sustained drug release, antibacterial properties, and self-healing capabilities, making them promising for bone regeneration. | [64] |
| Injectable hydrogel nanomaterials (PNIPAAM with CS, APS and cross-linked with PEGDMA) | Continuous subcutaneous insulin infusion (CSII) showed better blood glucose control and lower incidence of hypoglycemia compared with multiple daily injections (MDI) in children with type 1 diabetes mellitus (T1DM). | [68] |
| Injectable liquid metal nanoflake hydrogel | The LM-doxorubicin nanoflake hydrogel with pH-triggered drug release shows enhanced therapeutic efficacy in preventing postoperative tumor relapse. | [72] |
| Bio-inspired fluorescent nano-injectable hydrogel prepared by copolymerization of N-isopropylacrylamide (NIPAM) and acrylic functionalized nucleobase (adenine) | The injectable hydrogel with a phase-separated structure enables sequential release of different drugs and exhibits fluorescence characteristics, making it suitable for dual drug delivery and imaging. | [73] |
| Injectable micromotor@hydrogel system | The micromotor@hydrogel drug delivery system protects micromotors and enables sustained release of erythromycin, exhibiting excellent antibacterial effect for the treatment of bacterial infections. | [74] |
| Nano polydopamine crosslinked thiol-functionalized hyaluronic acid hydrogel | The hydrogel, crosslinked using polydopamine nanoparticles, shows good injectability, mechanical stability, sustained drug release, and enhanced endothelial cell behavior, making it suitable for angiogenic drug delivery and tissue engineering. | [75] |
| poly(lactic-co-glycolic acid) (PLGA) MS loaded with melatonin(Mel) + Laponite hydrogels | The injectable micro-gel compound and nano-PM compound based on sustained-release microspheres provide stable and prolonged drug release, repair neural function, and reduce biomaterial loss for the treatment of spinal cord injury. | [76] |
| Injectable thermosensitive hydrogel (poly(d,l-lactide)-poly(ethylene glycol)-poly(d,l-lactide))containing erlotinib-loaded hollow mesoporous silica nanoparticles | The injectable ERT@HMSNs/gel composite provides sustained release of erlotinib, improves efficacy against NSCLC, and demonstrates an impressive balance between antitumor efficacy and systemic safety. | [77] |
| Injectable PEG-induced silk nanofiber hydrogel | The injectable silk fibroin nanofiber hydrogel, prepared using a dissolving technique and PEG, exhibits fast gelation, amorphous structure, and superior antibacterial properties, making it suitable for vancomycin delivery in tissue engineering. | [78] |
| Injectable hydrogel and nanoparticle system | The injectable hydrogel-nanoparticle system provides a promising approach for delivering microRNAs to cardiac tissue, improving cardiac function after myocardial infarction. | [80] |

Table 1. Cont.

| Hydrogel Composition | Outcomes | Ref. |
|---|---|---|
| Sustained delivery system incorporating P24-loaded PLGA microspheres and nano-hydroxyapatite in composite hydrogel | The composite hydrogel with sustained P24 peptide release enhances bone tissue regeneration and shows potential for improving bone defect treatment in tissue engineering. | [81] |
| Injectable hydrogel (PLGA-PEG-PLGA) modified with hydroxyapatite particles | The hydrogel modified with nano- and core-shell hydroxyapatite particles enables controlled release of calcium cations, offering potential applications in bone regeneration. | [82] |
| Injectable thermo-sensitive hydrogel (hyaluronic acid-chitosan-g-poly(N-isopropylacrylamide) | The injectable hydrogel, combined with folic acid-conjugated graphene oxide (GOFA) nano-carrier, provides controlled and targeted intratumoral delivery of doxorubicin for breast cancer therapy. | [67] |
| Silica-triptorelin acetate depot | The silica-triptorelin acetate depot demonstrates sustained release of triptorelin, comparable to Pamorelin(R), and maintains equivalent pharmacodynamic effects with lower $C_{max}$ values, offering potential for prolonged therapeutic effects. | [83] |
| Injectable 3-D nano-scaffold hydrogel | Mixing peptide-amphiphile (PA) with BMP-2 formed a transparent hydrogel that induced significant ectopic bone formation, offering potential for tissue regeneration. | [84] |
| Multi-functional calcitriol delivery system for osteoporotic bone regeneration based on poly(D, L-lactide)-poly(ethylene glycol)-poly(D, L-lactide) hydrogel | PDLLA-PEG-PDLLA hydrogel integrated with HA-D and PCL-PEG-NH2 micelles enabled sustained release of calcitriol, promoting osteogenesis and bone regeneration both in vitro and in vivo. | [85] |
| FRET-enabled monitoring of thermosensitive micellar hydrogel assembly (poly(epsilon-caprolactone-co-1,4,8-trioxa[4.6]spiro-9-undecanone)-b-poly(ethylene glycol)-b-poly(epsilon-caprolactone-co-1,4,8-trioxa[4.6]spiro-9-undecanone) triblock copolymer. | PECT triblock copolymer facilitated hydrogel formation and sustained release of micelles, allowing precise imaging of the fate of macro biodegradable materials and potential for co-delivery of therapeutic agents. | [86] |
| Thermosensitive micellar hydrogel (PECT triblock copolymer) | Injectable MHg depot composed of PECT micelles immobilized DOX and I-131-HA, enabling localized delivery, sustained release, and enhanced antitumor effect with reduced side effects. | [87] |
| Cell penetrable nano-polyplex hydrogel | Protamine-conjugated poly(organo-phosphazene) hydrogel forms after injection, releasing nano-polyplexes for effective siRNA delivery and long-term gene silencing on target site. | [88] |
| Biopolymer nano-network (Chitosan and dextran sulfate) | Colloidal nano-network made of chitosan and dextran sulfate encapsulates PA-13 antimicrobial peptide, protecting it from degradation, and delivers it locally, eliminating bacteria without impacting bioactivity. | [89] |
| Graphene oxide-containing self-assembling peptide hybrid hydrogels | GO-reinforced peptide hydrogels promote high cell viability and metabolic activity, showing potential as injectable scaffolds for in vivo delivery of nucleus pulposus cells. | [91] |
| Alginate nanohydrogels | BMP-2@ANH system promotes proliferation and differentiation of human bone marrow stromal cells into osteoblasts, offering a potential method for facilitating stem cell differentiation in vivo. | [92] |
| Nano-hybrid oligopeptide hydrogel | Topical delivery of docetaxel using DTX-CTs/Gel inhibited post-surgical tumor recurrence and enhanced cell death, showing promise for cancer therapy. | [104] |

Table 1. *Cont.*

| Hydrogel Composition | Outcomes | Ref. |
|---|---|---|
| Chitosan-incorporated fatty acid vesicles hydrogel | Curcumin-loaded OCLAVs-CS hydrogel effectively reduced burst release, exhibited enhanced antioxidant activity, and can serve as an injectable or 3D printable drug delivery system. | [93] |
| Injectable two-step forming hydrogel (chitosan, collagen, hydroxypropyl-gamma-cyclodextrin and polyethylene glycol) | Hydrogel composed of chitosan, collagen, hydroxypropyl-gamma-cyclodextrin, and polyethylene glycol exhibited controlled release properties, adaptability for minimally invasive implantation, and support for cell proliferation. | [94] |
| Injectable poly(phosphazene) hydrogels with different anionic sidechains | Tunable hydrogel systems with optimized physical properties and BMP-2 release rates were identified, enabling effective bone regeneration in a critical-sized calvarial defect model. | [95] |
| Injectable hydrogel depot system using Exendin 4 (Ex-4) interactive and complex-forming polymeric ionic nanoparticles | The hydrogel system demonstrated prolonged release of Exendin 4 (Ex-4), offering potential as a long-term effective and reproducible treatment for type 2 diabetes mellitus. | [96] |
| Two-in-one injectable micelleplex-loaded thermogel system composed with polymerization of poly(ethylene glycol), poly(propylene glycol), and poly(3-hydroxybutyrate) | The novel nanoparticle-hydrogel system enabled prolonged release of pDNA micelleplexes, indicating its potential for sustained gene delivery applications. | [105] |
| Injectable alginate-based hydrogel cross-linked via the regulated release of divalent ions from the hydrolysis of D-glucono-delta-lactone | The hydrogel exhibited improved mechanical properties through the slow release of divalent ions from D-glucono-delta-lactone, making it suitable for bone tissue engineering applications. | [97] |
| Injectable bone regeneration composite (IBRC) with nano-hydroxyapatite/collagen particles in an alginate hydrogel carrier | The controlled release of rhBMP-2 from IBRC promoted bone formation, highlighting its potential as a bone defect repair material for clinical applications. | [98] |
| Moldable/injectable calcium phosphate cement (CPC) composite scaffolds | Strong, macroporous CPC scaffolds were developed, suitable for bone regeneration, cell delivery, and growth factor release, with potential applications in dental, craniofacial, and orthopedic reconstructions. | [100] |
| Injectable and quadruple-functional hydrogel (folate/polyethylenimine-conjugated poly(organophosphazene) polymer) encapsulated with siRNA and Au-$Fe_3O_4$ nanoparticles | The hydrogel-based delivery method improved tumor targeting efficiency compared with intravenous delivery, enabling sustained release, passive targeting, active targeting, and magnetic targeting for enhanced therapeutic effects. | [101] |
| Injectable thermosensitive polymeric hydrogel of poly(organophosphazene) combined with superparamagnetic iron oxide nanoparticles | The designed injectable hydrogel allowed controlled release of TRAIL/SPION nanocomplex under hyperthermia, resulting in enhanced cytotoxicity against TRAIL-resistant cancer cells and significant tumor reduction in vivo. | [102] |

Abbreviations: PNIPAAM, Poly(N-isopropylacrylamide); PEGDMA, Polyethylene glycol dimethacrylate; nHA, nano-hydroxyapatite; PLGA, poly(lactide-co-glycolide); PEG, poly(ethylene glycol); FRET, Fluorescence resonance energy transfer; PECT, poly(epsilon-caprolactone-co-1,4,8-trioxa[4.6]spiro-9-undecanone)-poly(ethyleneglycol)-poly(epsilon-caprolactone-co-1,4,8-trioxa[4.6]spiro-9-undecanone).

Injectable hydrogel systems have emerged as a highly promising platform for drug delivery, demonstrating notable advantages in achieving localized and sustained release, thereby facilitating targeted therapy. This feat is accomplished through the implementation of various strategies, such as the utilization of functional ions, nanomaterials, and responsive systems, which collectively contribute to the achievement of controlled release profiles. Furthermore, the integration of hybrid formulations, nanogels, liposome-in-gel systems, and thermo-responsive hydrogels effectively ensures prolonged drug release. Exciting prospects are also observed in combination therapies, where the amalgamation of drug delivery systems with complementary agents exhibits synergistic effects, promising enhanced therapeutic outcomes. In the specific context of bone tissue regeneration, hydrogels play a pivotal role by incorporating bioactive components such as peptides, calcium cations, micelles, and growth factors, thereby promoting osteogenesis and bone formation. Moreover, the tunability of mechanical properties in hydrogels facilitates cell proliferation and provides an adaptable environment for tissue engineering applications.

## 3. Tissue Engineering

In recent years, significant advancements have been made in the field of tissue engineering, particularly in the development of functional hydrogel-based constructs for tissue regeneration. Researchers have leveraged advancements in nano-based 3D and 4D scaffolds, stem cells, and biomaterial innovations to achieve promising results in various applications, including bone and cartilage tissue engineering [5]. Notably, electrospun nanofibrous scaffolds and hydrogel scaffolds that emulate the native extracellular matrix have substantially improved cell viability, adhesion, differentiation, and host integration in these areas [5]. Furthermore, aligned conductive core-shell biomimetic scaffolds have emerged as potential tools for peripheral nerve tissue regeneration [106].

In the realm of bone regeneration, injectable thermosensitive hydrogels enriched with nano-hydroxyapatite have demonstrated potential as biocompatible alternatives [6]. Researchers have tackled challenges associated with natural hydrogels by developing efficient encapsulation systems employing bivalent cobalt-doped nano-hydroxyapatite and gum tragacanth for bone tissue engineering [107]. Additionally, injectable collagen-based hydrogels with controlled mechanical properties have shown promise in bone regeneration applications [7].

The biocompatibility and immune response of injectable collagen/nano-hydroxyapatite (Col/nHA) hydrogels have been investigated for hard tissue engineering [108]. Injectable gels composed of carboxymethyl-chitosan, gelatin, and nano-hydroxyapatite have also demonstrated potential in bone tissue engineering [8]. Similarly, hydrogel constructs incorporating chondroitin sulfate nanoparticles and nano-hydroxyapatite have exhibited superior properties for osteochondral regeneration [109]. Hydrogels have also been recognized as promising vehicles for cardiac tissue regeneration, offering minimally invasive administration and effective delivery of therapeutic agents [110].

Polymeric scaffolds, including hydrogels with nano-additives, have emerged as solutions to address limitations in bone/cartilage and neural tissue engineering [111]. For instance, the addition of zirconium oxide nanoparticles to alginate-gelatin hydrogels has enhanced their mechanical properties and regulation of biodegradation, rendering them suitable for cartilage tissue engineering [112]. Injectable alginate-O-carboxymethyl chitosan/nano fibrin composite hydrogels have been developed for adipose tissue engineering, supporting stem cell proliferation and differentiation [113]. Carrageenan nanocomposite hydrogels incorporating whitlockite nanoparticles and an angiogenic drug have also shown promise for bone tissue engineering [114].

Poly(ethylene glycol)-poly(epsilon-caprolactone)-poly(ethylene glycol) nanocomposites with nano-hydroxyapatite demonstrate thermoresponsivity and favorable gelation properties, making them viable options for orthopedic tissue engineering [115]. Improving the mechanical strength and bioactivity of chitosan/collagen hydrogels through the integration of functionalized single-wall carbon nanotubes holds promise for bone regeneration [116]. Composite hydrogels incorporating nano-hydroxyapatite, glycol chitosan, and hyaluronic acid present porous structures and cytocompatibility, making them attractive for bone tissue engineering [117]. Similarly, a composite hydrogel combining laponite nanoparticles and silated hydroxypropylmethyl cellulose exhibits improved mechanical properties and cytocompatibility for cartilage tissue engineering [118]. Additionally, a biohydrogel incorporating nano SIM@ZIF-8 demonstrates osteogenic differentiation and lipid-lowering abilities, offering potential for bone repair [119].

In the field of tissue engineering, researchers have developed innovative approaches for fabricating injectable hydrogels with diverse applications [120–128]. One noteworthy example is an osteogenic hydrogel composed of gelatin-methacryloyl pre-polymer (GelMA) and nano silicate (SN) (Figure 3A). This hydrogel has shown promising results in in vitro studies on SDF-1α release and in vivo studies on a rat calvaria defect model (Figure 3B,C) [120]. The GelMA-SN-SDF-1α hydrogel exhibits injectability, controlled release of SDF-1α, and the ability to stimulate mesenchymal stem cell migration and expression of osteogenic-related biomarkers.

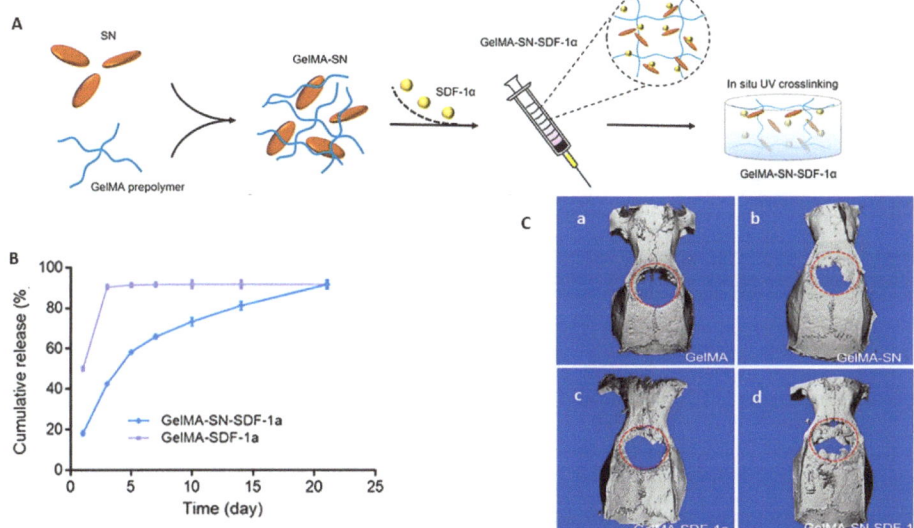

**Figure 3.** (**A**) Schematic presentation of GelMA-SN-SDF-1α hydrogel fabrication; (**B**) In vitro study of release profile of SDF-1α from GelMA-SDF-1α and GelMA-SN- SDF-1α and (**C**) Micro-CT scanning results of the bone healing of calvaria defects rats treated with GelMA, GelMA-SN, GelMA-SDF-1α, and GelMA-SN-SDF-1α hydrogel for 6 weeks. Adopted with permission [120].

The field of regenerative medicine has witnessed notable advancements in nanoengineered biomimetic hydrogels, particularly in the realm of 3D printing. These hydrogels exhibit improved mechanical properties and create an interactive environment that facilitates favorable outcomes in tissue regeneration [121]. Furthermore, the utilization of mineralized heparin-gelatin nanoparticles has shown promise in bone tissue engineering, serving as versatile fillers or multifunctional devices for nanotherapeutic approaches [122].

In the context of hair follicle tissue engineering, the application of GelMA/chitosan-microcarriers loaded with platelet-rich plasma and dermal papilla cells has yielded encouraging results, promoting hair follicle growth and vascularization to enable hair regeneration [123]. Injectable hydrogel composites based on polysaccharides and incorporating nano-hydroxyapatite have been developed as scaffolds for bone tissue engineering, exhibiting remarkable efficacy in bone repair [124]. Similarly, bioactive glass nanoparticle-reinforced injectable hydrogels composed of PEG and pNVC copolymers have demonstrated improved properties and enhanced osteogenesis, positioning them as suitable grafting materials for orthopedic reconstructive surgeries [125]. Additionally, an injectable hydrogel composed of collagen fibrils and a glycol-chitosan matrix has been tailored to address the unique challenges of mechanically strained tissues in soft tissue engineering [126].

Researchers have underscored the significance of integrating stem cells, nanotechnology, and biomaterial innovations in the development of functional hydrogel-based constructs for tissue regeneration [127]. Notably, visible-light-mediated nano-biomineralization has emerged as a rapid fabrication method for customizable biomineralized tough hydrogels, exhibiting enhanced mechanical and biological properties. These hydrogels hold great promise for applications in skin repair and bone regeneration [128].

The collective efforts in hydrogel development offer significant potential for advancing the field of tissue engineering and regenerative medicine. For a comprehensive overview of the hydrogel materials utilized in tissue engineering, please refer to Table 2.

Table 2. Different hydrogel materials used for tissue engineering.

| Hydrogel Composition | Outcomes | Ref. |
|---|---|---|
| Core-shell scaffold based on aligned conductive nanofiber yarns (NFYs) within a methacrylated gelatin (GelMA) hydrogel | Aligned nanofiber yarns within a hydrogel scaffold induce neurite alignment and extension, promoting the alignment and elongation of nerve cells, offering potential for nerve tissue engineering applications. | [106] |
| In situ forming thermosensitive chitosan-glycerol phosphate hydrogel loaded with risedronate and nano-hydroxyapatite | The prepared hydrogel formulation with risedronate and nano-hydroxyapatite showed sustained drug release, enhanced Saos-2 cell proliferation, alkaline phosphatase activity, and calcium deposition, making it a promising option for bone tissue engineering. | [6] |
| Protein-based hydrogels derived from natural tissues | Investigating the nano-/micro-structure and composition of protein-based hydrogels derived from natural tissues is crucial for their widespread use in tissue engineering and regenerative medicine. | [129] |
| Calcium alginate-gum tragacanth hydrogels incorporated with cobalt-doped nano-hydroxyapatite | The hydrogels exhibited enhanced swelling, degradation, diffusion, long-term viability of encapsulated cells, osteogenic differentiation, and angiogenic properties, making them suitable for bone tissue engineering applications. | [107] |
| Chemically crosslinked collagen/chitosan/hyaluronic acid hydrogels | Optimization of the hydrogel composition showed that using high concentrations of crosslinking agent and adjusting the hyaluronic acid content resulted in hydrogels with compact structure, good mechanical properties, prolonged degradation profile, and suitable biocompatibility for bone regeneration applications. | [7] |
| Injectable PCL-PEG-PCL-Col/nHA hydrogels | PCL-PEG-PCL-Col/nHA hydrogels showed successful integration of collagen and nano-hydroxyapatite, delayed biodegradation rate, no prominent pro-inflammatory response, and increased expression of CD31 and IL-10, indicating biocompatibility for hard tissue regeneration. | [108] |
| Enzymatically crosslinked CMC/gelatin/nHAp injectable gels | The enzymatically crosslinked injectable gels exhibited rigidity, adjustable crosslinking degree and strength, increased pore sizes with higher gelatin concentration, and support for osteoblast cell proliferation and differentiation, making them suitable for in situ bone tissue engineering applications. | [8] |
| Injectable semi-interpenetrating network hydrogel with chondroitin sulfate nanoparticles (ChS-NP)s and nanohydroxyapatite (nHA) | The gradient hydrogel construct demonstrated mineralized subchondral and chondral zones, higher osteoblast proliferation in the subchondral zone, porous structure with gradient interface, layer-specific retention of cells, and in vivo osteochondral regeneration with hyaline cartilage formation and subchondral bone integration. | [109] |

Table 2. Cont.

| Hydrogel Composition | Outcomes | Ref. |
|---|---|---|
| Alginate dialdehyde-gelatin scaffolds with zirconium oxide nanoparticles | Incorporation of $ZrO_2$ nanoparticles into alginate-gelatin hydrogels enhances mechanical and chemical properties. Nanocomposite hydrogels exhibit improved swelling behavior, controlled biodegradation, cell viability, and attachment, making them suitable for cartilage tissue regeneration. | [112] |
| Alginate-O-carboxymethyl chitosan/nano fibrin composite hydrogels | Alginate/O-CMC hydrogel blend demonstrated superior properties for tissue engineering applications, supporting the survival, adhesion, proliferation, and differentiation of adipose-derived stem cells. | [113] |
| Injectable carrageenan nanocomposite hydrogel | Carrageenan nanocomposite hydrogel incorporated with whitlockite nanoparticles and an angiogenic drug promoted osteogenesis and angiogenesis in vitro, showing potential for bone tissue engineering. | [114] |
| Injectable thermosensitive hydrogel made of poly(ethylene glycol)-poly(epsilon-caprolactone)-poly(ethylene glycol) (PECE) and nanohydroxyapatite (n-HA) | Thermosensitive hydrogel nanocomposites exhibited good thermosensitivity, injectability, and 3D network structure, making them promising for injectable orthopedic tissue engineering. | [115] |
| Chitosan/collagen hydrogels nano-engineered with functionalized single-wall carbon nanotubes | Integration of COOH-SWCNTs into chitosan and collagen hydrogels increased mechanical strength, bioactivity, and potential for bone tissue engineering and regenerative medicine. | [116] |
| Nano-hydroxyapatite/glycol chitosan/hyaluronic acid composite hydrogel | Composite hydrogel exhibited porous structure, enzymatic degradation, and cytocompatibility, making it suitable for bone tissue engineering applications. | [117] |
| Laponite nanoparticle-associated silated hydroxypropylmethyl cellulose hydrogel | Incorporation of laponites into silated hydroxypropylmethyl cellulose hydrogel resulted in an interpenetrating network that improved mechanical properties without compromising cytocompatibility, oxygen diffusion, or chondrogenic cell functionality. | [118] |
| Nano SIM@ZIF-8-modified injectable high-intensity biohydrogel composed of composed of poly (ethylene glycol) diacrylate (PEGDA) and sodium alginate (SA) + nano simvastatin-laden zeolitic imidazolate framework-8 | nSZPS hydrogel stimulates osteogenic differentiation, inhibits adipogenic differentiation, exhibits excellent injectability, mechanical strength, and promotes bone regeneration in hyperlipidemic microenvironments. | [119] |
| Nano-silicate-reinforced and SDF-1alpha-loaded gelatin-methacryloyl hydrogel | GelMA-SN-SDF-1alpha hydrogel demonstrates injectability, controlled release of SDF-1alpha, MSC migration and homing, and excellent bone regeneration ability in critical-sized calvaria defects. | [120] |
| Succinylated gelatin cross-linked with aldehyde heparin formed nanoparticles, which were mineralized with hydroxyapatite (mineralized heparin-gelatin nanoparticles) | These nanoparticles may enhance the mechanical properties of injectable hydrogels for bone regeneration. | [122] |
| Injectable platelet-rich plasma (PRP)/cell-laden microcarrier/hydrogel composite system | Gelatin methacryloyl (GelMA) and chitosan hydrogels were used to prepare scalable interpenetrating network GelMA/chitosan-microcarriers (IGMs) loaded with PRP and dermal papilla cells (DPCs). The composite system promoted DPC viability, hair inducibility, and hair follicle regeneration. | [123] |
| Polysaccharide-based injectable hydrogel compositing nano-hydroxyapatite | N-carboxyethyl chitosan (NCEC) and oxidized dextran (ODex) were cross-linked via Schiff base linkage to form an injectable hydrogel. The hydrogel, composited with nano-hydroxyapatite (nHAP), exhibited interconnected porous structure and showed excellent bone repair effect in vivo. | [124] |

**Table 2.** *Cont.*

| Hydrogel Composition | Outcomes | Ref. |
|---|---|---|
| Bioactive glass nanoparticle-incorporated triblock copolymeric injectable hydrogel | Injectable hydrogel with bioactive glass nanoparticles showed good gelling and injectability properties, excellent swelling properties, enhanced bone cell proliferation, ALP activity, and apatite mineralization for accelerated in vitro osteogenesis. | [125] |
| Nano-fibrillar hybrid injectable hydrogel with heterotypic collagen fibrils | Injectable hydrogel with semi-interpenetrating networks of heterotypic collagen fibrils in a glycol-chitosan matrix showed nano-fibrillar porous structure, mechanical stability, prolonged half-life, and support for cell implantation. | [126] |
| Visible-light-mediated nano-biomineralization of customizable tough hydrogels | Rapid preparation of biomineralized tough hydrogels with improved mechanical and biological properties under visible light irradiation, suitable for customizable skin repair and bone regeneration. | [128] |

Abbreviations: PCL-PEG-PCL-Col/nHA, Poly($\varepsilon$-caprolactone)-poly(ethylene glycol)-poly($\varepsilon$-caprolactone)/collagen/nano-hydroxyapatite; CMC, carboxymethyl-chitosan; SDF-1alpha stromal cell-derived factor-1 alpha.

These studies underscore the versatility and potential of hydrogels in tissue engineering and regenerative medicine. By customizing the composition, structure, and properties of hydrogels, researchers can design materials that promote cell alignment, controlled release of bioactive substances, integration of nanoparticles, and interpenetrating network structures, all of which contribute to their suitability for various tissue engineering applications.

## 4. Bone Repair

In recent years, considerable research endeavors have been devoted to the advancement of biomaterials tailored specifically for bone regeneration purposes. The primary objective of these biomaterials is to promote bone formation and address various bone defects, thereby offering potential solutions for a wide range of clinical applications. Among the emerging avenues in this field, composite biomaterials and injectable hydrogels have garnered significant attention.

Composite biomaterials, such as the combination of RADA16 peptide hydrogel with porous calcium sulfate/nano-hydroxyapatite ($CaSO_4$/HA) cement, have emerged as a promising approach. Notably, this particular composite has demonstrated improved osteogenic differentiation and enhanced bone formation in femoral condyle defects [9]. Another notable study developed an injectable hydrogel for craniofacial bone regeneration, incorporating bioglass or whitlockite nanoparticles with FGF-18 into a chitin-PLGA hydrogel. This hydrogel exhibited sustained release of FGF-18, resulting in near-complete bone regeneration in craniofacial bone defects [10].

The development of injectable hydrogels has also shown promise as a viable strategy for bone regeneration. For instance, a GelMA-HAMA/nHAP composite hydrogel encapsulating human-urine-derived stem cell exosomes (USCEXOs) exhibited controlled-release properties, fostering osteogenesis and angiogenesis in vitro, while significantly enhancing cranial bone defect repair in a rat model [11]. Additionally, injectable bone regeneration composites composed of nano-hydroxyapatite/collagen (nHAC) particles within an alginate hydrogel carrier demonstrated controllable degradation, biocompatibility, and great potential for bone repair and tissue engineering [12].

Noteworthy advancements in the realm of biomaterials for bone regeneration also encompass injectable hydrogels with dual functionality. One particular study focused on tumor microenvironment-modulated hydrogels (TME), which incorporated bio-responsive drug-loaded mesoporous bioactive glass nanoparticles (MTX-ss-MBGN), gelatin, and oxidized chondroitin sulfate (OCS) for the treatment of tumor-associated bone defects (Figure 4A). The injected Gel/OCS solution, along with MTX-ss-MBGN, rapidly formed a TME-modulated hydrogel, facilitating sustained drug-responsive release with targeted delivery to tumor cells through the depletion of protons and glutathione (GSH). Figure 4B,C depict the hydrogel formation process and the dual responsiveness, highlighting the controlled release capability of MTX-ss-MBGN GO hydrogels. Moreover, this hydrogel facilitated bone regeneration by transitioning into regenerative scaffolds (Figure 4D), effectively preventing tumor recurrence [17]. Similarly, an injectable hydrogel utilizing cisplatin and polydopamine-decorated nano-hydroxyapatite demonstrated tumor ablation, suppression of tumor growth, and improved adhesion and proliferation of bone mesenchymal stem cells [18].

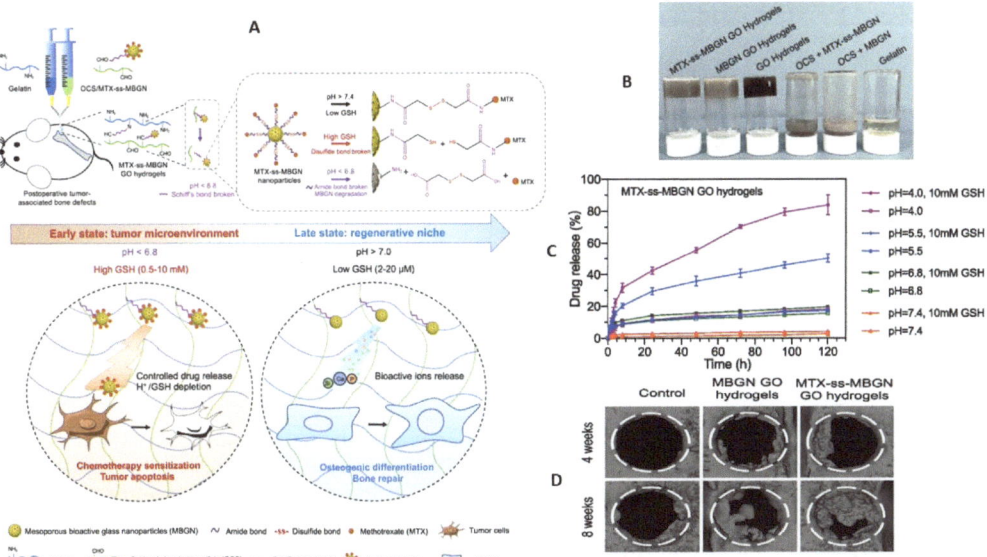

**Figure 4.** (**A**) Schematic representation of the synthesized injectable TME-modulated MTX-ss-MBGN GO hydrogels showing residual tumor apoptosis, restoring chemotherapy sensitivity and promoting bone regeneration for postoperative tumor-associated bone defect closed-loop management; (**B**) Observation study of hydrogels; (**C**) In vitro drug release study to confirm controlled release capability of MTX-ss-MBGN GO hydrogels under tumor-mimicking environment; (**D**) Micro-CT image of calvarial defect repair at 4 or 8 weeks under different treatments on healthy SD rats. Adopted with permission [17].

Recent scientific investigations have placed significant emphasis on the advancement of injectable gel systems and hydrogel composites designed for bone regeneration, with the aim of addressing a wide range of clinical applications. Noteworthy examples include the utilization of chitin-CaSO$_4$-nano-fibrin gel and a double cross-linked hydrogel consisting of gelatin, alginate dialdehyde, calcium ions, and nano-sized hydroxyapatite. These composite hydrogels have demonstrated improved angiogenesis, osteogenesis, and efficacy in bone defect repair [21,23].

Furthermore, the development of injectable hydrogel systems specifically tailored for bone regeneration has garnered considerable attention in research endeavors. One such system involves the incorporation of controlled-release microspheres of bone morphogenetic protein 2 (BMP-2) and 17 beta-estradiol within a composite hydrogel, showcasing uniform discharge and promising outcomes in tissue regeneration [24]. In another study, chitin nano-whiskers (CNWs) were integrated into a chitosan/beta-glycerophosphate disodium salt (CS/GP) hydrogel, resulting in enhanced mechanical properties, improved biocompatibility, and a controlled drug release rate [25].

Moreover, an injectable nano-composite hydrogel for bone regeneration was developed by facilitating the in-situ growth of calcium phosphate (CaP) nanoparticles. This approach exhibited superior mechanical properties, enhanced cell adhesion, osteodifferentiation, and noteworthy advancements in bone formation in vivo [26]. Similarly, an injectable bone regeneration composite (IBRC) employing a calcium alginate hydrogel matrix carrying nano-hydroxyapatite/collagen particles demonstrated structural homogeneity, histocompatibility, and bone healing capabilities [27].

The potential of injectable hydrogels in promoting bone formation, enhancing bone architecture, and addressing bone-related conditions has been substantiated by various studies. For instance, a rabbit-based investigation utilized a biomimetic/osteoinductive injectable hydrogel comprising hyaluronan and nano-hydroxyapatite crystals, leading to substantial improvements in bone density and architecture [130]. Another study compared the osteogenic potential of two injectable hydrogels, namely, demineralized dentin matrix (DDM) hydrogel and nano-hydroxyapatite (n-HA), with the DDM hydrogel demonstrating promising outcomes in promoting collagen-I gene expression and alkaline phosphatase activity [131]. Additionally, an injectable hydrogel based on gellan gum (GG) loaded with chlorhexidine (CHX) and nano-hydroxyapatite (nHA) exhibited a three-dimensional polymeric network, remarkable biocompatibility, and notable osteogenic properties, effectively inhibiting bacterial growth and presenting a potential treatment option for infectious bone defects [132]. A comprehensive summary of the hydrogel materials employed in tissue engineering can be found in Table 3.

Table 3. Different hydrogel materials used for bone repair and osteogenesis.

| Hydrogel Composition | Outcomes | Ref. |
| --- | --- | --- |
| RADA16 peptide hydrogel filled with porous calcium sulfate/nano-hydroxyapatite (CaSO$_4$/HA) composite biomaterial | Controlled and sustainable release of bFGF for more than 32 days from RADA16/CaSO$_4$/HA composite biomaterial, leading to enhanced osteogenic differentiation in vitro and improved bone formation in vivo. | [9] |
| Injectable chitin-PLGA hydrogel containing bioglass nanoparticles (nBG) or whitlockite nanoparticles (nWH) with FGF-18 | CGnWHF (nWH + FGF-18 containing CG) showed the highest osteogenic potential and near-complete bone regeneration in critical-sized defect region compared to other groups, indicating its potential for craniofacial bone defects. | [10] |
| GelMA-HAMA/nHAP composite hydrogel with human-urine-derived stem cell exosomes | Composite hydrogel with controlled release of USCEXOs promotes osteogenesis and angiogenesis, enhancing bone regeneration in vivo. | [11] |
| Injectable bone regeneration composite (IBRC) with nano-hydroxyapatite/collagen (nHAC) particles in alginate hydrogel carrier | IBRC exhibited controllable degradability and biocompatibility, making it a promising material for bone repair and tissue engineering. | [12] |
| poly(caprolactone)-poly(ethylene glycol)-poly(caprolactone) + gelatin and nano-hydroxyapatite | Hydrogels showed successful integration of Gel and nHA, lacked inflammation, and exhibited biocompatibility without toxic effects in in vivo conditions. | [13] |
| nano-hydroxyapatite hybrid methylcellulose hydrogel carrying bone mesenchymal stem cells | Addition of nHA to MC hydrogel enhances cell survival, osteogenic differentiation, and remediation efficiency in vivo. | [14] |
| Thermo-sensitive PEG-PCL-PEG copolymer/collagen/n-HA hydrogel composite | Composite hydrogel exhibits thermosensitivity, biocompatibility, and better performance in guided bone regeneration compared to self-healing processes. | [15] |

Table 3. Cont.

| Hydrogel Composition | Outcomes | Ref. |
|---|---|---|
| Injectable polysaccharide hydrogel-loaded nano-hydroxyapatite | Hydrogel/hydroxyapatite composite scaffold enhances new bone area and alveolar ridge promotion, while promoting soft tissue healing. | [16] |
| TME-modulated hydrogel (MBGN/Gel/OCS) | Hydrogel interferes with tumor microenvironment, overcomes cancer resistance, and promotes sustained drug release and osteogenesis. | [17] |
| Injectable hydrogel containing cisplatin (DDP) and polydopamine-decorated nano-hydroxyapatite (DDP/PDA/nHA) | Exhibits dual functions of tumor therapy and bone regeneration, effectively ablating tumor cells and inducing bone regeneration. | [18] |
| Light-cured hyaluronic acid composite hydrogels (nano-HA/chitosan) | Enhance mechanical properties and osteogenic potential, promising for bone regeneration applications. | [19] |
| Nanocellulose reinforced alginate hydrogel(AC) that carried beta-tricalcium phosphate (beta-TCP) nano-powder and liver-derived extracellular matrix (ECM) from porcine | ETAC Show enhanced cytocompatibility, accelerated bone regeneration, and improved healing quality compared to TAC and AC beads. | [20] |
| Chitin-CaSO$_4$-nFibrin gel | Demonstrates improved rheology, angiogenic potential, and osteo-regeneration compared to chitin control. | [21] |
| Silk-hydroxyapatite composite | Exhibits injectability, thixotropy, and osteodifferentiation potential, supporting improved osteogenesis and bone defect healing. | [22] |
| nHA@Gel/ADA hydrogel with gelatin, alginate dialdehyde, $Ca^{2+}$, borax, and nano-sized hydroxyapatite | Promotes efficient repair of critical-size skull bone defects and supports macrophage-BMSC crosstalk. | [23] |
| Composite hydrogel system incorporating PLGA-BMP-2 and PLA-17 beta-estradiol microspheres in a hydrogel core | Shows controlled release, refilling of bone defects, and regeneration in osteoporotic rats. | [24] |
| Injectable hydrogel with chitosan/beta-glycerophosphate disodium salt (CS/GP) and chitin nano-whiskers (CNWs) | Exhibits improved mechanical properties, gelation speed, and biocompatibility, suitable for tissue engineering scaffold applications. | [25] |
| PDH/mICPN hydrogel composed of DMAEMA, HEMA, CaP nanoparticles (ICPNs), and poly-L-glutamic acid (PGA) | Self-assembles in situ, demonstrating enhanced mechanical strength, cell adhesion, and osteodifferentiation for bone regeneration. | [26] |
| Injectable bone regeneration composite (IBRC) with calcium alginate hydrogel matrix carrying nano-hydroxyapatite/collagen | Demonstrates structural homogeneity, good biocompatibility, and the ability to promote bone healing. | [27] |
| Biomimetic/osteoinductive injectable hyaluronan-based hydrogel loaded with nano-hydroxyapatite crystals (Hya/HA) | Shows potential for enhancing bone architecture, with an osteoinductive effect and improved bone density and architecture in the rabbit distal femur. | [130] |

Table 3. *Cont.*

| Hydrogel Composition | Outcomes | Ref. |
|---|---|---|
| Demineralized dentin matrix hydrogel (DDMH) | Exhibits a porous structure and supports viability and differentiation of BMMSCs, with potential for promoting bone formation. A 50% concentration of DDMH shows promising results. | [131] |
| Gellan gum (GG)-based injectable hydrogel loaded with chlorhexidine (CHX) and nanohydroxyapatite (nHA) | Demonstrates superior biocompatibility, mechanical strength, osteogenic properties, and antibacterial effect against *E. faecalis*. Shows potential for treating infectious bone defects. | [132] |

Abbreviations: GelMA-HAMA/nHAP, Gelatin-methacrylate-Hyaluronic acid methacrylate/nano-hydroxyapatite; PEG-PCL-PEG, poly(ethylene glycol)-poly($\varepsilon$-caprolactone)-poly(ethylene glycol); TME, Tumor microenvironment; MBGN/Gel/OCS, modulated hydrogel composed of bio-responsive drug-loaded mesoporous bioactive glass nanoparticles/gelatin/oxidized chondroitin sulfate; DMAEMA, dimethylaminoethyl methacrylate; HEMA, 2-hydroxyethyl methacrylate.

These studies underscore the potential of hydrogels in bone tissue engineering and regeneration. By leveraging controlled-release mechanisms, ensuring biocompatibility and tissue integration, enhancing mechanical properties, and addressing specific clinical challenges such as bone defects and infections, hydrogels hold promise as materials for promoting bone healing, regeneration, and broader tissue engineering applications.

## 5. Wound Healing

Considerable advancements have been made in the realm of hydrogel-based materials for wound healing and therapy. These versatile hydrogels have garnered significant attention in the context of wound dressings due to their ability to address multiple aspects of wound management. A comprehensive review conducted in this domain highlights the numerous benefits associated with hydrogel dressings, including moisture retention, protective qualities, exudate absorption, as well as anti-inflammatory and antibacterial properties [133]. The review also delves into the feasibility and future trends concerning hydrogel dressing development, encompassing diverse preparation materials, cross-linking methods, and hydrogel types.

In a particular study, researchers explored the incorporation of catechol-modified quaternized chitosan (QCS-C) within a poly(d,l-lactide)-poly(ethylene glycol)-poly(d,l-lactide) (PLEL) hydrogel. This integration resulted in improved tissue adhesion and a reduced gelation temperature. Moreover, the inclusion of nano-scaled bioactive glass (nBG) further augmented angiogenesis and accelerated wound healing [134].

Chitosan-carboxymethyl cellulose (CMC)-based hydrogels loaded with nano-curcumin exhibited controlled release properties, cytocompatibility, and facilitated tissue regeneration, making them well-suited for diabetic wound repair [135].

A noteworthy study in the field involves the development of a black nano-titania thermogel, wherein nanosized black titania nanoparticles were integrated into a chitosan matrix. This composition highlighted simultaneous photothermal and photodynamic therapy effects. Furthermore, it supported normal skin cell functions and promoted the regeneration of skin tissue, thereby facilitating the healing of cutaneous tumor-induced wounds [136].

Regarding antibacterial activity and wound healing, an injectable silver-gelatin-cellulose hydrogel dressing, incorporated with silver nanoparticles, exhibited enhanced wound-healing properties, particularly in the context of infant nursing care [137]. Additionally, a berberine-modified ZnO nano-colloid hydrogel demonstrated exceptional moisturizing capabilities, anti-inflammatory effects, and notable wound healing abilities, positioning it as a promising candidate for diabetic wound healing [138]. Another intriguing hydrogel formulation involved the encapsulation of Ag-decorated polydopamine nanoparticles within a cationic guar gum hydrogel, exhibiting high photothermal conversion efficiency and potent antibacterial properties, thereby presenting a potential solution for wound healing and combatting bacterial infections [139].

Considerable advancements have been achieved in the realm of hydrogel-based materials specifically designed for wound-healing applications. Notably, the incorporation of nanoparticles has emerged as a valuable approach for enhancing the properties of these hydrogels. For instance, an alginate nanocomposite hydrogel incorporating nano-sized calcium fluoride particles has demonstrated enhanced bioactivity and antibacterial properties. This hydrogel formulation effectively promotes fibroblast cell proliferation, facilitates cell migration, and accelerates the process of wound healing [28].

Gellan gum methacrylate and laponite nanocomposite hydrogels have also exhibited notable improvements in mechanical properties and swelling behavior. These hydrogels hold potential as carriers for therapeutic agents, making them well-suited for the treatment of chronic infections in burn wounds [29]. In another study, researchers introduced an injectable nano-composite hydrogel composed of curcumin, N,O-carboxymethyl chitosan, and oxidized alginate. This hydrogel formulation demonstrates controlled release behavior and expedites wound healing in vivo, positioning it as a highly promising material for wound dressings [140].

Furthermore, a multifunctionalized injectable hydrogel, known as COA hydrogel, has garnered attention for its exceptional wound-repair capabilities. Comprising oxidized alginate/carboxymethyl chitosan (KA hydrogel) integrated with keratin nanoparticles (Ker NPs) and nanosized-EGCG covered with silver nanoparticles (AE NPs), this hydrogel displays promising outcomes. It promotes epithelization, scavenges radicals, and exhibits significant improvements in wound healing, as substantiated by an increase in epidermis thickness [141]. Please refer to Figure 5 for an illustration of the multifunctionalized injectable hydrogel (COA hydrogel) and Figure 6 for visual evidence of its efficacy in promoting wound healing.

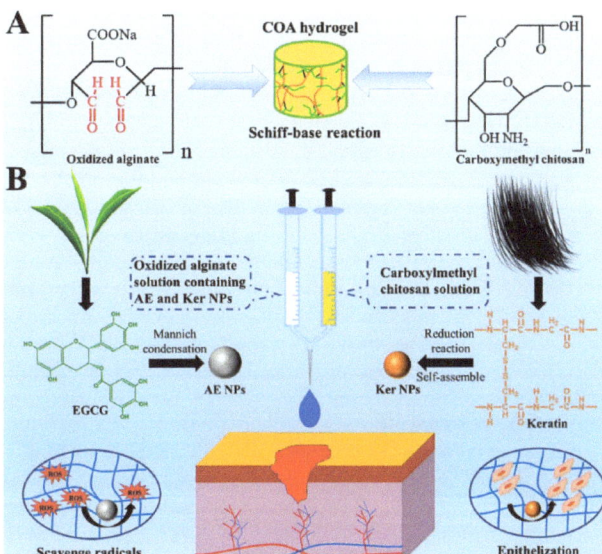

**Figure 5.** (**A**,**B**) Schematic representation of COA hydrogel fabrication and functionalization for wound repair. Adopted with permission [141].

An exemplary instance involves the utilization of an injectable hydrogel system that combines methacryloxylated silk fibroin, metformin-loaded mesoporous silica microspheres, and silver nanoparticles. This nano-dressing exhibits remarkable immunomodulatory effects, effectively inhibits bacterial growth, enhances fibroblast migration, and promotes angiogenesis. Consequently, it emerges as a highly promising solution for the treatment of diabetic wounds [142].

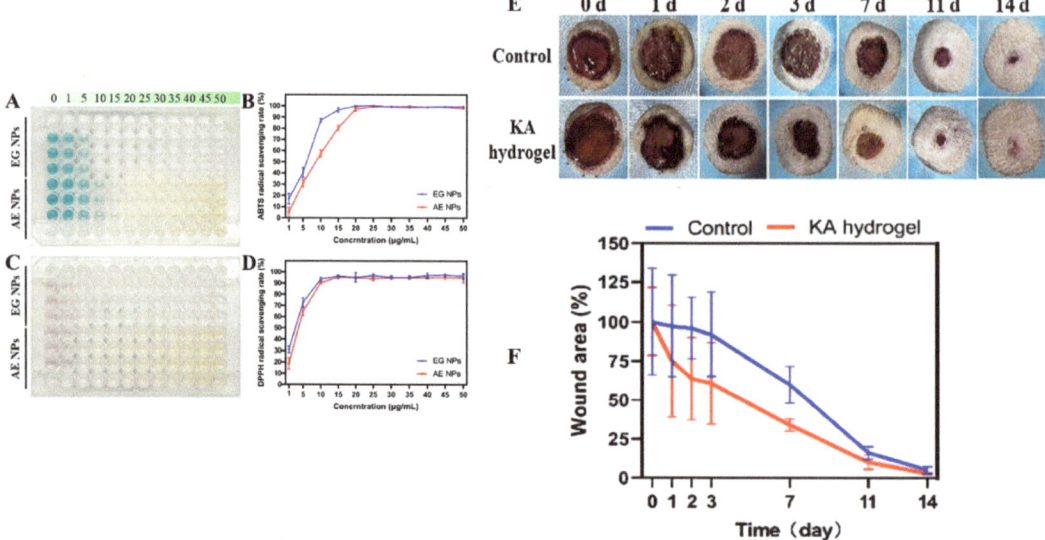

**Figure 6.** Images and quantitative antioxidant capacity of (**A,B**) ABTS and (**C,D**) DPPH radical scavenging assay of EG and AE NPs. (**E**) Photographs of the wound area of rats treated with or without KA hydrogel at different times. (**F**) Statistical results of wound area at different times where the wound area for KA hydrogel is 34% decreased than control group at day 7. Adopted with permission [141].

Injectable hydrogel dressings also show great potential for surgical applications. For instance, a hydrogel adhesive possessing rapid adhesion and anti-swelling properties holds promise for achieving prompt hemostasis and wound sealing [143]. Furthermore, a sodium alginate-based injectable hydrogel dressing, cross-linked with gallic acid-functionalized silver nanoparticles, exhibits sustained antimicrobial activity, reduces the inflammatory response, and accelerates wound healing in infected wounds [144].

Numerous novel injectable hydrogel dressings have been developed with outstanding antibacterial properties and enhanced wound-healing capabilities. These include injectable hydrogels incorporating Ag-doped Mo2C-derived polyoxometalate nanoparticles, fusiform-like zinc oxide nanorods, and mce-like Au-CuS heterostructural nanoparticles [145–147].

In a separate study, the therapeutic potential of an injectable hydrogel composed of nano-sized suspended formulations of human fibroblast-derived matrix (sFDM) is explored. This hydrogel demonstrates excellent biocompatibility, mechanical stability, and regenerative effects, effectively promoting wound healing, neovessel formation, and reducing necrosis and fibrosis in preclinical models [148].

The development of multifunctional hydrogels incorporating bioactive silver-lignin nanoparticles shows considerable promise for managing chronic wounds. These hydrogels possess commendable antimicrobial, antioxidant, and tissue remodeling properties [149]. Similarly, dual-functional hydrogels composed of guar gum-grafted-polyacrylamidoglycolic acid and silver nanocomposites exhibit self-healing capabilities, injectability, and bacterial inactivation properties, rendering them highly advantageous for wound-healing applications [150].

These innovative hydrogel-based materials present exciting opportunities for wound management and therapy. For further details and a comprehensive overview of various hydrogel materials and their outcomes, please refer to Table 4.

Table 4. Different hydrogel materials used for wound healing.

| Hydrogel Composition | Outcomes | Ref. |
|---|---|---|
| PLEL-nBG-QCS-C hydrogel: poly(d,l lactide)-poly(ethylene glycol)-poly(d,l-lactide) PLEL, nano-scaled bioactive glass (nBG), and catechol modified quaternized chitosan (QCS-C) | Exhibits thermo-sensitivity, antibacterial properties, tissue adhesion, and accelerates wound healing | [134] |
| Chitosan-CMC-g-PF127 injectable hydrogels loaded with nano-curcumin | Show controlled release, biocompatibility, and promote diabetic wound repair | [135] |
| BT-CTS thermogel: Injectable thermosensitive hydrogel with black titania nanoparticles (B-TiO$_2$-x) in chitosan matrix | Provides effective tumor therapy, wound closure, and tissue regeneration for skin tumors | [136] |
| Injectable silver-gelatin-cellulose ternary hydrogel dressing with aminated silver nanoparticles | Exhibits antibacterial properties and enhances cutaneous wound healing in infant nursing care | [137] |
| ZnO-Ber/H: Berberine-modified ZnO nano-colloids hydrogel | Promotes diabetic wound healing by enhancing wound healing rate, regulating antioxidant stress factors, downregulating inflammatory factors, and promoting the expression of vascular and epithelial tissue-related factors | [138] |
| CG/PDA@Ag hydrogel: Cationic guar gum hydrogel encapsulating Polydopamine NPs with Ag (PDA@Ag) | Combines high photothermal conversion efficiency and inherent antibacterial ability, demonstrating superior antibacterial efficacy for photothermal antibacterial therapy | [139] |
| Injectable alginate nanocomposite hydrogel containing nano-sized calcium fluoride particles | Enhances bioactivity, antibacterial property, cell proliferation, migration, and extracellular matrix deposition for accelerated wound healing | [28] |
| GG-MA/Laponite hydrogel: Gellan gum methacrylate (GG-MA) combined with laponite (R) XLG | Shows improved mechanical properties and potential as wound dressing materials for infected wounds | [29] |
| Nano-curcumin/CCS-OA hydrogel: In situ injectable hydrogel composed of curcumin, N,O-carboxymethyl chitosan, and oxidized alginate | Accelerates wound healing by promoting re-epithelialization and collagen deposition in rat dorsal wounds | [140] |
| KA hydrogel: Injectable oxidized alginate/carboxymethyl chitosan hydrogel functionalized with keratin nanoparticles (Ker NPs) and nanosized-EGCG covered with Ag nanoparticles (AE NPs) | Accelerates wound healing, particularly in the early stage, and improves the thickness of renascent epidermis | [141] |
| M@M-Ag-Sil-MA hydrogel: Photocurable methacryloxylated silk fibroin hydrogel (Sil-MA) co-encapsulated with metformin-loaded mesoporous silica microspheres (MET@MSNs) and silver nanoparticles (Ag NPs) | Resolves immune contradiction in diabetic wounds, promotes fibroblast migration and endothelial cell angiogenesis, and accelerates diabetic wound healing in a diabetic mouse model | [142] |
| RAAS hydrogel: Injectable hydrogel adhesive with rapid adhesion to wet tissues and anti-swelling properties. | Achieves rapid adhesion to wet tissues, exhibits excellent anti-swelling properties, and demonstrates fast hemostasis and stable adhesion strength in diverse hemorrhage models | [143] |
| GA@AgNPs-SA hydrogel: Injectable sodium alginate hydrogel loaded with gallic acid-functionalized silver nanoparticles (GA@AgNPs) | Exhibits long-term antimicrobial effect, reduces inflammatory response, and accelerates the repair of bacteria-infected wounds through sustained release of silver ions and promotion of angiogenesis | [144] |
| Injectable hydrogel with Ag-doped Mo2C-derived polyoxometalate (AgPOM) nanoparticles, urea, gelatin, and tea polyphenols (TPs) | Exhibits antibacterial activity, accelerates wound healing, and shows potential as a therapeutic agent for drug-resistant bacteria-infected wounds | [145] |
| CMCS-brZnO hydrogel: Injectable hydrogel synthesized by incorporating fusiform-like zinc oxide nanorods (brZnO) into carboxymethyl chitosan (CMCS) | Demonstrates injectability, self-healing, tissue adhesion, antibacterial activity, and promotion of wound healing through sustained release of antibacterial $Zn^{(2+)}$ ions | [146] |

Table 4. Cont.

| Hydrogel Composition | Outcomes | Ref. |
| --- | --- | --- |
| Silk fibroin-hyaluronic acid based injectable hydrogel incorporated with mace-like Au-CuS heterostructural nanoparticles (gAu-CuS HSs) | Enhances hemostasis, exhibits antibacterial activity, regulates cytokine expression, promotes angiogenesis, and accelerates wound healing, making it a promising strategy for diabetic wound healing | [147] |
| PH/sFDM hydrogel containing nano-sized suspended formulation and Pluronic F127/hyaluronic acid (HA) | Promotes neovessel formation, collagen deposition, blood reperfusion, and reduces necrosis and fibrosis in cutaneous wound and hindlimb ischemia models | [148] |
| Self-assembling hydrogels based on thiolated hyaluronic acid (HA-SH) and bioactive silver-lignin nanoparticles (Ag@Lig NPs) | Inhibits proteolytic enzymes, oxidative enzymes, and bacteria, while promoting tissue remodeling and skin integrity restoration in chronic wounds | [149] |
| Guar gum-grafted-polyacrylamidoglycolic acid (GG-g-PAGA) polymer-based silver nanocomposite (AgNC) hydrogels | Exhibits self-healing ability, injectability, stretchability, flowability, high swelling, porosity, mechanical behavior, and biodegradability, suitable for wound-healing applications | [150] |

Abbreviations: sFDM, Human fibroblast derived matrix in nano-sized suspended formulation.

The aforementioned studies serve as significant indicators of the potential that hydrogels hold in the realm of wound healing applications. With their inherent thermo-sensitivity, controlled release capabilities, antibacterial properties, and ability to facilitate various wound-healing processes, hydrogels offer a promising avenue for accelerating the healing of wounds, managing infections, and promoting tissue regeneration. The versatile formulation and properties of hydrogels further contribute to their efficacy in wound care. As research and development in this field progresses, the role of hydrogel-based materials in enhancing wound care and improving patient outcomes is expected to be further enhanced.

## 6. Photothermal

The use of injectable hydrogels has emerged as a promising approach for various modalities of cancer therapy. Specifically, studies have focused on developing injectable hydrogel systems for chemo-photothermal therapy, targeted drug delivery, and combination therapies.

In the realm of chemo-photothermal therapy, innovative hydrogel formulations have been developed to enhance therapeutic outcomes. For instance, a study investigated an in-situ-forming hydrogel incorporating dopamine-reduced graphene oxide (DOPA-rGO) and resveratrol for breast cancer therapy [30]. Another study developed an in-situ-injectable PEG hydrogel system loaded with albumin nanoparticles, demonstrating efficient singlet oxygen generation and hyperthermia, resulting in enhanced cancer cell killing [31]. Furthermore, injectable and biodegradable nano-photothermal DNA hydrogel nanoparticles were engineered to improve tumor cell sensitivity to photothermal and photodynamic treatments [32]. These studies highlight the potential of injectable hydrogels in chemo-photothermal therapy, providing targeted and effective options for cancer treatment.

Moreover, injectable hydrogels have been explored for targeted drug delivery and combination therapies. A self-healing nanocomposite hydrogel carrying graphene oxide (GO) and nano-hydroxyapatite (HAP) exhibited tumor inhibition and photothermal effects, offering a potential treatment for tumors without the side effects of chemotherapy [33]. An intelligent thermo-responsive hydrogel system loaded with berberine hydrochloride (BH) demonstrated enhanced anti-tumor activity when combined with laser irradiation [34].

Recent advancements have expanded the applications of injectable hydrogels in cancer therapy. For example, an injectable nano-composite hydrogel based on hyaluronic acid-chitosan derivatives demonstrated simultaneous photothermal-chemotherapy of cancer with anti-inflammatory capacity, exhibiting favorable tumor inhibition effects [35]. A silk fibroin nanofiber hydrogel system complexed with upconversion nanoparticles and nano-graphene oxide showed excellent biocompatibility and efficient cancer cell ablation through upconversion luminescence imaging and photothermal therapy [36]. Additionally, an injectable and near-infrared (NIR)/pH-responsive nanocomposite hydrogel incorporated gold nanorods, enabling sustained drug release and offering therapeutic potential for chemophotothermal synergistic cancer therapy [37].

Furthermore, the development of bifunctional biomaterials has been explored for bone tumor therapy, combining tumor photothermal therapy with enhanced bone regeneration [38]. The utilization of injectable thermosensitive hydrogels loaded with deferasirox nanoparticles presented a potential strategy for combined chemo-photothermal therapy in melanoma, offering localized drug delivery and photothermal therapy [39]. Please refer to Table 5 for a description of the hydrogel materials used in photothermal therapy.

**Table 5.** Different hydrogel materials used for photothermal therapy.

| Hydrogel Materials and Composition | Outcomes | Ref. |
|---|---|---|
| Chitosan-based injectable in-situ-forming hydrogels containing dopamine-reduced graphene oxide (DOPA-rGO) and resveratrol (RES) | Exhibits injectability, in situ gelation, suitable physicochemical properties, and good cytocompatibility, and significantly enhances the efficacy of chemo-photothermal therapy in breast cancer cells. | [30] |
| In situ injectable PEG hydrogel system formulated with albumin nanoparticles | Exhibits hyperthermia, singlet oxygen ($(1)O_{(2)}$) generation, and enhanced killing of tumor cells, showing potential for ablation of poorly responsive hypoxic tumors. | [31] |
| Injectable and biodegradable nano-photothermal DNA hydrogel | Exhibits improved penetration, sensitivity to photothermal therapy (PTT) and photodynamic treatment (PDT), easy cellular uptake, enhanced anti-tumor activity, and reduced drug resistance, providing a safe and efficient supplement for cancer therapy. | [32] |
| Injectable and self-healing nanocomposite hydrogel loaded with needle-like nano-hydroxyapatite (HAP) and graphene oxide (GO) | Effectively inhibits tumor cell proliferation, realizes the synergistic effect of photothermal therapy, and shows potential as an effective treatment approach for tumors. | [33] |
| Injectable in situ intelligent thermo-responsive hydrogel with glycyrrhetinic acid (GA)-conjugated nano graphene oxide (NGO) | Exhibits sustained and temperature-dependent drug release, enhanced anti-tumor activity when combined with laser irradiation, and shows potential for clinical treatment of malignant tumors. | [34] |
| Injectable nano-composite hydrogel based on hyaluronic acid-chitosan derivatives | Demonstrates tumor inhibition through a comprehensive approach of photothermal therapy, chemotherapy, and anti-inflammatory effects. | [35] |
| Silk fibroin nanofiber (SF) hydrogel system complexed with upconversion nanoparticles and nano-graphene oxide (SF/UCNP@NGO) | Shows potential for tumor imaging and therapy, with excellent biocompatibility, efficient cancer cell ablation, and outstanding antitumor efficacy. | [36] |
| Injectable, near-infrared (NIR)/pH-responsive nanocomposite hydrogel | Demonstrates potential as a long-term drug delivery platform for chemophotothermal synergistic cancer therapy, reducing adverse effects and enabling prolonged drug retention in the tumor region. | [37] |
| Thermosensitive TMPO-oxidized lignocellulose/cationic agarose hydrogel | Shows potential for photothermal therapy in melanoma, with short gelation time, high mechanical strength, efficient drug release, and reduced cytotoxicity with laser light irradiation. | [39] |

These studies highlight the potential of hydrogels in cancer therapy by providing a platform for synergistic treatments, controlled drug release, tumor ablation, and proliferation control. By leveraging these capabilities, hydrogels offer promising strategies for improving therapeutic efficacy and reducing adverse effects in cancer treatment.

## 7. Other Biomedical Applications

### 7.1. Angiogenesis

Injectable hydrogels have emerged as versatile biomedical tools with wide-ranging applications, including the treatment of ischemic brain injury, cardiovascular diseases, and cancer therapy. These hydrogels hold significant potential for addressing critical challenges in these fields.

In the context of ischemic brain injury, a promising approach involves the combination of bone marrow mesenchymal stem cells (BMSCs) with rigid-flexible composite scaffolds. This synergistic approach has demonstrated improved therapeutic effects by reducing brain edema, infarct volume, and neurological deficits, while promoting neuronal proliferation and vascular growth. Such advances in the use of injectable hydrogels with BMSCs hold promise for the treatment of brain injuries [42].

Furthermore, injectable hydrogels have shown promise in tissue regeneration. For example, a sulfated cellulose nanocrystal (CNC-S) hydrogel loaded with vascular endothelial growth factor (VEGF) has been developed to facilitate tissue regeneration by promoting cellular infiltration and angiogenesis. This innovative hydrogel-based approach offers a potential solution for promoting tissue regeneration in various clinical scenarios [43].

The use of injectable hydrogels in the treatment of cardiovascular diseases has also been explored. A comprehensive review highlights the potential of injectable hydrogels as minimally invasive therapies for cardiac applications. The review discusses various strategies, including the combination of hydrogels with stem cells, cytokines, nanoparticles, exosomes, and genetic material. These strategies present exciting possibilities for improving cardiac function and promoting tissue regeneration in the context of cardiovascular diseases [44].

Moreover, injectable hydrogels demonstrate promise in targeted cancer therapy. A thermo-responsive nano-hydrogel loaded with triptolide has exhibited localized and sustained-release treatment of breast cancer. This approach displays enhanced cytotoxicity and anti-angiogenesis effects, underscoring the potential of injectable hydrogels in targeted cancer therapy. Such advancements hold considerable potential for improving cancer treatment outcomes [45].

Collectively, these studies highlight the remarkable capability of injectable hydrogels to address critical challenges in ischemic brain injury, tissue regeneration in cardiovascular diseases, and targeted cancer therapy. The development and utilization of injectable hydrogels in these areas pave the way for significant advancements in biomedical applications, improving patient outcomes.

### 7.2. Antibacterial

Injectable hydrogels have emerged as a versatile tool in various biomedical applications, including active cargo delivery, tissue regeneration, antibacterial applications, and bone reconstruction. One notable example of the potential of injectable hydrogels is found in the development of an antimicrobial colloidal hydrogel. This hydrogel incorporates graphene oxide (GO), thermo-sensitive nanogels (tNG), and silver nanoparticles (AgNPs). The resulting hydrogel possesses several desirable features, including tunable mechanical strength, responsive drug release, high antibacterial activity, temperature responsiveness, and self-healing properties. Such multifunctional characteristics make this hydrogel suitable for scaffold-based applications and antibacterial therapies, positioning it as a promising material with broad biomedical applications [151,152].

In the realm of bone regeneration and reconstruction, injectable bone substitutes composed of carrageenan (CG), nano-hydroxyapatite (nHA), and polymethylmethacrylate (PMMA) bone cement have demonstrated significant promise. These substitutes exhibit exceptional osteoblast adhesion, tissue regeneration potential, antimicrobial properties, remineralization capacity, as well as favorable physicochemical and mechanical performance. By offering a versatile solution for bone tissue engineering that does not solely rely on pharmaceutical drugs, these injectable bone substitutes represent a significant advancement in the field [153,154].

These examples highlight the potential of injectable hydrogels in advancing biomedical applications. Through their unique properties and versatile formulations, injectable hydrogels continue to pave the way for groundbreaking developments in the fields of active cargo delivery, tissue regeneration, antibacterial applications, and bone reconstruction.

### 7.3. Immiunotherapy

The field of injectable biomaterials has witnessed notable progress, resulting in innovative strategies for personalized cancer immunotherapy and tumor treatment. These advancements offer promising avenues to augment immune responses and effectively combat tumors.

One significant study involves the utilization of a self-assembled nano-vaccine platform that combines a conjugate of Toll-like receptor 7/8 agonist and tumor epitope (TLR7/8a-epitope) [46]. This nano-vaccine has demonstrated the capacity to enhance CD8 T-cell immunity, positioning it as a promising candidate for personalized immunotherapy in the treatment of melanoma tumors.

Another promising development revolves around the design of injectable smart hydrogels (ISHs) as a robust cancer vaccine platform. These hydrogels have the ability to recruit dendritic cells (DCs) and elicit tumor-specific immune responses, resulting in the eradication of melanoma tumors in preclinical mouse models. The ISHs offer a targeted and efficient approach to stimulate immune responses against cancer [47].

The integration of nanotechnology and biomaterials has played a pivotal role in advancing injectable biomaterials for cancer immunotherapy and tumor treatment [48]. Organic and polymeric carriers have also emerged as valuable tools in localized tumor chemo-immunotherapy, contributing to enhanced treatment effectiveness [49]. These technologies find diverse applications in diagnostics, drug delivery, cancer treatment, and tissue engineering, facilitating the development of personalized therapeutic strategies.

Furthermore, an injectable immunotherapy system based on a sodium alginate hydrogel has exhibited promise in inhibiting tumor recurrence and metastasis [50]. This system presents a potential strategy for adjuvant immunotherapy, providing supplementary support in cancer treatment.

Collectively, these advancements underscore the substantial potential of injectable biomaterials in personalized cancer immunotherapy and tumor treatment. The utilization of nanovaccine platforms, ISHs, organic and polymeric carriers, and sodium alginate hydrogels exemplifies the broad spectrum of applications and advantages offered by injectable biomaterials in the fight against cancer. These innovative approaches pave the way for more targeted, efficient, and personalized therapies in cancer immunotherapy [46–50].

### 7.4. Cartilage Repair

Significant strides have been made in recent investigations aimed at the development of injectable hydrogels for cartilage regeneration, presenting promising strategies to tackle the challenges associated with cartilage tissue engineering.

One notable approach involves the formulation of an injectable double-crosslinked hydrogel functionalized with kartogenin (KGN)-conjugated polyurethane nanoparticles (PN-KGN) and transforming growth factor beta3 (TGF-beta3) [51]. This hydrogel effectively promotes the migration and chondrogenesis of endogenous mesenchymal stem cells (MSCs), offering a viable in situ strategy for cartilage regeneration.

Another study focuses on an injectable biphasic semi-interpenetrating polymer network (SIPN) hydrogel impregnated with chondroitin sulfate (ChS) nanoparticles and ChS-loaded zein nanoparticles [52]. These nanoparticles are dispersed within injectable SIPNs developed by blending alginate with poly(vinyl alcohol) and calcium crosslinking. This hydrogel exhibits compatibility with chondrocytes and stimulates cartilage-specific gene expression and protein synthesis, demonstrating promise for the regeneration of hyaline cartilage.

Furthermore, a novel injectable biphasic hydrogel composed of partially hydrolyzed polyacrylamide (HPAM) crosslinked with chromium acetate and incorporating nanocrystalline hydroxyapatite (nHAp) has been developed [53]. This hydrogel supports cell viability and differentiation, positioning it as a prospective candidate for applications in cartilage tissue engineering.

In aggregate, these studies contribute to the advancement of injectable hydrogels for cartilage regeneration. By facilitating cell migration, chondrogenesis, and the expression of cartilage-specific genes, these hydrogels offer promising solutions for effective cartilage repair and regeneration.

*7.5. Other Applications*

The integration of nanomaterials with injectable self-healing hydrogels has yielded significant progress in therapies and regenerative medicine, showcasing their potential across a wide range of therapeutic domains.

In the field of cancer therapy, researchers have explored the combination of self-healing hydrogels with nanomaterials, such as Mn-Zn ferrite@mesoporous silica nanospheres and DOX [155]. This integration allows for tumor imaging and synergistic magnetothermal-chemo-chemodynamic therapy, offering efficient diagnosis and treatment for tumors.

Nano-engineered materials have been employed to enhance artificial extracellular matrices (ECMs) for improved cell scaffolds [156]. By modifying the mechanical properties of ECMs and providing dynamic stimuli, these materials enable wireless monitoring of cell status within cultures, leading to advancements in artificial ECMs.

For neurodegenerative diseases, a bioactive self-healing hydrogel based on a tannic-acid-modified gold nano-crosslinker has emerged as a potential injectable brain implant for treating Parkinson's disease [157]. This hydrogel promotes neural stem cell proliferation and differentiation, possesses anti-inflammatory properties, and effectively restores motor function in a rat model of Parkinson's disease.

Biodegradable polymers (BDPs) have gained prominence in various biomedical applications, including ophthalmic drug delivery [158]. Leveraging BDP-based implants, microneedles, and injectable particles enables targeted drug delivery to the ocular posterior segment, enhancing drug retention and bioavailability for ophthalmic treatments.

Advancements in nano- and micro-technologies have empowered researchers to exert precise control over hydrogel properties and functionalities for regenerative engineering [159]. These advancements hold significant implications for tissue engineering, encompassing musculoskeletal, nervous, and cardiac tissues.

Additionally, the development of an injectable conductive hydrogel shows promise in myocardial infarction (MI) treatment [160]. This hydrogel exhibits comparable myocardial conductivity and anti-fatigue performance. When loaded with plasmid DNA encoding endothelial nitric oxide synthase (eNOs) and adipose-derived stem cells (ADSCs), it improves heart function, thereby presenting a potential therapeutic strategy for MI treatment.

In the realm of dermal fillers, hyaluronic acid (HAc)-hydroxyapatite (HAp) composite hydrogels have been formulated as injectable fillers with enhanced biostability and bioactivity [161]. These composite fillers stimulate dermis recovery, collagen synthesis, and elastic fiber formation, positioning them as attractive candidates for long-lasting and multifunctional soft tissue augmentation.

Nanotechnology-based delivery vehicles are being investigated for the treatment of erectile dysfunction (ED) [162]. These vehicles exhibit promise in topical drug delivery, injectable gels, hydrogels for nerve regeneration, and encapsulation of drugs to enhance erectile function. Basic science studies underscore the potential of nanotechnology in developing therapies for ED, highlighting its utility in addressing male sexual dysfunction.

Lastly, a dual-network hydrogel based on an ionic nano-reservoir (INR) has been developed for sealing gastric perforations [163]. This hydrogel displays exceptional adhesion and mechanical properties, offering a potential solution for biomedical challenges such as gastric perforation treatment.

Collectively, these studies and advancements underscore the diverse applications and potential benefits arising from the integration of nanomaterials with injectable hydrogels in various therapeutic areas.

*7.6. Perspective*

The literature review highlights several significant trends and perspectives in the fields of drug delivery, tissue engineering, wound healing, cancer therapy, and therapeutic applications of injectable hydrogel systems. Injectable hydrogels have gained popularity as versatile platforms for drug delivery due to their gelation upon injection, which enables localized and sustained drug release [4,55,56,60,63,64,73–78,80,103]. Achieving targeted and localized drug delivery to specific sites, such as tumor cells or pathological areas, has been a major focus of research, and various strategies have been explored to achieve this objective [40,41,59,62,73,75,77].

Another essential aspect of drug delivery systems is sustained and controlled drug release, which can enhance therapeutic efficacy and minimize side effects. Various techniques have been employed to achieve sustained drug release, including hybrid formulations, nanogels, liposome-in-gel systems, and thermo-responsive hydrogels [3,40,59,62,63,72,73,75,80]. Furthermore, the integration of drug delivery systems with other therapeutic agents or techniques, known as combination therapies, has shown promise in achieving synergistic effects and improving treatment outcomes [40,56,59,60,74,75,77,80].

In the field of bone tissue regeneration and treatment of bone defects, hydrogel-based drug delivery systems have demonstrated significant potential. Various agents, such as peptides, calcium cations, micelles, and growth factors, have been incorporated into hydrogels to promote osteogenesis, bone formation, and bone regeneration [81,82,85,95,97,98]. Additionally, efforts have been focused on tailoring the mechanical properties and adaptability of hydrogels for specific applications, with the aim of enhancing mechanical strength, macroporosity, and adaptability to facilitate cell proliferation, implantation, and tissue engineering [94,95,97,100].

In the field of tissue engineering, the reviewed literature reveals patterns that highlight the versatility and potential of hydrogels. Aligned nanofiber yarns within hydrogel scaffolds have shown potential in promoting the alignment and elongation of nerve cells and other cell types, which is beneficial for nerve and bone tissue engineering [106,107]. Controlled release of bioactive substances from hydrogels has been demonstrated, enabling sustained release and promoting cell proliferation and tissue regeneration [6]. The incorporation of nanoparticles into hydrogels improves their mechanical and chemical properties, leading to enhanced cell viability, attachment, and proliferation, making them suitable for various tissue regeneration applications [108,112,116,125]. Injectable and thermosensitive hydrogels have been developed, offering advantages such as adjustable crosslinking, injectability, and support for cell proliferation and differentiation [8,115]. Interpenetrating network structures in hydrogels show promise for bone tissue engineering by providing porous structures, improved mechanical properties, and support for cell implantation [117,118,126].

In the field of bone tissue engineering and regeneration using hydrogel systems, controlled release systems have shown potential in enhancing osteogenic differentiation and bone formation by delivering bioactive factors over time [9,11]. Hydrogels have demonstrated biocompatibility and the ability to integrate with native tissues, making them suitable for tissue engineering and implantation [13]. Composite hydrogels with improved mechanical properties have been developed, exhibiting osteogenic potential and facilitating bone healing [19,25]. Additionally, hydrogels have shown promise in the treatment of various bone defects, including critical-size defects and infectious bone defects [10,23,132]. Advancements in rheological properties, injectability, and self-assembly capabilities have further improved the ease of application and in situ gel formation [22,26].

In the field of wound healing, the literature review reveals important trends and potential applications of injectable hydrogel systems. Thermo-sensitive hydrogels with antibacterial properties and tissue adhesion have shown potential for wound healing [134,139,146]. Hydrogels with controlled release capabilities and biocompatibility have demonstrated effectiveness in diabetic wound repair [135,138,142,147]. Various hydrogel formulations have been developed to promote wound healing, tissue regeneration, and control infections [28,136,140,141,147,148]. Antibacterial hydrogels have shown efficacy in inhibiting bacterial growth and managing drug-resistant infections [137,144,145,149]. Certain hydrogels also exhibit hemostatic abilities, contributing to efficient wound closure [143,147]. Injectable, self-healing, and versatile hydrogels have been designed for easy application and adaptation to different wound types [146,150].

In the field of cancer therapy, hydrogel systems have shown promise in improving therapeutic outcomes. The combination of multiple treatment modalities within hydrogels has been shown to have synergistic effects, leading to improved outcomes [30,32,35]. Hydrogels with controlled and temperature-dependent drug release capabilities have demonstrated targeted delivery and enhanced anti-tumor activity [34]. Additionally, hydrogels have shown potential in tumor ablation and inhibition through hyperthermia, singlet oxygen generation, and comprehensive treatment approaches [31,33,35]. The physicochemical properties and cytocompatibility of hydrogels further contribute to their clinical applicability in cancer therapy [30].

The integration of nanomaterials with injectable hydrogels opens up promising possibilities for therapeutic applications [49,156]. These hybrid systems have shown therapeutic effects in various diseases and conditions, including ischemic insult, CNS diseases, cancer, neurodegenerative diseases, ophthalmic treatments, tissue engineering, myocardial infarction, dermal fillers, erectile dysfunction, and gastric perforations. Nanomaterial-integrated hydrogels serve as versatile platforms for targeted drug delivery, tissue regeneration, immunotherapy, and implantable devices in biomedical contexts [49,156]. The combination of nanomaterials with injectable hydrogels enables synergistic treatment approaches, leading to improved outcomes in diagnosis, drug release, and therapeutic efficacy. The biocompatibility and biodegradability of these hydrogels make them suitable for long-term implantation or drug release without adverse effects [156,159]. Furthermore, injectable hydrogels integrated with nanomaterials hold promise for personalized therapies, tailoring treatments to individual patients and enhancing therapeutic effects against tumors [46,47].

It is also important to note that while the literature review provides valuable insights, further research and development are still necessary to overcome challenges and optimize the use of injectable hydrogel systems in clinical settings. Future studies could focus on refining the design and properties of hydrogels, exploring new combinations of therapeutic agents, advancing the understanding of their interactions with biological systems, and conducting rigorous preclinical and clinical trials to establish their safety and efficacy.

## 8. Conclusions

In conclusion, the development of nanostructured injectable hydrogel systems has advanced biomedical applications and expanded therapeutic possibilities. These materials play a crucial role in drug delivery, tissue engineering, wound management, cancer therapy, and bone regeneration. Injectable hydrogels provide precise control over drug release, targeted delivery, and improved mechanical properties. They show promise in various areas such as cardiac regeneration, joint diseases, ocular disorder treatment, and gynecological drug delivery. Furthermore, they have potential applications in diabetes treatment, wound healing, cancer therapy, and neuroregeneration. The integration of nanomaterials with injectable hydrogels has further broadened their use in advanced therapies and regenerative medicine. Ongoing advancements in hydrogel-based systems drive innovation in drug delivery and tissue engineering, leading to personalized treatments and improved patient outcomes. With their versatility, controlled release capabilities, and biocompatibility, injectable hydrogels offer a promising platform for addressing clinical challenges and revolutionizing medical treatments. They hold great potential for the future of regenerative medicine and contribute to the advancement of healthcare.

Disclosure: The authors partly used OpenAI's large-scale language-generation model. The authors reviewed, revised, and edited the document for accuracy and take full responsibility for the content of this publication. The authors used Bing AI image creator to draw the graphical abstract.

Authors used Web of Science and PubMed to conduct literature search, used EndNote and over 20 essential keywords to screen the number of references down to 163.

**Author Contributions:** The authors confirm contribution to the paper as follows: Conceptualization, writing, review and editing, H.O.; Investigation, review and editing, S.D.C. All authors have read and agreed to the published version of the manuscript.

**Funding:** This research received no external funding.

**Institutional Review Board Statement:** Not applicable.

**Informed Consent Statement:** Not applicable.

**Data Availability Statement:** Not applicable.

**Conflicts of Interest:** The authors declare no conflict of interest.

# References

1. Bar, A.; Cohen, S. Inducing Endogenous Cardiac Regeneration: Can Biomaterials Connect the Dots? *Front. Bioeng. Biotechnol.* **2020**, *8*, 126. [CrossRef] [PubMed]
2. Bruno, M.C.; Cristiano, M.C.; Celia, C.; D'Avanzo, N.; Mancuso, A.; Paolino, D.; Wolfram, J.; Fresta, M. Injectable Drug Delivery Systems for Osteoarthritis and Rheumatoid Arthritis. *ACS Nano* **2022**, *16*, 19665–19690. [CrossRef]
3. Chen, Y.L.; Lai, J.Y.; Zhou, R.F.; Ouyang, Y.F.; Fu, H. The Analgesic Effect of Dexmedetomidine Loaded with Nano-Hydrogel as a Novel Nano-Drug Delivery System for Thoracic Paravertebral Block After Thoracic Surgery. *J. Biomed. Nanotechnol.* **2022**, *18*, 1604–1612. [CrossRef]
4. Du Toit, L.C.; Choonara, Y.E.; Pillay, V. An Injectable Nano-Enabled Thermogel to Attain Controlled Delivery of p11 Peptide for the Potential Treatment of Ocular Angiogenic Disorders of the Posterior Segment. *Pharmaceutics* **2021**, *13*, 176. [CrossRef]
5. Qasim, M.; Chae, D.S.; Lee, N.Y. Advancements and frontiers in nano-based 3D and 4D scaffolds for bone and cartilage tissue engineering. *Int. J. Nanomed.* **2019**, *14*, 4333–4351. [CrossRef] [PubMed]
6. Morsi, N.M.; Nabil Shamma, R.; Osama Eladawy, N.; Abdelkhalek, A.A. Bioactive injectable triple acting thermosensitive hydrogel enriched with nano-hydroxyapatite for bone regeneration: In-vitro characterization, Saos-2 cell line cell viability and osteogenic markers evaluation. *Drug Dev. Ind. Pharm.* **2019**, *45*, 787–804. [CrossRef] [PubMed]
7. Gilarska, A.; Lewandowska-Lancucka, J.; Horak, W.; Nowakowska, M. Collagen/chitosan/hyaluronic acid—Based injectable hydrogels for tissue engineering applications—Design, physicochemical and biological characterization. *Colloids Surf. B Biointerfaces* **2018**, *170*, 152–162. [CrossRef]
8. Mishra, D.; Bhunia, B.; Banerjee, I.; Datta, P.; Dhara, S.; Maiti, T.K. Enzymatically crosslinked carboxymethyl-chitosan/gelatin/nano-hydroxyapatite injectable gels for in situ bone tissue engineering application. *Mater. Sci. Eng. C* **2011**, *31*, 1295–1304. [CrossRef]
9. He, B.; Zhang, M.Z.; Yin, L.F.; Quan, Z.X.; Ou, Y.S.; Huang, W. bFGF-incorporated composite biomaterial for bone regeneration. *Mater. Des.* **2022**, *215*, 110469. [CrossRef]
10. Amirthalingam, S.; Lee, S.S.; Pandian, M.; Ramu, J.; Iyer, S.; Hwang, N.S.; Jayakumar, R. Combinatorial effect of nano whitlockite/nano bioglass with FGF-18 in an injectable hydrogel for craniofacial bone regeneration. *Biomater. Sci.* **2021**, *9*, 2439–2453. [CrossRef]
11. Lu, W.; Zeng, M.; Liu, W.B.; Ma, T.L.; Fan, X.L.; Li, H.; Wang, Y.A.; Wang, H.Y.; Hu, Y.H.; Xie, J. Human urine-derived stem cell exosomes delivered via injectable GelMA templated hydrogel accelerate bone regeneration. *Mater. Today Bio* **2023**, *19*, 100569. [CrossRef]
12. Tan, R.W.; Feng, Q.L.; She, Z.D.; Wang, M.B.; Jin, H.; Li, J.Y.; Yu, X. In vitro and in vivo degradation of an injectable bone repair composite. *Polym. Degrad. Stab.* **2010**, *95*, 1736–1742. [CrossRef]
13. Alipour, M.; Ashrafihelan, J.; Salehi, R.; Aghazadeh, Z.; Rezabakhsh, A.; Hassanzadeh, A.; Firouzamandi, M.; Heidarzadeh, M.; Rahbarghazi, R.; Aghazadeh, M.; et al. In vivo evaluation of biocompatibility and immune modulation potential of poly(caprolactone)-poly(ethylene glycol)-poly(caprolactone)-gelatin hydrogels enriched with nano-hydroxyapatite in the model of mouse. *J. Biomater. Appl.* **2021**, *35*, 1253–1263. [CrossRef]
14. Deng, L.Z.; Liu, Y.; Yang, L.Q.; Yi, J.Z.; Deng, F.L.; Zhang, L.M. Injectable and bioactive methylcellulose hydrogel carrying bone mesenchymal stem cells as a filler for critical-size defects with enhanced bone regeneration. *Colloids Surf. B Biointerfaces* **2020**, *194*, 111159. [CrossRef]

15. Fu, S.Z.; Ni, P.Y.; Wang, B.Y.; Chu, B.Y.; Zheng, L.; Luo, F.; Luo, J.C.; Qian, Z.Y. Injectable and thermo-sensitive PEG-PCL-PEG copolymer/collagen/n-HA hydrogel composite for guided bone regeneration. *Biomaterials* **2012**, *33*, 4801–4809. [CrossRef] [PubMed]
16. Pan, Y.S.; Zhao, Y.; Kuang, R.; Liu, H.; Sun, D.; Mao, T.J.; Jiang, K.X.; Yang, X.T.; Watanabe, N.; Mayo, K.H.; et al. Injectable hydrogel-loaded nano-hydroxyapatite that improves bone regeneration and alveolar ridge promotion. *Mater. Sci. Eng. C* **2020**, *116*, 111158. [CrossRef]
17. Cai, M.; Li, X.J.; Xu, M.; Zhou, S.Q.; Fan, L.; Huang, J.Y.; Xiao, C.R.; Lee, Y.C.; Yang, B.; Wang, L.; et al. Injectable Tumor Microenvironment-Modulated Hydrogels with Enhanced Chemosensitivity and Osteogenesis for Tumor-Associated Bone Defects Closed-Loop Management. *Chem. Eng. J.* **2022**, *450*, 138086. [CrossRef]
18. Luo, S.Y.; Wu, J.; Jia, Z.R.; Tang, P.F.; Sheng, J.; Xie, C.M.; Liu, C.; Gan, D.L.; Hu, D.; Zheng, W.; et al. An Injectable, Bifunctional Hydrogel with Photothermal Effects for Tumor Therapy and Bone Regeneration. *Macromol. Biosci.* **2019**, *19*, 1900047. [CrossRef]
19. Abdul-Monem, M.M.; Kamoun, E.A.; Ahmed, D.M.; El-Fakharany, E.M.; Al-Abbassy, F.H.; Aly, H.M. Light-cured hyaluronic acid composite hydrogels using riboflavin as a photoinitiator for bone regeneration applications. *J. Taibah Univ. Med. Soc.* **2021**, *16*, 529–539. [CrossRef]
20. Rahaman, M.S.; Park, S.S.; Kang, H.J.; Sultana, T.; Gwon, J.G.; Lee, B.T. Liver tissue-derived ECM loaded nanocellulose-alginate-TCP composite beads for accelerated bone regeneration. *Biomed. Mater.* **2022**, *17*, 055016. [CrossRef]
21. Kumar, A.; Sivashanmugam, A.; Deepthi, S.; Bumgardner, J.D.; Nair, S.V.; Jayakumar, R. Nano-fibrin stabilized $CaSO_4$ crystals incorporated injectable chitin composite hydrogel for enhanced angiogenesis & osteogenesis. *Carbohydr. Polym.* **2016**, *140*, 144–153. [CrossRef]
22. Ding, Z.Z.; Han, H.Y.; Fan, Z.H.; Lu, H.J.; Sang, Y.H.; Yao, Y.L.; Cheng, Q.Q.; Lu, Q.; Kaplan, D.L. Nanoscale Silk-Hydroxyapatite Hydrogels for Injectable Bone Biomaterials. *ACS Appl. Mater. Interfaces* **2017**, *9*, 16914–16922. [CrossRef]
23. Zhou, X.H.; Sun, J.W.; Wo, K.Q.; Wei, H.J.; Lei, H.Q.; Zhang, J.Y.; Lu, X.F.; Mei, F.; Tang, Q.M.; Wang, Y.F.; et al. nHA-loaded gelatin/alginate hydrogel with combined physical and bioactive features for maxillofacial bone repair. *Carbohydr. Polym.* **2022**, *298*, 120127. [CrossRef] [PubMed]
24. Garcia-Garcia, P.; Reyes, R.; Segredo-Morales, E.; Perez-Herrero, E.; Delgado, A.; Evora, C. PLGA-BMP-2 and PLA-17 β-Estradiol Microspheres Reinforcing a Composite Hydrogel for Bone Regeneration in Osteoporosis. *Pharmaceutics* **2019**, *11*, 648. [CrossRef]
25. Wang, Q.Q.; Chen, S.Y.; Chen, D.J. Preparation and characterization of chitosan based injectable hydrogels enhanced by chitin nano-whiskers. *J. Mech. Behav. Biomed. Mater.* **2017**, *65*, 466–477. [CrossRef]
26. Kuang, L.J.; Ma, X.Y.; Ma, Y.F.; Yao, Y.; Tariq, M.; Yuan, Y.; Liu, C.S. Self-Assembled Injectable Nanocomposite Hydrogels Coordinated by in Situ Generated CaP Nanoparticles for Bone Regeneration. *ACS Appl. Mater. Interfaces* **2019**, *11*, 17234–17246. [CrossRef]
27. Tan, R.W.; Feng, Q.L.; Jin, H.; Li, J.Y.; Yu, X.; She, Z.D.; Wang, M.B.; Liu, H.Y. Structure and Biocompatibility of an Injectable Bone Regeneration Composite. *J. Biomater. Sci. Polym. Ed.* **2011**, *22*, 1861–1879. [CrossRef]
28. Shin, D.Y.; Cheon, K.H.; Song, E.H.; Seong, Y.J.; Park, J.U.; Kim, H.E.; Jeong, S.H. Fluorine-ion-releasing injectable alginate nanocomposite hydrogel for enhanced bioactivity and antibacterial property. *Int. J. Biol. Macromol.* **2019**, *123*, 866–877. [CrossRef]
29. Pacelli, S.; Paolicelli, P.; Moretti, G.; Petralito, S.; Di Giacomo, S.; Vitalone, A.; Casadei, M.A. Gellan gum methacrylate and laponite as an innovative nanocomposite hydrogel for biomedical applications. *Eur. Polym. J.* **2016**, *77*, 114–123. [CrossRef]
30. Melo, B.L.; Lima-Sousa, R.; Alves, C.G.; Moreira, A.F.; Correia, I.J.; de Melo-Diogo, D. Chitosan-based injectable in situ forming hydrogels containing dopamine-reduced graphene oxide and resveratrol for breast cancer chemo-photothermal therapy. *Biochem. Eng. J.* **2022**, *185*, 108529. [CrossRef]
31. Lee, W.T.; Yoon, J.; Kim, S.S.; Kim, H.; Nguyen, N.T.; Le, X.T.; Lee, E.S.; Oh, K.T.; Choi, H.G.; Youn, Y.S. Combined Antitumor Therapy Using in Situ Injectable Hydrogels Formulated with Albumin Nanoparticles Containing Indocyanine Green, Chlorin e6, and Perfluorocarbon in Hypoxic Tumors. *Pharmaceutics* **2022**, *14*, 148. [CrossRef]
32. Zhou, L.P.; Pi, W.; Hao, M.D.; Li, Y.S.; An, H.; Li, Q.C.; Zhang, P.X.; Wen, Y.Q. An injectable and biodegradable nano-photothermal DNA hydrogel enhances penetration and efficacy of tumor therapy. *Biomater. Sci.* **2021**, *9*, 4904–4921. [CrossRef] [PubMed]
33. Qi, Y.J.; Qian, Z.Y.; Yuan, W.Z.; Li, Z.H. Injectable and self-healing nanocomposite hydrogel loading needle-like nano-hydroxyapatite and graphene oxide for synergistic tumour proliferation inhibition and photothermal therapy. *J. Mater. Chem. B* **2021**, *9*, 9734–9743. [CrossRef]
34. Wang, H.H.; Wang, B.Y.; Wang, S.S.; Chen, J.Q.; Zhi, W.W.; Guan, Y.B.; Cai, B.R.; Zhu, Y.H.; Jia, Y.Y.; Huang, S.N.; et al. Injectable in situ intelligent thermo-responsive hydrogel with glycyrrhetinic acid-conjugated nano graphene oxide for chemo-photothermal therapy of malignant hepatocellular tumor. *J. Biomater. Appl.* **2022**, *37*, 151–165. [CrossRef]
35. Rong, L.D.; Liu, Y.; Fan, Y.; Xiao, J.; Su, Y.H.; Lu, L.G.; Peng, S.J.; Yuan, W.Z.; Zhan, M.X. Injectable nano-composite hydrogels based on hyaluronic acid-chitosan derivatives for simultaneous photothermal-chemo therapy of cancer with anti-inflammatory capacity. *Carbohydr. Polym.* **2023**, *310*, 120721. [CrossRef]
36. He, W.; Li, P.; Zhu, Y.; Liu, M.M.; Huang, X.N.; Qi, H. An injectable silk fibroin nanofiber hydrogel hybrid system for tumor upconversion luminescence imaging and photothermal therapy. *New J. Chem.* **2019**, *43*, 2213–2219. [CrossRef]
37. Xu, X.Y.; Huang, Z.Y.; Huang, Z.Q.; Zhang, X.F.; He, S.Y.; Sun, X.Q.; Shen, Y.F.; Yan, M.N.; Zhao, C.S. Injectable, NIR/pH-Responsive Nanocomposite Hydrogel as Long-Acting Implant for Chemophotothermal Synergistic Cancer Therapy. *ACS Appl. Mater. Interfaces* **2017**, *9*, 20361–20375. [CrossRef]

38. Liao, J.F.; Han, R.X.; Wu, Y.Z.; Qian, Z.Y. Review of a new bone tumor therapy strategy based on bifunctional biomaterials. *Bone Res.* **2021**, *9*, 18. [CrossRef]
39. Veisi, H.; Varshosaz, J.; Rostami, M.; Mirian, M. Thermosensitive TMPO-oxidized lignocellulose/cationic agarose hydrogel loaded with deferasirox nanoparticles for photothermal therapy in melanoma. *Int. J. Biol. Macromol.* **2023**, *238*, 124126. [CrossRef]
40. Gangrade, A.; Mandal, B.B. Injectable Carbon Nanotube Impregnated Silk Based Multifunctional Hydrogel for Localized Targeted and On-Demand Anticancer Drug Delivery. *ACS Biomater. Sci. Eng.* **2019**, *5*, 2365–2381. [CrossRef]
41. Gil, M.S.; Thambi, T.; Phan, V.H.G.; Kim, S.H.; Lee, D.S. Injectable hydrogel-incorporated cancer cell-specific cisplatin releasing nanogels for targeted drug delivery. *J. Mater. Chem. B* **2017**, *5*, 7140–7152. [CrossRef]
42. Pei, Y.H.; Huang, L.F.; Wang, T.; Yao, Q.H.; Sun, Y.R.; Zhang, Y.; Yang, X.M.; Zhai, J.L.; Qin, L.H.; Xue, J.J.; et al. Bone marrow mesenchymal stem cells loaded into hydrogel/nanofiber composite scaffolds ameliorate ischemic brain injury. *Mater. Today Adv.* **2023**, *17*, 100349. [CrossRef]
43. Min, K.; Tae, G. Cellular infiltration in an injectable sulfated cellulose nanocrystal hydrogel and efficient angiogenesis by VEGF loading. *Biomater. Res.* **2023**, *27*, 28. [CrossRef] [PubMed]
44. Liao, X.S.; Yang, X.S.; Deng, H.; Hao, Y.T.; Mao, L.Z.; Zhang, R.J.; Liao, W.Z.; Yuan, M.M. Injectable Hydrogel-Based Nanocomposites for Cardiovascular Diseases. *Front. Bioeng. Biotechnol.* **2020**, *8*, 251. [CrossRef] [PubMed]
45. Luo, Y.Y.; Li, J.J.; Hu, Y.C.; Gao, F.; Leung, G.P.H.; Geng, F.N.; Fu, C.M.; Zhang, J.M. Injectable thermo-responsive nano-hydrogel loading triptolide for the anti-breast cancer enhancement via localized treatment based on "two strikes" effects. *Acta Pharm. Sin. B* **2020**, *10*, 2227–2245. [CrossRef]
46. Song, H.J.; Su, Q.; Shi, W.F.; Huang, P.S.; Zhang, C.N.; Zhang, C.; Liu, Q.; Wang, W.W. Antigen epitope-TLR7/8a conjugate as self-assembled carrier-free nanovaccine for personalized immunotherapy. *Acta Biomater.* **2022**, *141*, 398–407. [CrossRef]
47. Duong, H.T.T.; Thambi, T.; Yin, Y.; Kim, S.H.; Nguyen, T.L.; Phan, V.H.G.; Kim, J.; Jeong, J.H.; Lee, D.S. Degradation-regulated architecture of injectable smart hydrogels enhances humoral immune response and potentiates antitumor activity in human lung carcinoma. *Biomaterials* **2020**, *230*, 119599. [CrossRef]
48. Cellesi, F.; Tirelli, N. *Injectable Nanotechnology*; Woodhead Publ Ltd.: Cambridge, UK, 2011; pp. 298–322.
49. Bai, Y.T.; Wang, T.R.; Zhang, S.L.; Chen, X.S.; He, C.L. Recent advances in organic and polymeric carriers for local tumor chemo-immunotherapy. *Sci. China Technol. Sci.* **2022**, *65*, 1011–1028. [CrossRef]
50. Zhang, Y.Y.; Wang, T.G.; Zhuang, Y.P.; He, T.D.; Wu, X.L.; Su, L.; Kang, J.; Chang, J.; Wang, H.J. Sodium Alginate Hydrogel-Mediated Cancer Immunotherapy for Postoperative In Situ Recurrence and Metastasis. *ACS Biomater. Sci. Eng.* **2021**, *7*, 5717–5726. [CrossRef]
51. Fan, W.; Yuan, L.; Li, J.; Wang, Z.; Chen, J.; Guo, C.; Mo, X.; Yan, Z. Injectable double-crosslinked hydrogels with kartogenin-conjugated polyurethane nano-particles and transforming growth factor β3 for in-situ cartilage regeneration. *Mater. Sci. Eng. C* **2020**, *110*, 110705. [CrossRef]
52. Radhakrishnan, J.; Subramanian, A.; Sethuraman, S. Injectable glycosaminoglycan-protein nano-complex in semi-interpenetrating networks: A biphasic hydrogel for hyaline cartilage regeneration. *Carbohydr. Polym.* **2017**, *175*, 63–74. [CrossRef]
53. Koushki, N.; Tavassoli, H.; Katbab, A.A.; Katbab, P.; Bonakdar, S. A New Injectable Biphasic Hydrogel Based on Partially Hydrolyzed Polyacrylamide and Nano Hydroxyapatite, Crosslinked with Chromium Acetate, as Scaffold for Cartilage Regeneration. In Proceedings of the 30th International Conference of the Polymer-Processing-Society (PPS), Cleveland, OH, USA, 6–12 June 2014.
54. Chyzy, A.; Tomczykowa, M.; Plonska-Brzezinska, M.E. Hydrogels as Potential Nano-, Micro- and Macro-Scale Systems for Controlled Drug Delivery. *Materials* **2020**, *13*, 188. [CrossRef]
55. Holyoak, D.T.; Wheeler, T.A.; van der Meulen, M.C.H.; Singh, A. Injectable mechanical pillows for attenuation of load-induced post-traumatic osteoarthritis. *Regen. Biomater.* **2019**, *6*, 211–219. [CrossRef]
56. Li, X.Q.; Shi, Y.L.; Xu, S.X. Local delivery of tumor-targeting nano-micelles harboring GSH-responsive drug release to improve antitumor efficiency. *Polym. Adv. Technol.* **2022**, *33*, 2835–2844. [CrossRef]
57. Gosecka, M.; Gosecki, M. Antimicrobial Polymer-Based Hydrogels for the Intravaginal Therapies—Engineering Considerations. *Pharmaceutics* **2021**, *13*, 1393. [CrossRef]
58. GuhaSarkar, S.; More, P.; Banerjee, R. Urothelium-adherent, ion-triggered liposome-in-gel system as a platform for intravesical drug delivery. *J. Control. Release* **2017**, *245*, 147–156. [CrossRef]
59. GuhaSarkar, S.; Pathak, K.; Sudhalkar, N.; More, P.; Goda, J.S.; Gota, V.; Banerjee, R. Synergistic locoregional chemoradiotherapy using a composite liposome-in-gel system as an injectable drug depot. *Int. J. Nanomed.* **2016**, *11*, 6435–6448. [CrossRef]
60. Liu, C.J.; Guo, X.L.; Ruan, C.P.; Hu, H.L.; Jiang, B.P.; Liang, H.; Shen, X.C. An injectable thermosensitive photothermal-network hydrogel for near-infrared-triggered drug delivery and synergistic photothermal-chemotherapy. *Acta Biomater.* **2019**, *96*, 281–294. [CrossRef]
61. Malekmohammadi, S.; Sedghi Aminabad, N.; Sabzi, A.; Zarebkohan, A.; Razavi, M.; Vosough, M.; Bodaghi, M.; Maleki, H. Smart and Biomimetic 3D and 4D Printed Composite Hydrogels: Opportunities for Different Biomedical Applications. *Biomedicines* **2021**, *9*, 1537. [CrossRef] [PubMed]
62. Nam, K.; Kim, Y.M.; Choi, I.; Han, H.S.; Kim, T.; Choi, K.Y.; Roh, Y.H. Crystallinity-tuned ultrasoft polymeric DNA networks for controlled release of anticancer drugs. *J. Control. Release* **2023**, *355*, 7–17. [CrossRef]

63. Parameswaran-Thankam, A.; Parnell, C.M.; Watanabe, F.; RanguMagar, A.B.; Chhetri, B.P.; Szwedo, P.K.; Biris, A.S.; Ghosh, A. Guar-Based Injectable Thermoresponsive Hydrogel as a Scaffold for Bone Cell Growth and Controlled Drug Delivery. *ACS Omega* **2018**, *3*, 15158–15167. [CrossRef]
64. Ren, B.W.; Chen, X.Y.; Du, S.K.; Ma, Y.; Chen, H.A.; Yuan, G.L.; Li, J.L.; Xiong, D.S.; Tan, H.P.; Ling, Z.H.; et al. Injectable polysaccharide hydrogel embedded with hydroxyapatite and calcium carbonate for drug delivery and bone tissue engineering. *Int. J. Biol. Macromol.* **2018**, *118*, 1257–1266. [CrossRef] [PubMed]
65. Salehi, S.; Naghib, S.M.; Garshasbi, H.R.; Ghorbanzadeh, S.; Zhang, W. Smart stimuli-responsive injectable gels and hydrogels for drug delivery and tissue engineering applications: A review. *Front. Bioeng. Biotechnol.* **2023**, *11*, 1104126. [CrossRef] [PubMed]
66. El-Sherbiny, I.; Khalil, I.; Ali, I.; Yacoub, M. Updates on smart polymeric carrier systems for protein delivery. *Drug Dev. Ind. Pharm.* **2017**, *43*, 1567–1583. [CrossRef] [PubMed]
67. Fong, Y.T.; Chen, C.H.; Chen, J.P. Intratumoral Delivery of Doxorubicin on Folate-Conjugated Graphene Oxide by In-Situ Forming Thermo-Sensitive Hydrogel for Breast Cancer Therapy. *Nanomaterials* **2017**, *7*, 388. [CrossRef]
68. Sun, Y.; Niu, H.S.; Wang, Z.X.; Wang, Y.; Li, X.C.; Hao, J.L. Comparative analysis of different drug delivery methods of injectable hydrogel nanomaterials of insulin biomaterials via multiple daily injections and continuous subcutaneous insulin infusion in the treatment of type 1 diabetes mellitus in children. *Mater. Express* **2021**, *11*, 1154–1160. [CrossRef]
69. Tong, S.; Li, Q.Y.; Liu, Q.Y.; Song, B.; Wu, J.Z. Recent advances of the nanocomposite hydrogel as a local drug delivery for diabetic ulcers. *Front. Bioeng. Biotechnol.* **2022**, *10*, 14. [CrossRef]
70. Wanakule, P.; Roy, K. Disease-Responsive Drug Delivery: The Next Generation of Smart Delivery Devices. *Curr. Drug Metab.* **2012**, *13*, 42–49. [CrossRef]
71. Wu, M.; Chen, J.S.; Huang, W.J.; Yan, B.; Peng, Q.Y.; Liu, J.F.; Chen, L.Y.; Zeng, H.B. Injectable and Self-Healing Nanocomposite Hydrogels with Ultrasensitive pH-Responsiveness and Tunable Mechanical Properties: Implications for Controlled Drug Delivery. *Biomacromolecules* **2020**, *21*, 2409–2420. [CrossRef]
72. Xiong, J.J.; Yan, J.J.; Li, C.; Wang, X.Y.; Wang, L.Z.; Pan, D.H.; Xu, Y.P.; Wang, F.; Li, X.X.; Wu, Q.; et al. Injectable liquid metal nanoflake hydrogel as a local therapeutic for enhanced postsurgical suppression of tumor recurrence. *Chem. Eng. J.* **2021**, *416*, 129092. [CrossRef]
73. Xu, Y.; Yang, M.M.; Ma, Q.Y.; Di, X.; Wu, G.L. A bio-inspired fluorescent nano-injectable hydrogel as a synergistic drug delivery system. *New J. Chem.* **2021**, *45*, 3079–3087. [CrossRef]
74. Yang, S.H.; Ren, J.Y.; Wang, H. Injectable Micromotor@Hydrogel System for Antibacterial Therapy. *Chem. Eur. J.* **2022**, *28*, 6. [CrossRef] [PubMed]
75. Yegappan, R.; Selvaprithiviraj, V.; Mohandas, A.; Jayakumar, R. Nano polydopamine crosslinked thiol-functionalized hyaluronic acid hydrogel for angiogenic drug delivery. *Colloids Surf. B Biointerfaces* **2019**, *177*, 41–49. [CrossRef]
76. Zhang, M.; Bai, Y.; Xu, C.; Lin, J.T.; Jin, J.K.; Xu, A.K.; Lou, J.N.; Qian, C.; Yu, W.; Wu, Y.L.; et al. Novel optimized drug delivery systems for enhancing spinal cord injury repair in rats. *Drug Deliv.* **2021**, *28*, 2548–2561. [CrossRef]
77. Zhou, X.H.; He, X.L.; Shi, K.; Yuan, L.P.; Yang, Y.; Liu, Q.Y.; Ming, Y.; Yi, C.; Qian, Z.Y. Injectable Thermosensitive Hydrogel Containing Erlotinib-Loaded Hollow Mesoporous Silica Nanoparticles as a Localized Drug Delivery System for NSCLC Therapy. *Adv. Sci.* **2020**, *7*, 2001442. [CrossRef]
78. Sun, Y.S.; Zhang, P.; Zhang, F.; Pu, M.Y.; Zhong, W.T.; Zhang, Y.; Shen, Y.C.; Zuo, B.Q. Injectable PEG-induced silk nanofiber hydrogel for vancomycin delivery. *J. Drug Deliv. Sci. Technol.* **2022**, *75*, 103596. [CrossRef]
79. Rodriguez-Velazquez, E.; Alatorre-Meda, M.; Mano, J.F. Polysaccharide-Based Nanobiomaterials as Controlled Release Systems for Tissue Engineering Applications. *Curr. Pharm. Des.* **2015**, *21*, 4837–4850. [CrossRef]
80. Bheri, S.; Davis, M.E. Nanoparticle-Hydrogel System for Post-myocardial Infarction Delivery of MicroRNA. *ACS Nano* **2019**, *13*, 9702–9706. [CrossRef] [PubMed]
81. Cai, Q.; Qiao, C.Y.; Ning, J.; Ding, X.X.; Wang, H.Y.; Zhou, Y.M. A Polysaccharide-based Hydrogel and PLGA Microspheres for Sustained P24 Peptide Delivery: An In vitro and In vivo Study Based on Osteogenic Capability. *Chem. Res. Chin. Univ.* **2019**, *35*, 908–915. [CrossRef]
82. Chamradova, I.; Vojtova, L.; Castkova, K.; Divis, P.; Peterek, M.; Jancar, J. The effect of hydroxyapatite particle size on viscoelastic properties and calcium release from a thermosensitive triblock copolymer. *Colloid Polym. Sci.* **2017**, *295*, 107–115. [CrossRef]
83. Forsback, A.P.; Noppari, P.; Viljanen, J.; Mikkola, J.; Jokinen, M.; Leino, L.; Bjerregaard, S.; Borglin, C.; Halliday, J. Sustained In-Vivo Release of Triptorelin Acetate from a Biodegradable Silica Depot: Comparison to Pamorelin® LA. *Nanomaterials* **2021**, *11*, 1578. [CrossRef] [PubMed]
84. Hosseinkhani, H.; Hosseinkhani, M.; Khademhosseini, A.; Kobayashi, H. Bone regeneration through controlled release of bone morphogenetic protein-2 from 3-D tissue engineered nano-scaffold. *J. Control. Release* **2007**, *117*, 380–386. [CrossRef]
85. Hu, Z.C.; Tang, Q.; Yan, D.Y.; Zheng, G.; Gu, M.B.; Luo, Z.C.; Mao, C.; Qian, Z.Y.; Ni, W.F.; Shen, L.Y. A multi-functionalized calcitriol sustainable delivery system for promoting osteoporotic bone regeneration both in vitro and in vivo. *Appl. Mater. Today* **2021**, *22*, 100906. [CrossRef]
86. Huang, P.S.; Song, H.J.; Zhang, Y.M.; Liu, J.J.; Cheng, Z.; Liang, X.J.; Wang, W.W.; Kong, D.L.; Liu, J.F. FRET-enabled monitoring of the thermosensitive nanoscale assembly of polymeric micelles into macroscale hydrogel and sequential cognate micelles release. *Biomaterials* **2017**, *145*, 81–91. [CrossRef]

87. Huang, P.S.; Zhang, Y.M.; Wang, W.W.; Zhou, J.H.; Sun, Y.; Liu, J.J.; Kong, D.L.; Liu, J.F.; Dong, A.J. Co-delivery of doxorubicin and I-131 by thermosensitive micellar-hydrogel for enhanced in situ synergetic chemoradiotherapy. *J. Control. Release* 2015, *220*, 456–464. [CrossRef] [PubMed]
88. Kim, Y.M.; Park, M.R.; Song, S.C. An injectable cell penetrable nano-polyplex hydrogel for localized siRNA delivery. *Biomaterials* 2013, *34*, 4493–4500. [CrossRef] [PubMed]
89. Klubthawee, N.; Bovone, G.; Marco-Dufort, B.; Guzzi, E.A.; Aunpad, R.; Tibbitt, M.W. Biopolymer Nano-Network for Antimicrobial Peptide Protection and Local Delivery. *Adv. Healthc. Mater.* 2022, *11*, e2101426. [CrossRef]
90. Kumar, P.; Choonara, Y.E.; Modi, G.; Naidoo, D.; Pillay, V. Multifunctional Therapeutic Delivery Strategies for Effective Neuro-Regeneration Following Traumatic Spinal Cord Injury. *Curr. Pharm. Des.* 2015, *21*, 1517–1528. [CrossRef]
91. Ligorio, C.; Zhou, M.; Wychowaniec, J.K.; Zhu, X.Y.; Bartlam, C.; Miller, A.F.; Vijayaraghavan, A.; Hoyland, J.A.; Saiani, A. Graphene oxide containing self-assembling peptide hybrid hydrogels as a potential 3D injectable cell delivery platform for intervertebral disc repair applications. *Acta Biomater.* 2019, *92*, 92–103. [CrossRef]
92. Lim, H.J.; Do Ghim, H.; Choi, J.H.; Chung, H.Y.; Lim, J.O. Controlled Release of BMP-2 from Alginate Nanohydrogels Enhanced Osteogenic Differentiation of Human Bone Marrow Stromal Cells. *Macromol. Res.* 2010, *18*, 787–792. [CrossRef]
93. Liu, H.; Meng, X.Y.; Li, L.; Xia, Y.M.; Hu, X.Y.; Fang, Y. The incorporated hydrogel of chitosan-oligoconjugated linoleic acid vesicles and the protective sustained release for curcumin in the gel. *Int. J. Biol. Macromol.* 2023, *227*, 17–26. [CrossRef]
94. Perez-Herrero, E.; Garcia-Garcia, P.; Gomez-Morales, J.; Llabres, M.; Delgado, A.; Evora, C. New injectable two-step forming hydrogel for delivery of bioactive substances in tissue regeneration. *Regen. Biomater.* 2019, *6*, 149–162. [CrossRef] [PubMed]
95. Seo, B.B.; Koh, J.T.; Song, S.C. Tuning physical properties and BMP-2 release rates of injectable hydrogel systems for an optimal bone regeneration effect. *Biomaterials* 2017, *122*, 91–104. [CrossRef] [PubMed]
96. Seo, B.B.; Park, M.R.; Song, S.C. Sustained Release of Exendin 4 Using Injectable and Ionic-Nano-Complex Forming Polymer Hydrogel System for Long-Term Treatment of Type 2 Diabetes Mellitus. *ACS Appl. Mater. Interfaces* 2019, *11*, 15201–15211. [CrossRef]
97. Sun, X.J.; Li, Z.Y.; Cui, Z.D.; Wu, S.L.; Zhu, S.L.; Liang, Y.Q.; Yang, X.J. Preparation and physicochemical properties of an injectable alginate-based hydrogel by the regulated release of divalent ions via the hydrolysis of d-glucono-δ-lactone. *J. Biomater. Appl.* 2020, *34*, 891–901. [CrossRef] [PubMed]
98. Tan, R.W.; She, Z.D.; Wang, M.B.; Yu, X.; Jin, H.; Feng, Q.L. Repair of rat calvarial bone defects by controlled release of rhBMP-2 from an injectable bone regeneration composite. *J. Tissue Eng. Regen. Med.* 2012, *6*, 614–621. [CrossRef]
99. Vong, L.B.; Nagasaki, Y. Nitric Oxide Nano-Delivery Systems for Cancer Therapeutics: Advances and Challenges. *Antioxidants* 2020, *9*, 791. [CrossRef]
100. Xu, H.H.K.; Weir, M.D.; Simon, C.G. Injectable and strong nano-apatite scaffolds for cell/growth factor delivery and bone regeneration. *Dent. Mater.* 2008, *24*, 1212–1222. [CrossRef]
101. Zhang, Z.Q.; Kim, Y.M.; Song, S.C. Injectable and Quadruple-Functional Hydrogel as an Alternative to Intravenous Delivery for Enhanced Tumor Targeting. *ACS Appl. Mater. Interfaces* 2019, *11*, 34634–34644. [CrossRef]
102. Zhang, Z.Q.; Song, S.C. Multiple hyperthermia-mediated release of TRAIL/SPION nanocomplex from thermosensitive polymeric hydrogels for combination cancer therapy. *Biomaterials* 2017, *132*, 16–27. [CrossRef]
103. Boffito, M.; Pontremoli, C.; Fiorilli, S.; Laurano, R.; Ciardelli, G.; Vitale-Brovarone, C. Injectable Thermosensitive Formulation Based on Polyurethane Hydrogel/Mesoporous Glasses for Sustained Co-Delivery of Functional Ions and Drugs. *Pharmaceutics* 2019, *11*, 501. [CrossRef] [PubMed]
104. Liu, C.D.; Ma, Y.D.; Guo, S.; He, B.F.; Jiang, T.Y. Topical delivery of chemotherapeutic drugs using nano-hybrid hydrogels to inhibit post-surgical tumour recurrence. *Biomater. Sci.* 2021, *9*, 4356–4363. [CrossRef] [PubMed]
105. Soh, W.W.M.; Teoh, R.Y.P.; Zhu, J.L.; Xun, Y.R.; Wee, C.Y.; Ding, J.; San Thian, E.; Li, J. Facile Construction of a Two-in-One Injectable Micelleplex-Loaded Thermogel System for the Prolonged Delivery of Plasmid DNA. *Biomacromolecules* 2022, *23*, 3477–3492. [CrossRef] [PubMed]
106. Wang, L.; Wu, Y.B.; Hu, T.L.; Ma, P.X.; Guo, B.L. Aligned conductive core-shell biomimetic scaffolds based on nanofiber yarns/hydrogel for enhanced 3D neurite outgrowth alignment and elongation. *Acta Biomater.* 2019, *96*, 175–187. [CrossRef]
107. Kulanthaivel, S.; Agarwal, T.; Rathnam, V.S.S.; Pal, K.; Banerjee, I. Cobalt doped nano-hydroxyapatite incorporated gum tragacanth-alginate beads as angiogenic-osteogenic cell encapsulation system for mesenchymal stem cell based bone tissue engineering. *Int. J. Biol. Macromol.* 2021, *179*, 101–115. [CrossRef]
108. Hassanzadeh, A.; Ashrafihelan, J.; Salehi, R.; Rahbarghazi, R.; Firouzamandi, M.; Ahmadi, M.; Khaksar, M.; Alipour, M.; Aghazadeh, M. Development and biocompatibility of the injectable collagen/nano-hydroxyapatite scaffolds as in situ forming hydrogel for the hard tissue engineering application. *Artif. Cells Nanomed. Biotechnol.* 2021, *49*, 136–146. [CrossRef]
109. Radhakrishnan, J.; Manigandan, A.; Chinnaswamy, P.; Subramanian, A.; Sethuraman, S. Gradient nano-engineered in situ forming composite hydrogel for osteochondral regeneration. *Biomaterials* 2018, *162*, 82–98. [CrossRef]
110. Saludas, L.; Pascual-Gil, S.; Prosper, F.; Garbayo, E.; Blanco-Prieto, M. Hydrogel based approaches for cardiac tissue engineering. *Int. J. Pharm.* 2017, *523*, 454–475. [CrossRef]
111. Niemczyk-Soczynska, B.; Zaszczynska, A.; Zabielski, K.; Sajkiewicz, P. Hydrogel, Electrospun and Composite Materials for Bone/Cartilage and Neural Tissue Engineering. *Materials* 2021, *14*, 6899. [CrossRef]

112. Ghanbari, M.; Salavati-Niasari, M.; Mohandes, F.; Firouzi, Z.; Mousavi, S.D. The impact of zirconium oxide nanoparticles content on alginate dialdehyde-gelatin scaffolds in cartilage tissue engineering. *J. Mol. Liq.* **2021**, *335*, 116531. [CrossRef]
113. Jaikumar, D.; Sajesh, K.M.; Soumya, S.; Nimal, T.R.; Chennazhi, K.P.; Nair, S.V.; Jayakumar, R. Injectable alginate-O-carboxymethyl chitosan/nano fibrin composite hydrogels for adipose tissue engineering. *Int. J. Biol. Macromol.* **2015**, *74*, 318–326. [CrossRef] [PubMed]
114. Yegappan, R.; Selvaprithiviraj, V.; Amirthalingam, S.; Mohandas, A.; Hwang, N.S.; Jayakumar, R. Injectable angiogenic and osteogenic carrageenan nanocomposite hydrogel for bone tissue engineering. *Int. J. Biol. Macromol.* **2019**, *122*, 320–328. [CrossRef] [PubMed]
115. Fu, S.Z.; Gun, G.; Gong, C.Y.; Zeng, S.; Liang, H.; Luo, F.; Zhang, X.N.; Zhao, X.; Wei, Y.Q.; Qian, Z.Y. Injectable Biodegradable Thermosensitive Hydrogel Composite for Orthopedic Tissue Engineering. 1. Preparation and Characterization of Nanohydroxyapatite/Poly(ethylene glycol)-Poly($\varepsilon$-caprolactone)-Poly(ethylene glycol) Hydrogel Nanocomposites. *J. Phys. Chem. B* **2009**, *113*, 16518–16525. [CrossRef]
116. Kaur, K.; Paiva, S.S.; Caffrey, D.; Cavanagh, B.L.; Murphy, C.M. Injectable chitosan/collagen hydrogels nano-engineered with functionalized single wall carbon nanotubes for minimally invasive applications in bone. *Mater. Sci. Eng. C* **2021**, *128*, 112340. [CrossRef] [PubMed]
117. Huang, Y.X.; Zhang, X.L.; Wu, A.M.; Xu, H.Z. An injectable nano-hydroxyapatite (n-HA)/glycol chitosan (G-CS)/hyaluronic acid (HyA) composite hydrogel for bone tissue engineering. *RSC Adv.* **2016**, *6*, 33529–33536. [CrossRef]
118. Boyer, C.; Figueiredo, L.; Pace, R.; Lesoeur, J.; Rouillon, T.; Visage, C.L.; Tassin, J.F.; Weiss, P.; Guicheux, J.; Rethore, G. Laponite nanoparticle-associated silated hydroxypropylmethyl cellulose as an injectable reinforced interpenetrating network hydrogel for cartilage tissue engineering. *Acta Biomater.* **2018**, *65*, 112–122. [CrossRef]
119. Qiao, M.X.; Xu, Z.Y.; Pei, X.B.; Liu, Y.H.; Wang, J.; Chen, J.Y.; Zhu, J.; Wan, Q.B. Nano SIM@ZIF-8 modified injectable High-intensity biohydrogel with bidirectional regulation of osteogenesis and Anti-adipogenesis for bone repair. *Chem. Eng. J.* **2022**, *434*, 134583. [CrossRef]
120. Shi, Z.; Xu, Y.; Mulatibieke, R.; Zhong, Q.; Pan, X.; Chen, Y.; Lian, Q.; Luo, X.; Shi, Z.; Zhu, Q. Nano-Silicate-Reinforced and SDF-1$\alpha$-Loaded Gelatin-Methacryloyl Hydrogel for Bone Tissue Engineering. *Int. J. Nanomed.* **2020**, *15*, 9337–9353. [CrossRef]
121. Cernencu, A.I.; Dinu, A.I.; Stancu, I.C.; Lungu, A.; Iovu, H. Nanoengineered biomimetic hydrogels: A major advancement to fabricate 3D-printed constructs for regenerative medicine. *Biotechnol. Bioeng.* **2022**, *119*, 762–783. [CrossRef]
122. Yang, Y.; Tang, H.H.; Kowitsch, A.; Mader, K.; Hause, G.; Ulrich, J.; Groth, T. Novel mineralized heparin-gelatin nanoparticles for potential application in tissue engineering of bone. *J. Mater. Sci. Mater. Med.* **2014**, *25*, 669–680. [CrossRef]
123. Zhang, Y.F.; Yin, P.J.; Huang, J.F.; Yang, L.N.; Liu, Z.; Fu, D.L.; Hu, Z.Q.; Huang, W.H.; Miao, Y. Scalable and high-throughput production of an injectable platelet-rich plasma (PRP)/cell-laden microcarrier/hydrogel composite system for hair follicle tissue engineering. *J. Nanobiotechnol.* **2022**, *20*, 22. [CrossRef]
124. Cao, Z.; Bai, X.; Wang, C.B.; Ren, L.L.; Ma, D.Y. A simple polysaccharide based injectable hydrogel compositing nano-hydroxyapatite for bone tissue engineering. *Mater. Lett.* **2021**, *293*, 129755. [CrossRef]
125. Pal, A.; Das Karmakar, P.; Vel, R.; Bodhak, S. Synthesis and Characterizations of Bioactive Glass Nanoparticle-Incorporated Triblock Copolymeric Injectable Hydrogel for Bone Tissue Engineering. *ACS Appl. Bio Mater.* **2023**, *6*, 445–457. [CrossRef] [PubMed]
126. Latifi, N.; Asgari, M.; Vali, H.; Mongeau, L. A tissue-mimetic nano-fibrillar hybrid injectable hydrogel for potential soft tissue engineering applications. *Sci. Rep.* **2018**, *8*, 1047. [CrossRef] [PubMed]
127. Dubey, S.K.; Alexander, A.; Sivaram, M.; Agrawal, M.; Singhvi, G.; Sharma, S.; Dayaramani, R. Uncovering the Diversification of Tissue Engineering on the Emergent Areas of Stem Cells, Nanotechnology and Biomaterials. *Curr. Stem Cell Res. Ther.* **2020**, *15*, 187–201. [CrossRef]
128. Wei, H.Q.; Zhang, B.; Lei, M.; Lu, Z.; Liu, J.P.; Guo, B.L.; Yu, Y. Visible-Light-Mediated Nano-biomineralization of Customizable Tough Hydrogels for Biomimetic Tissue Engineering. *ACS Nano* **2022**, *16*, 4734–4745. [CrossRef]
129. Jabbari, E. Challenges for Natural Hydrogels in Tissue Engineering. *Gels* **2019**, *5*, 30. [CrossRef]
130. Nageeb, M.; Nouh, S.R.; Bergman, K.; Nagy, N.B.; Khamis, D.; Kisiel, M.; Engstrand, T.; Hilborn, J.; Marei, M.K. Bone Engineering by Biomimetic Injectable Hydrogel. *Mol. Cryst. Liq. Cryst.* **2012**, *555*, 177–188. [CrossRef]
131. Sultan, N.; Jayash, S.N. Evaluation of osteogenic potential of demineralized dentin matrix hydrogel for bone formation. *BMC Oral Health* **2023**, *23*, 247. [CrossRef] [PubMed]
132. Xu, L.J.; Bai, X.; Yang, J.J.; Li, J.S.; Xing, J.Q.; Yuan, H.; Xie, J.; Li, J.Y. Preparation and characterisation of a gellan gum-based hydrogel enabling osteogenesis and inhibiting Enterococcus faecalis. *Int. J. Biol. Macromol.* **2020**, *165*, 2964–2973. [CrossRef]
133. Zeng, D.; Shen, S.H.; Fan, D.D. Molecular design, synthesis strategies and recent advances of hydrogels for wound dressing applications. *Chin. J. Chem. Eng.* **2021**, *30*, 308–320. [CrossRef]
134. Zheng, Z.Q.; Bian, S.Q.; Li, Z.Q.; Zhang, Z.Y.; Liu, Y.; Zhai, X.Y.; Pan, H.B.; Zhao, X.L. Catechol modified quaternized chitosan enhanced wet adhesive and antibacterial properties of injectable thermo-sensitive hydrogel for wound healing. *Carbohydr. Polym.* **2020**, *249*, 116826. [CrossRef] [PubMed]
135. Shah, S.A.; Sohail, M.; Karperien, M.; Johnbosco, C.; Mahmood, A.; Kousar, M. Chitosan and carboxymethyl cellulose-based 3D multifunctional bioactive hydrogels loaded with nano-curcumin for synergistic diabetic wound repair. *Int. J. Biol. Macromol.* **2023**, *227*, 1203–1220. [CrossRef] [PubMed]

136. Wang, X.C.; Ma, B.; Xue, J.M.; Wu, J.F.; Chang, J.; Wu, C.T. Defective Black Nano-Titania Thermogels for Cutaneous Tumor-Induced Therapy and Healing. *Nano Lett.* **2019**, *19*, 2138–2147. [CrossRef]
137. Gou, L.; Xiang, M.L.; Ni, X.L. Development of wound therapy in nursing care of infants by using injectable gelatin-cellulose composite hydrogel incorporated with silver nanoparticles. *Mater. Lett.* **2020**, *277*, 128340. [CrossRef]
138. Yin, X.C.; Fan, X.Y.; Zhou, Z.P.; Li, Q. Encapsulation of berberine decorated ZnO nano-colloids into injectable hydrogel using for diabetic wound healing. *Front. Chem.* **2022**, *10*, 14. [CrossRef]
139. Qi, X.L.; Huang, Y.J.; You, S.Y.; Xiang, Y.J.; Cai, E.Y.; Mao, R.T.; Pan, W.H.; Tong, X.Q.; Dong, W.; Ye, F.F.; et al. Engineering Robust Ag-Decorated Polydopamine Nano-Photothermal Platforms to Combat Bacterial Infection and Prompt Wound Healing. *Adv. Sci.* **2022**, *9*, 2106015. [CrossRef] [PubMed]
140. Li, X.Y.; Chen, S.; Zhang, B.J.; Li, M.; Diao, K.; Zhang, Z.L.; Li, J.; Xu, Y.; Wang, X.H.; Chen, H. In situ injectable nano-composite hydrogel composed of curcumin, N,O-carboxymethyl chitosan and oxidized alginate for wound healing application. *Int. J. Pharm.* **2012**, *437*, 110–119. [CrossRef] [PubMed]
141. Ma, L.; Tan, Y.F.; Chen, X.Y.; Ran, Y.Q.; Tong, Q.L.; Tang, L.W.; Su, W.; Wang, X.L.; Li, X.D. Injectable oxidized alginate/carboxylmethyl chitosan hydrogels functionalized with nanoparticles for wound repair. *Carbohydr. Polym.* **2022**, *293*, 119733. [CrossRef]
142. Mei, J.W.; Zhou, J.; Kong, L.T.; Dai, Y.; Zhang, X.Z.; Song, W.Q.; Zhu, C. An injectable photo-cross-linking silk hydrogel system augments diabetic wound healing in orthopaedic surgery through spatiotemporal immunomodulation. *J. Nanobiotechnol.* **2022**, *20*, 232. [CrossRef] [PubMed]
143. Bian, S.Q.; Hao, L.Z.; Qiu, X.; Wu, J.; Chang, H.; Kuang, G.M.; Zhang, S.; Hu, X.H.; Dai, Y.K.; Zhou, Z.Y.; et al. An Injectable Rapid-Adhesion and Anti-Swelling Adhesive Hydrogel for Hemostasis and Wound Sealing. *Adv. Funct. Mater.* **2022**, *32*, 2207741. [CrossRef]
144. Hu, Q.S.; Nie, Y.; Xiang, J.; Xie, J.W.; Si, H.B.; Li, D.H.; Zhang, S.Y.; Li, M.; Huang, S.S. Injectable sodium alginate hydrogel loaded with plant polyphenol-functionalized silver nanoparticles for bacteria-infected wound healing. *Int. J. Biol. Macromol.* **2023**, *234*, 123691. [CrossRef] [PubMed]
145. Huang, H.; Su, Y.; Wang, C.X.; Lei, B.; Song, X.J.; Wang, W.J.; Wu, P.; Liu, X.Y.; Dong, X.C.; Zhong, L.P. Injectable Tissue-Adhesive Hydrogel for Photothermal/Chemodynamic Synergistic Antibacterial and Wound Healing Promotion. *ACS Appl. Mater. Interfaces* **2023**, *15*, 2714–2724. [CrossRef] [PubMed]
146. Hu, T.; Wu, G.P.; Bu, H.T.; Zhang, H.Y.; Li, W.X.; Song, K.; Jiang, G.B. An injectable, adhesive, and self-healable composite hydrogel wound dressing with excellent antibacterial activity. *Chem. Eng. J.* **2022**, *450*, 138201. [CrossRef]
147. Wang, L.; Hussain, Z.; Zheng, P.H.; Zhang, Y.J.; Cao, Y.; Gao, T.; Zhang, Z.Z.; Zhang, Y.H.; Pei, R.J. A mace-like heterostructural enriched injectable hydrogel composite for on-demand promotion of diabetic wound healing. *J. Mater. Chem. B* **2023**, *11*, 2166–2183. [CrossRef]
148. Ha, S.S.; Kim, J.H.; Savitri, C.; Choi, D.; Park, K. Nano-Sized Extracellular Matrix Particles Lead to Therapeutic Improvement for Cutaneous Wound and Hindlimb Ischemia. *Int. J. Mol. Sci.* **2021**, *22*, 13265. [CrossRef]
149. Perez-Rafael, S.; Ivanova, K.; Stefanov, I.; Puiggali, J.; del Valle, L.J.; Todorova, K.; Dimitrov, P.; Hinojosa-Caballero, D.; Tzanov, T. Nanoparticle-driven self-assembling injectable hydrogels provide a multi-factorial approach for chronic wound treatment. *Acta Biomater.* **2021**, *134*, 131–143. [CrossRef]
150. Palem, R.R.; Madhusudana Rao, K.; Kang, T.J. Self-healable and dual-functional guar gum-grafted-polyacrylamidoglycolic acid-based hydrogels with nano-silver for wound dressings. *Carbohydr. Polym.* **2019**, *223*, 115074. [CrossRef]
151. Cheng, W.H.; Chen, Y.H.; Teng, L.J.; Lu, B.H.; Ren, L.; Wang, Y.J. Antimicrobial colloidal hydrogels assembled by graphene oxide and thermo-sensitive nanogels for cell encapsulation. *J. Colloid Interface Sci.* **2018**, *513*, 314–323. [CrossRef]
152. Niu, Y.L.; Guo, T.T.; Yuan, X.Y.; Zhao, Y.H.; Ren, L.X. An injectable supramolecular hydrogel hybridized with silver nanoparticles for antibacterial application. *Soft Matter* **2018**, *14*, 1227–1234. [CrossRef]
153. Ocampo, J.I.G.; Bassous, N.; Orozco, C.P.O.; Webster, T.J. Evaluation of cytotoxicity and antimicrobial activity of an injectable bone substitute of carrageenan and nano hydroxyapatite. *J. Biomed. Mater. Res. Part A* **2018**, *106*, 2984–2993. [CrossRef]
154. Wang, M.; Sa, Y.; Li, P.; Guo, Y.R.; Du, Y.M.; Deng, H.B.; Jiang, T.; Wang, Y.N. A versatile and injectable poly(methyl methacrylate) cement functionalized with quaternized chitosan-glycerophosphate/nanosized hydroxyapatite hydrogels. *Mater. Sci. Eng. C* **2018**, *90*, 264–272. [CrossRef] [PubMed]
155. Wang, C.Y.; Zhao, N.U.; Huang, Y.X.; He, R.; Xu, S.C.; Yuan, W.Z. Coordination of injectable self-healing hydrogel with Mn-Zn ferrite@mesoporous silica nanospheres for tumor MR imaging and efficient synergistic magnetothermal-chemo-chemodynamic therapy. *Chem. Eng. J.* **2020**, *401*, 126100. [CrossRef]
156. Jooken, S.; Deschaume, O.; Bartic, C. Nanocomposite Hydrogels as Functional Extracellular Matrices. *Gels* **2023**, *9*, 153. [CrossRef]
157. Xu, J.P.; Chen, T.Y.; Tai, C.H.; Hsu, S.H. Bioactive self-healing hydrogel based on tannic acid modified gold nano-crosslinker as an injectable brain implant for treating Parkinson's disease. *Biomater. Res.* **2023**, *27*, 8. [CrossRef] [PubMed]
158. Osi, B.; Khoder, M.; Al-Kinani, A.A.; Alany, R.G. Pharmaceutical, biomedical and ophthalmic applications of biodegradable polymers (BDPs): Literature and patent review. *Pharm. Dev. Technol.* **2022**, *27*, 341–356. [CrossRef]
159. Guan, X.F.; Avci-Adali, M.; Alarcin, E.; Cheng, H.; Kashaf, S.S.; Li, Y.X.; Chawla, A.; Jang, H.L.; Khademhosseini, A. Development of hydrogels for regenerative engineering. *Biotechnol. J.* **2017**, *12*, 1600394. [CrossRef]

160. Wang, W.; Tan, B.Y.; Chen, J.R.; Bao, R.; Zhang, X.R.; Liang, S.; Shang, Y.Y.; Liang, W.; Cui, Y.L.; Fan, G.W.; et al. An injectable conductive hydrogel encapsulating plasmid DNA-eNOs and ADSCs for treating myocardial infarction. *Biomaterials* **2018**, *160*, 69–81. [CrossRef]
161. Jeong, S.H.; Fan, Y.F.; Baek, J.U.; Song, J.; Choi, T.H.; Kim, S.W.; Kim, H.E. Long-lasting and bioactive hyaluronic acid-hydroxyapatite composite hydrogels for injectable dermal fillers: Physical properties and in vivo durability. *J. Biomater. Appl.* **2016**, *31*, 464–474. [CrossRef]
162. Wang, A.Y.; Podlasek, C.A. Role of Nanotechnology in Erectile Dysfunction Treatment. *J. Sex. Med.* **2017**, *14*, 36–43. [CrossRef]
163. Yuan, Y.H.; Wu, H.; Ren, X.Y.; Wang, J.W.; Liu, R.Q.; Hu, B.H.; Gu, N. Dual-network hydrogel based on ionic nano-reservoir for gastric perforation sealing. *Sci. China Mater.* **2022**, *65*, 827–835. [CrossRef]

**Disclaimer/Publisher's Note:** The statements, opinions and data contained in all publications are solely those of the individual author(s) and contributor(s) and not of MDPI and/or the editor(s). MDPI and/or the editor(s) disclaim responsibility for any injury to people or property resulting from any ideas, methods, instructions or products referred to in the content.

Review

# Mucoadhesive Polymers and Their Applications in Drug Delivery Systems for the Treatment of Bladder Cancer

Caroline S. A. de Lima [1,*], Justine P. R. O. Varca [1], Victória M. Alves [1], Kamila M. Nogueira [1], Cassia P. C. Cruz [1], M. Isabel Rial-Hermida [2], Sławomir S. Kadłubowski [3], Gustavo H. C. Varca [1] and Ademar B. Lugão [1]

[1] Nuclear and Energy Research Institute, IPEN-CNEN/SP—University of São Paulo, Av. Prof. Lineu Prestes, No. 2242, Cidade Universitária, São Paulo 05508-000, Brazil
[2] I+D Farma Group (GI-1645), Departamento de Farmacología, Farmacia y Tecnología Farmacéutica, Facultad de Farmacia, Instituto de Materiales (iMATUS) and Health Research Institute of Santiago de Compostela (IDIS), Universidade de Santiago de Compostela, 15782 Santiago de Compostela, Spain
[3] Institute of Applied Radiation Chemistry (IARC), Lodz University of Technology, Wroblewskiego No. 15, 93-590 Lodz, Poland
* Correspondence: caroline.lima@usp.br

Citation: de Lima, C.S.A.; Varca, J.P.R.O.; Alves, V.M.; Nogueira, K.M.; Cruz, C.P.C.; Rial-Hermida, M.I.; Kadłubowski, S.S.; Varca, G.H.C.; Lugão, A.B. Mucoadhesive Polymers and Their Applications in Drug Delivery Systems for the Treatment of Bladder Cancer. Gels 2022, 8, 587. https://doi.org/10.3390/gels8090587

Academic Editor: Jordi Puiggali

Received: 1 August 2022
Accepted: 8 September 2022
Published: 15 September 2022

Publisher's Note: MDPI stays neutral with regard to jurisdictional claims in published maps and institutional affiliations.

Copyright: © 2022 by the authors. Licensee MDPI, Basel, Switzerland. This article is an open access article distributed under the terms and conditions of the Creative Commons Attribution (CC BY) license (https://creativecommons.org/licenses/by/4.0/).

**Abstract:** Bladder cancer (BC) is the tenth most common type of cancer worldwide, affecting up to four times more men than women. Depending on the stage of the tumor, different therapy protocols are applied. Non-muscle-invasive cancer englobes around 70% of the cases and is usually treated using the transurethral resection of bladder tumor (TURBIT) followed by the instillation of chemotherapy or immunotherapy. However, due to bladder anatomy and physiology, current intravesical therapies present limitations concerning permeation and time of residence. Furthermore, they require several frequent catheter insertions with a reduced interval between doses, which is highly demotivating for the patient. This scenario has encouraged several pieces of research focusing on the development of drug delivery systems (DDS) to improve drug time residence, permeation capacity, and target release. In this review, the current situation of BC is described concerning the disease and available treatments, followed by a report on the main DDS developed in the past few years, focusing on those based on mucoadhesive polymers as a strategy. A brief review of methods to evaluate mucoadhesion properties is also presented; lastly, different polymers suitable for this application are discussed.

**Keywords:** mucoadhesion; drug release; bladder tumor; polymeric hydrogels; intravesical therapy

## 1. Overview

One of the most probable causes of mortality in the worldwide population is cancer. The prevalence of this set of diseases seems to be decreasing very slowly due to enhancements in early detection and better treatments. Nevertheless, cancer remains a major problem concerning public health systems [1]. Above 18 million new cases are diagnosed each year, and one in every five people develops this condition before the age of 75 years old. Subsequently, around 10 million people die from cancer per year [2].

Bladder cancer is the tenth most common type, representing 3% of the new diagnoses and 2.1% of cancer deaths [2,3]. Focusing on gender, men present three to four times more chances to develop bladder cancer than women [4].

Moreover, there is a significant variance concerning occurrence in the geographical regions; higher rates are observed in Europe and North America, while a lower percentage of cases can be found in Latin America and Northern Africa (Figure 1). The registers concerning bladder cancer vary around the world and are more easily found in European countries and Australia. Developing countries usually lack registers of regional recurrence of cases, in addition to being deficient in providing access to care and diagnostic procedures. However, the differences in recurrence are mostly due to differences in exposure to risk factors such as cigarette smoking, chemical carcinogens, chemotherapy, pelvic radiotherapy, traces of arsenic in drinking water, or endemic chronic urinary infections caused by *Schistosoma haematobium* [3,5].

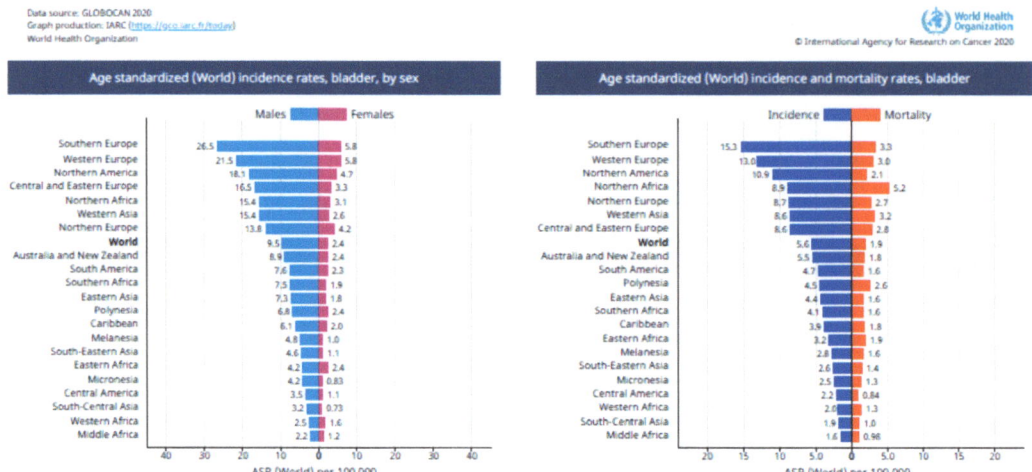

**Figure 1.** Age-standardized incidence rates of bladder cancer in the world according to the World Health Organization. Reprinted from Bladder Cancer, The Global Cancer Observatory, Copyright (2020) [2].

Smoking is the main risk factor for bladder cancer and is related to up to 50% of all cases, particularly urothelial tumors. On the other hand, the infection by *Schistosoma haematobium* is usually related to squamous cell carcinoma. For example, in Egypt, where there was an endemic scenario related to schistosomiasis, there is a dominance of this type of bladder cancer [3].

The symptoms that could potentially indicate the existence of the tumor are the presence of blood in the urine, irritative voiding symptoms such as urgency to urinate frequently, and repetitive urinary infections. Furthermore, 75% of bladder tumors are non-muscle-invasive (urothelial) and, therefore, less aggressive, while the other 25% are muscle-invasive or metastatic diseases. The stage of urothelial carcinoma is the most important prognostic factor, which is based on cytologic atypia.

The tumors are classified according to the TNM scale, which describes tumor size/depth and nodal or metastatic spread (Ta, T1, T2, T3, or T4—Figure 2), and the muscle-invasive forms are those above T2. However, T1 tumors must receive significant attention because they affect the lamina propria, which indicates their potential to become invasive [6,7]. Depending on the disease stage, different protocols of therapies are required. Superficial tumors are often treated with single instillation of mitomycin C (MMC), epirubicin, gemcitabine, or BCG (bacillus Calmette–Guérin), while more invasive tumors may demand the combination of more than one drug for chemotherapy [8,9].

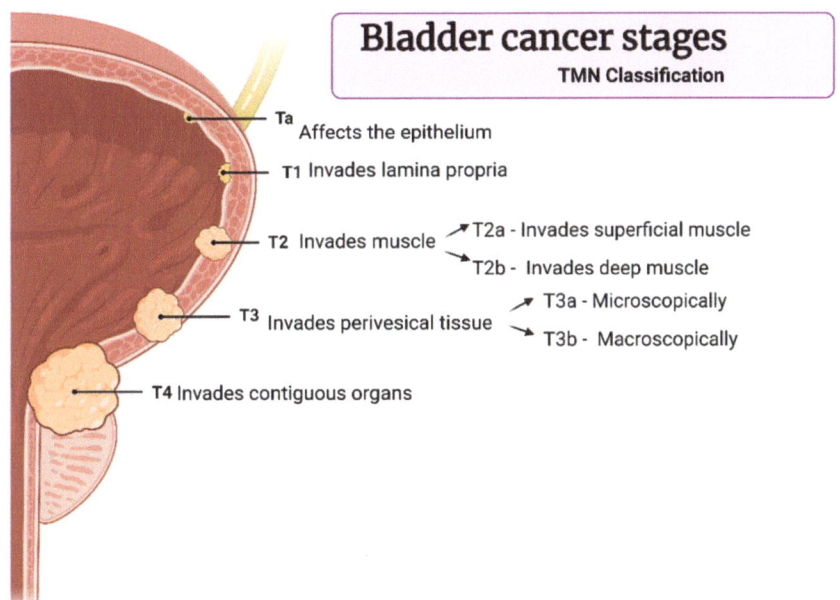

**Figure 2.** Bladder cancer stages according to TNM classification. Created with BioRender.com adapted from Surgery, 34:10, Down et al., Bladder Cancer, Pages 532–539, Copyright (2016), with permission from Elsevier [7].

## 2. Fundamentals

### 2.1. Mucosa Structure

The urinary bladder wall is composed of the urothelium (inner layer), the lamina propria (submucosal connective tissue layer), the muscular layer, and the serosal layer covering it on the outer layer (Figure 3). Usually, women present $3.0 \pm 1.0$ mm of bladder wall thickness, while men present $3.3 \pm 1.1$ mm [10].

The internal face of the bladder is covered by a mucosa composed of transitional epithelium known as the urothelium, basement membrane, and sub-urothelium. The urothelium is a specialized epithelium coated with mucopolysaccharide and glycosaminoglycan, with an important function of protection from the urine. The structure of this mucosal surface is wrinkled, which allows the cycles of filling and voiding the vesicle without compromising the barrier function. The urothelium has three layers; the first one is made of basal cells attached to a basement membrane. The superficial layers, i.e., the second and the third ones, are made of large hexagonal cells, the umbrella cells. One of the main functions of the urothelium is to isolate the urine from the underlying tissues, which is possible due to the tight junctions between umbrella cells. Between the urothelium and the detrusor layer, there is a layer composed of an extracellular matrix known as the "sub-urothelium". This layer contains fibril-shaped or bundle-shaped collagens (type I and III), an elastin fibrous network, interstitial cells, fibroblasts, adipocytes, afferent and efferent endings, blood vessels, and a muscular layer called *muscularis mucosae* [10,11].

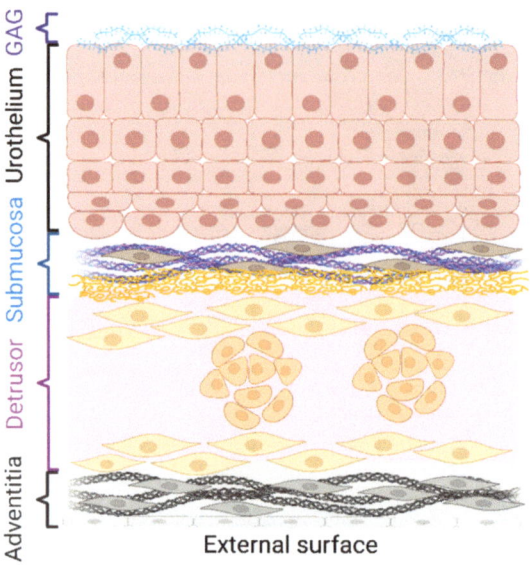

**Figure 3.** Layers that compose the bladder wall. The inner layer is called the urothelium, and its luminal surface is covered with glycosaminoglycan (GAG). The urothelium is composed of three sublayers. The first one is made of basal cells attached to a basement membrane; the superficial layers, i.e., the second and the third ones, are made of large hexagonal cells, the umbrella cells. Under the urothelium, the submucosa or lamina propria is composed of fibril-shaped or bundle-shaped collagens (type I and III), an elastin fibrous network, interstitial cells, fibroblasts, adipocytes, afferent and efferent endings, and blood vessels. Lastly, the outer layer is the muscular one (detrusor) covered by the adventitia. Created with BioRender.com, adapted from open access Wang et al., Pharmaceutics; published by MDPI, 2021 [12].

*2.2. Available Treatments*

Superficial bladder cancer is the most frequent form of bladder cancer. Usually, it is treated by telescopic removal of the tumor using a technique called transurethral resection of bladder tumor (TURBIT), followed by the instillation of chemotherapy or immunotherapy. The monitoring of the bladder is conducted by cystoscopy. Around 70% of the patients diagnosed with bladder cancer present the non-muscle-invasive kind. The American Urological Association (AUA) classifies patients into low-, intermediate-, and high-risk categories, according to factors such as age, smoking history, and symptoms (Figure 4) [9]. Low-risk tumors (low-grade Ta) are treated by TUBIT followed by a single instillation of a chemotherapy drug (MMC, epirubicin, or gemcitabine, for example) [13]. High-grade Ta and T1 just treated with TURBIT present a 50% chance of recurrence; thus, it is common to combine TURBIT with MMC or BCG. Chemotherapy is used when there is a greater risk of progression or recurrence in non-muscle-invasive types.

When the patient presents a high risk, major surgery to remove the organ (cystectomy) could be considered. In cases where the tumor has invaded the bladder muscle, the cure may be achieved via treatment with chemotherapy, radiotherapy, or cystectomy. On the other hand, when the tumor presents too high a grade and, therefore, does not have a perspective of cure, treatment with radiotherapy and chemotherapy are extremely recommended [14].

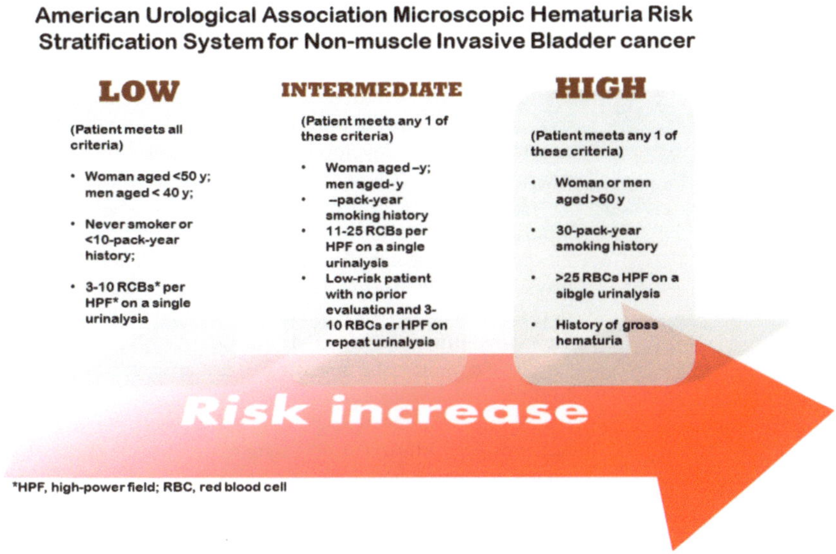

**American Urological Association Microscopic Hematuria Risk Stratification System for Non-muscle Invasive Bladder cancer**

**LOW**
(Patient meets all criteria)
- Woman aged <50 y; men aged < 40 y;
- Never smoker or <10-pack-year history;
- 3-10 RCBs* per HPF* on a single urinalysis

**INTERMEDIATE**
(Patient meets any 1 of these criteria)
- Woman aged –y; men aged- y
- –pack-year smoking history
- 11-25 RCBs per HPF on a single urinalysis
- Low-risk patient with no prior evaluation and 3-10 RBCs er HPF on repeat urinalysis

**HIGH**
(Patient meets any 1 of these criteria)
- Woman or men aged >60 y
- 30-pack-year smoking history
- >25 RBCs HPF on a sibgle urinalysis
- History of gross hematuria

**Risk increase**

*HPF, high-power field; RBC, red blood cell

**Figure 4.** Stratification system for superficial bladder cancer.

BCG was developed as a vaccine for the prevention of tuberculosis disease. However, it has been applied in oncology as an immunotherapeutic for several types of cancer, including non-muscle-invasive bladder cancer. Since 1976, BCG has been recommended for superficial bladder cancer treatment for patients who present a high risk for recurrence and progression as it is the best option to delay them [6]. The vaccine acts by recruiting different types of cells in the tumor microenvironment, such as $CD4^+$ and $CD8^+$ lymphocytes, granulocytes, macrophages, and dendritic cells, which lead the tumor cells to apoptosis [8].

Non-metastasized bladder tumors are treated with intravesical chemotherapy. According to the tumor progression, a single chemotherapeutic or a combination may be used. Gemcitabine and cisplatin, dose-dense methotrexate, vinblastine, doxorubicin (DOX) and cisplatin (MVAC), cisplatin, methotrexate, and vinblastine (CMV), and gemcitabine and paclitaxel are the common combination drugs [8].

Intravesical formulations to treat bladder cancer must attend to some specific features such as having the ability to overcome the urothelium wall, molecular weight under or equal to 200 Da, pH between 6 and 7, and aqueous/organic phase partition coefficient from $-0.4$ to $-0.2$ or from $-7.5$ to $-8.0$. In addition, the presence of charge in eventual drug nanoparticles may help with cellular uptake. Positively charged particles are more rapidly absorbed into tissues in comparison with anionic or neutral particles [15,16].

Usually, the volume of drug formulation instilled into the urine bladder is around 50 mL. Afterward, micturition is prevented for 1–2 h. Regardless of the preparation of the patient, residual urine is often present in the human urinary bladder, which causes dilution of the drug or even washing it out. These circumstances demand frequent catheter insertion with a reduced interval between doses and even irritation of the urothelial lining or urinary tract infection. To overcome that, new formulations with improved mucoadhesion, targeting, and controlled delivery have been studied in the past years [15].

### 2.3. Drug Delivery Systems (DDS)

One of the main advantages of drug delivery systems is the release of the drug in a targeted location, increasing the absorption by the organism. Accordingly, drug delivery systems are an option to increase treatment efficacy.

In DDS, there are generally two factors used to evaluate the efficacy of the system: the quantity of the drug loading and the duration of the presence of the drug in the organism.

Consequently, the design of drug delivery systems involves the chemical formulation of the drugs, the route of administration, the form of dosage, and the use of supplemental medical devices [17].

Researchers aim to develop a target-specific, effective, and safe drug delivery system to boost therapeutic actions and reduce side effects. Advances in drug delivery studies can facilitate the development of an active carrier for targeted action with improved pharmacokinetic behavior [18].

Drug properties can vary significantly when used to treat the same symptoms, depending on the chemical composition, size, hydrophilicity, and ability to bind to a specific receptor. The drugs can suffer from insufficient bioavailability due to insolubility in physiological fluids and low permeability of different organs. Therefore, the therapeutic performance is dependent not only on the activity of the applied drug but also on the bioavailability of the targeted site [19].

The delivery of a specific drug at a programmed rate over a prolonged period is a topic of interest in the drug delivery field. The strategy of releasing drugs at slower rates is very useful for pharmaceutical ingredients that are either subject to a fast metabolism and eliminated from the organism quickly after administration or for providing extended pharmacological action. Sustained drug delivery can be reached by preventing the drug molecules from completely entering the aqueous environment for a viable period; it can be achieved by adjusting the degradation speed of the carrier or by adjusting the diffusion rate of the drug over an insoluble polymer matrix or shell. A constant dosage of drug within the therapeutic window is beneficial to counter the side-effects related, for example, to chemotherapy [20].

The ideal drug carrier should have the following characteristics: good biocompatibility and specific release of drugs at the lesion tissue or targeted cells. Even though no clinical formulation possesses all these characteristics, researchers continue to design smart DDSs with multiple functions, aiming to explore improved strategies for the treatment of diseases and to obtain promising formulations for clinical translation [21]. Below, several DDS implemented for the treatment of BC are briefly explained.

Thiolated chitosan has been frequently explored for intravesical DDS due to its high mucoadhesive properties, but usually, this kind of system is used for the delivery of hydrophilic drugs. To allow the administration of a lipophilic drug in such a system, a self-emulsifying drug delivery system (SEDDS) decorated with thiolated chitosan was prepared. Formulations composed of S-protected chitosan complexed with sodium dodecyl sulfate (SEDDS-CS-MNA-SDS) were those able to be retained in the porcine mucosa for a longer period after voiding several times [22].

A new intravesical DDS composed of a Foley-type catheter (FT-C), which contains an inflation balloon at the tip, was developed by replacing the impermeable silicone rubber of the balloon with a permeable membrane made of interpenetrating polymer network (IPN). The system allowed the diffusion of water-soluble drugs such as MMC, providing prolonged drug release into the bladder. Drug release and anti-carcinoma cell efficacy were investigated. In vitro results showed a sustained release of MMC for up to 12 days with an inhibitory effect against HTB-9 (ATCC bladder carcinoma cell line), but that time could be extended once the drug reservoir can be reloaded without removing the catheter. In vivo short-term studies were also performed in porcine models, and the therapeutic MMC concentration was released after 2 h. However, optimization of the system and longer pre-clinical studies are needed, as little or no MMC tissue uptake was observed [23].

Kaldybekov and coworkers [24] developed maleimide-functionalized PEGylated liposomes (PED-Mal) for intravesical drug delivery. The liposomes presented good adhesion to the bladder mucosa, with a retention of 32% of the formulation after 50 min washing. Fluorescence microscopy assays revealed that the PEGylated liposomes presented a higher capacity of permeation than conventional and PEG-Mal ones. This is mainly because the maleimide-functionalized PEG liposomes formed strong covalent bonds to the mucosa slowing down their penetration. Concerning drug release, PEG-Mal liposomes presented a sustained release through 8 h while conventional and PEGylated ones presented faster release, in 2 h and 4 h, respectively.

New approaches to drug design based on nanoparticles and nanostructures for effective drug delivery are crucial for the future of medical treatment, especially for cancer therapy. Nanotechnology associated with the appropriate material may present great potential in increasing drug delivery efficiency. Furthermore, for biomedical applications, they must be biodegradable, have a prolonged circulation half-life, not be inclined to aggregate or cause an inflammatory response in the organism, and be cost-effective. The efficacy of those structures is very dependent on their chemical properties, as well as on their size, charge, shape, surface modifications, and loading methods [25].

Nanoparticles (NPs) or nanocarriers (NCs) are increasingly considered candidates to safely carry therapeutic agents into selected sections in the body, such as a cell or a particular tissue. Various nano or micro delivery systems are designed to encapsulate the active agent such as polymeric nanocapsules, dendrimers, and liposomes. They can conceal some of the adverse biopharmaceutical properties of the molecule and replace them with properties of materials used for nano-delivery systems. Furthermore, advances in the nanomedical field are also applied for site-specific drug delivery [26]. The benefits of nanoscience and nanotechnology progress and their application in therapeutic drug delivery are huge, aiming to overcome the undesirable effects of previous therapies and develop treatments for several diseases. As a result, the pharmacokinetics can be modulated, and the transport and specific targeting through the controlled drug release with reduced dosing can be achieved. In addition, the solubility, biodistribution, and in vivo stability can be increased [25].

For instance, mesoporous silica nanoparticles were modified superficially with poly (amidoamine) (PAMAM) dendrimers through a layer-by-layer method for the delivery of DOX in the treatment of bladder cancer. The number of PAMAM dendrimer layers was capable of controlling the release rate of the system which was triggered by acid pH. The mucoadhesion was increased by enhancing the number of amino groups in the PAMAM. This was concluded after observing the increase in the hydrodynamic size of nanoparticles after electrostatic interactions between their positive charges and the mucin's negative charges. However, there was no significant difference between two and three layers of the polymer, indicating its biding saturation [27].

Self-immolative systems (SIS) are systems capable of being activated by a stimulus that will initiate spontaneous intramolecular disassembling, breaking them into their building blocks and, therefore, releasing the drug encapsulated in the structure. This feature has received attention as it is possible to program the drug release according to the specific environment found in diseased tissues such as different pH, reductive conditions, or enzyme expression [28]. The synthesis of a macromolecular system with high renal clearance efficiency and activatable near-infrared fluorescence was reported as a self-immolative system with the potential for real-time noninvasive imaging of orthotopic bladder cancer (Figure 5). The aminopeptidase N enzyme (APN) has paramount importance in the processes of tumor invasion, angiogenesis, and metastasis; accordingly, it is overexpressed in BC, in a way that its levels indicate the tumor size, lymph node metastasis, and metastasis stage. In addition, APN is considered a reliable urinary biomarker for BC detection. However, the challenge related to the optical imaging of BC relies on the ability of probes to go through renal clearance to reach the bladder and to present high specific signals concerning BC-associated biomarkers. The preclinical results of this study showed a renal clearance efficiency of 94% ID 24 h post-injection, and the synthesized macromolecule was effectively able to detect the APN levels related to bladder cancer [29].

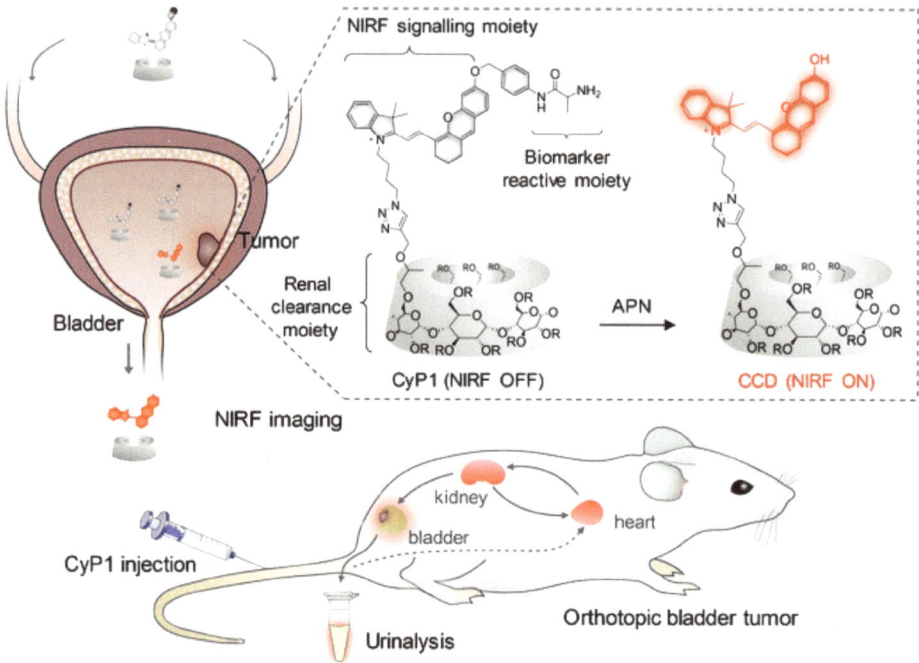

**Figure 5.** Scheme representing the design and mechanisms of the renal-clearable macromolecular reporter (CyP1) for NIRF imaging and urinalysis of BC in living mice. Chemical structures of CyP1 and its activated form as CCD in response to APN (R = H or $CH_2CHOHCH_3$) are also represented. Huang et al.: A Renal-Clearable Macromolecular Reporter for Near-Infrared Fluorescence Imaging of Bladder Cancer. Angewandte Chemie. 2020. 59. P. 4416. Copyright Wiley-VCH GmbH. Reproduced with permission [29].

## 2.4. Adhesion/Bioadhesion/Mucoadhesion

Defined as the formation of an attachment between a biological material and an artificial substrate, bioadhesion is of interest in the development of drug delivery systems. Usually, biopolymers show bioadhesive properties and are used for diverse therapeutic purposes. The bioadhesive polymers can be classified into two groups: (i) specific, with the ability to adhere to certain chemical structures within the biological molecules, and (ii) nonspecific, with the capacity of binding the cell surfaces and the mucosal layer [30].

The direct contact between a delivery vehicle and an absorptive epithelium can result in the intensified specificity and therapeutic effectiveness of delivered compounds [31]. Bioadhesive pharmaceutical formulations are usually designed to enhance drug bioavailability by increasing the residence time of drug compounds and localizing the effect at the targeted site. Simultaneously, they contribute to local drug delivery formulation design, improving bioavailability by avoiding metabolic pathways [32,33]:

The current mucoadhesive formulations deal, primarily, with adhesion force-mediated transmucosal drug delivery. Those adhesion forces are generated from the epithelial cell layer, the mucus layer, or a combination of both. There are various mucoadhesive formulations, such as gel, tablet, ointment, powder, and film agent, and their sites of absorption are various mucosal epidermis cells, including buccal and nasal mucosa or ocular surfaces [34].

While mucoadhesive materials can increase contact with a specific site or tissue, the mucoadhesion of a system can be compromised by the natural defenses of the body against the deposition of impurities onto the mucous membrane. Hence, there is a need to possess suitable features to help the maintenance of effective drug concentration at the action site, control the drug release, enable a decrease in drug administration frequency, and increase patient compliance to the therapy [35].

## 3. Mucoadhesive Polymeric Drug Delivery Systems

Mucoadhesive DDS may be of diverse types such as particles with surface modification (cationic or thiolated particulate system) and chemical or physical drug entrapment (system composed of nanoparticles and hydrogels). In a system composed of nanoparticles and hydrogels, it is still possible to obtain non-floating in situ gelations (non-floating composite system of polymeric nanoparticles and hydrogels) [15]. Therefore, to improve drug residence in the urothelial wall of the bladder, after urination, strategic and advanced formulations (Table 1) have been developed through intelligent drug carriers, which combine characteristics of solubility, permeability, adhesion, loading, release, and cytotoxic effect against cancer cells. The mucoadhesion of a system relies on the interaction between the hydrogel and the mucosa structure. The swelling of the hydrogel during contact with the mucosa promotes bioadhesion with physical or chemical interactions and helps to induce cellular uptake. The interactions between the hydrogel and mucosa depend on features such as molecular weight of the polymer, hydration, hydrogen bonding capacity, chain flexibility, charge, and biological environmental factors [36]. In Table 2, some advantages and disadvantages of mucoadhesive DDS are listed.

**Table 1.** Mucoadhesive polymeric systems for controlled drug release in the bladder—advanced formulations.

| Drug | Carrier | Polymer | Cancer Cells | Encapsulation Efficiency (%) | Reference |
|---|---|---|---|---|---|
| Doxorubicin | Nanodiamonds with surface modification | Chitosan | HT-1197 | >90 | [37] |
| Doxorubicin | Nanoparticles with surface modification | Poly(amidoamine) | UMUC3 | >90 | [27] |
| Gambogic acid | Nanoparticles with surface modification | Chitosan | MB49 and MH-3T3 | - | [38] |
| Docetaxel | Nanogel | Polyacrylamide | UMC3 and T24 | >90 | [39] |
| MMC | Gel | Chitosan/$\beta$-glycerophosphate | - | - | [40] |
| Fluorescein diacetate | Micro and nanoparticles | CH glycol (GCH), N-acetylcysteine (NAC), and glutadione (GSH) | - | 12.2–100% | [41] |
| Gemcitabine hydrochloride | Microspheres | Carbopol 2020 NF, Eudragit E100 (EE100), poloxamer and chitosan | T24 (ATCC HTB4TM) and RT4 (ATCC HTB2TM) | >80 | [42] |
| Paclitaxel | Nanoparticles | Gelatin | - | 0.52 | [43] |
| Doxorubicin and peptide-modified cisplatin | Nanocapsules | Chitosan, polymethacrylic acid | UMUC3 | >80 | [44] |
| Paclitaxel | Liposomes in a gel system | Gellan gum | NBT-II and T24 (ATCC, USA) | >90 | [45] |

**Table 2.** Mucoadhesive DDS: advantages and disadvantages [46].

| Advantages | Disadvantages |
| --- | --- |
| Prolong drug residence time at the tumor site | Dislodgement of the formulation may happen |
| Increase drug bioavailability | Overhydration may compromise the formulation structure |
| Reduce dosing frequency | |
| Improve drug permeability | |
| Reduce the dose of drug administered | |
| Fast onset of action | |

An emergent delivery system, based on nanodiamonds (NDs) with a massive surface-to-area ratio, was developed to improve bladder cancer treatment. In this study, chitosan (CH) was used to attribute adhesiveness, increasing the electrostatic force between the positive charge of the polymer and the negative charge of the mucosal wall. Furthermore, the polymer provided steric stability for the colloid, preventing rapid aggregation of the NDs through ion release when in contact with physiological media. To increase the stability of NDs loaded with DOX and coated with CH (CH-NDx) in a neutral pH medium, the CH-NDx were coated with polyanionic molecules (dextran sulfate—DS and pentasodium tripolyphosphate—TPP). Greater release and greater retention of the drug were identified in the TPPCH-NDx formulation (75% and 45%, respectively). TPPCH-NDx also proved to be efficient in the cytotoxic effect against cancer cells, with a lower $IC_{50}$ compared to the $IC_{50}$ of the free drug and the drug trapped only in uncoated NDs. The NDs system proposed by the authors presented encouraging results, highlighting it as a possible option for a more efficient bladder cancer treatment [37].

Studies by Wang et al. [27] focused on the surface modification of mesoporous silica nanoparticles (MSNPs) loaded with DOX, through coating with poly(amidoamine) (PAMAM). In this study, the potential of loading efficiency, release, mucoadhesion, and cytotoxic effect could be controlled by the number of PAMAM dendrimer layers on the surface of MSNPs. Therefore, MSNPs formulations containing zero (G0), one (G1), two (G2), and three (G3) layers of PAMAM were created for investigation. Mucoadhesion assays with mucin particles and in ex vivo porcine bladders showed that the MSNPs-G2 formulation showed greater adhesion to the urothelial mucosa compared to MSNPs-G0. This result is associated with the capacity for electrostatic interaction between positively charged PAMAM and negatively charged urothelium mucin. The loading efficiency and 100 h DOX release for MSNPs-G2 were 95.5% and 65.3% vs. 94.7% and 93.8% for MSNPs-G0, respectively. The presence of PAMAM dendrimer decreased the release rate of DOX, but increased the mucoadhesion of the system, allowing sustained release. The authors also observed that the release of DOX@MSNPs-G2 is responsive to pH, with values of 89.6% (pH 4.5), 67.7% (pH 6.1), and 34.3% (pH 4. 5). The study of cytotoxic effects had an $IC_{50}$ of 1.07 ug/mL for free DOX and 5.63 ug/mL for DOX@MSNPs-G2. The obtention of a formulation with sustained release is a common problem due to the frequent burst behavior of hydrogels. In this research, this problem was fixed by upgrading the functionalities of the materials, such as improvement of mucoadhesion and pH responsiveness.

Xu and coworkers [38] developed mucoadhesive CH nanoparticles modified with mono-benzylic acid (CB) responsive (SSCB) and nonresponsive (CCCB) carriers to intracellular glutathione (GSH) loaded with the gambogic acid drug (GA) activated by ROS (reactive oxygen species). In this study, drug release could be triggered by increased GSH concentration and ROS levels within bladder cancer cells. To efficiently encapsulate hydrophobic charges, CH was modified by mono-benzylic acid, to create hydrophobicity in the amino groups of CH. Part of these amino groups was modified by hydrophobic portions of the benzyl group. In terms of cytotoxic effect against MB49 (mouse bladder cancer cells) and NIH-3T3 (mouse fibroblast cells), the $IC_{50}$ of GA (free drug) was lower than that of GB (drug activated with ROS). For both cells, the best $IC_{50}$ was in the order GA < SSCBGA < CCCBGA. The study of mucoadhesion was carried out in vivo, with fluorescence imaging testing using mice. The tested formulations were responsive (SS-CBGB) and non-responsive (CCCBGB) GB, with Cy7.5 as a positive control. After 60 h of incubation, SSCBGB and CCCBGB presented enhanced adhesion when compared to free control. In short, cationic CH promoted the mucoadhesion of the prodrug and ROS-related properties gave SSCBGB the precision to deliver GA to bladder cancer cells. This system presented, thus, a great contribution to bladder treatment studies, as it represents a system with improved residence time and with targeting features.

Another system, composed of a polyacrylamide nanogel (PAm) functionalized with a cationic amine group ($NH_2$) and loaded with the hydrophobic drug docetaxel (DTX), was prepared as a mucoadhesive platform for cancer treatment. The interaction between the hydrophilic matrix and the hydrophobic drug constituted a slow and sustained release of up to 9 days; after 9 days of experimentation, 76% of DTX was released from the PAm-NH2-DTX nanogels. The authors carried out a study of the mucoadhesion of PAm-NH2-DTX nanogels comparing them with free CH. It was identified that PAm-NH2-DTX nanogels were more mucoadhesive than CH. The cell viability in human urothelial carcinoma cell lines, UMCU3 and T24, for PAm-N H2-DTX and PAm-DTX was similar to free DTX, but PAm-NH2-DTX exhibited superior inhibition compared to UMUC3 cells when compared to T24 cells, with a minimal inhibitory concentration ($IC_{50}$) of 5.6 ng/mL versus 535.6 ng/mL, respectively. The $IC_{50}$ difference became even more pronounced as exposure time increased. Thus, this study represented an important option of polymer functionalization to increase mucoadhesive properties, extending the options for systems that present electrostatic interaction with urothelium beyond those containing CH [39].

Studies by Kolawole et al. emphasized the use of CH with β-glycerophosphate (CHGP) for its ability to form a thermosensitive in situ gelling system, with physical crosslinking at 37 °C and pH 6. The study related the thermogelling behavior, drug release, retention, and mucoadhesive properties with the molecular weight of chitosan. In this way, CH of low (LCH), medium (MCH), and high (HCH) molecular weight was used for the formulations. HCH in the presence of GP (HCH-GP) showed a higher gelation index at 37 °C of 138.09 PA, compared to the 59.7 and 30.2 PA of MCH-GP and LCH-GP, respectively. In addition to gelling, mucosal retention was also favored by the high weight of CH. In contrast, formulations with HCH-GP showed less release of MMC, compared to formulations of MCH-GP and LCH-GP; therefore, a higher molecular weight of CH led to the lower release of the drug. The release was still favored in the presence of GP; for all molecular weights of CH, the formulations that contained GP showed greater release of the drug compared to the formulations without GP. In terms of mucoadhesion, the presence of GP in the formulations reduced mucosal adhesion, but the formulations still presented satisfactory adhesion results. Mediating all results, the HCH-GP formulation was identified as promising for intravesical application [40]. In this study, the influence of the molar mass of the polymer chosen on the construction of the system, as well as its behavior concerning mucoadhesion and drug release, was clear.

The mucoadhesion of CH for intravesical administration through the thiolation process has been also investigated and reported in the literature. One of the studies addressed the conjugation of CH glycol (GCH) with thiolated polymers N-acetylcysteine (NAC) and glutadione (GSH) to produce microparticles (MP) and nanoparticles (NP). The degree of thiolation by quantifying thiol and disulfide bonds was verified using Ellman's method. In particular, GSH- and NAC-GC were characterized by 3.6 and 6.3 mmol of immobilized free thiol groups and 0.2 and 0.8 mmol of disulfide bonds per gram of polymer conjugate, respectively. Using the mucin particle method, mucoadhesive properties were measured as a function of turbidimetry and zeta potential in artificial urine with pH 5.0 and 7.0. Conjugates of GC with NAC (NAC-GC) and with GSH (GSH-GC), at both pH, showed greater mucoadhesion compared to unconjugated GC. In all profiles, the NAC-GC formula had greater adhesion strength. Thus, it is possible to observe that the preservation of the thiol groups from oxidation and the greater formation of disulfide bonds resulted in the best mucoadhesion property. Polymer thiolation has been a frequent strategy for mucoadhesion improvement [41]. Moreover, studies conducted by Kolawole et al. (2018) showed that, for CH, mucoadhesion is linked to the degree of methacrylation of H-CH (370 kDa) with anhydrous methacrylate. Different levels of the degree of methacrylation in CH were approached as low (LMe-CH) and high (HMe-CHi) molecular weight. Through the ex vivo study in porcine bladder with sodium fluorescein (FS), the high degree of methacrylation formula revealed better mucoadhesion, due to the presence of a higher percentage of unsaturated methacrylate groups that form covalent bonds with thiols present on the mucosal surface.

The mucoadhesion of high-molecular-weight CH was also evaluated under the influence of the degree of boronation, conjugated by reaction with 4-carboxyphenylboronic. This study pointed out that boron CH can interact with the mucosal surface through three mechanisms: (i) covalent bonding of phenylboronic acid with mucosal sialic acid, (ii) hydrogen bonds with glycoproteins, and (iii) electrostatic interaction between cationic polymer and sialic acid. Therefore, using the traction method, it was possible to observe that the mucoadhesive property was improved in the formulation containing a high degree of boronation (HBChi), due to the greater presence of boronate groups interacting with the mucosal surface. This represents another possibility to confer enhanced mucoadhesion features to polymers by modifying their structures [47].

*3.1. Theories and Mechanisms*

Mucoadhesion could be explained by some theories that include the electronic theory, the wetting theory, the adsorption theory, the diffusion theory, the mechanical theory, the cohesive theory, and the fracture theory [48,49].

Despite the complexity of the mechanism responsible for the formation of mucoadhesive forces, there are two general steps involved in the process, namely, contact and consolidation (Figure 6). In the contacting step, the adhesive polymer and the mucous initiate free contact with each other, sometimes influenced by external forces such as the peristaltic movements of the gastrointestinal tract, motions of organic fluids, or Brownian movements. Thus, both attractive and repulsive forces act and the adhesion process initiates only when the repulsive forces are surpassed. This initial step can be correlated to wetting, electronic, adsorption, and mechanical theories [48].

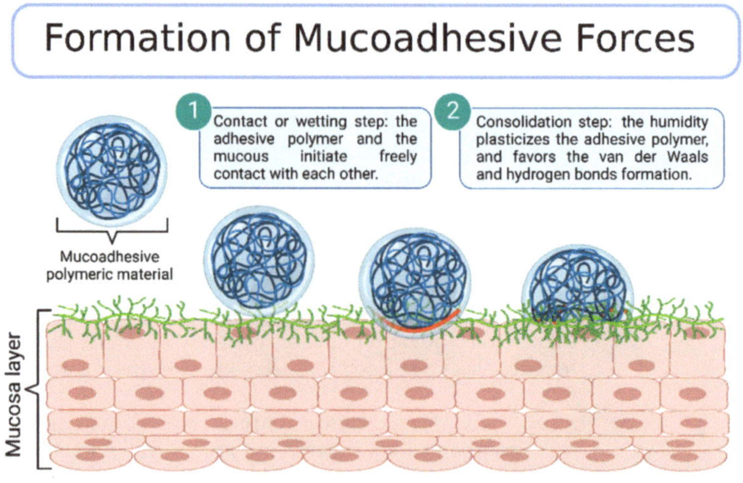

**Figure 6.** Formation of mucoadhesive forces scheme. Created with BioRender.com Adapted from Biomaterials and Bionanotechnology, Tekade et al., Thiolated-Chitosan: A Novel Mucoadhesive Polymer for Better-Targeted Drug Delivery Muktika, pages 459–493, Copyright (2019), with permission from Elsevier [50].

In the consolidation step, the humidity plasticizes the adhesive polymer and favors the formation of van der Waals and hydrogen bonds. Initially, the adhesive polymer and glycoproteins from the mucous interact by interdiffusion and form secondary bonds. After that, the adhesive polymer forms an instantaneous gel hydrated by the aqueous environment [48]. This step is regarded as the diffusion and the cohesive theories, which enforce the idea that the mucoadhesion mechanism is better explained by combining all the mentioned theories [49]. This section describes the various theories and mechanisms involved in the mucoadhesion process.

### 3.1.1. Electronic Theory

The electronic theory (Figure 7) explains the presence of attractive forces between the biological and the adhesive system surfaces due to the formation of an electrical double layer produced from the electron transfer among the surfaces [51].

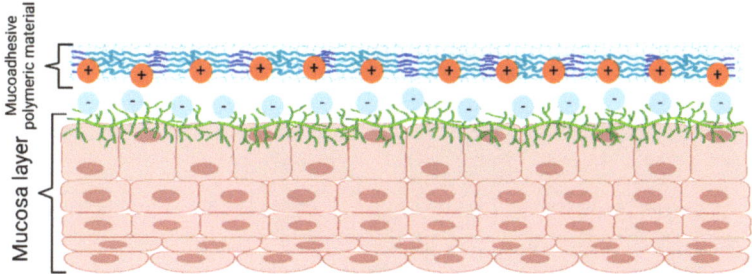

**Figure 7.** Mucoadhesion mechanism according to electronic theory. Created with BioRender.com Adapted from Biomaterials and Bionanotechnology, Tekade et al., Thiolated-Chitosan: A Novel Mucoadhesive Polymer for Better-Targeted Drug Delivery Muktika, pages 459–493, Copyright (2019), with permission from Elsevier [50].

### 3.1.2. Wetting Theory

The wetting theory is applied to adhesive systems with low viscosity and high affinity to the substrate. It correlates the adhesion strength to the contact angle of the low-viscosity system (Figure 8), the spreadability coefficient (the difference in the surface energies between the biological surface and the liquid), and the work of adhesion (the energy needed to separate the two phases). In general, at contact angles close to zero, the adhesion strength is benefited due to the increased contact area. In addition, higher individual surface energies are correlated with a better adhesive strength of the interface [49].

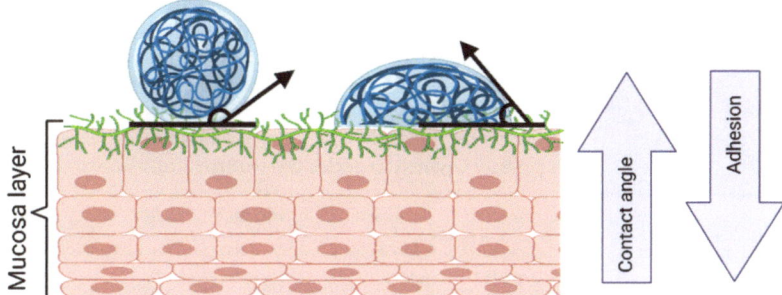

**Figure 8.** The wetting theory according to contact angle. Created with BioRender.com Adapted from Biomaterials and Bionanotechnology, Tekade et al., Thiolated-Chitosan: A Novel Mucoadhesive Polymer for Better-Targeted Drug Delivery Muktika, pages 459–493, Copyright (2019), with permission from Elsevier [50].

### 3.1.3. Adsorption Theory

The adsorption theory approaches the presence of intermolecular forces, namely, hydrogen bonding and Van der Waals forces, that act between the biological substrate and the adhesive material [51]. Despite the isolated interaction being weak, the combined effect of several forces could lead to strong interactions [48].

### 3.1.4. Diffusion Theory

The diffusion theory presumes the polymer chain interpenetration through the substrate surface, forming a network structure (Figure 9). The depth of penetration depends on the polymer diffusion coefficient, flexibility and mobility of the mucin structure, the polymer–substrate contact time, the mutual solubility, and the similarity in the chemical structures [51].

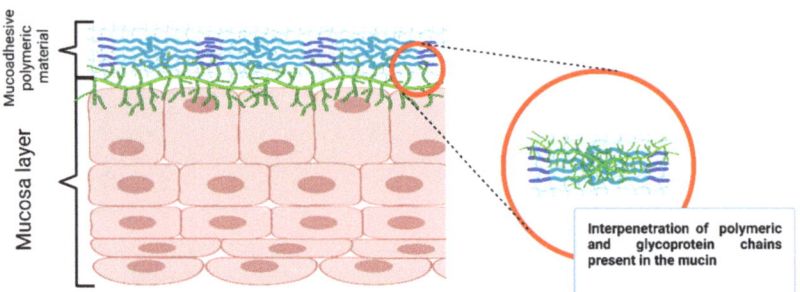

**Figure 9.** Diffusion theory mechanism. Created with BioRender.com.

### 3.1.5. Mechanical Theory

The mechanical theory assumes the diffusion of the low-viscosity polymeric system to an irregular and rough biological surface which must increase the surface area available for interaction, forming an interlocked structure that benefits the adhesion process, as well as viscoelastic and plastic dissipation of energies [48].

### 3.1.6. Cohesive or Fracture Theory

The cohesive theory postulates that the mucoadhesion phenomenon occurs mostly due to the intermolecular interactions among the polymer molecules and biomolecules present in the mucus [52].

This theory considers the strength required for detaching two surfaces after adhesion. Therefore, the adhesion is described by the force required for rupture, in addition to the factors that promote the adhesive interaction, considering the other theories to explain the mucoadhesion process. Moreover, the rupture of the adhesive bonds occurs through the interface, mostly at the weakest point of interaction between the surfaces. This theory is mainly applicable to rigid and semirigid bioadhesive materials [53,54].

Based on previous theories, mucoadhesion can be generally classified into two categories: (i) chemical, which comprehends the electronic and the adsorption theories, and (ii) physical, which includes the wetting, diffusion, and cohesive theories.

## 3.2. Methods to Evaluate Mucoadhesivity of Polymers

There are several in vivo, ex vivo, and in vitro methods for assessing the efficacy of the mucoadhesion capacity of a polymer system. In vitro tests are firstly performed to screen potential bioadhesives and are usually accomplished using mechanical and rheological tests. Some typical tests include tensile strength, shear strength, rheological methods, colloidal gold staining method, mechanical spectroscopic method, and falling liquid film method. This section describes the tests mentioned above.

### 3.2.1. In Vitro and Ex Vivo Methods

Tensile Strength

This method measures the force required to break the adhesive between the mucoadhesive polymer system and the mucous using testers and balances [48]. For example, Ferreira and colleagues [55] evaluated their adhesive formulations using a texture analyzer.

Shear Strength

The shear strength is determined by measuring the required force for a mucoadhesive polymer system to slide from the mucous surface considering the parallel direction to the plane of the contact area. For instance, Silva and collaborators [56] evaluated the synergism between their polymeric adhesive system and mucin using a controlled stress rheometer.

Rheological Methods

The viscometrical method is used to measure the mucin–mucoadhesive polymer bond strength. Thus, the force of mucoadhesion is measured concerning the rheological changes of the polymer–mucin mixtures. The mucoadhesive polymers/mucin mixtures tend to present a higher viscosity than the sum of viscosities of the individual components of the mixture. The interaction between the mucin and the mucoadhesive polymer commonly generates improved viscosity depending on the polymer applied to the system [48].

Washability Test

The washability test is a modification of the Franz diffusion cell to concurrently measure the drug released by diffusion from a mucoadhesive polymeric system and the amount washed by a tangential flow. This method adopts the use of a modified donor that has the chamber closed with two sideways that enable a thermostatic buffer to stream over the sample. The buffer simulates the physiologic conditions and is fluxed over the mucoadhesive polymeric system at a constant rate, being gathered in a beaker under constant stirring. Moreover, the tracer or drug washed is quantified by a proper analytical method [56].

Colloidal Gold Staining Method

The colloidal gold staining method is based on the measurement of the change in the color intensity of the mucin molecules promoted by the red colloidal gold particles resulting from the interaction between the mucoadhesive polymer system and the mucin gold conjugates, which tends to develop a red color on the mucoadhesive surface [48]. Huang and colleagues [57] improved the classic colloidal gold staining technique developed by Park in 1989 by developing the Pt-staining method based on test strips to create platinum nanoshells on the surface of colloidal gold. This method not only retains the original advantages of colloidal gold with easy synthesis and bonding but inserts Pt. nanoparticles with excellent catalytic activity as a signal marker to reach sensitive quantitative detection [57,58].

Mechanical Spectroscopic Method

The mechanical spectroscopic method consists of the study of the effect of pH and polymer chain length over mucoadhesion. The difference between the storage modulus of the mucoadhesive polymer system and the individual components at the same concentration is evidenced by the magnitude of the interaction between the polymer and mucin. A higher difference indicates a stronger presumed interaction [48].

Falling Liquid Film Method

The falling liquid film method is an ex vivo test proposed by Teng and Ho who set small intestine sections from rats on an inclined Tygon tube flute (angle of 45°) [59]. The particle suspension migrates over the mucous interface, and the adhesion strength is determined by the particle portion adhered to the mucous surface [48]. Efiana and collaborators adapted the method for porcine intestinal mucosa, and Sudan Red G was used as a label signal for absorbance measurement [60].

Biacore System

The Biacore is an instrument based on the principle of an optical phenomenon, namely, surface plasmon resonance (SPR), which expresses a response of the measurement of the refractive index that varies as a function of the solute content present in a solution that contacts the sensor chip. During the detection process, the polymer molecules are retained on the sensor chip surface, and the mucin suspension migrates through the sensor chip. When the analyte, mucin particle, links to the ligand molecule, the polymer on the sensor chip surface, the solute concentration, and the refractive index on the surface change, increasing the resonance unit response. Once they dissociate, the resonance unit response falls [30,60].

Confocal Laser Scanning Microscopy (CLSM) Method

Confocal laser scanning microscopy (CLSM or LSCM) is a high-resolution and high-selectivity technique. The main aspect of this optical imaging technique is the possibility of obtaining in-focus images from selected depths. CLSM combines the laser scanning method with the 3D detection of biological objects marked with fluorescent signalers. For example, Dyawanapelly and collaborators [61] evaluated the mucoadhesion properties of the CH oligosaccharide surface-modified polymer nanoparticles developed for mucosal delivery of proteins.

3.2.2. In Vivo Methods

In vivo mucoadhesion evaluation is generally based on residence time or relative bioavailability assays. Due to the cost, time, and ethical concerns, there are few studies described in the literature. Accordingly, these techniques are sparse in comparison with in vitro and ex vivo methods. In vivo methods are a greater indicator of clinical performance, although they cannot distinguish between mucoadhesion and other factors that influence the residence time besides commonly present high standard deviations [51].

The in vivo data can usually be correlated to in vitro analysis. Low in vitro/in vivo correlation of the mucoadhesive strength indicates that a polymeric system with strong mucoadhesion properties in vitro might not reach longer mucosal residence in vivo, which may be explained by the different environments in vitro and in vivo [51]. Gamma scintigraphy is a technique used for the diagnosis of diseases in neurology, oncology, and cardiology based on the use of gamma-ray emitting radioisotopes that allow following in real time the path of the labeled formulation in the body after administration. Therefore, a formulation under investigation may be tracked inside the body via a gamma-emitting radionuclide label. Moreover, it is also possible to obtain quantitative information about the molecules in organs by counting the radioactivity in them. The radiolabeling of the location/organ of interest takes place by using a short-lived radioisotope that can emit gamma rays, such as $^{99}$Tc [62].

*3.3. Mucoadhesive Polymers Suitable for the Development of DDs for Bladder Cancer Treatment*

Several different polymers have been used to receive materials suitable for the treatment of bladder cancer. According to the origin of the macromolecules, they could be synthetic (poly(N-vinyl-2-pyrrolidone), poly(vinyl alcohol), poly(2-hydroxyethyl methacrylate), Carbopol, Pluronic®, etc.) or natural (gellan gum, sodium alginate, hydroxypropyl cellulose, or carboxymethylcellulose). Below, general information on the selected polymers can be found.

Poly(N-vinyl-2-pyrrolidone) (PVP) was developed in 1939 by Walter Reppe at BASF. Subsequently, it started being utilized in a multiplicity of sectors: pharmaceutical, cosmetic, and detergent industries. PVP is now being used in several technical applications such as membranes, glue sticks, hot-melt adhesives, and crop protection. Due to its versatile features (such as water solubility, film and complex formation, and adhesive and bonding power), as well as its toxicological harmlessness, PVP is one of the most interesting technical specialty polymers in the field of chemistry [63]. Grant and colleagues recently prepared electrospun nanofibrous mats of CS and PVP for the delivery of the chemotherapy drug 5-fluorouracil (5-Fu) to treat lung cancer. The developed material demonstrated efficiency in killing cells for over 24 h and, therefore, presented potential as a DDS for the application proposed [64]. Nanocomposites of PVP combined with alginate and polydopamine were also prepared and studied regarding their potential for cancer treatment purposes. The release studies showed the capacity of delivering the DOX drug for 50 h and the possibility to combine a photothermal treatment with chemotherapy. The results confirmed the efficacy of the combined therapy, lowering the cell activity to only 13.2% [65].

Poly(vinyl alcohol) (PVA) is a water-soluble, biodegradable (under both aerobic and anaerobic conditions) [66], and biocompatible polymer that is obtained from poly(vinyl acetate) through alkaline hydrolysis. It presents a high ability to form films, with high surface stabilization and chelation properties [67]. This polymer is one of the most important synthetic polymers used in commercial, industrial, medical, and nutraceutical applications [68]. Furthermore, several studies have used PVA in the development of improved cancer therapies [67]. Ullah and coauthors prepared formulations composed of carboxymethyl chitosan and PVA for the delivery of oxaliplatin in the treatment of colorectal cancer. In their studies, they were able to develop a pH-responsive hydrogel which, together with the concentration variation of both polymers, allowed tailoring the delivery properties of the material [69].

Poly(2-hydroxyethyl methacrylate) (PHEMA) is a stable, optically transparent, hydrophilic methacrylate polymer. In the dry state, the material is hard and glassy; however, in polar media, the pendant hydroxyethyl group can extend outward, and the material becomes soft and flexible. Due to its good biocompatibility, PHEMA has been extensively researched for biomedical applications [70], such as hydrogel systems for drug delivery or scaffolds for tissue engineering. PHEMA-based hydrogels can be engineered to possess similar water content and mechanical properties as tissue and exhibit excellent cytocompatibility. The most prominent example of a biomedical device based on pHEMA may be the very first modern soft contact lenses developed by Otto Wichterle around 1960 [71].

Carbopol (Carbomer) is a high-molecular-weight, acrylic acid-based polymer crosslinked with allyl sucrose or allyl pentaerythritol that contains between 56% and 68% $w/w$ carboxylic acid groups [72]. Carbopol polymers were first described and patented in 1957 [73]. Since then, several release tablet formulations, which involve carbomer matrices, have been patented [74]. Today, Carbopol polymers are widely accepted ingredients in pharmaceutical dosage systems of almost every form, from controlled-release tablets to oral suspensions and other novel delivery systems, as well as a variety of topical products. Carbomers demonstrate good mucoadhesion, particularly at low pH values where they are present in a protonated state [75]. Carbopol 940 has been combined with micelles containing paclitaxel for the local treatment of melanoma. The studies showed that the formulation proposed was capable of increasing the retention the permeability of the drug into the skin. One of the reasons for this behavior was the positive charges presented in the polymer that were also helpful to promote melanoma cellular uptake and improve the in vitro cytotoxicity [76].

Pluronic® (Poloxamer) is a synthetic amphiphilic copolymer based on hydrophilic poly(ethylene oxide) (PEO) blocks and hydrophobic poly(propylene oxide) (PPO) blocks organized in a triblock structure PEO–PPO–PEO. PEG refers to polyols of Mw below 20,000 Da, while PEO is relevant to polyols with higher molecular weight [77]. The properties of the Pluronic® copolymers can be changed by adjusting the molar mass ratio between the PEO and PPO blocks [78]. In an aqueous environment, these block copolymers self-assemble into micelles with a hydrophilic PEO outer shell that interfaces with water. Since these micelles are amphiphilic, they could accommodate lipophilic molecules in the central hydrophobic core area. Consequently, Pluronic® micelles are effectively used as drug carriers because their assemblies can act as passive drug containers [79]. Researchers have developed a system composed of chitosan thioglycolic acid nanoparticles loaded with gemcitabine HCl and dispersed into a bioadhesive CH gel or in an in situ gelling poloxamer solution as potential formulations for the treatment of superficial bladder cancer. Both formulations presented mucoadhesive properties and the capacity of enhancing drug residence time; however, poloxamer lost its gelling property when diluted in artificial urine at body temperature. Thus, it would be recommended to empty the bladder of the patient before application of this formulation to guarantee its gelling capacity and, therefore, good mucoadhesion and a sustained release [80].

*3.4. In Situ Gelling Polymers*

In situ gelling systems are polymeric formulations in solution before entering the body, where, under physiological conditions, they change into a gel form. This can occur through different types of devices correlated with the properties of the polymers used in the delivery system and due to physical or chemical crosslinking that can be triggered by factors such as changes in temperature, pH, and the presence of ions. The sol-gel transition is very common for thermosetting polymers, i.e., those that present an upper-critical solution temperature or lower-critical solution temperature, in which, according to temperature changes, the gel-forming units interact with each other via physical (van der Waals and electrostatic) or covalent bonds, forming a gel network. These systems are characterized by being easy to administer, presenting the sustained release of the drug at the target, with the possibility to be administered via different routes to obtain local or systemic effects of the loaded drug [81,82].

The pH-specific polymers have the characteristics of structures with ionizable groups—weakly basic or acidic. Changing the pH will produce changes in the ionization state, as well as in the solubility and conformation that result in polymer gelation [83].

Thermosensitive in situ gelling systems have sol-gel transition triggered at temperatures close to physiological ones (32–37 °C). This transition occurs via a change in the aqueous solubility of the polymers, characterized by structures of hydrophobic and hydrophilic groups. With the increase in temperature, there is a rearrangement of polymer–water interactions, which is responsible for the polymer separation and dehydration of the solvated polymer chains in a rapid way. These polymers that have hydrophobic and hydrophilic segments form self-assembled micelles that, at higher temperatures, cause their packing and, thus, the change of the solution into gel form [84]. Some polymers are sensitive to ions such as alginate, gellan gum, and pectin, and crosslinking occurs due to some monovalent or divalent cations that are present in physiological fluids, such as tears and saliva. The viscosity of the gel obtained depends on the cation type and its concentration [85].

3.4.1. Thermo-Responsive Systems

In situ gelation triggered by temperature occurs in thermo-responsive polymers, with sol-gel transitions directly linked to specific temperature limits. When a polymer solidification occurs above a temperature limit, the system presents a "lowest critical temperature" [86]. Thus, it is understood that, at low or room temperatures (20–25 °C), these thermo-responsive polymers have a fluid aspect, whereas, under physiological conditions with a temperature between 35–37 °C, they present a gel behavior (Figure 10). These gels sensitive to in situ temperature have high fluidity and low viscosity, providing easy application [87]. Solutions that show a sol-gel change when cooling present a "higher critical temperature" due to micellar growth, hydrophobic interaction, and transition from the coil to a helix. This phenomenon is observed in gelatin or carrageenan solutions, where a random coil shape occurs in the solution, thus generating a continuous network via partial helix formation after cooling [88].

Some polymers present gelation at a higher temperature, due to the gradual loss of water when the temperature rises, thus increasing the intermolecular interaction and aggregation of the network structure, leading to the gelation of the system [89]. For example, hydroxypropylmethylcellulose presents the lowest critical temperature of the solution between 75 and 90 °C, while methylcellulose presents it at 40 to 50 °C [90].

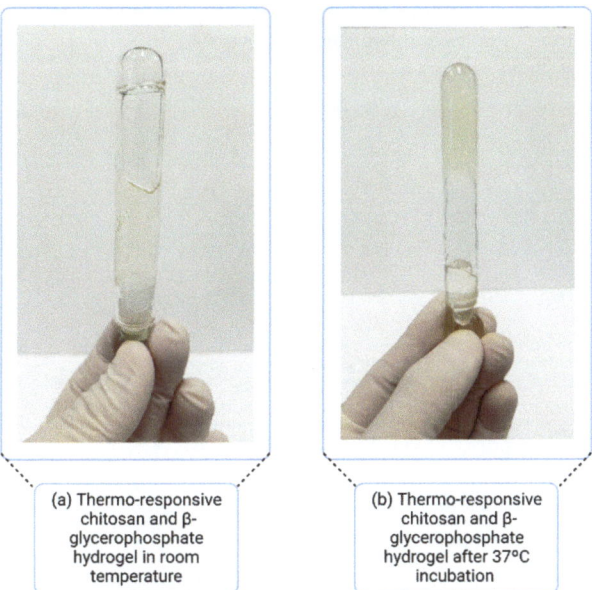

(a) Thermo-responsive chitosan and β-glycerophosphate hydrogel in room temperature

(b) Thermo-responsive chitosan and β-glycerophosphate hydrogel after 37°C incubation

**Figure 10.** Thermo-responsive gel system with lowest critical temperature.

3.4.2. pH-Responsive Systems

The pH is considered an important factor governing the degree of ionization of polymeric systems and their solubility in water. Repulsive electrostatic forces and osmotic forces in the presence of ions cause the polymer to swell pH-dependently or to disintegrate the gel [91].

Some polymers undergo gelling triggered by changes in pH, based on ionizing proportions. Polymers such as polyacrylic acid and carbopols have high molecular weight and a large number of carboxylic acid groups. These polymers show little swelling at low pH, as they have low concentrations of dissociated acid portions. When there is an increase in pH, an electrostatic repulsion triggered by additional charges leads to an expansion of these polymers, promoting their gelation [92]. If acidic pH is proposed to develop in situ gelling formulations using these polymers to maintain low viscosity, the stability could be affected, especially in pH-sensitive drugs [86].

Another polymer that has pH-sensitive properties is CH, which is a polymer rich in amine bonds. In acidic pH, it is soluble due to electrostatic repulsion, whereas, at pH above 6.2, it forms a dissociated precipitate [92]. Studies with a CH derivative containing palmitic acid linkages in its free amines (N-palmitoyl CH) showed gelling properties at physiological pH. This allowed applying the liquid solution at pH 6.5, which became solid after reaching pH 7.4 [93]. At low pH, free amines are protonated, and electrostatic forces block the cohesive attraction between polymer chains; when there is an increase in pH, the hydrophobic interactions between palmitoyl groups are more significant, boosting the condensation of molecules of N-palmitoyl CH.

3.4.3. Ionic-Responsive Systems

In situ gelation triggered by ions occurs through the interaction of anionic fractions present in the molecule and cations, such as calcium ($Ca^{2+}$), usually present in body fluids such as vaginal, nasal, or lacrimal [94].

Ionic-sensitive polymers commonly used for in situ gelling preparation are natural polysaccharides such as alginate, gellan gum (Figure 11), and pectin, which are generated by gels triggered by the interaction between carboxylic acid residues present in the polymeric structure and surrounding cations. Monovalent cations weaken the electrostatic repulsion, promoting hydrophobic interactions, while divalent cations cause the association of helical sections of polymer chains, generating junction zones. It should be noted that the variation in ion concentration can lead to different mechanical properties, creating an 'egg box' structure responsible for gelling, resulting in heterogeneous gels [95].

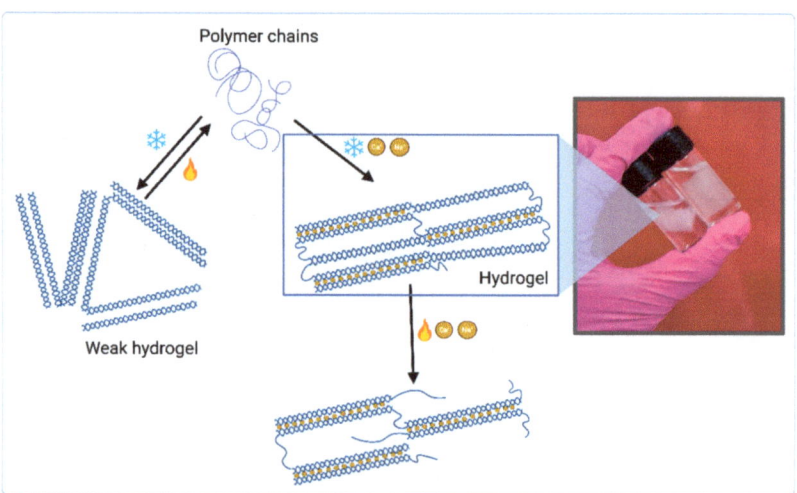

**Figure 11.** Gellan gum hydrogel system triggered by the presence of cations and temperature. Above 50 °C, gellan gum polymer chains are disordered, but the colling process of the solution induces the formation of double helices stabled by hydrogen bonds. The presence of cations in the solution allows the interconnection of these helixes and the formation of a 3D matrix. Created with BioRender.com.

### 3.4.4. In Situ Gelation Triggered by Genipin

Less toxic alternatives than glutaraldehyde for polymer crosslinking have been studied, and genipin, a compound of natural origin, was demonstrated to be a promising alternative. The pH of the mixture influences the reaction mechanism, giving genipin a ring-opening polymerization under basic pH conditions, while a Schiff reaction together with primary amines gives rise to the formation of crosslinked networks under acidic or neutral conditions. In these cases, a nucleophilic attack occurs on genipin, resulting in the formation of heterocyclic amines and, thus, crosslinking via genipin bounds [96]. Researchers investigated the in situ gelation of the collagen–genipin mixture in treatments for gastrointestinal ulcers, with good results under physiological conditions [97].

### 3.5. Rheological Aspects

Rheology is the study of the material flow and its deformation behavior, which can be measured by applying force to a sample. Combined with formulation viscosity, plasticity, and elasticity, the rheological behavior may impact product manufacturing, long-term stability, appearance, dispensing, sensory properties, packaging, and in vivo performance [98].

One essential property of semisolids and viscoelastic materials is the rheological behavior. Gels can be cited as a typical example of a pharmaceutical semisolid that behaves in a non-Newtonian manner. The viscosity of a gel is defined as a flow curve reflecting the shear stress as a function of shear rate or strain, and the viscoelastic properties are presented as a frequency sweep reflecting the moduli at increasing frequencies [99].

Mathematical modeling can be applied in rheology to reliably predict the rheological properties of concentrated or diluted polymeric liquids. With those models applied in the assessment of hydrogel networks and their rheological characteristics, it is possible to identify key parameters for the process, formulation, and mechanisms of drug delivery. Consequently, the mathematical understanding of the gel-forming material properties and of the way the formulation and process parameters interact can facilitate the intelligent design of a hydrogel network [100].

The understanding of the rheological properties of hydrogels is key to gaining insight into the mechanical properties, viscoelastic behavior, and interactions between the hydrogel components. Those properties are the fusion of multiple factors such as the structure and nature of polymers, temperature, ionic strength, pH, concentration, and crosslinking of polymers components within the hydrogel. Further knowledge about those properties helps the determination of possible industrial applications for the synthesized hydrogel [101].

A thorough rheological analysis helps in understanding the properties of any synthesized material. The rheological study helps in perceiving the viscosity, elasticity, crosslinking, flow, and mechanical behavior of the material, in response to an applied strain or stress. Those kinds of properties are also known to vary with changes in the molecular network, and they play a crucial role in the determination of the field of application of the synthesized material [102].

When it comes to developing formulations for intravesical applications, rheology is a key factor to allow instillation via a catheter and to understand its behavior inside the bladder concerning bioadhesion, stability, and drug release.

## 4. Conclusions and Future Perspectives

Bladder cancer's survival rate is high when compared to other types of tumors such as pancreatic and glioblastoma. However, it affects considerably the quality of life of the patients. Additionally, many of them present the progression of superficial cancer, pass through chemoresistance, and end up developing muscle-invasive bladder cancer [8]. Depending on the tumor stage, treatment may englobe intravesical chemotherapy, which usually demands frequent catheter insertions due to the reduced residence time of the drug. Several procedures of chemotherapy instillations may irritate the urinary tract and other side-effects, which have encouraged researchers to develop drug delivery systems for bladder cancer treatment.

Drug delivery systems must present some important properties such as specific release at the lesion tissue or targeted cells and biocompatibility. Concerning nanotechnologies, they may be very helpful in targeting nonspecific drug release, via passive and active approaches, as an option to solve side-effects related to chemotherapy [8].

Bioadhesion is an important strategy to enhance drug residence time and target delivery. Among other key points to evaluate, mucosal thickness, low absorptive surface area, mucosal microbiome, and mucosal secretion are challenges that must be taken into account when developing a mucoadhesive system with optimum therapeutic response [103]. Natural and synthetic polymers have been explored due to their properties of mucoadhesion with different mechanisms as possible vehicles for drug delivery in bladder cancer treatment. However, most of the research in progress is still in need of further in vitro and in vivo studies. Moreover, therapeutic outcomes and topics concerning safety must be evaluated by clinical trials, as in vitro and in vivo results provide limited information. The progress in these aspects of new and sophisticated drug delivery systems will allow the improvement of bladder cancer treatment [104].

**Author Contributions:** Conceptualization, C.S.A.d.L. and G.H.C.V.; writing—original draft preparation, C.S.A.d.L., J.P.R.O.V., V.M.A., K.M.N., C.P.C.C. and S.S.K.; writing—review and editing, C.S.A.d.L., M.I.R.-H. and G.H.C.V.; supervision, M.I.R.-H. and A.B.L.; project administration, A.B.L.; funding acquisition, A.B.L. All authors have read and agreed to the published version of the manuscript.

**Funding:** This research was funded by the São Paulo Research Foundation (FAPESP), grant number 2017/50332-0.

**Institutional Review Board Statement:** Not applicable.

**Informed Consent Statement:** Not applicable.

**Data Availability Statement:** Not applicable.

**Acknowledgments:** Caroline Santos Alves de Lima acknowledges the São Paulo Research Foundation (FAPESP) for her PhD fellowships, grant numbers 2019/01315-1 and 2021/09636-1; M. I. Rial-Hermida acknowledges Xunta de Galicia for her postdoctoral grant (grant ED481B 2018/009). Authors also acknowledge the International Agency for Research on Cancer ("IARC") for Figure 1 reprint.

**Conflicts of Interest:** The authors declare no conflict of interest.

## References

1. Siegel, R.L.; Miller, K.D.; Fuchs, H.E.; Jemal, A. Cancer Statistics, 2021. *CA Cancer J. Clin.* **2021**, *71*, 7–33. [CrossRef] [PubMed]
2. International Agency for Research on Cancer; GLOBOCAN Global Bladder Cancer Statistics. 2020, pp. 1–2. Available online: https://gco.iarc.fr/today/data/factsheets/cancers/30-Bladder-fact-sheet.pdf (accessed on 9 September 2022).
3. Richters, A.; Aben, K.K.H.; Kiemeney, L.A.L.M. The Global Burden of Urinary Bladder Cancer: An Update. *World J. Urol.* **2020**, *38*, 1895–1904. [CrossRef] [PubMed]
4. Degeorge, K.C.; Holt, H.R.; Hodges, S.C. Bladder Cancer: Diagnosis and Treatment. *Am. Fam. Physician* **2017**, *96*.
5. Cassell, A.; Yunusa, B.; Jalloh, M.; Mbodji, M.M.; Diallo, A.; Ndoye, M.; Diallo, Y.; Labou, I.; Niang, L.; Gueye, S.M. Non-Muscle Invasive Bladder Cancer: A Review of the Current Trend in Africa. *World J. Oncol.* **2019**, *10*, 123–131. [CrossRef]
6. Martinez Rodriguez, R.H.; Buisan Rueda, O.; Ibarz, L. Bladder Cancer: Present and Future. *Med. Clin. (Barc).* **2017**, *149*, 449–455. [CrossRef] [PubMed]
7. Down, C.J.; Nair, R.; Thurairaja, R. Bladder Cancer. *Surgery* **2016**, *34*, 532–539. [CrossRef]
8. Jain, P.; Kathuria, H.; Momin, M. Clinical Therapies and Nano Drug Delivery Systems for Urinary Bladder Cancer. *Pharmacol. Ther.* **2021**, *226*. [CrossRef] [PubMed]
9. Lenis, A.T.; Lec, P.M.; Chamie, K. Bladder Cancer a Review. *JAMA - J. Am. Med. Assoc.* **2020**, *324*, 1980–1991. [CrossRef]
10. Ajalloueian, F.; Lemon, G.; Hilborn, J.; Chronakis, I.S.; Fossum, M. Bladder Biomechanics and the Use of Scaffolds for Regenerative Medicine in the Urinary Bladder. *Nat. Rev. Urol.* **2018**, *15*, 155–174. [CrossRef]
11. Fry, C.H.; Vahabi, B. The Role of the Mucosa in Normal and Abnormal Bladder Function. *Basic Clin. Pharmacol. Toxicol.* **2016**, *119*, 57–62. [CrossRef]
12. Wang, S.; Jin, S.; Shu, Q.; Wu, S. Strategies to Get Drugs across Bladder Penetrating Barriers for Improving Bladder Cancer Therapy. *Pharmaceutics* **2021**, *13*, 166. [CrossRef] [PubMed]
13. Kamat, A.M.; Hahn, N.M.; Efstathiou, J.A.; Lerner, S.P.; Malmström, P.U.; Choi, W.; Guo, C.C.; Lotan, Y.; Kassouf, W. Bladder Cancer. *Lancet* **2016**, *388*, 2796–2810. [CrossRef]
14. Care, P. Bladder Cancer: Diagnosis and Management of Bladder Cancer: © NICE (2015) Bladder Cancer: Diagnosis and Management of Bladder Cancer. *BJU Int.* **2017**, *120*, 755–765. [CrossRef]
15. Kolawole, O.M.; Lau, W.M.; Mostafid, H.; Khutoryanskiy, V.V. Advances in Intravesical Drug Delivery Systems to Treat Bladder Cancer. *Int. J. Pharm.* **2017**, *532*, 105–117. [CrossRef] [PubMed]
16. GuhaSarkar, S.; Banerjee, R. Intravesical Drug Delivery: Challenges, Current Status, Opportunities and Novel Strategies. *J. Control. Release* **2010**, *148*, 147–159. [CrossRef] [PubMed]
17. Edgar, J.Y.C.; Wang, H. Introduction for Design of Nanoparticle Based Drug Delivery Systems. *Curr. Pharm. Des.* **2016**, *23*. [CrossRef]
18. Shende, P.; Basarkar, V. Recent Trends and Advances in Microbe-Based Drug Delivery Systems. *DARU, J. Pharm. Sci.* **2019**, *27*, 799–809. [CrossRef]
19. Laffleur, F.; Keckeis, V. Advances in Drug Delivery Systems: Work in Progress Still Needed? *Int. J. Pharm. X* **2020**, *2*, 100050. [CrossRef]
20. Sharif, S.; Abbas, G.; Hanif, M.; Bernkop-Schnürch, A.; Jalil, A.; Yaqoob, M. Mucoadhesive Micro-Composites: Chitosan Coated Halloysite Nanotubes for Sustained Drug Delivery. *Colloids Surfaces B Biointerfaces* **2019**, *184*. [CrossRef]
21. Lian, Z.; Ji, T. Functional Peptide-Based Drug Delivery Systems. *J. Mater. Chem. B* **2020**, *8*, 6517–6529. [CrossRef]
22. Lupo, N.; Jalil, A.; Nazir, I.; Gust, R.; Bernkop-Schnürch, A. In Vitro Evaluation of Intravesical Mucoadhesive Self-Emulsifying Drug Delivery Systems. *Int. J. Pharm.* **2019**, *564*, 180–187. [CrossRef] [PubMed]
23. Stærk, K.; Hjelmager, J.S.; Alm, M.; Thomsen, P.; Andersen, T.E. A New Catheter-Integrated Drug-Delivery System for Controlled Intravesical Mitomycin C Release. *Urol. Oncol. Semin. Orig. Investig.* **2022**, *40*, 409.e19–409.e26. [CrossRef]
24. Kaldybekov, D.B.; Tonglairoum, P.; Opanasopit, P.; Khutoryanskiy, V.V. Mucoadhesive Maleimide-Functionalised Liposomes for Drug Delivery to Urinary Bladder. *Eur. J. Pharm. Sci.* **2018**, *111*, 83–90. [CrossRef] [PubMed]

25. Nikezić, A.V.V.; Bondžić, A.M.; Vasić, V.M. Drug Delivery Systems Based on Nanoparticles and Related Nanostructures. *Eur. J. Pharm. Sci.* **2020**, *151*. [CrossRef] [PubMed]
26. Colone, M.; Calcabrini, A.; Stringaro, A. Drug Delivery Systems of Natural Products in Oncology. *Molecules* **2020**, *25*, 4560. [CrossRef]
27. Wang, B.; Zhang, K.; Wang, J.; Zhao, R.; Zhang, Q.; Kong, X. Poly(Amidoamine)-Modified Mesoporous Silica Nanoparticles as a Mucoadhesive Drug Delivery System for Potential Bladder Cancer Therapy. *Colloids Surf. B Biointerfaces* **2020**, *189*, 110832. [CrossRef]
28. Gonzaga, R.V.; do Nascimento, L.A.; Santos, S.S.; Machado Sanches, B.A.; Giarolla, J.; Ferreira, E.I. Perspectives About Self-Immolative Drug Delivery Systems. *J. Pharm. Sci.* **2020**, *109*, 3262–3281. [CrossRef]
29. Huang, J.; Jiang, Y.; Li, J.; He, S.; Huang, J.; Pu, K. A Renal-Clearable Macromolecular Reporter for Near-Infrared Fluorescence Imaging of Bladder Cancer. *Angew. Chemie Int. Ed.* **2020**, *59*, 4415–4420. [CrossRef]
30. Kumar, K.; Dhawan, N.; Sharma, H.; Vaidya, S.; Vaidya, B. Bioadhesive Polymers: Novel Tool for Drug Delivery. *Artif. Cells Nanomedicine Biotechnol.* **2014**, *42*, 274–283. [CrossRef]
31. Estrellas, K.M.; Fiecas, M.; Azagury, A.; Laulicht, B.; Cho, D.Y.; Mancini, A.; Reineke, J.; Furtado, S.; Mathiowitz, E. Time-Dependent Mucoadhesion of Conjugated Bioadhesive Polymers. *Colloids Surf. B Biointerfaces* **2019**, *173*, 454–469. [CrossRef]
32. Laurén, P.; Paukkonen, H.; Lipiäinen, T.; Dong, Y.; Oksanen, T.; Räikkönen, H.; Ehlers, H.; Laaksonen, P.; Yliperttula, M.; Laaksonen, T. Pectin and Mucin Enhance the Bioadhesion of Drug Loaded Nanofibrillated Cellulose Films. *Pharm. Res.* **2018**, *35*. [CrossRef] [PubMed]
33. Khutoryanskiy, V.V. Advances in Mucoadhesion and Mucoadhesive Polymers. *Macromol. Biosci.* **2011**, *11*, 748–764. [CrossRef] [PubMed]
34. Jiao, Y.; Pang, X.; Liu, M.; Zhang, B.; Li, L.; Zhai, G. Recent Progresses in Bioadhesive Microspheres via Transmucosal Administration. *Colloids Surf. B Biointerfaces* **2016**, *140*, 361–372. [CrossRef]
35. Bruschi, M.L.; de Francisco, L.M.B.; Borghi, F.B. An Overview of Recent Patents on Composition of Mucoadhesive Drug Delivery Systems. *Recent Pat. Drug Deliv. Formul.* **2015**, *9*, 79–87. [CrossRef]
36. Hanafy, N.A.N.; Leporatti, S.; El-Kemary, M.A. Mucoadhesive Hydrogel Nanoparticles as Smart Biomedical Drug Delivery System. *Appl. Sci.* **2019**, *9*, 825. [CrossRef]
37. Ali, M.S.; Metwally, A.A.; Fahmy, R.H.; Osman, R. Chitosan-Coated Nanodiamonds: Mucoadhesive Platform for Intravesical Delivery of Doxorubicin. *Carbohydr. Polym.* **2020**, *245*, 116528. [CrossRef]
38. Xu, X.; Liu, K.; Jiao, B.; Luo, K.; Ren, J.; Zhang, G.; Yu, Q.; Gan, Z. Mucoadhesive Nanoparticles Based on ROS Activated Gambogic Acid Prodrug for Safe and Efficient Intravesical Instillation Chemotherapy of Bladder Cancer. *J. Control. Release* **2020**, *324*, 493–504. [CrossRef]
39. Lu, S.; Neoh, K.G.; Kang, E.-T.; Mahendran, R.; Chiong, E. Mucoadhesive Polyacrylamide Nanogel as a Potential Hydrophobic Drug Carrier for Intravesical Bladder Cancer Therapy. *Eur. J. Pharm. Sci.* **2015**, *72*, 57–68. [CrossRef]
40. Kolawole, O.M.; Lau, W.M.; Khutoryanskiy, V.V. Chitosan/β-Glycerophosphate in Situ Gelling Mucoadhesive Systems for Intravesical Delivery of Mitomycin-C. *Int. J. Pharm. X* **2019**, *1*, 100007. [CrossRef]
41. Denora, N.; Lopedota, A.; Perrone, M.; Laquintana, V.; Iacobazzi, R.M.; Milella, A.; Fanizza, E.; Depalo, N.; Cutrignelli, A.; Lopalco, A.; et al. Spray-Dried Mucoadhesives for Intravesical Drug Delivery Using N-Acetylcysteine- and Glutathione-Glycol Chitosan Conjugates. *Acta Biomater.* **2016**, *43*, 170–184. [CrossRef]
42. Karavana, S.Y.; Şenyiğit, Z.A.; Çalışkan, Ç.; Sevin, G.; Özdemir, D.İ.; Erzurumlu, Y.; Şen, S.; Baloğlu, E. Gemcitabine Hydrochloride Microspheres Used for Intravesical Treatment of Superficial Bladder Cancer: A Comprehensive in Vitro/Ex Vivo/in Vivo Evaluation. *Drug Des. Devel. Ther.* **2018**, *12*, 1959–1975. [CrossRef] [PubMed]
43. Lu, Z.; Yeh, T.; Wang, J.; Chen, L.; Lyness, G.; Xin, Y.; Wientjes, M.G.; Bergdall, V.; Couto, G.; Alvarez-berger, F.; et al. Paclitaxel Gelatin Nanoparticles for Intravesical Bladder Cancer Therapy. *J. Urol.* **2011**, *185*, 1478–1483. [CrossRef] [PubMed]
44. Lu, S.; Xu, L.; Tang, E.; Mahendran, R.; Chiong, E.; Gee, K. Co-Delivery of Peptide-Modified Cisplatin and Doxorubicin via Mucoadhesive Nanocapsules for Potential Synergistic Intravesical Chemotherapy of Non-Muscle-Invasive Bladder Cancer. *Eur. J. Pharm. Sci.* **2016**, *84*, 103–115. [CrossRef] [PubMed]
45. GuhaSarkar, S.; More, P.; Banerjee, R. Urothelium-Adherent, Ion-Triggered Liposome-in-Gel System as a Platform for Intravesical Drug Delivery. *J. Control. Release* **2017**, *245*, 147–156. [CrossRef] [PubMed]
46. Khan, A.B.; Mahamana, R.; Pal, E. Review on Mucoadhesive Drug Delivery System: Novel Approaches in Modern Era. *RGUHS J. Pharm. Sci.* **2014**, *4*, 128–141. [CrossRef]
47. Kolawole, O.M.; Lau, W.M.; Khutoryanskiy, V.V. Synthesis and Evaluation of Boronated Chitosan as a Mucoadhesive Polymer for Intravesical Drug Delivery. *J. Pharm. Sci.* **2019**, *108*, 3046–3053. [CrossRef]
48. Muppalaneni, S.; Mastropietro, D.; Omidian, H. Mucoadhesive Drug Delivery Systems. In *Engineering Polymer Systems for Improved Drug Delivery*; Bader, R.A., Putnam, D.A., Eds.; John Wiley & Sons, Inc.: Hoboken, NJ, USA, 2013; ISBN 9781118747896.
49. Chatterjee, B.; Amalina, N.; Sengupta, P.; Mandal, U.K. Mucoadhesive Polymers and Their Mode of Action: A Recent Update. *J. Appl. Pharm. Sci.* **2017**. [CrossRef]
50. Tekade, M.; Maheshwari, N.; Youngren-Ortiz, S.R.; Pandey, V.; Chourasiya, Y.; Soni, V.; Deb, P.K.; Sharma, M.C. *Thiolated-Chitosan: A Novel Mucoadhesive Polymer for Better-Targeted Drug Delivery*; Elsevier Inc.: London, UK, 2019; ISBN 9780128144282.

51. Bassi da Silva, J.; de Ferreira, S.B.S.; de Freitas, O.; Bruschi, M.L. A Critical Review about Methodologies for the Analysis of Mucoadhesive Properties of Drug Delivery Systems. *Drug Dev. Ind. Pharm.* **2017**, *43*, 1053–1070. [CrossRef]
52. Roy, S.; Pal, K.; Anis, A.; Pramanik, K.; Prabhakar, B. Polymers in Mucoadhesive Drug-Delivery Systems: A Brief Note. *Des. Monomers Polym.* **2009**, *12*, 483–495. [CrossRef]
53. Singh, I.; Rana, V. Techniques for the Assessment of Mucoadhesion in Drug Delivery Systems: An Overview. *J. Adhes. Sci. Technol.* **2012**, *26*, 2251–2267. [CrossRef]
54. Roy, S.K.; Prabhakar, B. Bioadhesive Polymeric Platforms for Transmucosal Drug Delivery Systems—A Review. *Trop. J. Pharm. Res.* **2010**, *9*, 91–104. [CrossRef]
55. De Souza Ferreira, S.B.; Moço, T.D.; Borghi-Pangoni, F.B.; Junqueira, M.V.; Bruschi, M.L. Rheological, Mucoadhesive and Textural Properties of Thermoresponsive Polymer Blends for Biomedical Applications. *J. Mech. Behav. Biomed. Mater.* **2016**, *55*, 164–178. [CrossRef]
56. Bassi da Silva, J.; Ferreira, S.; Reis, A.; Cook, M.; Bruschi, M. Assessing Mucoadhesion in Polymer Gels: The Effect of Method Type and Instrument Variables. *Polymers* **2018**, *10*, 254. [CrossRef]
57. Huang, D.; Lin, B.; Song, Y.; Guan, Z.; Cheng, J.; Zhu, Z.; Yang, C. Staining Traditional Colloidal Gold Test Strips with Pt Nanoshell Enables Quantitative Point-of-Care Testing with Simple and Portable Pressure Meter Readout. *ACS Appl. Mater. Interfaces* **2019**, *11*, 1800–1806. [CrossRef]
58. PARK, K. A New Approach to Study Mucoadhesion: Colloidal Gold Staining. *Int. J. Pharm.* **1989**, *53*, 209–217. [CrossRef]
59. Teng, C.L.C.; Ho, N.F.H. Mechanistic Studies in the Simultaneous Flow and Adsorption of Polymer-Coated Latex Particles on Intestinal Mucus I: Methods and Physical Model Development. *J. Control. Release* **1987**, *6*, 133–149. [CrossRef]
60. Efiana, N.A.; Mahmood, A.; Lam, H.T.; Zupančič, O.; Leonaviciute, G.; Bernkop-Schnürch, A. Improved Mucoadhesive Properties of Self-Nanoemulsifying Drug Delivery Systems (SNEDDS) by Introducing Acyl Chitosan. *Int. J. Pharm.* **2017**, *519*, 206–212. [CrossRef]
61. Dyawanapelly, S.; Koli, U.; Dharamdasani, V.; Jain, R.; Dandekar, P. Improved Mucoadhesion and Cell Uptake of Chitosan and Chitosan Oligosaccharide Surface-Modified Polymer Nanoparticles for Mucosal Delivery of Proteins. *Drug Deliv. Transl. Res.* **2016**, *6*, 365–379. [CrossRef]
62. Kumari, P.V.K.; Anitha, S.; Rao, Y.S. Review on Pharmacoscintigraphy. *J. Pharm. Res. Int.* **2020**, *32*, 23–32. [CrossRef]
63. Fischer, F.; Bauer, S. Polyvinylpyrrolidon. Ein Tausendsassa in Der Chemie. *Chemie unserer Zeit* **2009**, *43*, 376–383. [CrossRef]
64. Grant, J.J.; Pillai, S.C.; Perova, T.S.; Hehir, S.; Hinder, S.J.; McAfee, M.; Breen, A. Electrospun Fibres of Chitosan/PVP for the Effective Chemotherapeutic Drug Delivery of 5-Fluorouracil. *Chemosensors* **2021**, *9*, 70. [CrossRef]
65. Xu, Y.; Zhao, J.; Zhang, Z.; Zhang, J.; Huang, M.; Wang, S.; Xie, P. Preparation of Electrospray ALG/PDA–PVP Nanocomposites and Their Application in Cancer Therapy. *Soft Matter* **2020**, *16*, 132–141. [CrossRef] [PubMed]
66. Marušincová, H.; Husárová, L.; Růžička, J.; Ingr, M.; Navrátil, V.; Buňková, L.; Koutny, M. Polyvinyl Alcohol Biodegradation under Denitrifying Conditions. *Int. Biodeterior. Biodegradation* **2013**, *84*, 21–28. [CrossRef]
67. Rivera-Hernández, G.; Antunes-Ricardo, M.; Martínez-Morales, P.; Sánchez, M.L. Polyvinyl Alcohol Based-Drug Delivery Systems for Cancer Treatment. *Int. J. Pharm.* **2021**, *600*. [CrossRef]
68. Ben Halima, N. Poly(Vinyl Alcohol): Review of Its Promising Applications and Insights into Biodegradation. *RSC Adv.* **2016**, *6*, 39823–39832. [CrossRef]
69. Ullah, K.; Sohail, M.; Murtaza, G.; Khan, S.A. Natural and Synthetic Materials Based CMCh/PVA Hydrogels for Oxaliplatin Delivery: Fabrication, Characterization, In-Vitro and In-Vivo Safety Profiling. *Int. J. Biol. Macromol.* **2019**, *122*, 538–548. [CrossRef]
70. Peppas, N.A.; Hilt, J.Z.; Khademhosseini, A.; Langer, R. Hydrogels in Biology and Medicine: From Molecular Principles to Bionanotechnology. *Adv. Mater.* **2006**, *18*, 1345–1360. [CrossRef]
71. WICHTERLE, O.; LÍM, D. Hydrophilic Gels for Biological Use. *Nature* **1960**, *185*, 117–118. [CrossRef]
72. Brady, J.; Dürig, t.; Lee, P.I.; Li, J.X. Chapter 7—Polymer Properties and Characterization. In *Developing Solid Oral Dosage Forms*; Academic Press: Cambridge, MA, USA, 2017; pp. 181–223. ISBN 9780128024478.
73. Brown, H.P. Carboxylic Polymers 1957. US Patent US2798053A, 2 July 1957.
74. B F Goodrich Bulletin. *Sustained Release Patents Using Carbopol Resin*; B F Goodrich Bulletin: Cleveeland, OH, USA, 1987.
75. Panzade, P.; Puranik, P.K. Carbopol Polymers: A Versatile Polymer for Pharmaceutical Applications. *Res. J. Pharm. Technol.* **2010**, *3*, 672–675.
76. Xu, H.; Wen, Y.; Chen, S.; Zhu, L.; Feng, R.; Song, Z. Paclitaxel Skin Delivery by Micelles-Embedded Carbopol 940 Hydrogel for Local Therapy of Melanoma. *Int. J. Pharm.* **2020**, *587*, 119626. [CrossRef]
77. Harris, J.M. *Poly (Ethylene Glycol) Chemistry: Biotechnical and Biomedical Applications*; Springer Science & Business Media: New York, NY, USA, 2013.
78. Pitto-Barry, A.; Barry, N.P.E. Pluronic®Block-Copolymers in Medicine: From Chemical and Biological Versatility to Rationalisation and Clinical Advances. *Polym. Chem.* **2014**, *5*, 3291–3297. [CrossRef]
79. Ottenbrite, R.M.; Javan, R. *Encyclopedia of Condensed Matter Physics*; Academic Press: Cambridge, MA, USA, 2005; pp. 99–108.
80. Şenyiğit, Z.A.; Karavana, S.Y.; İlem-Özdemir, D.I.; Çalışkan, C.; Waldner, C.; Şen, S.; Bernkop-Schnürch, A.; Baloğlu, E. Design and Evaluation of an Intravesical Delivery System for Superficial Bladder Cancer: Preparation of Gemcitabine HCl-Loaded Chitosan-Thioglycolic Acid Nanoparticles and Comparison of Chitosan/Poloxamer Gels as Carriers. *Int. J. Nanomedicine* **2015**, *10*, 6493–6507. [CrossRef] [PubMed]

81. Ajazuddin, A.A.; Khan, J.; Giri, T.K.; Tripathi, D.K.; Saraf, S.; Saraf, S. Advancement in Stimuli Triggered in Situ Gelling Delivery for Local and Systemic Route. *Expert Opin. Drug Deliv.* **2012**, *9*, 1573–1592. [CrossRef] [PubMed]
82. Suman, K.; Shanbhag, S.; Joshi, Y.M. Phenomenological Model of Viscoelasticity for Systems Undergoing Sol-Gel Transition. *Phys. Fluids* **2021**, *33*. [CrossRef]
83. Tang, H.; Zhao, W.; Yu, J.; Li, Y.; Zhao, C. Recent Development of PH-Responsive Polymers for Cancer Nanomedicine. *Molecules* **2018**, *24*, 4. [CrossRef]
84. Caramella, C.M.; Rossi, S.; Ferrari, F.; Bonferoni, M.C.; Sandri, G. Mucoadhesive and Thermogelling Systems for Vaginal Drug Delivery. *Adv. Drug Deliv. Rev.* **2015**, *92*, 39–52. [CrossRef]
85. Al-Kinani, A.A.; Zidan, G.; Elsaid, N.; Seyfoddin, A.; Alani, A.W.G.; Alany, R.G. Ophthalmic Gels: Past, Present and Future. *Adv. Drug Deliv. Rev.* **2018**, *126*, 113–126. [CrossRef]
86. Zahir-Jouzdani, F.; Wolf, J.D.; Atyabi, F.; Bernkop-Schnürch, A. In Situ Gelling and Mucoadhesive Polymers: Why Do They Need Each Other? *Expert Opin. Drug Deliv.* **2018**, *15*, 1007–1019. [CrossRef]
87. Lihong, W.; Xin, C.; Yongxue, G.; Yiying, B.; Gang, C. Thermoresponsive Ophthalmic Poloxamer/Tween/Carbopol in Situ Gels of a Poorly Water-Soluble Drug Fluconazole: Preparation and in Vitro – in Vivo Evaluation. *Drug Dev. Ind. Pharm.* **2014**, *40*, 1402–1410. [CrossRef]
88. Matanović, M.R.; Kristl, J.; Grabnar, P.A. Thermoresponsive Polymers: Insights into Decisive Hydrogel Characteristics, Mechanisms of Gelation, and Promising Biomedical Applications. *Int. J. Pharm.* **2014**, *472*, 262–275. [CrossRef]
89. Park, H.; Kim, M.H.; Yoon, Y.I.; Park, W.H. One-Pot Synthesis of Injectable Methylcellulose Hydrogel Containing Calcium Phosphate Nanoparticles. *Carbohydr. Polym.* **2017**, *157*, 775–783. [CrossRef] [PubMed]
90. Sarkar, N. Thermal Gelation Properties of Methyl and Hydroxypropyl Methylcellulose. *J. Appl. Polym. Sci.* **1979**, *24*, 1073–1087. [CrossRef]
91. Sharma, M.; Deohra, A.; Reddy, K.R.; Sadhu, V. Biocompatible In-Situ Gelling Polymer Hydrogels for Treating Ocular Infection. In *Methods in Microbiology*; Academic Press: Cambridge, MA, USA, 2019; pp. 93–114.
92. Gupta, P.; Vermani, K.; Garg, S. Hydrogels: From Controlled Release to PH-Responsive Drug Delivery. *Drug Discov. Today* **2002**, *7*, 569–579. [CrossRef]
93. Gupta, S. Carbopol/Chitosan Based PH Triggered In Situ Gelling System for Ocular Delivery of Timolol Maleate. *Sci. Pharm.* **2010**, *78*, 959–976. [CrossRef] [PubMed]
94. Vigani, B.; Rossi, S.; Sandri, G.; Bonferoni, M.C.; Caramella, C.M.; Ferrari, F. Recent Advances in the Development of In Situ Gelling Drug Delivery Systems for Non-Parenteral Administration Routes. *Pharmaceutics* **2020**, *12*, 859. [CrossRef]
95. CAO, S.; REN, X.; ZHANG, Q.; CHEN, E.; XU, F.; CHEN, J.; LIU, L.; JIANG, X. In Situ Gel Based on Gellan Gum as New Carrier for Nasal Administration of Mometasone Furoate. *Int. J. Pharm.* **2009**, *365*, 109–115. [CrossRef]
96. Mi, F.-L.; Shyu, S.-S.; Peng, C.-K. Characterization of Ring-Opening Polymerization of Genipin and PH-Dependent Cross-Linking Reactions between Chitosan and Genipin. *J. Polym. Sci. Part A Polym. Chem.* **2005**, *43*, 1985–2000. [CrossRef]
97. Narita, T.; Yunoki, S.; Ohyabu, Y.; Yahagi, N.; Uraoka, T. In Situ Gelation Properties of a Collagen–Genipin Sol with a Potential for the Treatment of Gastrointestinal Ulcers. *Med. Devices Evid. Res.* **2016**, *9*, 429–439. [CrossRef]
98. Simões, A.; Miranda, M.; Cardoso, C.; Vitorino, F. Rheology by Design: A Regulatory Tutorial for Analytical Method Validation. *Pharmaceutics* **2020**, *12*, 820. [CrossRef]
99. Qwist, P.K.; Sander, C.; Okkels, F.; Jessen, V.; Baldursdottir, S.; Rantanen, J. On-Line Rheological Characterization of Semi-Solid Formulations. *Eur. J. Pharm. Sci.* **2019**, *128*, 36–42. [CrossRef]
100. Ghica, M.V.; Hîrjău, M.; Lupuleasa, D.; Dinu-Pîrvu, C.-E. Flow and Thixotropic Parameters for Rheological Characterization of Hydrogels. *Mol. 2016, Vol. 21, Page 786* **2016**, *21*, 786. [CrossRef] [PubMed]
101. Kalia, S.; Choudhury, A.R. Synthesis and Rheological Studies of a Novel Composite Hydrogel of Xanthan, Gellan and Pullulan. *Int. J. Biol. Macromol.* **2019**, *137*, 475–482. [CrossRef] [PubMed]
102. Choudhury, A.R. PH Mediated Rheological Modulation of Chitosan Hydrogels. *Int. J. Biol. Macromol.* **2020**, *156*, 591–597. [CrossRef] [PubMed]
103. Kumar, A.; Naik, P.K.; Pradhan, D.; Ghosh, G.; Rath, G. Mucoadhesive Formulations: Innovations, Merits, Drawbacks, and Future Outlook. *Pharm. Dev. Technol.* **2020**, *25*, 797–814. [CrossRef] [PubMed]
104. Choi, Y.W.; Yoon, H.Y.; Yang, H.M.; Kim, C.H.; Goo, Y.T.; Kang, M.J.; Lee, S. Current Status of the Development of Intravesical Drug Delivery Systems for the Treatment of Bladder Cancer. *Expert Opin. Drug Deliv.* **2020**, *17*, 1555–1572. [CrossRef]

MDPI AG
Grosspeteranlage 5
4052 Basel
Switzerland
Tel.: +41 61 683 77 34

*Gels* Editorial Office
E-mail: gels@mdpi.com
www.mdpi.com/journal/gels

Disclaimer/Publisher's Note: The statements, opinions and data contained in all publications are solely those of the individual author(s) and contributor(s) and not of MDPI and/or the editor(s). MDPI and/or the editor(s) disclaim responsibility for any injury to people or property resulting from any ideas, methods, instructions or products referred to in the content.

www.ingramcontent.com/pod-product-compliance
Lightning Source LLC
LaVergne TN
LVHW072324090526
838202LV00019B/2345